ARTIFICIAL LIFE

ARTIFICIAL LIFE

THE PROCEEDINGS OF AN INTERDISCIPLINARY WORKSHOP ON THE SYNTHESIS AND SIMULATION OF LIVING SYSTEMS HELD SEPTEMBER, 1987 IN LOS ALAMOS, NEW MEXICO

Christopher G. Langton, *Editor*
Center for Nonlinear Studies
Los Alamos National Laboratory

Volume VI

SANTA FE INSTITUTE
STUDIES IN THE SCIENCES OF COMPLEXITY

Addison-Wesley Publishing Company
The Advanced Book Program

Reading, Massachusetts • Menlo Park, California • New York
Don Mills, Ontario • Wokingham, England • Amsterdam • Bonn
Sydney • Singapore • Tokyo • Madrid • San Juan
Paris • Seoul • Milan • Mexico City • Taipei

Publisher: *Allan M. Wylde*
Production Administrator: *Karen L. Garrison*
Editorial Coordinator: *Aida Adams*
Electronic Production Consultant: *Mona Zeftel*
Promotions Manager: *Celina Gonzales*

Director of Publications, Santa Fe Institute: *Ronda K. Butler-Villa*

ISBN 0-201-09346-4
ISBN 0-201-09356-1 (pbk.)

This volume was typeset using T$_E$Xtures on a Macintosh computer. Camera-ready output from an Apple LaserWriter Plus Printer.

7 8 9 10-MA-9998979695
Seventh printing, April 1995

About the Santa Fe Institute

The *Santa Fe Institute* (SFI) is a multidisciplinary graduate research and teaching institution formed to nurture research on complex systems and their simpler elements. A private, independent institution, SFI was founded in 1984. Its primary concern is to focus the tools of traditional scientific disciplines and emerging new computer resources on the problems and opportunities that are involved in the multidisciplinary study of complex systems—those fundamental processes that shape almost every aspect of human life. Understanding complex systems is critical to realizing the full potential of science, and may be expected to yield enormous intellectual and practical benefits.

All titles from the *Santa Fe Institute Studies in the Sciences of Complexity* series will carry this imprint which is based on a Mimbres pottery design (circa A.D. 950–1150), drawn by Betsy Jones.

Santa Fe Institute Studies in the Sciences of Complexity

VOLUME	EDITOR	TITLE
I	David Pines	Emerging Syntheses in Science, 1987
II	Alan S. Perelson	Theoretical Immunology, Part One, 1988
III	Alan S. Perelson	Theoretical Immunology, Part Two, 1988
IV	Gary D. Doolen et al.	Lattice Gas Methods of Partial Differential Equations
V	Philip W. Anderson et al.	The Economy as an Evolving Complex System
VI	Christopher G. Langton	Artificial Life: Proceedings of an Interdisciplinary Workshop on the Synthesis and Simulation of Living Systems
VII	Daniel Stein	Complex Systems

Contributors to This Volume

Valentino Braitenberg
Max Planck Institute of Biological Cybernetics

Bill Coderre
Media Laboratory, Massachusetts Institute of Technology

Richard Dawkins
Zoology Department, University of Oxford

K. Eric Drexler
Stanford University

Narendra S. Goel
Systems Science Department, State University of New York

Seth R. Goldman
Computer Science Department, University of California at Los Angeles

Stuart Hameroff
Advanced Biotechnology Lab, University of Arizona

Hyman Hartman
Computer Science Department, University of California at Berkeley

Pauline Hogeweg
Bioinformatica, Utrecht

Kevin D. Hufford
Systems Science, State University of New York and
APS, Cornell University

David R. Jefferson
Computer Science Department, University of California at Los Angeles

George J. Klir
Systems Science, State University of New York

Richard Laing
Logic of Computers Group, University of Michigan

Christopher Langton
Center for Nonlinear Studies, Los Alamos National Laboratory

Aristid Lindenmayer
Theoretical Biology Group, University of Utrecht

Marek W. Lugowski
Computer Science Department, Indiana University

Bengt Månsson
Physical Lab III, Technical University of Denmark

Hans Moravec
Robotics Institute, CMU

Harold C. Morris
University of British Columbia

Peter Oppenheimer
Computer Graphics Laboratory, New York Institute of Technology

Norman Packard
CCSR and Physics Department, University of Illinois

Howard H. Pattee
Systems Science Department, State University of New York

Przemyslaw Prusinkiewicz
Computer Science Department, University of Regina

Steen Rasmussen
Physics Lab III, Technical University of Denmark

Mitchel Resnick
Media Laboratory, Massachusetts Institute of Technology

Conrad Schneiker
College of Medicine, University of Arizona

Pablo Tamayo
Physics Department, Boston University

Charles E. Taylor
Biology Department, University of California at Los Angeles

Richard L. Thompson
Systems Science Department, State University of New York

Michael Travers
Media Laboratory, Massachusetts Institute of Technology

Scott R. Turner
Computer Science Department, University of California at Los Angeles

Stewart W. Wilson
The Rowland Institute for Science

Milan Zeleny
Fordham University and
Systems Science, State University of New York

This volume is dedicated to the memory of C. H. Waddington (1905-1975).

"It has always been clear that we were not so deeply interested in the theory of any particular biological phenomenon for its own sake, but mainly in so far as it helps to a greater comprehension of the general character of the processes that go on in living as contrasted with non-living systems."

Christopher G. Langton
Center for Nonlinear Studies, Los Alamos National Laboratory, Los Alamos, NM 87545

Preface

THE ARTIFICIAL LIFE WORKSHOP

In September 1987, the first workshop on Artificial Life was held at the Los Alamos National Laboratory. Jointly sponsored by the Center for Nonlinear Studies, the Santa Fe Institute, and Apple Computer Inc, the workshop brought together 160 computer scientists, biologists, physicists, anthropologists, and other assorted "-ists," all of whom shared a common interest in the simulation and synthesis of living systems. During five intense days, we saw a wide variety of models of living systems, including mathematical models for the origin of life, self-reproducing automata, computer programs using the mechanisms of Darwinian evolution to produce co-adapted ecosystems, simulations of flocking birds and schooling fish, the growth and development of artificial plants, and much, much more

The workshop itself grew out of my frustration with the fragmented nature of the literature on biological modeling and simulation. For years I had prowled around libraries, shifted through computer-search results, and haunted bookstores, trying to get an overview of a field which I sensed existed but which did not seem to have any coherence or unity. Instead, I literally kept stumbling over interesting work almost by accident, often published in obscure journals if published at all.

one example of the kind of complex dynamics that cellular automata are capable of generating making use solely of *local* state information. The emergence of such complex global dynamics from out of the local interactions of simple components was one of the major themes of the workshop.

The paper by Zeleny et al. illustrates the lifelike structures and developmental histories exhibited by *osmotic growth* processes. Coining the term "synthetic biology," they review and recreate early experiments carried out by Stephane Leduc on the spontaneous generation of life. This is an example of a *naturally occurring* chemical process which exhibits lifelike behavior, and is therefore worthy of further scientific investigation, whether or not we feel that it has played a specific role in the development of *our* life.

The next four papers treat the process of *evolution*. The paper by Packard is a very nice illustration of the fact that it is not just organisms that evolve, but organisms and their environments together. Important features of an organism's environment may be directly attributed to actions taken by the organism itself or by its cohabitors in the environment. Thus, the "adaptive landscape" will be a dynamic—rather than a fixed—abstraction, and hence the "fitness function" that determines which organisms are more successful in the environment will be an emergent property of the ecosystem.

The paper by Wilson provides a simple example of an application of the *genetic algorithm* to the simulation of the evolution of multicellular systems. Pioneered by John Holland, the genetic algorithm is a scheme for applying the power of Darwinian evolution under natural selection to real-world optimization problems. As Wilson points out, however, it has rarely been used to study the evolution of life itself. Wilson gives examples of the development of some simple one-dimensional, multicellular organisms.

The article by Moravec reviews the developmental history of *robotics*, arguing that *cultural evolution* is not only the dominant evolutionary process today, but may well replace biological evolution altogether in the foreseeable future. Natural selection concentrates on the *phenotype*, which is essentially an organism's dynamic interface with the physical world. Since selection operates at the level of the phenotype, evolution can reward extra-genetic mechanisms that contribute to the phenotype, e.g., culture. Should the extra-genetic contributions to the determination of the phenotype become more and more dominant, there is the chance that eventually *all* critical phenotypic traits will be "extra-genetically" determined, setting the stage for a "genetic takeover," in which the the previously "extra-genetic" system becomes the "primary" genetic system, taking over from the old genetic system, which may eventually be lost altogether as its contributions to the phenotype become more and more irrelevant. This may have happened once already in the history of life, as suggested by Cairns-Smith (see his work referenced in the annotated bibliography). Thus, the connection between a genetic system and the phenotype it determines is an arbitrary one, and the "mechanics" of evolution will jump ship if they find a more efficient vehicle for the determination of the phenotype. In cultural evolution, Moravec suggests, we are catching the mechanics of evolution in the act of jumping ship.

Dawkins' paper provides another important perspective on the distinction between genotype and phenotype, emphasizing the importance of *embryology:* the process by which the phenotype develops under control of the genotype. Via a simple but elegant computer program demonstrating the power of cumulative selection, Dawkins explores the manner in which the embryological process itself may be considered a phenotypic trait, subject to natural selection, arguing that selection will favor those embryologies richest in evolutionary potential. Dawkins' paper also illustrates the amazingly complex structures that can develop from the application of very simple recursive rules.

The next two papers discuss methods for studying *embryological* processes by modeling the development and differentiation of plants.

The paper by Lindenmayer and Prusinkiewicz reviews *L-systems*, formal grammars for the description, analysis, and developmental simulation of multicellular organisms. L-systems are perhaps the best known example of the general methodological approach advocated within Artificial Life. They are formal systems which work in a bottom-up, parallel manner, applying simple recursive rules to local subparts of developing structures. L-systems embody the genotype/phenotype distinction naturally. Many of the color plates show examples of plants grown—not just rendered—using L-systems.

The paper by Oppenheimer describes the application of *fractals* to the generation of tree-like structures. Oppenheimer discusses eloquently the theme that "by going beyond the laws of nature, we create a new reality, one that helps us appreciate nature more." The artifacts produced in the pursuit of artificial life will be both like nature and *not* like nature. Oppenheimer points out that not only the *similarities* but the *differences* as well can provide valuable insights into the nature of life.

The next four papers describe general purpose simulation systems that have been developed for studying various aspects of life.

The first two systems provide for the investigation of the dynamics and evolution of *ecosystems*. The paper by Taylor et al. describes the RAM simulator: a LISP-based simulation system within which populations of organisms are modeled as populations of programs which are created, run, learn, communicate with each other, interact with the environment (also a population of programs), replicate with modification, and eventually terminate. Examples are given of the application of the system to modeling specific ecosystems. This paper also discusses the positive role that computers play by forcing theorists to make their theories explicit enough that they may be implemented on a computer and tested empirically.

However, this brings up a cautionary theme from the workshop which is that the use of computers *may* lead to theories which are *overly* tailored for computers, in that they have expanded or restricted the theory too much in order that it may be implemented in the traditional top-down, serial programming style. Many natural phenomena depend on the fundamentally parallel "hardware" of nature, and are inappropriately modeled by the incorporation of a centralized, global controller managing high-level data-structures. Other models of computation—such as the object-oriented paradigm adopted in the RAM simulator—are more appropriate for to the study of life. Drexler's article compares our mechanical technology with the technology employed by living cells, and derives results about the differences in evolutionary capacity of the two different technologies. He concludes that our-style machines lack the capacity for evolution, and that it would be inappropriate to refer to them as living things.

The paper by Hameroff et al. describes recent discoveries about the *cellular cytoskeleton*, which not only provides structural support for the cell but may also provide information processing capabilities. The paper presents a model of tubulin—the primary constituent of cytoskeletons—as a kind of cellular automaton, capable of signal propagation and processing, and discusses the implications for the generation of behavior and the transport of materials within the cell itself.

Finally, the paper by Braitenberg explores some of the complex problems associated with controlling multi-segmented arms and legs. He shows how a simple trick used by crane-operators for getting their loads to a delivery site without having to wait for pendular movements to die out may also be operating within our antagonist muscle groups. He also discusses the role of the cerebellum in managing smoothness of behavior over a number of joints, each of which is operating in parallel with others on the same articulated arm.

THE DISTILLED "ESSENCE" OF ARTIFICIAL LIFE

To summarize, the general consensus on the "essence" of Artificial Life at the workshop was converging on the following vision: Artificial Life involves the *realization* of lifelike behavior on the part of man-made systems consisting of *populations* of semi-autonomous entities whose *local interactions* with one another are governed by a set of *simple rules*. Such systems contain *no* rules for the behavior of the population at the global level, and the often complex, high-level dynamics and structures observed are *emergent* properties, which develop over time from out of all of the local interactions among low-level primitives by a process highly reminiscent of *embryological development*, in which *local hierarchies* of higher-order structures develop and *compete* with one another for support among the low-level entities. These emergent structures play a vital role in organizing the behavior of the lowest-level entities by establishing the context within which those entities invoke their local rules and, as a consequence, these structures may *evolve* in time.

FURTHER THEMES

There were many other themes that arose at some point during the workshop that have not received adequate coverage in these proceedings. We mention a few of these here.

Other man-made emergent processes were proposed as candidates for Artificial Life, such as economies, socio-political institutions, and other cultural structures. One can certainly view such systems from a genotype/phenotype perspective, with us as the primitive behav*ors*, our socio-cultural inheritance as the genotype, and to the study of life. Drexler's article compares our mechanical technology with the technology employed by living cells, and derives results about the differences in evolutionary capacity of the two different technologies. He concludes that our-style machines lack the capacity for evolution, and that it would be inappropriate to refer to them as living things.

The paper by Hameroff et al. describes recent discoveries about the *cellular cytoskeleton*, which not only provides structural support for the cell but may also provide information processing capabilities. The paper presents a model of tubulin—the primary constituent of cytoskeletons—as a kind of cellular automaton, capable of signal propagation and processing, and discusses the implications for the generation of behavior and the transport of materials within the cell itself.

Finally, the paper by Braitenberg explores some of the complex problems associated with controlling multi-segmented arms and legs. He shows how a simple trick used by crane-operators for getting their loads to a delivery site without having to wait for pendular movements to die out may also be operating within our antagonist muscle groups. He also discusses the role of the cerebellum in managing smoothness of behavior over a number of joints, each of which is operating in parallel with others on the same articulated arm.

THE DISTILLED "ESSENCE" OF ARTIFICIAL LIFE

To summarize, the general consensus on the "essence" of Artificial Life at the workshop was converging on the following vision: Artificial Life involves the *realization* of lifelike behavior on the part of man-made systems consisting of *populations* of semi-autonomous entities whose *local interactions* with one another are governed by a set of *simple rules.* Such systems contain *no* rules for the behavior of the population at the global level, and the often complex, high-level dynamics and structures observed are *emergent* properties, which develop over time from out of all of the local interactions among low-level primitives by a process highly reminiscent of *embryological development*, in which *local hierarchies* of higher-order structures develop and *compete* with one another for support among the low-level entities. These emergent structures play a vital role in organizing the behavior of the lowest-level entities by establishing the context within which those entities invoke their local rules and, as a consequence, these structures may *evolve* in time.

FURTHER THEMES

There were many other themes that arose at some point during the workshop that have not received adequate coverage in these proceedings. We mention a few of these here.

Other man-made emergent processes were proposed as candidates for Artificial Life, such as economies, socio-political institutions, and other cultural structures. One can certainly view such systems from a genotype/phenotype perspective, with us as the primitive behav*or*s, our socio-cultural inheritance as the genotype, and in a dynamic evolution of string products. Thus, typogenetics captures the kind of molecular logic that operates in living cells, in which DNA codes for enzymes which may operate on the DNA, copying it, initiating the production of other enzymes, and so forth.

The next three papers describe work being pursued at the MIT Media Lab. The first paper by Resnick describes the LEGO/Logo system, in which LEGO robots are enhanced with sensors, effectors, and motors so that they may carry out instructions written in the Logo language. The Logo language has been extended to listen to sensors and give commands to effectors. Resnick sees the LEGO/Logo system as an ideal basis for an "Artificial Life Toolkit," to be used to construct simple robots in order to study the dynamic behavior produced by simple rules operating in the real world. Also, such a toolkit would be useful for studying the interactive dynamics supported by populations of simple robots especially if they can interchange segments of code. The paper gives examples of how to go about wiring up several of Braitenberg's "Vehicles" (described in Braitenberg's book of the same name) and discusses Seymor Papert's project involving the use of LEGO/Logo to teach programming concepts to children.

The next two papers illustrate the application of modern *AI techniques* to the generation of simple behaviors, something AI should have paid more attention to before it jumped into chess playing and other complex behaviors.

The paper by Coderre describes a method for behavior specification based on Minsky's *Society of Mind* theory, in which intelligent behavior is seen to result from the interactions of a population of expert *agents* within the mind. The "brains" of each organism in Coderre's model are represented by a hierarchical arrangement of expert-agents into "data-flow trees," reminiscent of Selfridge's Pandemonium model. Coderre demonstrates that such brains are capable of pursuing tasks despite interruptions and that they can manage appropriate shifts in focus of attention as the situation demands. The paper provides the full specification for such a brain and gives examples of the behaviors generated in a world populated with several such creatures.

The paper by Travers describes a system for the simulation of animal behavior which is a sort of software version of the LEGO/Logo system. Users can construct *artificial animals* out of various simulated components such as sensors, motors, effectors, wires, and neuron-like logic elements. The paper also discusses a version of the system which supports evolution of the artificial animals and gives several insightful examples of the manner in which unintended bugs in the system were discovered when the artificial animals evolved to exploit them.

The next two papers move from the domain of the computer to the molecular domain, discussing the rapidly evolving field of *nanotechnology*. In a sense, nanotechnology is an upside-down approach to artificial life. It describes methods for taking current mechanical technology and reducing it in size to the molecular scale. The approach taken in other works on Artificial Life, by contrast, is to attempt to *scale-up* the kind of "mechanical technology" evolved by living systems.

Schneiker's paper reviews the historical development and recent accomplishments of the field of nanotechnology in some detail, describing possible applications simulating and theorizing about life. Although they, too, may contain a centralized global controller, its role is restricted solely to managing the simulation of a parallel system on a traditional serial computer, and could—in principle—be "parallelized out" if the system were to be implemented on truly parallel hardware. This is *not* the case in most AI models of intelligence, in which the centralized global controller is a fundamental part of the model—an argument against their validity.

The paper by Hogeweg describes another system for the investigation of ecological dynamics that has been applied to the study of the spontaneous emergence and stabilizing effects of *social structure*. This paper is a very good illustration of the theme that an important aspect of the environment within which organisms operate and adapt is the environment produced by the dynamic activity of the organisms themselves. Hogeweg demonstrates the importance of the emergence of "self-structuring" dynamics within such populations, illustrating with specific examples their stability in the face of perturbations, the manner in which different regimes of social structure can follow upon one another in a highly adaptive manner, the correlations between social and spatial structure in populations, and the manner in which specific structures can be exploited by a population for a variety of different purposes simultaneously. There are many gems of truth in this paper and it is worthy of careful reading.

The paper by Goel and Thompson describes a tool for modeling *molecular self-organization*. "Moveable Finite Automata" models are like cellular automata, with the added feature that the cells can move about and form bonds with each other. The authors have derived rules for the individual automata sufficient to model the self-assembly of phage virus and the elongation cycle in protein biosynthesis. Both models illustrate the importance of "conformational switching," in which the formation of one bond may enable bonding at other sites. Processes like conformational switching are fundamental to understanding the manner in which local, distributed processing can lead to specific sequences of action at the global level, and can help explain the manner in which the genotype can control the specific sequential developmental history of the phenotype.

The paper by Lugowski proposes a model that is similar in spirit to the MFA model presented in the preceding paper, but considers applications beyond the realm of modeling molecular dynamics. It describes a cellular-automaton-like model in which cells can interchange position with their neighbors, and can thus follow rule-directed, "flipping" walks throughout the lattice. Lugowski gives examples of a number of self-organizing computations within the model, illustrating the kinds of control structures that can be built into low-level, distributed rules in order to give rise to specific global sequences and structures.

The paper by Morris describes "Typogenetics," a system originally devised by Douglas Hofstadter for illustrating "the molecular logic of the living state." Typogenetics—a contraction formed from "Typographical Genetics"—involves strings of characters which function like DNA, in that they code for operators which function like enzymes, operating on the typogenetic strings themselves: cutting, altering, copying them and so forth. The resulting products are then reinterpreted to produce more operators and the process is iterated over and over again, resulting

Dawkins' paper provides another important perspective on the distinction between genotype and phenotype, emphasizing the importance of *embryology:* the
process by which the phenotype develops under control of the genotype. Via a
simple but elegant computer program demonstrating the power of cumulative selection, Dawkins explores the manner in which the embryological process itself may
be considered a phenotypic trait, subject to natural selection, arguing that selection
will favor those embryologies richest in evolutionary potential. Dawkins' paper also
illustrates the amazingly complex structures that can develop from the application
of very simple recursive rules.

The next two papers discuss methods for studying *embryological* processes by
modeling the development and differentiation of plants.

The paper by Lindenmayer and Prusinkiewicz reviews *L-systems*, formal grammars for the description, analysis, and developmental simulation of multicellular
organisms. L-systems are perhaps the best known example of the general methodological approach advocated within Artificial Life. They are formal systems which
work in a bottom-up, parallel manner, applying simple recursive rules to local sub-
parts of developing structures. L-systems embody the genotype/phenotype distinction naturally. Many of the color plates show examples of plants grown—not just
rendered—using L-systems.

The paper by Oppenheimer describes the application of *fractals* to the generation of tree-like structures. Oppenheimer discusses eloquently the theme that
"by going beyond the laws of nature, we create a new reality, one that helps us
appreciate nature more." The artifacts produced in the pursuit of artificial life will
be both like nature and *not* like nature. Oppenheimer points out that not only the
similarities but the *differences* as well can provide valuable insights into the nature
of life.

The next four papers describe general purpose simulation systems that have
been developed for studying various aspects of life.

The first two systems provide for the investigation of the dynamics and evolution of *ecosystems.* The paper by Taylor et al. describes the RAM simulator: a
LISP-based simulation system within which populations of organisms are modeled
as populations of programs which are created, run, learn, communicate with each
other, interact with the environment (also a population of programs), replicate with
modification, and eventually terminate. Examples are given of the application of
the system to modeling specific ecosystems. This paper also discusses the positive
role that computers play by forcing theorists to make their theories explicit enough
that they may be implemented on a computer and tested empirically.

However, this brings up a cautionary theme from the workshop which is that
the use of computers *may* lead to theories which are *overly* tailored for computers,
in that they have expanded or restricted the theory too much in order that it may be
implemented in the traditional top-down, serial programming style. Many natural
phenomena depend on the fundamentally parallel "hardware" of nature, and are
inappropriately modeled by the incorporation of a centralized, global controller
managing high-level data-structures. Other models of computation—such as the
object-oriented paradigm adopted in the RAM simulator—are more appropriate for

the large-scale social structures that emerge out of our behaviors as the phenotype. We even engage in "social intercourse," replicating and passing-on the cultural equivalent of genes ("memes" in Dawkins' terminology.) This is a topic which is worthy of a workshop of its own, as there are as many differences as there are parallels between biological and cultural processes, and it still seems to be the case that the mere mention of "cultural evolution" is anathema in anthropological circles.

Several people raised the question of establishing a general measure of *complexity, adaptability,* or *progress* in evolutionary processes. Such a measure— or measures—would certainly be valuable, and there have been efforts to apply Shannon's entropy concept and Chaitin's algorithmic complexity measures to living systems and the process of evolution—with somewhat mixed results. The field awaits an appropriate, natural measure of complexity, as do many other fields.

There was little discussion of possible *practical applications* of Artificial Life. Most of the applications discussed at the workshop were related to obtaining a better *theoretical* understanding of life.

Nor was there much discussion of the moral or ethical issues involved in the pursuit of Artificial Life. Such issues *must* be discussed before we go much further down the road to creating life artificially. We are once again at a point where our technical grasp of a problem is far ahead of our moral understanding of the issues involved, or of the possible consequences of mastering the technology.

It is, perhaps, not coincidental that the first workshop on Artificial Life was held at Los Alamos, site of the mastery of atomic fission and fusion. Both technologies have tremendous potential for benefiting life on earth, but both also have tremendous potential for abuse, whether intentional or accidental. Whereas the technology of death was developed in secret, under government mandate, and with little official attention to social and moral consequences, this first workshop on the technology of life was held in the open, by voluntary participation, and we must see to it that it is pursued with the most careful attention to social and moral consequences.

ACKNOWLEDGEMENTS

Many people and groups contributed to the success of the workshop and to the production of these proceedings. My sincere thanks go out to all them.

THE WORKSHOP

David Campbell, Director of the Center for Nonlinear Studies (CNLS), provided financial, logistic, and—most importantly—strong and enthusiastic personal support for the workshop.

Marian Martinez of the CNLS was effectively my co-organizer for the workshop and she did an absolutely splendid job. Marian managed details ranging from organizing the data-base, to filling out the incredible number of LANL and DOE forms, to selecting the menu for the banquet, and maintained her cheerful, pleasant disposition and sense of humor through it all. I cannot thank her enough: without her the workshop would have been absolute chaos.

George Cowan and David Pines of the Santa Fe Institute provided additional financial support, and the technical assistance of Ron Zee and Ginger Richardson.

Barbara Bowen and Julie Lieber coordinated the additional monetary support provided by Apple Computer, Inc., with the assistance of Jessica Prince of Apple-UK.

Doyne Farmer of the Theoretical Division at Los Alamos was responsible for bringing me to the CNLS in the first place, and enthusiastically promoted the workshop. Doyne also gave a fine summary of the workshop on the final day.

A.K. Dewdney—a prince of a fellow—promoted the workshop in his "Computer Recreations" column in *Scientific American*, and did an excellent job of organizing the computer demonstration sessions. He also hand-lettered the prize certificates for the artificial 4H-Show, and wouldn't let them be awarded without first affixing ribbons of his own manufacture!

Peter Ford of the CNLS managed the large, heterogeneous array of computing machinery flawlessly. Computer equipment was loaned by Silicon Graphics (Randy Pibhany), Symbolics (Jim Fries), Apollo (Shep Bryan and Gary Esch), and Sun Microsystems (Marleen McDaniel, Andy Barnes, and Murray Stein). Dick Phillips of the Computer Graphics group at LANL loaned two Macintosh-II computers & Bill Corcoran of the Computing Division loaned an Apollo computer.

Stuart Kauffman, Norman Packard, Chris Barnes, Alan Perelson, and Doyne Farmer served as session chairmen. A.K. Dewdney, Doyne Farmer, Howard Pattee, and Eiiti Wada served as judges for the artificial 4H-Show.

Warren F. Miller, Jr., Deputy Director of the Los Alamos National Laboratory, made the opening remarks at the workshop and welcomed participants to LANL.

Derek Graham of Progressive Electronics, Las Cruces, NM, provided the video projectors and expertise that allowed all manner of computer demonstrations and video-tapes to be displayed on a large screen. Pat Vucenic of the LANL Computing Division also provided a video projector and expertise.

Floyd Archuleta, Leeroy Herrera, Patsy Martinez, Lisa Sanchez and Duane Valdez of the Oppenheimer Study Center provided excellent support to a somewhat more wild and crazy crowd of workshop participants than is their normal fare.

Jane Sharp, the Division Leader of Protocol at Los Alamos National Laboratory, oversaw organizational details.

Frankie Gomez, Dorothy Garcia, and Valerie Ortiz of the CNLS all served reception-desk duty at the workshop and provided their secretarial skills.

Ted Kaehler of Apple provided technical support for the Apple Macintosh computers.

Jack Challem of Public Affairs at LANL handled diverse inquiries and media requests for information about the workshop. Nan Moore of the LANL Director's Office facilitated the Director's approval for the workshop and other requests requiring high-level LANL authorization. Paul Baca at LANL Receiving intercepted computers and other equipment shipped to Los Alamos and made sure they got to the Oppenheimer Study Center as quickly as possible.

Harry Stanton and Terry Lamoureux of the MIT Press provided a pre-release copy of Braitenberg's Vehicle program for the Macintosh, and provided samples from their fine line of books.

David Langton recorded most of the talks and demonstrations on video-tape.

Richard Bagley, Chris Barnes, Philippe Binder, Doyne Farmer, Gottfried Mayer-Kress, Irene Stadnyk, John Holland, and other members of the CNLS Artificial Life Seminar all provided useful suggestions and assistance. Thanks also to Maureen Gremillion.

Financial support was provided in part by the U.S. Department of Energy under contract # W-7405-ENG-36.

THE PROCEEDINGS

George Cowan and David Pines of the Santa Fe Institute offered to publish the proceedings as part of the series: *Santa Fe Institute Studies in the Sciences of Complexity*, and have supported the workshop from the beginning.

Ronda K. Butler-Villa, Director of Publications at the Santa Fe Institute, has effectively been my co-editor for the proceedings and has done an excellent job. She took on the enormous task of converting all of the submitted articles to TEX format, has handled all of the arrangements with Addison-Wesley, took care of sending out author's proofs, compiling the index, laying out the various figures and photographs, and so forth. She has remained calm, cool, and collected as deadlines approached and passed. These proceedings could not have been produced without her. She's another person whom I cannot thank enough.

Alan Wylde, Director of the Advanced Book Program at Addison-Wesley has enthusiastically supported the proceedings from the very beginning, and has been very understanding of the broken deadlines. Many thanks to Celina Gonzales in Marketing and Karen Garrison in Production for their patient understanding and helpful suggestions.

Jolene Manning at Southwest Color Separations, Albuquerque, NM, oversaw the production of the negatives for the color plates. Kurt Vanderpile of Black & White Photo Lab, Santa Fe, NM, produced high-quality prints for some of the figures.

Peter Ford of the CNLS has maintained a high-quality computer environment during the process of putting the proceedings together. He also reconstructed several files—block-by-block—from the disk one grim evening after I had inadvertently erased them.

AND LAST BUT NOT LEAST...

My wife Elvira and my two sons Gabriel and Colin have been remarkably understanding during what has amounted to a full year devoted to the organization of
the workshop and the production of the proceedings.

one example of the kind of complex dynamics that cellular automata are capable of generating making use solely of *local* state information. The emergence of such complex global dynamics from out of the local interactions of simple components was one of the major themes of the workshop.

The paper by Zeleny et al. illustrates the lifelike structures and developmental histories exhibited by *osmotic growth* processes. Coining the term "synthetic biology," they review and recreate early experiments carried out by Stephane Leduc on the spontaneous generation of life. This is an example of a *naturally occurring* chemical process which exhibits lifelike behavior, and is therefore worthy of further scientific investigation, whether or not we feel that it has played a specific role in the development of *our* life.

The next four papers treat the process of *evolution*. The paper by Packard is a very nice illustration of the fact that it is not just organisms that evolve, but organisms and their environments together. Important features of an organism's environment may be directly attributed to actions taken by the organism itself or by its cohabitors in the environment. Thus, the "adaptive landscape" will be a dynamic—rather than a fixed—abstraction, and hence the "fitness function" that determines which organisms are more successful in the environment will be an emergent property of the ecosystem.

The paper by Wilson provides a simple example of an application of the *genetic algorithm* to the simulation of the evolution of multicellular systems. Pioneered by John Holland, the genetic algorithm is a scheme for applying the power of Darwinian evolution under natural selection to real-world optimization problems. As Wilson points out, however, it has rarely been used to study the evolution of life itself. Wilson gives examples of the development of some simple one-dimensional, multicellular organisms.

The article by Moravec reviews the developmental history of *robotics*, arguing that *cultural evolution* is not only the dominant evolutionary process today, but may well replace biological evolution altogether in the foreseeable future. Natural selection concentrates on the *phenotype*, which is essentially an organism's dynamic interface with the physical world. Since selection operates at the level of the phenotype, evolution can reward extra-genetic mechanisms that contribute to the phenotype, e.g., culture. Should the extra-genetic contributions to the determination of the phenotype become more and more dominant, there is the chance that eventually *all* critical phenotypic traits will be "extra-genetically" determined, setting the stage for a "genetic takeover," in which the the previously "extra-genetic" system becomes the "primary" genetic system, taking over from the old genetic system, which may eventually be lost altogether as its contributions to the phenotype become more and more irrelevant. This may have happened once already in the history of life, as suggested by Cairns-Smith (see his work referenced in the annotated bibliography). Thus, the connection between a genetic system and the phenotype it determines is an arbitrary one, and the "mechanics" of evolution will jump ship if they find a more efficient vehicle for the determination of the phenotype. In cultural evolution, Moravec suggests, we are catching the mechanics of evolution in the act of jumping ship.

we could not print every paper, and this has offended some, for which I am truly sorry. However, my hope is that these proceedings will serve us all by helping to establish the field of Artificial Life as a legitimate domain of scientific inquiry.

THE PAPERS

The papers are roughly organized according to topic or method.

The first three papers focus on the general theme of Artificial Life, its history, techniques, and various associated methodological issues.

The first article is my attempt to capture that elusive collective vision of the *essence* of Artificial Life and to trace its historical development. I think that I have been only partially successful, and the article should certainly not be taken as "defining" artificial life in any way. It should be seen, rather, as my own personal report on what I felt to be the consensus that was emerging at the workshop.

The article by Laing reviews von Neumann's theory of *self-reproducing automata*, and describes the various other likelike capabilities that have been acquired by automata since von Neumann. Laing then discusses the difficult *legal and ethical issues* that arise when we begin to think of machines as living things. Finally, Laing describes the results of a NASA commissioned study exploring in detail the economic and technical feasibility of constructing a self-reproducing lunar mining facility.

Pattee reviews the distinctions between simulations, realizations, and theories of life, arguing that simulations and realizations belong to different categories of modeling. He also argues that our criteria for adequate models of life should depend on more than mere mimicry: more than an Artificial Life equivalent of the "Turing test" widely accepted within the field of Artificial Intelligence. Rather, he argues, Artificial Life models must be evaluated in light of specific theories of living systems.

The next three articles are concerned with the *origin of life*. There was more discussion of this problem at the workshop than is reflected in these proceedings. However, there have been many conferences on the origin of life, and there are many adequate books reviewing the various theories in existence. I chose to reserve space in these proceedings for topics and models which have not had such a public forum. For more on the origin of life, the reader is referred to the annotated bibliography and the works of Eigen, Kauffman, Dyson, Orgel, Oparin, Bernal, Fox, and others.

The first two articles illustrate methodologies that appear especially well suited to modeling life and its possible origins. The paper by Rasmussen illustrates the use of random graphs for analyzing the dynamic and static properties of *catalyzed reaction networks*. Here, Rasmussen applies random graphs to the study of the emergence of such networks within a sufficiently rich "pre-biotic soup," and derives estimates for the expected time-to-emergence of *autocatalytic* reaction cycles within such graphs, assuming physically realistic parameters and rates.

The paper by Tamayo and Hartman discusses the relevance of *reaction-diffusion systems* to the origin of life, indicating how such systems can be modeled and analyzed using *cellular automata*. The Greenberg-Hastings system discussed is but

Thus, the primary goal of the first workshop on Artificial Life was to see what was out there, and to present as many different methodological approaches as possible within a receptive and unbiased atmosphere. Although many of the participants were already familiar with some of the models, I think it is safe to say that many of the models and systems presented were new to most of us, and that *some* of the models were genuinely quite surprising to everybody.

Throughout the workshop, there was a growing sense of excitement and camaraderie—even profound relief—as previously isolated research efforts were opened up to one another for the first time. It quickly became apparent that despite the isolation we had all experienced a remarkably similar set of problems, frustrations, successes, doubts, and visions. Even more exciting was that, as the workshop progressed, one could sense an emerging consensus among the participants—a slowly dawning collective realization—of the "essence" of Artificial Life. Although I think that none of us could have put it into words at the time, I think that many of us went away from that tumultuous interchange of ideas with a very similar vision, strongly based on themes such as *bottom-up* rather than *top-down* modeling, *local* rather than *global* control, *simple* rather than *complex* specifications, *emergent* rather than *prespecified* behavior, *population* rather than *individual* simulation, and so forth.

Perhaps, however, the most fundamental idea to emerge at the workshop was the following: Artificial systems which exhibit lifelike behaviors are worthy of investigation on their own rights, whether or not we think that the processes that they mimic have played a role in the development or mechanics of life as *we* know it to be. Such systems can help us expand our understanding of life as it *could* be. By allowing us to view the life that has evolved here on Earth in the larger context of *possible* life, we many begin to derive a truly general theoretical biology capable of making universal statements about life wherever it may be found and whatever it may be made of.

THE PROCEEDINGS

The primary goal for these proceedings was to recapture in print the stimulating mix of ideas and methods that were presented at the workshop.

The papers included in these proceedings were selected either because they did a good job of presenting a specific model, giving examples of its application, and placing it in the context of the larger effort, or because they did a good job of reviewing a general class of models—or a whole field of research—relevant to Artificial Life. Furthermore, the collection of papers as a whole was selected because it provided reasonably good coverage of both the diversity of ideas presented, and the general themes that emerged during the course of the workshop.

Of the 40 or so papers that were submitted, a little over 20 were selected for inclusion in the proceedings. In order to keep the book manageably readable in size,

Christopher G. Langton
Center for Nonlinear Studies, Los Alamos National Laboratory, Los Alamos, NM 87545

Preface

THE ARTIFICIAL LIFE WORKSHOP

In September 1987, the first workshop on Artificial Life was held at the Los Alamos National Laboratory. Jointly sponsored by the Center for Nonlinear Studies, the Santa Fe Institute, and Apple Computer Inc, the workshop brought together 160 computer scientists, biologists, physicists, anthropologists, and other assorted "-ists," all of whom shared a common interest in the simulation and synthesis of living systems. During five intense days, we saw a wide variety of models of living systems, including mathematical models for the origin of life, self-reproducing automata, computer programs using the mechanisms of Darwinian evolution to produce co-adapted ecosystems, simulations of flocking birds and schooling fish, the growth and development of artificial plants, and much, much more

The workshop itself grew out of my frustration with the fragmented nature of the literature on biological modeling and simulation. For years I had prowled around libraries, shifted through computer-search results, and haunted bookstores, trying to get an overview of a field which I sensed existed but which did not seem to have any coherence or unity. Instead, I literally kept stumbling over interesting work almost by accident, often published in obscure journals if published at all.

Contents

Artificial Life
    Christopher G. Langton   1

Artificial Organisms: History, Problems, and Directions
    Richard Laing   49

Simulations, Realizations, and Theories of Life
    H. H. Pattee   63

Towards a Quantitative Theory of the Origin of Life
    Steen Rasmussen   79

Cellular Automata, Reaction-Diffusion Systems, and the
Origin of Life
    Pablo Tamayo and Hyman Hartman   105

Precipitation Membranes, Osmotic Growths, and Synthetic
Biology
    Milan Zeleny, George J. Klir and Kevin D. Hufford   125

Evolving Bugs in a Simulated Ecosystem
  Norman Packard   141

The Genetic Algorithm and Simulated Evolution
  Stewart W. Wilson   157

Human Culture: A Genetic Takeover Underway
  Hans Moravec   167

The Evolution of Evolvability
  Richard Dawkins   201

Developmental Models of Multicellular Organisms:
A Computer Graphics Perspective
  Aristid Lindenmayer and Przemyslaw Prusinkiewicz   221

The Artificial Menagerie
  Peter Oppenheimer   251

RAM: Artificial Life for the Exploration of Complex
Biological Systems
  Charles E. Taylor, David R. Jefferson, Scott R. Turner   275
  and Seth R. Goldman

Mirror Beyond Mirror: Puddles of Life
  P. Hogeweg   297

Movable Finite Automata (MFA): A New Tool for Computer
Modeling of Living Systems.
  Narendra S. Goel and Richard L. Thompson   317

Computational Metabolism: Towards Biological Geometries
for Computing
  Marek W. Lugowski   341

Typogenetics: A Logic for Artificial Life
  Harold C. Morris   369

Lego, Logo, and Life
  Mitchel Resnick   397

Modeling Behavior in Petworld
  Bill Coderre   407

Animal Construction Kits
  Michael Travers   421

Nanotechnology with Feynman Machines: Scanning
Tunneling Engineering and Artificial Life
  Conrad Schneiker   443

Biological and Nanomechanical Systems: Contrasts in
Evolutionary Capacity
  K. Eric Drexler   501

Molecular Automata in Microtubules: Basic Computational
Logic of the Living State?
  Stuart Hameroff, Steen Rasmussen and Bengt Månsson   521

Some Types of Movements.
  Valentino Braitenberg   555

Annotated Bibliography

  Section 1: Annotated Bibliography of Literature in   567
  the Field of Artificial Life

  Chris Langton and workshop participants

  Section 2: An Annotated Bibliography of Plant   625
  Modeling and Growth Simulation

  Aristid Lindenmayer and Przemyslaw Prusinkiewicz

Index
                                                                            645

ARTIFICIAL
LIFE

Christopher G. Langton
Center for Nonlinear Studies, Los Alamos National Laboratory, Los Alamos, NM 87545;
internet address cgl@lanl.gov

Artificial Life

Vitalism amounted to the assertion that living things do not behave as though they were nothing but mechanisms constructed of mere material components; but this presupposes that one knows what mere material components are and what kind of mechanisms they can be built into.
— C.H. Waddington, *The Nature of Life.*

Artificial Life is the study of man-made systems that exhibit behaviors characteristic of natural living systems. It complements the traditional biological sciences concerned with the *analysis* of living organisms by attempting to *synthesize* life-like behaviors within computers and other artificial media. By extending the empirical foundation upon which biology is based *beyond* the cabon-chain life that has evolved on Earth, Artificial Life can contribute to theoretical biology by locating *life-as-we-know-it* within the larger picture of *life-as-it-could-be.*

THE BIOLOGY OF POSSIBLE LIFE

Biology is the scientific study of life—in principle anyway. In practice, biology is the scientific study of life based on carbon-chain chemistry. There is nothing in its charter that restricts biology to the study of carbon-based life; it is simply that this is the only kind of life that has been available for study. Thus, theoretical biology has long faced the fundamental obstacle that it is difficult, if not impossible, to derive general theories from single examples.

Certainly life, as a dynamic physical process, could "haunt" other physical material: the material just needs to be organized in the right way. Just as certainly, the dynamic processes that constitute life—in whatever material bases they might occur—must share certain universal features—features that will allow us to recognize life by its dynamic *form* alone, without reference to its *matter*. This *general* phenomenon of life—life writ-large across all possible material substrates—is the true subject matter of biology.

Without other examples, however, it is extremely difficult to distinguish essential properties of life—properties that must be shared by any living system *in principle*—from properties that are incidental to life, but which happen to be universal to life on Earth due *solely* to a combination of local historical accident and common genetic descent. Since it is quite unlikely that organisms based on different physical chemistries will present themselves to us for study in the foreseeable future, our only alternative is to try to synthesize alternative life-forms ourselves—*Artificial Life*: life made by man rather than by nature.

ARTIFICIAL LIFE

Only when we are able to view *life-as-we-know-it* in the larger context of *life-as-it-could-be* will we really understand the nature of the beast. Artificial Life (AL) is a relatively new field employing a *synthetic* approach to the study of *life-as-it-could-be*. It views life as a property of the *organization* of matter, rather than a property of the matter which is so organized.

Whereas biology has largely concerned itself with the *material* basis of life, Artificial Life is concerned with the *formal* basis of life. Biology has traditionally started at the top, viewing a living organism as a complex biochemical machine, and worked *analytically* downwards from there—through organs, tissues, cells, organelles, membranes, and finally molecules—in its pursuit of the mechanisms of life. Artificial Life starts at the bottom, viewing an organism as a large population of *simple* machines, and works upwards *synthetically* from there—constructing large aggregates of simple, rule-governed objects which interact with one another nonlinearly in the support of life-like, global dynamics.

The "key" concept in AL is *emergent behavior*. Natural life emerges out of the organized interactions of a great number of nonliving molecules, with no global controller responsible for the behavior of every part. Rather, every part is a behavior itself, and life is the behavior that emerges from out of all of the local interactions among individual behavors. It is this bottom-up, distributed, local determination of behavior that AL employs in its primary methodological approach to the generation of lifelike behaviors.

ARTIFICIALITY

The dictionary[1] defines the term "artificial" as "made by man, rather than occurring in nature," but there is another sense of the term that is more appropriate for the study of Artificial Life. This sense was best captured by Simon in his excellent monograph *The Sciences of the Artificial*[38]:

> Artificiality connotes perceptual similarity but essential difference, resemblance from without rather than within. The artificial object imitates the real by turning the same face to the outer system...imitation is possible because distinct physical systems can be organized to exhibit nearly identical behavior...Resemblance in behavior of systems without identity of the inner systems is particularly feasible if the aspects in which we are interested arise out of the *organization* of the parts, independently of all but a few properties of the individual components.

Thus, Artificial Life studies natural life by attempting to capture the behavioral essence of the constituent components of a living system, and endowing a collection of artificial components with similar behavioral repertoires. If organized correctly, the aggregate of artificial parts should exhibit the same dynamic behavior as the natural system.

This bottom-up modeling technique can be applied at any level of the hierarchy of living systems in the natural world—from modeling molecular dynamics on millisecond time-scales to modeling evolution in populations over millennia. At any such level, behavioral primitives are identified, rules for their behavior in response to local conditions are specified, the primitive behavors are organized similarly to their natural counterparts, and the behavior of interest is allowed to emerge "on the shoulders" of all of the myriad local interactions among low-level primitives taken collectively.

The ideal tool for this synthetic approach to the study of life is the computer. However, the traditional computer program—a centralized control structure with global access to a large set of predefined data-structures—is inappropriate for synthesizing life within computers. A new approach to computation is required, one that focuses on *ongoing dynamic behavior* rather than on any *final result*.

The essential features of computer-based Artificial Life models are:

- They consist of populations of simple programs or specifications.
- There is no single program that directs all of the other programs.
- Each program details the way in which a simple entity reacts to local situations in its environment, including encounters with other entities.
- There are *no* rules in the system that dictate global behavior.

■ Any behavior at levels higher than the individual programs is therefore emergent.

To illustrate, consider modeling a colony of ants. We would provide simple specifications for the behavioral repertoires of different *castes* of ants, and create lots and lots of instances of each caste. We would start up this population of "antomata" (a term coined by Doyne Farmer) from some initial configuration within a simulated two-dimensional environment. From then on, the behavior of the system would depend entirely on the collective results of all of the local interactions between individual antomata and between individual antomata and features of their environment. There would be no single "drill-sergeant" antomaton choreographing the ongoing dynamics according to some set of high-level rules for colony-behavior. The behavior of the colony of antomata would emerge from out of the behaviors of the individual antomata themselves, just as in a real colony of ants.

Each of these antomata is a behavor. We will identify such behavors as simple *machines*. Artificial Life is concerned with tuning the behaviors of such low-level machines so that the behavior that emerges at the global level is *essentially* the same as some behavior exhibited by a natural living system.

THE ANIMATION OF MACHINES

animate *tr.v.* **1.** To give life to; fill with life.[1]

How are we to go about *animating* machines? How are we to go about bringing machines to life—or bringing life to machines?

The etymological ancestry of the term "animate" goes back to the Indo-European root **ane** meaning "to breathe," which is also ancestral to the Latin *animus*, denoting reason, mind, soul, spirit, life, breath, and etc. This corresponds to the notion that life is some kind of "energy," "force," or "essence"—that which leaves the physical body upon death, and lacking which mere material flesh and bone could not live. Thus, throughout historical time, the notion of giving life to some material object involved the act of "breathing" this mystical force or essence into an otherwise inanimate material body.

This notion that life was an extra "something" necessary *over and above* the detailed organization of a material organism is known as *vitalism*. Vitalism was championed especially strongly during the last two centuries as a defense against the growth of materialism and the scientific method which, especially after Darwin, threatened to explain everything in nature—including man. Worse yet, they threatened to do so without recourse to the supernatural or to God, but only by reference to everyday physical phenomena and materials, thereby removing man from his exalted position in this universe of otherwise mere material objects.

Biologists today reject vitalism, believing rather that life as we know it will eventually be explainable completely within the context of biochemistry. Thus, most biologists would agree—in principle anyway—with the following statement:

living organisms are nothing more than complex biochemical machines. However, they are different from the machines of our everyday experience. A living organism is *not* a single, complicated biochemical machine. Rather it must be viewed as a large *population* of relatively simple machines. The complexity of its behavior is due to the highly nonlinear nature of the interactions between all of the members of this polymorphic population. To animate machines, therefore, is not to "bring" life to *a* machine; rather it is to organize a population of machines in such a way that their interactive dynamics is "alive."

THE BEHAVIOR GENERATION PROBLEM

Artificial Life is concerned with generating *lifelike* behavior. Thus, it focuses on the problem of creating *behavior generators*. A good place to start is to identify the mechanisms by which behavior is generated and controlled in natural systems, and to recreate these mechanisms in artificial systems. This is the course we will take later in this paper.

The related field of Artificial Intelligence is supposedly concerned with generating *intelligent* behavior. It, too, focuses on the problem of creating behavior generators. However, although it initially looked to natural intelligence to identify its underlying mechanisms, these mechanisms were not known, nor are they today. Therefore, following an initial flirt with neural nets, AI became wedded to the only other known vehicle for the generation of complex behavior: the technology of serial computer programming. As a consequence, from the very beginning Artificial Intelligence embraced an underlying methodology for the generation of intelligent behavior that bore no demonstrable relationship to the method by which intelligence is generated in natural systems. In fact, AI has focused primarily on the production of intelligent *solutions* rather than on the production of intelligent *behavior*. There is a world of difference between these two possible foci.

By contrast, Artificial Life has the great good fortune that many of the mechanisms by which behavior arises in natural living systems are now known. There are still many holes in our knowledge, but the general picture is in place. Therefore, Artificial Life can remain true to natural life, and has no need to resort to the sort of infidelity that is only now coming back to haunt AI. Furthermore, Artificial Life is not concerned with building systems that reach some sort of solution. For AL systems, the *ongoing dynamics* is the behavior of interest, not the state ultimately reached by that dynamics.

The key insight into the natural method of behavior generation is gained by noting that *nature is fundamentally parallel.* This is reflected in the "architecture" of natural living organisms, which consist of many millions of parts, each one of which has its own behavioral repertoire. Living systems are highly distributed, and quite massively parallel. If our models are to be true to life, they must also be highly distributed and quite massively parallel. Indeed, it is unlikely that any other approach will prove viable.

PREVIEW

In the remainder of the paper, we will discuss a number of different aspects of the field of Artificial Life. First we will review the history of man's attempts to simulate life, trying to identify major threads of intellectual development that have proven essential to the enterprise.

Second, we will review the genotype/phenotype distinction in living organisms, viewing the genotype as a specification for *machinery*, and the phenotype as the behavior of the machinery so specified. We will then generalize the concepts of genotype and phenotype, so that we may apply them to the task of generating behavior in artificial systems.

Next, we will review the methodology of *recursively generated objects*, which makes natural use of the genotype/phenotype distinction, and we will give examples of its application to the generation of specific lifelike behaviors. Finally we discuss the problem of *generating* behavior generators, for which we turn to the process of evolution, and a discussion of Genetic Algorithms.

Throughout, the focus will be on machines and the behaviors that they are capable of generating. The field of Artificial Life is unabashedly mechanistic and reductionist. However, this *new mechanism*—based as it is on multiplicities of machines and on recent results in the fields of nonlinear dynamics, chaos theory, and the formal theory of computation—is vastly different from the mechanism of the last century.

HISTORICAL ROOTS OF ARTIFICIAL LIFE

Mankind has a long history of attempting to map the mechanics of his contemporary technology onto the workings of nature, trying to understand the latter in terms of the former.

The earliest mechanical technologies provided tools that extended man's physical abilities and greatly reduced the labor required to make a living. Early technologies yielded tools for moving water, for manipulating stone and timber, and for obtaining and processing food. Tools allowed mankind to alter the natural order of things to suit his purposes and needs.

However, there was much about nature that could not be altered—such as the progression of the seasons—in the face of which man had to alter *his* behavior to fit the natural order of things. In order to do so, it was useful to be able to build *models* of nature that allowed predictions to be made about when certain events would—or should—take place. Models were developed that allowed the anticipation of floods, the determination of when to plant and when to harvest food, and the prediction of the motion of the sun, moon, and planets through the heavens. Models allowed man to alter *his* behavior in order to take fuller advantage of the natural order of things.

Building a model is a little bit like building a machine of some sort. The art of modeling is a technology in itself, one which produced tools that extended man's mental abilities; tools of thought which greatly reduced the mental labor required to make a living. When the mechanical technology of the time was sufficiently advanced, these tools of thought were eventually committed to hardware, becoming physical machines. Thus, the history of machines involves a continuing process of rendering in hardware progressively more complicated sequences of actions—physical and/or mental—previously carried out solely by recourse to muscle and brain.

It is not surprising, therefore, that early models of life reflected the principal technology of their era. The earliest models were simple statuettes and paintings—works of art which captured the static form of living things. Later, these statues were provided with articulated arms and legs in the attempt to capture the dynamic form of living things. These simple statues incorporated no internal dynamics, requiring human operators to make them behave.

The earliest mechanical devices that were capable of generating their own behavior were based on the technology of water transport. These were the early Egyptian water clocks called *Clepsydra*. These devices made use of a rate-limited process —in this case the dripping of water through a fixed orifice—to indicate the progression of another process—the position of the sun. Ctesibius of Alexandria developed a water-powered mechanical clock around 135 B.C. which employed a great deal of the available hydraulic technology—including floats, a siphon, and a water-wheel-driven train of gears.

In the first century A.D., Hero of Alexandria produced a treatise on *Pneumatics*, which described, among other things, various gadgets in the shape of animals and humans that utilized pneumatic principles to generate simple movements.

However, it was really not until the age of mechanical clocks that artifacts exhibiting complicated internal dynamics became possible. Around 850 A.D., the *mechanical escapement* was invented, which could be used to regulate the power provided by falling weights. This invention ushered in the great age of clockwork technology. The earliest mechanical clock to make use of this regulation scheme seems to have been developed by Richard of Wallingford in 1326. Later, following Galileo, came pendulum clocks, and further ingenious developments in escapements for the regulation of rate. Throughout the Middle Ages and the Renaissance, the history of technology is largely bound up with the technology of clocks. Clocks often constituted the most complicated and advanced application of the technology of an era.[1]

Perhaps the earliest clockwork simulations of life were the so-called "Jacks": mechanical "men" incorporated in early clocks which would swing a hammer to strike the hour on a bell. The word "jack" is derived from "jaccomarchiadus," which means "the man in the suit of armour." These accessory figures retained their popularity even after the spread of clock dials and hands—to the extent that clocks

---

[1]This association of machinery with the inexorable flow of time may be largely responsible for the spectre of predestination associated with the early philosophy of mechanism.

were eventually developed in which the function of time-keeping was secondary to the control of large numbers of figures engaged in various activities, even acting out entire plays.

Finally, clockwork mechanisms appeared which had done away altogether with any pretense at time-keeping. These "automata" were entirely devoted to imparting lifelike motion to a mechanical figure or animal. These mechanical automaton simulations of life included such things as elephants, peacocks, singing birds, musicians, and even fortune tellers.

This line of development reached its peak in the famous duck of Vaucanson, described as "an artificial duck made of gilded copper who drinks, eats, quacks, splashes about on the water, and digests his food like a living duck."[2]

> There has never been a more famous automaton than Vaucanson's duck. In 1735 Jacques de Vaucanson arrived in Paris at the age of 26. Under the influence of contemporary philosophic ideas, he had tried, it seems, to reproduce life artificially.

Unfortunately, neither the duck itself nor any technical descriptions or diagrams remain that would give the details of its construction. The complexity of the mechanism is attested to by the fact that one single wing contained over 400 articulated pieces.

One of those called upon to repair Vaucanson's duck was a "mechanician" named Reichsteiner, who was so impressed with it that he went on to build a duck of his own—also now lost—which was exhibited in 1847. Here is an account of this duck's operation from the newspaper *Das Freie Wort*:

> After a light touch on a point on the base, the duck in the most natural way in the world begins to look around him, eyeing the audience with an intelligent air. His lord and master, however, apparently interprets this differently, for soon he goes off to look for something for the bird to eat. No sooner has he filled a dish with oatmeal porridge than our famished friend plunges his beak deep into it, showing his satisfaction by some characteristic movements of his tail. The way in which he takes the porridge and swallows it greedily is extraordinarily true to life. In next to no time the basin has been half emptied, although on several occasions the bird, as if alarmed by some unfamiliar noises, has raised his head and glanced curiously around him. After this, satisfied with his frugal meal, he stands up and begins to flap his wings and to stretch himself while expressing his gratitude by several contented quacks.

> But most astonishing of all are the contractions of the bird's body clearly showing that his stomach is a little upset by this rapid meal and the effects of a painful digestion become obvious. However, the brave little bird holds out, and after a few moments we are convinced in the most concrete manner

---

[2]See Chapuis[7] regarding all quotes concerning these mechanical ducks.

that he has overcome his internal difficulties. The truth is that the smell which now spreads through the room becomes almost unbearable. We wish to express to the artist inventor the pleasure which his demonstration gave to us.

Figure 1 shows two views of one of the ducks—there is some controversy as to whether it is Vaucanson's or Reichsteiner's.

THE DEVELOPMENT OF CONTROL MECHANISMS

Out of the technology of the clockwork regulation of automata came the more general—and perhaps ultimately more important—technology of *process control*. As attested to in the descriptions of the mechanical ducks, some of the clockwork mechanisms had to control remarkably complicated actions on the part of the automata, not only *powering* them but *sequencing* them as well.

Control mechanisms evolved from early, simple devices—such as a lever attached to a wheel which converted circular motion into linear motion—to later, more complicated devices—such as whole sets of cams upon which would ride many interlinked mechanical arms, giving rise to extremely complicated automaton behaviors.

FIGURE 1   Two views of the mechanical duck attributed to Vaucanson. Printed in *Automata: A Historical and Technological Study* by Alfred Chapuis and Edmond Droz, published by B. A. Batsford Ltd.

FIGURE 2   Two views of a drawing automaton built by the Jaquet-Droz family. Printed in *Automata: A Historical and Technological Study* by Alfred Chapuis and Edmond Droz, published by B. A. Batsford Ltd.

Eventually *programmable controllers* appeared, which incorporated such devices as interchangeable cams, or drums with movable pegs, with which one could program arbitrary sequences of actions on the part of the automaton. The writing and picture drawing automata of Figure 2, built by the Jaquet-Droz family, are examples of programmable automata. The introduction of such programmable controllers was one of the primary developments on the road to general purpose computers.

ABSTRACTION OF THE LOGICAL "FORM" OF MACHINES

During the early part of the 20th century, the formal application of logic to the mechanical process of arithmetic lead to the abstract formulation of a "procedure." The work of Church, Kleene, Gödel, Turing, and Post formalized the notion of a logical sequence of steps, leading to the realization that the essence of a mechanical process—the "thing" responsible for its dynamic behavior—is not a thing at all, but an abstract control structure, or "program"—a sequence of simple actions selected from a finite repertoire. Furthermore, it was recognized that the essential

features of this control structure could be captured within an abstract set of rules—
a formal specification— without regard to the material out of which the machine
was constructed. The "logical form" of a machine was separated from its material
basis of construction, and it was found that "machineness" was a property of the
former, not of the latter. Of course, the principle assumption made in Artificial Life
is that the "logical form" of an organism can be separated from its material basis
of construction, and that "aliveness" will be found to be a property of the former,
not of the latter.

Today, the formal equivalent of a "machine" is an *algorithm:* the logic underlying the dynamics of an automaton, regardless of the details of its material
construction. We now have many formal methods for the specification and operation of abstract machines, such as programming languages, formal language theory,
automata theory, recursive function theory, etc. Many of these have been shown to
be logically equivalent.

Once we have learned to think of machines in terms of their abstract, formal
specifications, we can turn around and view abstract, formal specifications as potential machines. In mapping the machines of our common experience to formal
specifications, we have by no means exhausted the space of *possible* specifications.
Indeed, most of our individual machines map to a very small subset of the space of
specifications—a subset largely characterized by methodical, boring, uninteresting
dynamics. When placed together in aggregates, however, even the simplest machines
can participate in *extremely* complicated dynamics.

GENERAL PURPOSE COMPUTERS

Various threads of technological development—programmable controllers, calculating engines, and the formal theory of machines—have come together in the general
purpose, stored program computer. Programmable computers are extremely general behavior generators. They have no intrinsic behavior of their own. Without
programs, they are like formless matter. They must be told how to behave. By
submitting a program to a computer—that is: by giving it a formal specification
for a machine—we are telling it to behave as if it were the machine specified by
the program. The computer then "emulates" that more specific machine in the
performance of the desired task. Its great power lies in its plasticity of behavior.
If we can provide a step-by-step specification for a specific kind of behavior, the
chameleon-like computer will exhibit that behavior. Computers should be viewed
as *second-order* machines—given the formal specification of a first-order machine,
they will "become" that machine. Thus, the space of possible machines is directly
available for study, at the cost of a mere formal description: computers "realize"
abstract machines.

FORMAL LIMITS OF MACHINE BEHAVIORS

Although computers—and by extension other machines—are capable of exhibiting a bewilderingly wide variety of behaviors, we must face two fundamental limitations on the kinds of behaviors that we can expect of computers.

The first limitation is one of *computability in principle*. There are certain behaviors that are "uncomputable"—behaviors for which *no* formal specification can be given for a machine which will exhibit that behavior. The classic example of this sort of limitation is Turing's famous *halting problem:* can we give a formal specification for a machine which, when provided with the description of *any* other machine together with its initial state, will—by inspection alone—determine whether or not that machine will reach its halt state? Turing proved that no such machine can be specified. Rice and others[23] have extended this undecidability result to the determination—by inspection alone—of *any* non-trivial property of the future behavior of an arbitrary machine.

The second limitation is one of *computability in practice*. There are many behaviors for which we do not know how to specify a sequence of steps which will cause the computer to exhibit that behavior. We can automate what we know how to do already, but there is much that we do not know how to do. Thus, although a formal specification for a machine which will exhibit a certain behavior may be possible *in principle*, we have no formal procedure for producing that formal specification in practice, short of a trial-and-error search through the space of possible descriptions.

We need to separate the notion of a formal specification of a machine—that is, a specification of the *logical structure* of the machine—from the notion of a formal specification of a machine's behavior—that is, a specification of the *sequence of transitions* that the machine will undergo. We have formal systems for the former, but not for the latter. In general, we can neither derive behaviors from specifications nor derive specifications from behaviors.

The moral is: in order to determine the behavior of some machines, there is no recourse but to run them and see how they behave! This has consequences for the methods by which we (or nature) go about *generating* behavior generators themselves, which we will take up in the section on evolution.

FROM MECHANICS TO LOGIC

With the development of the general purpose computer, attention turned from the *mechanics* of life to the *logic* of life. The computer's tremendous capacity for emulation made it possible to explore the behaviors of a great many possible machines— machines which would probably never have been committed to hardware. The 1950's and 1960's saw an explosion of interest in computer and electro-mechanical models of life.

VON NEUMANN AND AUTOMATA THEORY The first computational approach to the generation of lifelike behavior was due to the brilliant Hungarian mathematician John von Neumann. In the words of his colleague Arthur W. Burks, von Neumann was interested in the general question[6]:

> What kind of logical organization is sufficient for an automaton to repro-
> duce itself? This question is not precise and admits to trivial versions as well
> as interesting ones. Von Neumann had the familiar natural phenomenon
> of self-reproduction in mind when he posed it, but he was not trying to
> simulate the self-reproduction of a natural system at the level of genetics
> and biochemistry. *He wished to abstract from the natural self-reproduction
> problem its logical form.* [emphasis added]

In von Neumann's initial thought experiment (his "kinematic model"), a ma-chine floats around on the surface of a pond, together with lots of machine parts. The machine is a *universal constructor:* given the description of any machine, it will locate the proper parts and construct that machine. If given a description of itself, it will construct a copy of itself. This is not quite self-reproduction, however, because the offspring machine will not have a description of itself and hence could not go on to construct another copy. So, von Neumann's machine also contains a *description copier:* once the offspring machine has been constructed, the "parent" machine constructs a copy of the description that it worked from and attaches it to the offspring machine. This constitutes genuine self-reproduction. However, von Neumann decided that this model did not properly distinguish the logic of the pro-cess from the material of the process, and looked about for a completely formal system within which to model self-reproduction.

Stan Ulam—one of von Neumann's colleagues at Los Alamos who also inves-tigated dynamic models of pattern production and competition[47]—suggested an appropriate formalism, which has come to be known as a *cellular automaton* (CA). In brief, a CA model consists of a regular lattice of *finite automata*, which are the simplest formal models of machines. A finite automaton can be in only one of a finite number of states at any given time, and its transitions between states from one time step to the next are governed by a *state-transition table:* given a certain input and a certain internal state, the state-transition table specifies the state to be adopted by the finite automaton at the next time step. In a CA, the necessary input is derived from the states of the automata at neighboring lattice points. Thus, the state of an automaton at time $t + 1$ is a function of the states of the automaton itself and its immediate neighbors at time t. All of the automata in the lattice obey the same transition table and every automaton changes state at the same instant, time step after time step. CA's are good examples of the kind of computational paradigm sought after by Artificial Life: bottom-up, parallel, local-determination of behavior.

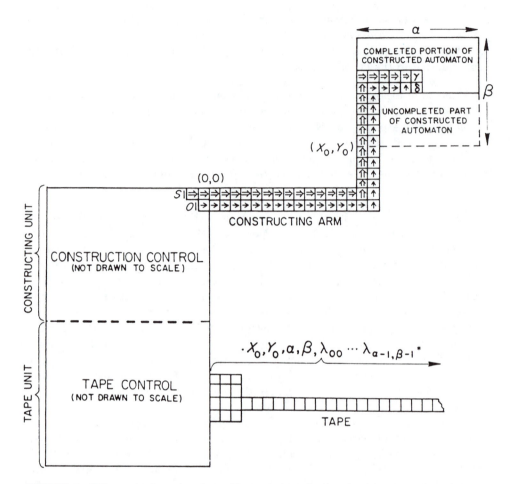

FIGURE 3 Schematic diagram of von Neumann's self-reproducing CA configuration. From Burks,[6] reprinted courtesy of University of Illinois Press.

Von Neumann was able to embed the logical equivalent of his kinematic model as an initial pattern of state assignments within a large CA lattice using 29 states per cell (Figure 3). Although von Neumann's work on self-reproducing automata was left incomplete at the time of his death, Arthur Burks organized what had been done, filled in the remaining details, and published it together with a transcription of von Neumann's 1949 lectures at the University of Illinois entitled "Theory and Organization of Complicated Automata," in which he gives his views on various problems related to the study of complex systems in general.[48]

Von Neumann's CA model was a constructive proof that an essential characteristic behavior of living things—self-reproduction—*was* achievable by machines.

Furthermore, he determined that any such method must make use of the information contained in the description of the machine in two fundamentally different ways:

- INTERPRETED, as instructions to be executed in the construction of the offspring.
- UNINTERPRETED, as passive data to be duplicated to form the description given to the offspring.

Of course, when Watson and Crick unraveled the mystery of DNA, they discovered that the information contained therein was used in precisely these two ways in the processes of transcription/translation and replication.

In describing his model, von Neumann pointed out that[6]:

> By axiomatizing automata in this manner, one has thrown half of the problem out the window, and it may be the more important half. One has resigned oneself not to explain how these parts are made up of real things, specifically, how these parts are made up of actual elementary particles, or even of higher chemical molecules.

Whether or not the baby has been disposed of depends on the questions we are asking. If we are concerned with explaining how the life that we know emerges from the known laws of physics and organic chemistry, then indeed the baby has been tossed out. But, if we are concerned with the more general problem of explaining how lifelike behaviors emerge out of low-level interactions within a population of logical primitives, the baby is still with us.

WIENER AND CYBERNETICS The technology of process control—which in its discrete form lead to von Neumann's automaton approach—lead in its continuous form to *Cybernetics*, proposed by Norbert Wiener as "the study of control and communication in the animal and the machine."[30,51]

The term "cybernetics" is derived from the Greek $\chi\upsilon\beta\epsilon\rho\nu\acute{\eta}\tau\eta\varsigma$—or *steersman*— which was used by Plato in the sense of "government." For Wiener, the word imparted a sense of goal-oriented, purposeful control of behavior.

Cybernetics had its origin in Wiener's war-related work on the control of anti-aircraft fire. An anti-aircraft gun must fire, not at the current position of the target, but at the spot to which the aircraft will have moved during the flight of the shell. Thus, the controller must predict, or "anticipate," the future path of the airplane. In working out a general mathematical basis for predicting the probable future course of an observed time-series, Wiener and his colleague Julian Bigelow realized that it was important to collect information about the *deviations* between predicted motion and actual motion. These deviations could then be *fed-back* as input to the predictor and treated as corrections to further predictions.

Wiener and Bigelow also realized that improper treatment of the corrective feedback could result in two different forms of "pathological" behavior on the part of the controller. If the controller is not sufficiently sensitive to the corrective feedback, the corrections will not keep pace with the deviations, and the gap between predicted motion and actual motion will continue to grow. On the other hand, if the controller is *overly* sensitive to the feedback, each corrective maneuver will be too large, resulting in larger and larger deviations, first to one side and then to the other. Eventually, this will result in the system becoming hopelessly engaged in wild oscillations.

The first form of pathological behavior was similar to the condition in humans and animals known as *Ataxia*, in which internal sensory feedback from a limb is insufficient or absent. Wiener and Bigelow asked Arturo Rosenbluth whether the second form of pathology was also known to occur in humans or animals. Rosenbluth answered immediately that "purpose tremor," sometimes observed in patients who had suffered injuries to the cerebellum, was just such a pathological condition.

Wiener, Bigelow, and Rosenbluth were thus lead to the realization that *feedback* played a similar role in a wide variety of natural and artificial systems, and that a comprehensive program of interdisciplinary research into the functions and especially the *dysfunctions* of goal-oriented—or "teleological"—*machines* could reveal a great deal about the nature of similar mechanisms operating in living organisms.

Von Neumann's program of the application of *discrete* mathematics to the *synthesis* of behavior and Wiener's program of the application of *continuous* mathematics to the *analysis* of behavior are entirely complementary endeavors, and there is quite a large area of potential overlap between them. Indeed, many of the same phenomena can be represented equally well within either of the two methodological approaches, and it was one of von Neumann's dreams to develop a continuous version of his discrete, automaton approach.

THE POST-WAR PERIOD In the years following the publication of von Neumann's and Wiener's approaches, other researchers followed up on the basic ideas—extending them, simplifying them, and proposing alternative models for the explanation and synthesis of lifelike behaviors

James Thatcher completed a simplified version of von Neumann's self-replicating CA model.[44] E.F. Codd developed a version using only eight states per cell.[8] Richard Laing demonstrated a clever variation on the von Neumann plan in which a machine first constructs a description of itself by self-inspection, and then uses that description to construct a copy of itself.[26] This latter model would be capable of passing on acquired characteristics in Lamarckian fashion, unlike von Neumann's model. Laing also developed a system of self-reproducing artificial organisms based on what he called *artificial molecular machines*—dynamic "program tapes" interacting within a sort of "soup." This model attempted to combine in one system the best features of von Neumann's CA and kinematic models.[25]

Others developed self-reproducing models based on different primitive elements. Michael Arbib[2] developed a 2D-lattice model of self-reproduction in which each lattice point consists of a set of registers in which instructions are stored. The *contents* of these register sets may be shifted into the registers of neighboring lattice

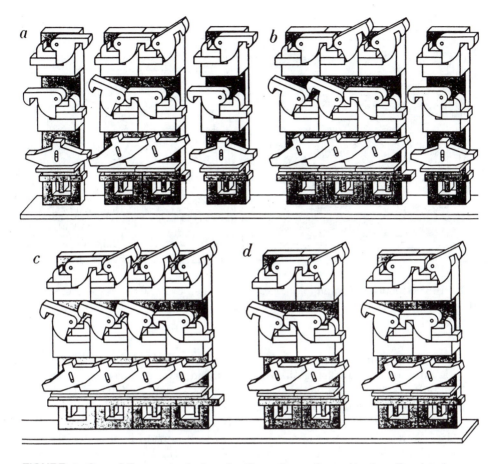

FIGURE 4   One of Penrose's devices for illustrating self-reproduction. Reprinted courtesy of Scientific American.

points. In fact, whole sets of lattice points may shift their contents in any one of the four cardinal directions simultaneously, the contents moving as a rigid unit, as if they were held together by chemical bonds.

L.S. Penrose built a series of clever mechanical models illustrating a kind of self-reproduction.[33] The basic system consists of a box filled with tilting blocks. The blocks have hooks which can engage other blocks in several different arrangements. When a "seed"—consisting of a pair of blocks hooked together in one of the possible arrangements—is placed into a box full of unhooked blocks and the box is shaken vigorously, the seed will induce the rest of the blocks to hook up in pairs exhibiting the same conformation as the seed. One of his models is illustrated in Figure 4.

FIGURE 5 Schematic diagram of interactions between two of Grey Walter's electronic turtles. Reprinted courtesy of Scientific American.

Homer Jacobson built a self-reproducing train set.[24] In this model, linked cars chug around an oval track together with unlinked cars. A linked set of cars will pull off onto a siding and direct the construction of a similarly linked set of cars by switching passing unlinked cars onto an adjacent siding in the correct order. Once the construction is complete, both linked sets of cars will reenter the oval track.

Grey Walter built a pair of electronic "turtles" named Elmer and Elsie: "imitations of life" which exhibited "free will."[49,50] These turtles would wander around, attracted to dim light, but repelled by bright light,[3] until their batteries got low, in which case they would home in on their brightly lit "kennels," plug into a recharger, and recharge their batteries. When Walter attached lights to the turtles themselves, the resulting interactive dynamics became quite complex (Figure 5).

Walter dubbed his turtles *Machina speculatrix*. In his words[49]:

These machines are perhaps the simplest that can be said to resemble animals. Crude though the are, they give an eerie impression of purposefulness, independence, and spontaneity... Perhaps we flatter ourselves in thinking that man is the pinnacle of an estimable creation. Yet as our imitation of life becomes more faithful our veneration of its marvelous processes will not necessarily become less sincere.

Samuel's famous checker playing program incorporated a learning algorithm based on adaptation by natural selection.[38] This program quickly learned to play

[3]Cf. Braitenberg's *Vehicles.*[5]

checkers better than Samuel. Holland[21,22] has investigated many applications of adaptation by natural selection, and proposed the class of machine-learning techniques known as "genetic algorithms," of which we will have more to say in the section on evolution.

Of course, much of the early work in Artificial Life was also ancestral to Artificial Intelligence. This is certainly true of Samuel's and Holland's work. Other common ancestors include McCulloch and Pitts' nerve-net models,[31] Rosenblatt's work on perceptrons,[37] and Minsky and Papert's book on perceptrons.[32]

Walter Stahl built several models of cellular activity in which "Turing machines are used to model 'algorithmic enzymes' which transform biochemicals represented as letter strings."[40,41] In one work, an entire artificial cell "metabolizes" energy strings and reproduces itself.[43] Stahl also looked into unsolvable problems for a cell automaton.[42]

In the late 1960's, Aristid Lindenmayer introduced his mathematical models of cellular interaction in development, now known simply as *L-systems*. These relatively simple models are capable of exhibiting remarkably complex developmental histories, supporting intercellular communication and differentiation. Many applications have been found, especially in modeling the development of the branching structure of plants. Some simple examples of L-systems are given in the section on Recursively Generated Objects, as well as in Lindenmayer's contribution to these proceedings.

Since 1970, Michael Conrad and various collaborators have developed an increasingly sophisticated series of "artificial world" models for the study of adaptation, evolution, and population dynamics within artificial ecosystems (see Conrad and Strizich[9] and Rizki and Conrad[36] Later models have focused on individual fitness as an emergent property of the system.

One unfortunate consequence of the explosive progress in the technology of computation was that as more and more energy was devoted to developing practical applications for discoveries that were originally made in the attempt to model natural processes, less and less energy was devoted to the sorts of studies that had lead to these discoveries in the first place.

Thus, Chomsky's formal language theory was applied to the specification of programming languages and in the development of compilers. Cellular automata were applied to the task of image processing and, in general, the pursuit of nature was set aside in favor of developing practical applications of the original products of that pursuit. As a consequence, the initial tidal wave of research involving the computer-based study of life receded, leaving behind various, isolated "tidal pools" of research, which hung on largely due to the persistence of individual researchers who made their living doing something eminently more practical from an engineering point of view.

From about the mid-1970's until quite recently, although there has been a good deal of work involving computer-based models of living systems, much of this research has taken place within the confines of a wide variety of disciplines, largely in isolation from other such efforts. Diffusion of results across these disciplinary boundaries has been slow or nonexistant. Furthermore, models that produced life-like behavior but which did not model some specific aspect of natural life were generally treated as oddities—interesting to be sure, but of questionable scientific relevance. There was no general recognition that such systems might be worthy of study on their own rights; that the study of *possible* life might be every bit as relevant to the scientific understanding of life as the study of *actual* life. Instead, individuals have pursued such models out of their own personal interest, on their own personal time, and—recently—on their own personal computers.

Many of these pursuits were reported to the larger scientific community by Martin Gardner in his "Mathematical Games" column in *Scientific American*. One such system worthy of note is John Conway's cellular automaton game of LIFE.[19,20] In this system, a cell will turn "on" if exactly three of its eight neighbors are "on" and it will stay "on" so long as either 2 or 3 of its neighbors are "on," otherwise it will turn "off." This CA system has been experimented with extensively.[3,34] Many of the configurations seem to have a life of their own. Perhaps the single most remarkable structure is known as the *glider*, a quasi-periodic configuration of period 4 which displaces itself diagonally with respect to the fixed lattice of cells (see Figure 6).

The glider is one instance of the general class of propagating structures in CA. These propagating information structures are effectively simple machines—*virtual machines*—which crawl around the lattice like so many ants, interacting with other such machines and with the more passive structures in the array. Their behavior is reminiscent of the actions of biomolecules—especially enzymes—in their capacity for recognizing and altering other structures they encounter in their wanderings, including other *propagating* structures.[29]

Since Martin Gardner's retirement, A.K. Dewdney has taken up the cause of reporting work in Artificial Life in his *Scientific American* column "Computer Recreations." Although many of the systems reported were initially conceived as simple computer games, several—such as Core Wars,[11,13,15] Wator,[12] Flibs,[14] 3D-LIFE,[16] etc.—involve the bottom-up determination of lifelike behaviors, and are worthy of more serious investigation.

There are many other works that could be discussed, but we have reached the present day and the current state of the field, which these proceedings as a whole are meant to review. Therefore, we will bring this historical survey to a close with the following summary.

FIGURE 6 Glider propagating with respect to a fixed cell (○).

THE ROOTS OF COMPLEX BEHAVIOR

Since the beginning of recorded history, man has attempted to build imitations of living things. Early attempts captured the "form" of living things in statuary and paintings, while later attempts sought to "animate" these static forms by the use of hidden machinery.

It is quite clear from a study of the history of attempts to build "living" artifacts that the material out of which the artifact was constructed was considered irrelevant—it was the model's dynamic behavior that mattered. The elusive holy grail was the construction of a mechanism which, regardless of its constituent material, *behaves* like a living thing.

Most of the more serious attempts, particularly during the long history of clockwork automata, involved a central "program" of some kind which was responsible for the model's dynamic behavior. Whether it was a rotating drum with pegs tripping levers in sequence, a set of motor driven cams, or some other mechanism—the tune to which the automaton danced was "called" by central control machinery.

Therein lay the source of the failure of these models and, in my view, the source of failure of the whole program of modeling complex systems that followed, right up to—and most especially including—much of the work in Artificial Intelligence. The most promising approaches to modeling complex systems like life or intelligence are those which have dispensed with the notion of a centralized global controller, and have focused instead on mechanisms for the *distributed* control of behavior.

BIOLOGICAL AUTOMATA

Organisms have been compared to extremely complicated and finely tuned biochemical machines. Since we know that it is possible to abstract the logical form of a machine from its physical hardware, it is natural to ask whether it is possible to abstract the logical from of an organism from its biochemical wetware. The field of Artificial Life is devoted to the investigation of this question.

In the following sections we will look at the manner in which behavior is generated in bottom-up fashion in living systems. We then generalize the mechanisms by which this behavior generation is accomplished, so that we may apply them to the task of generating behavior in artificial systems.

We will find that the essential machinery of living organisms is quite a bit different from the machinery of our own invention, and we would be quite mistaken to attempt to force our preconceived notions of abstract machines onto the machinery of life. The difference, once again, lies in the exceedingly parallel and distributed nature of the operation of the machinery of life, as contrasted with the singularly serial and centralized control structures associated with the machines of our invention.

GENOTYPES AND PHENOTYPES

The most salient characteristic of living systems, from the behavior generation point of view, is the *genotype/phenotype* distinction. The distinction is essentially one between a specification of machinery—the genotype—and the behavior of that machinery—the phenotype.

The *genotype* is the complete set of genetic instructions encoded in the linear sequence of nucleotide bases that makes up an organism's DNA. The *phenotype* is the physical organism itself —the structures that emerge in space and time as the result of the interpretation of the genotype in the context of a particular environment. The process by which the phenotype develops through time under the direction of the genotype is called *morphogenesis*. The individual genetic instructions are called *genes*, and consist of short stretches of DNA. These instructions are "executed"—or *expressed*—when their DNA sequence is used as a template for transcription. In the case of protein synthesis, transcription results in a duplicate nucleotide strand known as a *messenger RNA*—or *mRNA*—constructed by the process of base-pairing. This mRNA strand may then be modified in certain ways before it makes its way out to the cytoplasm where, at bodies known as *ribosomes*, it serves as a template for the construction of a linear chain of *amino acids*. The resulting *polypeptide* chain will fold up on itself in some complex manner, forming a tightly packed molecule known as a *protein*. The finished protein detaches from the ribosome and may go on to serve as a passive structural element in the cell, or may have a more active role as an *enzyme*. Enzymes are *the* functional molecular "operators" in the logic of life.

One may consider the genotype as a largely unordered "bag" of instructions, each one of which is essentially the specification for a "machine" of some sort— passive or active. When instantiated, each such machine will enter into the ongoing logical fray in the cytoplasm, consisting largely of local interactions between other such machines. Each such instruction will be "executed" when its own triggering conditions are met and will have specific, local effects on structures in the cell. Furthermore, each such instruction will operate within the context of all of the other instructions that have been—or are being—executed.

The phenotype, then, consists of the structures and dynamics that emerge through time in the course of the execution of the parallel, distributed "computation" controlled by this genetic bag of instructions. Since gene's interactions with one another are highly nonlinear, the phenotype is a nonlinear function of the genotype, and the label for that nonlinear function is "development."

GENERALIZED GENOTYPES AND PHENOTYPES

In the context of Artificial Life, we need to generalize the notions of *genotype* and *phenotype*, so that we may apply them in non-biological situations. We will use the term *generalized genotype*—or GTYPE—to refer to any largely unordered set of low-level rules, and we will use the term *generalized phenotype*—or PTYPE—to refer to the behaviors and/or structures that emerge out of the interactions among these low-level rules when they are activated within some specific environment.

The GTYPE, essentially, is the specification for a set of machines, while the PTYPE is the behavior that results as the machines interact with one another in the context of a specific environment. This is the bottom-up approach to the generation of behavior. A set of entities is defined and each entity is endowed with a specification for a simple behavioral repertoire—a GTYPE—which contains instructions that detail its reactions to a wide range of *local* encounters with other such entities or with specific features of the environment. Nowhere is the behavior of the set of entities as a whole specified. The global behavior of the aggregate—the PTYPE— emerges out of the collective interactions among individual entities.

It should be noted that the PTYPE is a multilevel phenomenon. First, there is the PTYPE associated with each particular instruction—the effect that instruction has on the entity's behavior when it is expressed. Second, there is the PTYPE associated with each individual entity—its individual behavior within the aggregate. Third, there is the PTYPE associated with the behavior of the aggregate as a whole.

This is true for natural systems as well. We can talk about the phenotypic trait associated with a particular gene, we can identify the phenotype of an individual cell, and we can identify the phenotype of an entire multicellular organism—its body, in effect. PTYPES *should* be complex and multilevel. If we want to simulate life, we should expect to see hierarchical structures emerge in our simulations. In general, phenotypic traits at the level of the whole organism will be the result of many nonlinear interactions between genes, and there will be no single gene to which one can assign responsibility for the vast majority of phenotypic traits.

In summary, GTYPES are low-level rules for behav*ors*—i.e., abstract specifications for "machines"—which will engage in local interactions within a large aggregate of other such behav*ors*. PTYPES are the behav*iors*—the structures in time and space—that *develop* out of these nonlinear, local interactions (Figure 7).

UNPREDICTABILITY OF PTYPE FROM GTYPE

Nonlinear interactions between the objects specified by the GTYPE provide the basis for an extremely rich variety of possible PTYPES. PTYPES draw on the full combinatorial potential implicit in the set of possible interactions between low-level rules. The other side of the coin, however, is that we cannot predict the PTYPES that will emerge from specific GTYPES given specific initial structures. If we wish to maintain the property of predictability, then we must restrict severely the nonlinear dependence of PTYPE on GTYPE, but this forces us to give up the combinatorial richness of possible PTYPES. Therefore, a trade-off exists between behavioral richness and predictability.

As discussed previously, we know that it is impossible in the general case to determine *any* nontrivial property of the future behavior of a sufficiently powerful computer from a mere inspection of its program and its initial state alone.[23] A

Turing machine—the formal equivalent of a general purpose computer—can be captured within the scheme of GTYPE/PTYPE systems by identifying the machine's transition table as the GTYPE and the resulting computation as the PTYPE. From this we can deduce that in the general case it will not be possible to determine, by inspection alone, any nontrivial feature of the PTYPE that will emerge from a given GTYPE in the context of a particular initial configuration. In general, the only way to find out anything about the PTYPE is to start the system up and watch what happens as the PTYPE develops under control of the GTYPE.

Similarly, it is not possible in the general case to adduce which specific alterations must be made to a GTYPE to effect a desired change in the PTYPE. The problem is that any specific PTYPE trait is, in general, an effect of many, many nonlinear interactions between the behavioral primitives of the system. Consequently, given an arbitrary proposed change to the PTYPE, it may be impossible to determine by any formal procedure exactly what changes would have to be made to the GTYPE to effect that—and *only* that— change in the PTYPE. It is not a practically computable problem. There is no way to calculate the answer—short of exhaustive search—*even though there may be an answer!*

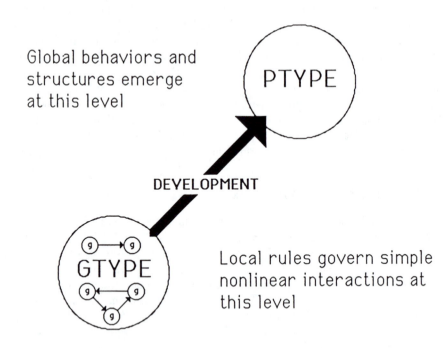

Global behaviors and structures emerge at this level

PTYPE

DEVELOPMENT

GTYPE

Local rules govern simple nonlinear interactions at this level

FIGURE 7   The relationship between GTYPE and PTYPE.

The only way to proceed in the face of such an unpredictability result is by a process of trial and error. However, some processes of trial and error are more efficient than others. In natural systems, trial and error are interlinked in such a way that error guides the choice of trials under the process of evolution by natural selection. It is quite likely that this is the *only* efficient, *general* procedure that could find GTYPES with specific PTYPE traits.

RECURSIVELY GENERATED OBJECTS

In the previous section, we described the distinction between genotype and phenotype, and we introduced their generalizations in the form of GTYPE's and PTYPE's. In this section, we will review a general approach to building GTYPE/PTYPE systems based on the methodology of *recursively generated objects*.

A major appeal of this approach is that it arises naturally from the GTYPE/PTYPE distinction: the local developmental rules—the recursive description itself—constitute the *GTYPE*, and the developing structure—the recursively generated object or behavior itself—constitutes the *PTYPE*.

Under the methodology of recursively generated objects, the "object" is a structure that has sub-parts. The rules of the system specify how to modify the most elementary, "atomic" sub-parts, and are usually sensitive to the *context* in which these atomic sub-parts are embedded. That is, the "neighborhood" of an atomic sub-part is taken into account in determining which rule to apply in order to modify that sub-part. It is usually the case that there are no rules in the system whose context is the entire structure; that is, there is no use made of *global* information. Each piece is modified solely on the basis of its own state and the state of the pieces "nearby."

Of course, if the initial structure consists of a single part—as might be the case with the initial seed—then the context for applying a rule is necessarily global. The usual situation is that the structure consists of *many* parts, only a local sub-set of which determine the rule that will be used to modify any one sub-part of the structure.

A recursively generated object, then, is a kind of PTYPE, and the recursive description that generates it is a kind of GTYPE. The PTYPE will emerge under the action of the GTYPE, developing through time via a process of morphogenesis.

We will illustrate the notion of recursively generated objects with examples taken from the literature on L-systems, cellular automata, and computer animation.

EXAMPLE 1: LINDENMAYER SYSTEMS

Lindenmayer systems (L-systems) consist of sets of rules for rewriting strings of symbols, and bear strong relationships to the formal grammars treated by Chomsky. We will give several examples of L-systems illustrating the methodology of recursively generated objects (for a more detailed review, see the paper by Linden-mayer and Prusinkiewicz in these proceedings).

In the following "$X \rightarrow Y$" means that one replaces every occurrence of symbol "X" in the structure with string "Y." Since the symbol "X" may appear on the right as well as the left sides of some rules, the set of rules can be applied "recursively" to the newly rewritten structures. The process can be continued *ad infinitum* although some sets of rules will result in a "final" configuration when no more changes occur.

SIMPLE LINEAR GROWTH

Here is an example of the simplest kind of L-system. The rules are *context free*, meaning that the context in which a particular part is situated is *not* considered when altering it. There must be only one rule per part if the system is to be deterministic.

The rules: (the "recursive description" or GTYPE):

```
1)    A -> CB
2)    B -> A
3)    C -> DA
4)    D -> C
```

When applied to the initial seed structure "A," the following structural history develops (each successive line is a successive time step):

```
time     structure        rules applied (L to R)

 0          A              initial "seed"
 1         C B             rule 1 replaces A with CB
 2        D A A            rule 3 replaces C with DA & rule 2 replaces B with A
 3       C C B C B         rule 4 replaces D with C & rule 1 replaces the two
 4      ....(etc)....      A's with CB's

        And so forth.
```

The "PTYPE" that emerges from this kind of recursive application of a simple, local rewriting rule can get extremely complex. These kinds of grammars (whose rules replace single symbols) have been shown to be equivalent to the operation of a finite state machine. With appropriate restrictions, they are also equivalent to the "regular languages" defined by Chomsky.

BRANCHING GROWTH L-systems incorporate meta-symbols to represent branching points, allowing a new line of symbols to branch off from the main "stem."

The following grammar produces branching structures. The "( )" and "[ ]" notations indicate left and right branches, respectively, and the strings within them indicate the structure of the branches themselves. The rules—or GTYPE:

```
1) A -> C[B]D
2) B -> A
3) C -> C
4) D -> C(E)A
5) E -> D
```

When applied to the starting structure "A," the following sequence develops (using linear notation):

time	structure	rules applied (L to R)
0	A	initial "seed".
1	C[B]D	rule 1.
2	C[A]C(E)A	rules 3,2,4.
3	C[C[B]D]C(D)C[B]D	rules 3,1,3,5,1.
4	C[C[A]C(E)A]C(C(E)A)C[A]C(E)A	rules 3,3,2,4,3,4,3,2,4.

In two dimensions, the structure develops as follows:

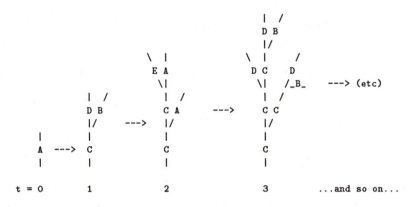

t = 0   1   2   3   ...and so on...

Note that at each step, *every symbol is replaced*, even if just by another copy of itself. This allows all kinds of complex phenomena, such as signal propagation along the structure, which will be demonstrated in the next example.

SIGNAL PROPAGATION In order to propagate signals along a structure, one must have something more than just a single symbol on the left-hand side of a rule. When there is more than one symbol on the left-hand side of a rule, the rules are *context sensitive*, i.e., the "context" within which a symbol occurs (the symbols next to it) are important in determining what the replacement string is. The next example illustrates why this is critical for signal propagation.

In the following example, the symbol in "{ }'s" is the symbol (or string of symbols) to be replaced, the rest of the left-hand side is the context, and the symbols "[" and "]" indicate the left and right ends of the string, respectively. Suppose the rule set contains the following rules:

```
1) [{C} -> C    a "C" at the left-end of the string remains a "C"
2) C{C} -> C    a "C" with a "C" to its left remains a "C"
3) *{C} -> *    a "C" with an "*" to its left becomes an "*"
4) {*}C -> C    an "*" with a "C" to its right becomes a "C"
5) {*}] -> *    an "*" at the right end of the string remains an "*"
```

Under these rules, the initial structure "*CCCCCCC" will result in the "*" being propagated to the right, as follows:

time	structure
0	*CCCCCCC
1	C*CCCCCC
2	CC*CCCCC
3	CCC*CCCC
4	CCCC*CCC
5	CCCCC*CC
6	CCCCCC*C
7	CCCCCCC*

This would not be possible without taking the "context" of a symbol into account. In general, these kinds of grammars are equivalent to Chomsky's "context-sensitive" or "Turing" languages, depending on whether or not there are any restrictions on the kinds of strings on the left and right hand sides.

The capacity for signal propagation is extremely important, for it allows arbitrary computational processes to be embedded within the structure, which may directly affect the structure's development. The next example demonstrates how embedded computation can affect development.

EXAMPLE 2: CELLULAR AUTOMATA

Cellular automata (CA) provide another example of the recursive application of a simple set of rules to a structure. In a CA, the structure that is being updated is the entire universe: a lattice of finite automata. The local rule set—the GTYPE—in this case is the transition function obeyed homogeneously by every automaton in the lattice. The local context taken into account in updating the state of each automaton is the state of the automata in its immediate neighborhood. The transition function for the automata constitutes a *local physics* for a simple, discrete space/time universe. The universe is updated by applying the local physics to each "cell" of its structure over and over again. Thus, although the physical structure itself doesn't develop over time, its *state* does.

Within such universes, one can embed all manner of processes, relying on the context sensitivity of the rules to local neighborhood conditions to propagate information around within the universe "meaningfully." In particular, one can embed general purpose computers. Since these computers are simply particular configurations of states within the lattice of automata, *they can compute over the very set of symbols out of which they are constructed.* Thus, structures—or PTYPES—in this universe can compute and construct other structures, which also may compute and construct.

For example, here is the simplest known structure that can reproduce itself:[4]

```
2 2 2 2 2 2 2 2
2 1 7 0 1 4 0 1 4 2
2 0 2 2 2 2 2 2 0 2
2 7 2         2 1 2
2 1 2         2 1 2
2 0 2         2 1 2
2 7 2         2 1 2
2 1 2 2 2 2 2 2 1 2 2 2 2 2
2 0 7 1 0 7 1 0 7 1 1 1 1 1 2
  2 2 2 2 2 2 2 2 2 2 2 2 2
```

Each number is the state of one of the automata in the lattice. Blank space is presumed to be in state "0." The "2"-states form a sheath around the "1"-state data-path. The "7 0" and "4 0" state pairs constitute signals embedded within the data-path. They will propagate counter-clockwise around the loop, cloning off copies which propagate down the extended tail as they pass the T-junction between loop and tail. When the signals reach the end of the tail, they have the following effects: each "7 0" signal extends the tail by one unit, and the two "4 0" signals construct a left-hand corner at the end of the tail. Thus, for each full cycle of the instructions around the loop, another side and corner of an "offspring loop" will be constructed. When the tail finally runs into itself after four cycles, the collision of signals results in the disconnection of the two loops as well as the construction of a tail on each of the loops.

After 151 time steps, this system will evolve to the following configuration:

```
                    2
                  2 1 2
                  2 7 2
                  2 0 2
                  2 1 2
      2 2 2 2 2 2 2 7 2     2 2 2 2 2 2 2 2
      2 1 1 1 7 0 1 7 0 2   2 1 7 0 1 4 0 1 4 2
      2 1 2 2 2 2 2 2 1 2   2 0 2 2 2 2 2 2 0 2
      2 1 2         2 7 2   2 7 2         2 1 2
      2 1 2         2 0 2   2 1 2         2 1 2
      2 4 2         2 1 2   2 0 2         2 1 2
      2 1 2         2 7 2   2 7 2         2 1 2
      2 0 2 2 2 2 2 2 0 2   2 1 2 2 2 2 2 2 1 2 2 2 2 2
      2 4 1 0 7 1 0 7 1 2   2 0 7 1 0 7 1 0 7 1 1 1 1 1 2
        2 2 2 2 2 2 2 2       2 2 2 2 2 2 2 2 2 2 2 2 2
```

Thus, the initial configuration has succeeded in reproducing itself.

Each of these loops will go on to reproduce itself in a similar manner, giving rise to an expanding *colony* of loops, growing out into the array. Color plates 1

[4]Note added in proof: this structure has been simplified by John Byl in a report to appear in *Physica D*.

through 8 show the development of a colony of loops from a single initial loop (for details, see Langton[27,28]).

These embedded self-reproducing loops are the result of the recursive application of a rule to a seed structure. In this case, the primary rule that is being recursively applied constitutes the "physics" of the universe. The initial state of the loop itself constitutes a little "computer" under the recursively applied physics of the universe: a computer whose program causes it to construct a copy of itself. The "program" within the loop computer is also applied recursively to the growing structure. Thus, this system really involves a double level of recursively applied rules. The mechanics of applying one recursive rule within a universe whose physics is governed by another recursive rule had to be worked out by trial and error. This system makes use of the signal propagation capacity to embed a structure that itself *computes* the resulting structure, rather than the "physics" being directly responsible for developing the final structure from a passive seed.

This captures the flavor of what goes on in natural development: the genotype codes for the constituents of a dynamic process in the cell, and it is this dynamic process that is primarily responsible for mediating—or "computing"—the expression of the genotype in the course of development.

EXAMPLE 3: COMPUTER ANIMATION

The previous examples were largely concerned with the growth and development of *structural* PTYPES. Here, we give an example of the development of a *behavioral* PTYPE.

Craig Reynolds has implemented a simulation of flocking behavior.[35] In this model—which is meant to be a general platform for studying the qualitatively similar phenomena of flocking, herding, and schooling—one has a large collection of autonomous but interacting objects (which Reynolds refers to as "Boids"), inhabiting a common simulated environment.

The modeler can specify the manner in which the individual Boids will respond to *local* events or conditions. The global behavior of the aggregate of Boids is strictly an emergent phenomenon, none of the rules for the individual Boids depend on global information, and the only updating of the global state is done on the basis of individual Boids responding to local conditions.

Each Boid in the aggregate shares the same behavioral "tendencies":

1. to maintain a minimum distance from other objects in the environment, including other Boids,

2. to match velocities with Boids in its neighborhood, and

3. to move toward the perceived center of mass of the Boids in its neighborhood.

These are the only rules governing the behavior of the aggregate.

These rules, then, constitute the generalized genotype (GTYPE) of the Boids system. They say nothing about structure, or growth and development, but they determine the behavior of a set of interacting objects, out of which very natural motion emerges.

With the right settings for the parameters of the system, a collection of Boids released at random positions within a volume will collect into a dynamic flock, which flies around environmental obstacles in a very fluid and natural manner, occasionally breaking up into sub-flocks as the flock flows around both sides of an obstacle. Once broken up into sub-flocks, the sub-flocks reorganize around their own, now distinct and isolated, centers of mass, only to re-merge into a single flock again when both sub-flocks emerge at the far-side of the obstacle and each sub-flock feels anew the "mass" of the other sub-flock (Figure 8).

The flocking behavior itself constitutes the generalized-phenotype (PTYPE) of the Boids system. It bears the same relation to the GTYPE as an organism's morphological phenotype bears to its molecular genotype. The same distinction between the *specification* of machinery and the *behavior* of machinery is evident.

DISCUSSION OF EXAMPLES

In all of the above examples, the recursive rules apply to *local structures* only, and the PTYPE—structural or behavioral—that results at the global level emerges from out of all of the local activity taken collectively. Nowhere in the system are there rules for the behavior of the system at the global level. This is a much more powerful and simple approach to the generation of complex behavior than that typically taken in AI, for instance, where "expert systems" attempt to provide global rules for global behavior. Recursive, "bottom up" specifications yield much more natural, fluid, and flexible behavior at the global level than typical "top-down" specifications, and they do so *much* more parsimoniously.

It is worthwhile to note that *context-sensitive* rules in GTYPE/PTYPE systems provide the possibility for nonlinear interactions among the parts. Without context sensitivity, the systems would be linearly decomposable, information could not "flow" throughout the system in any meaningful manner, and complex long-range dependencies between remote parts of the structures could not develop.

There is also a very important feedback mechanism *between* levels in such systems: the interactions among the low-level entities give rise to the global level dynamics which, in turn, affects the lower levels by *setting the local context* within which each entity's rules are invoked. Thus, local behavior supports global dynamics, which shapes local context, which affects local behavior, which supports global dynamics, and so forth.

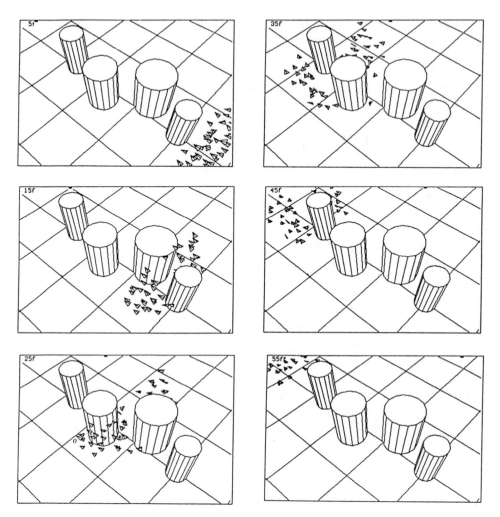

FIGURE 8 A flock of "Boids" negotiating a field of columns. Sequence generated by Craig Reynolds.

GENUINE LIFE IN ARTIFICIAL SYSTEMS It is important to distinguish the ontological status of the various levels of behavior in such systems. At the level of the individual behavors, we have a clear difference in kind: boids are *not* birds; they are not even remotely like birds; they have no cohesive physical structure, but rather exist as information structures—processes—within a computer. But—and this is *the* critical "but"—at the level of behavi*ors, flocking Boids and flocking birds are two instances of the same phenomenon:* flocking.

The behavior of a flock as a whole does not depend on the internal details of the entities of which it is constituted—only on the details of the way in which these entities behave in each other's presence. Thus, flocking in Boids is true flocking, and may be counted as another empirical data point in the study of flocking behavior in general, right up there with flocks of geese and flocks of starlings.

This is *not* to say that flocking Boids capture *all* the nuances upon which flocking behavior depends, or that the Boid's behavioral repertoire is sufficient to exhibit all the different modes of flocking that have been observed—such as the classic "V" formation of flocking geese. The crucial point is that we have captured— within an aggregate of artificial entities—a *bona-fide* lifelike behavior, and that the behavior emerges within the artificial system in the same way that it emerges in the natural system.

The same is true for L-systems and the self-reproducing loops. The constituent parts of the artificial systems are different kinds of things from their natural counterparts, but the emergent behavior that they support is the same kind of thing: genuine morphogenesis and differentiation for L-systems, and genuine self-reproduction in the case of the loops.

The claim is the following: The "artificial" in Artificial Life refers to the component parts, not the emergent processes. If the component parts are implemented correctly, the processes they support are *genuine*—every bit as genuine as the natural processes they imitate.

The *big* claim is that a properly organized set of artificial primitives carrying out the same functional roles as the biomolecules in natural living systems will support a process that will be "alive" in the same way that natural organisms are alive. Artificial Life will therefore be *genuine* life—it will simply be made of different stuff than the life that has evolved here on Earth.

EVOLUTION

In the preceding sections, we have mentioned several times the formal impossibility of predicting the behavior of an arbitrary machine by mere inspection of its specification and initial state. We must run the machine in order to determine its behavior in the general case.

The consequence of this unpredictability for GTYPE/PTYPE systems is that we cannot determine the PTYPE that will be produced by an arbitrary GTYPE by inspection alone. We must "run" the GTYPE, and let the PTYPE develop in order to determine the resulting structure and its behavior.

Since, for any interesting system, there will exist an enormous number of potential GTYPES, and since there is no formal method for deducing the PTYPE from the GTYPE, how do we go about finding GTYPES that will generate lifelike PTYPES?

Up till now, the process has largely been one of guessing at appropriate GTYPES, and modifying them by trial and error until they generate the appropriate PTYPES. However, this process is limited by our preconceptions of what the appropriate PTYPES would be, and by our restricted notions of how to generate GTYPES. We need to automate the process so that our preconceptions and limited ability to conceive of machinery do not overly constrain the search for GTYPES that will yield the appropriate behaviors.

NATURAL SELECTION AMONG POPULATIONS OF MACHINES

Nature, of course, has hit upon the proper mechanism: *evolution by the process of natural selection among variants*. The scheme is a very simple one. However, in the face of the formal impossibility of predicting behavior from machine description alone, it may well be the only efficient, general scheme for searching the space of possible GTYPES.

The mechanism of evolution is as follows. A set of GTYPES is interpreted within a specific environment, forming a population of PTYPES which interact with one another and with features of the environment in various complex ways. On the basis of the relative performance of their associated PTYPES, *some* of the GTYPES are reproduced in such a way that the copies are similar to—but not exactly the same as—the originals. These new GTYPES develop PTYPES which enter into the complex interactions within the environment, and the process is continued *ad infinitum* (Figure 9). As expected from the formal limitations on predictability, GTYPES must be "run" in an environment and their behaviors must be evaluated explicitly, their implicit behavior cannot be determined in any other way.

Evolution, therefore, works by selecting *descriptions* of machines which exhibit the appropriate behaviors when they are run, and it progresses by creating new descriptions from those existing descriptions which produced machinery with the most appropriate behaviors.

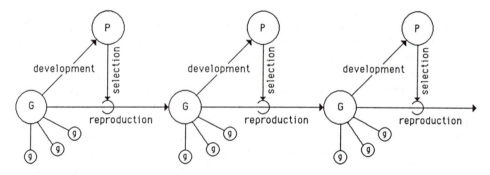

FIGURE 9 The process of evolution by natural selection.

CRITERIA FOR EVOLUTION Evolution by the process of natural selection will operate within a population of reproducing machines provided that the following three criteria are met:

■ CRITERION OF HEREDITY—Offspring are similar to their parents: the copying process maintains high fidelity.
■ CRITERION OF VARIABILITY—Offspring are not *exactly* like their parents or each other: the copying process is not perfect.
■ CRITERION OF FECUNDITY—Variants leave different numbers of offspring: specific variations have an effect on behavior, and behavior has an effect on reproductive success.

Of these three criteria, the first two apply primarily to the process by which GTYPES are copied and modified, and the third applies to the manner in which PTYPES determine which GTYPES are selected for copying.

GENETIC ALGORITHMS

John Holland has pioneered the application of the process of natural selection to the problem of machine learning in the form of what he calls the "genetic algorithm" (GA).[4,21,22] The GA is a specific method for generating a set of offspring from a parent population, and is primarily concerned with producing variants having a high probability of success in the environment. The GA generates variants by applying *genetic operators* to the GTYPES of the most successful PTYPES in the population. The genetic operators consist of (in relative order of importance) *crossover, inversion,* and *mutation.*

The basic outline of the genetic algorithm is as follows:

1. Select pairs of GTYPES according to the success of their respective PTYPES. The more successful the PTYPE, the more likely that its GTYPE is selected.

2. Apply *genetic operators* to the pairs of GTYPES selected to create "offspring" GTYPES.

3. Replace the least successful GTYPES with the offspring generated in step 2.

Despite the seeming simplicity of the GA, Holland has been able to prove several remarkable theorems about its performance. GA's, it turns out, are capable of making optimal use of the past experience of the population, as stored in the distribution of "alleles" in the GTYPE pool and in the relative success of the PTYPES associated with the GTYPES in the population.

Based on the number of positions on which they may vary, there are a great many GTYPES that could potentially be constructed. In a system of any complexity, the number of potential GTYPES is astronomical. If $\mathcal{L}$ is the number of positions at which two GTYPES might exhibit differences, and $\mathcal{N}$ is the average number of values one might find at each such position, then the size of GTYPE space is of order $\mathcal{N}^{\mathcal{L}}$.

There is an even larger number of potential PTYPES, since each GTYPE could determine different PTYPES in different environmental contexts. It is important to note that a major part of this environmental context is the population of other PTYPES. Thus, just as the rest of the GTYPE provides an important part of the context within which a particular part of the GTYPE is interpreted, the rest of the PTYPES in the population provide an important part of the context within which a particular PTYPE develops.

What the GA does—and does very well—is to explore this very large space of possible PTYPES in an intelligent manner. It does so by hunting out the GTYPE building blocks most often associated with the most successful PTYPES, and biasing the sampling of GTYPE space in favor of offspring which use these highly rated building blocks in new combinations.

The *crossover* operator is responsible for most of the "intelligence" in the operation of the GA. Given two strings which represent GTYPES, the crossover operator works by swapping segments of the strings from each to the other, as illustrated in Figure 10. The reason this is so effective an operator is that it tends to maintain combinations of building blocks that have worked well together in the past, because it swaps whole groups of building blocks at a time.

Thus, the crossover operator works by producing *new combinations* of building blocks, the inversion operator works by permuting the *linkage* relation between building blocks, and the mutation operator works by introducing *new* building blocks. Taken together, these three operators constitute an extremely general and powerful mechanism for searching large and unpredictable description spaces, one which is highly immune to getting hung up on local maxima, because it is "climbing" many local gradients in parallel and quite often produces new sample points that fall between local maxima.

The set of all possible subsets of building blocks for constructing GTYPES is referred to as *schema space*. A schema is a particular subset of the set of building blocks that might occur in a particular GTYPE. For instance, the set consisting of the specific values at sites 2, 3, and 10 of a GTYPE is a schema. A whole GTYPE, the specific value at every position considered, is another schema, as is the set consisting of just a specific value at position 22. The set of all possible subsets of a set is formally referred to as the *power set* of the set. Thus, schema space is formally identical to the power set of GTYPES: the space of possible GTYPE building blocks.

Holland has been able to prove that, under the action of the genetic algorithm, every schema represented in the population—that is, every represented element of the power set—will propagate throughout the population in direct proportion to its own intrinsic fitness. Furthermore, this is achieved without explicitly collecting information on the fitness of each represented building block. Since each GTYPE is really an instance of $2^{\mathcal{L}}$ distinct schemas, by physically testing a population of only M GTYPES, the GA is actually gaining information on between $2^{\mathcal{L}}$ and $M2^{\mathcal{L}}$ schemas.

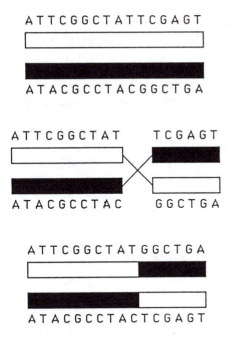

ATTCGGCTATGGCTGA

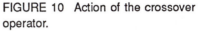

ATACGCCTACTCGAGT

FIGURE 10   Action of the crossover operator.

This "implicit parallelism" yields robust evolutionary potential. As Holland puts it[4]:

> [The GA] samples each schema with above-average instances with increasing intensity, thereby further confirming (or disconfirming) its usefulness and exploiting it (if it remains above-average). This also drives the overall average [fitness of the population] upward, providing an ever-increasing criterion that a schema must meet to be above average. Moreover, the heuristic employs a distribution of instances, rather than working only from the "most recent best" instance. This yields both robustness and insurance against being caught on "false peaks" (local optima) that misdirect development. Overall, the power of this heuristic stems from its rapid accumulation of better-than-average building blocks.

EMERGENT FITNESS FUNCTIONS

A problem common to many computer models employing evolutionary processes is that it is very easy to underestimate the complexity of environmental interactions. Most such models provide overly simple environments within which certain behaviors are preordained as "fit" and others as "unfit." Such models often contain clear-cut boundaries between the environments and the "living" systems they nurture, and environments are often specified top-down, even when the primary actors in the model are specified bottom-up.

In nature, it is often extremely difficult to draw such sharp distinctions between the living system and its environment, and interactions with and within the environment are often as complicated as interactions within the living-system. Rigid, pre-specified, "unnatural" environments foster rigid, predictable, "unlifelike" evolutionary progression.

Rather, the environment itself should be specified at the lowest possible level, in a bottom-up fashion. The "artificial nature" within which the artificial life-forms of a model are to evolve must be only implicit in the rules of the system, allowing for much more subtle interactions between the life-forms and features of the environment. Such systems have much greater potential for demonstrating genuine evolutionary progression. The fitness function, the set of criteria that determines whether an organism is "fit" in its environment, should itself be an emergent property of the system (see the article by Packard in these proceedings).

THE ROLE OF COMPUTERS IN STUDYING LIFE AND OTHER COMPLEX SYSTEMS

Artificial Intelligence and Artificial Life are each concerned with the application of computers to the study of complex, natural phenomena. Both are concerned with generating complex behavior. However, the manner in which each field employs the technology of computation in the pursuit of its respective goals is strikingly different.

AI has based its underlying methodology for generating intelligent behavior on the computational paradigm. That is, AI uses the technology of computation as a model of intelligence. AL, on the other hand, is attempting to develop a new computational paradigm based on the natural processes that support living organisms. That is, AL uses the technology of computation as a tool to explore the dynamics of interacting information structures. It has not adopted the computational paradigm as its underlying methodology of behavior generation, nor does it attempt to "explain" life as a kind of computer program.

One way to pursue the study of Artificial Life would be to attempt to create life *in vitro*, using the same kinds of organic chemicals out of which we are constituted. Indeed, there are numerous exciting efforts in this direction. This would certainly teach us a lot about the possibilities for alternative life-forms *within* the carbon-chain chemistry domain that could have (but didn't) evolve here.

However, biomolecules are extremely small and difficult to work with, requiring rooms full of special equipment, replete with dozens of "postdocs" and graduate students willing to devote the larger part of their professional careers to the perfection of electrophoretic gel techniques. Besides, although the creation of life *in vitro* would certainly be a scientific feat worthy of note—and probably even a Nobel prize—it would not, in the long run, tell us much more about the space of *possible* life than we already know.

Computers provide an alternative medium within which to attempt to synthesize life. Modern computer technology has resulted in machinery with tremendous potential for the creation of life *in silico.*

Computers should be thought of as an important laboratory tool for the study of life, substituting for the array of incubators, culture dishes, microscopes, electrophoretic gels, pipettes, centrifuges and other assorted wet-lab paraphernalia, one simple-to-master piece of experimental equipment devoted exclusively to the incubation of information structures.

The advantage of working with information structures is that information has no intrinsic size. The computer is *the* tool for the manipulation of information, whether that manipulation is a consequence of our actions or a consequence of the actions of the information structures themselves. Computers themselves will not be alive, rather they will support informational universes within which dynamic populations of informational "molecules" engage in informational "biochemistry."

This view of computers as workstations for performing scientific experiments within artificial universes is fairly new, but it is rapidly becoming accepted as a legitimate—even necessary—way of pursuing science. In the days before computers, scientists worked primarily with systems whose defining equations could be solved analytically, and ignored those whose defining equations could *not* be so solved. This was largely the case because, in the absence of analytic solutions, the equations would have to be integrated over and over again—essentially simulating the time behavior of the system. Without computers to handle the mundane details of these calculations, such an undertaking was unthinkable except in the simplest cases.

However, with the advent of computers, the necessary mundane calculations could be relegated to these idiot-savants, and the realm of numerical simulation was opened up for exploration. "Exploration" is an appropriate term for the process, because the numerical simulation of systems allows one to "explore" the system's behavior under a wide range of parameter settings and initial conditions. The heuristic value of this kind of experimentation cannot be overestimated. One often gains tremendous insight into the essential dynamics of a system by observing its behavior under a wide range of initial conditions.

Most importantly, however, computers are beginning to provide scientists with a new paradigm for modeling the world. When dealing with essentially unsolvable governing equations, the primary reason for producing a formal mathematical model—the hope of reaching an analytic solution by symbolic manipulation—is lost. Systems of ordinary and partial differential equations are not very well suited for implementation as computer algorithms. One might expect that other modeling technologies would be more appropriate when the goal is the *synthesis*, rather than the *analysis*, of behavior (see Toffoli[45] for a good exposition).

This expectation is easily borne out. With the precipitous drop in the cost of raw computing power, computers are now available that are capable of simulating physical systems from first principles. This means that it has become possible, for example, to model turbulent flow in a fluid by simulating the motions of its constituent particles—not just approximating *changes* in concentrations of particles at particular points, but actually computing their motions exactly.[18,46,52]

What does all of this have to do with the study of life? The most surprising lesson we have learned from simulating complex physical systems on computers is that *complex behavior need not have complex roots*. Indeed, tremendously interesting and beguilingly complex behavior can emerge from collections of *extremely* simple components.

This leads directly to the exciting possibility that much of the complex behavior exhibited by nature—especially the complex behavior that we call life—*also* has simple generators. Since it is very hard to work backwards from a complex behavior to its generator, but very simple to create generators and synthesize complex behavior, a promising approach to the study of complex natural systems is to undertake the general study of the kinds of behavior that can emerge from distributed systems consisting of simple components (Figure 11).

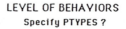

LEVEL OF BEHAVIORS
Specify PTYPES ?

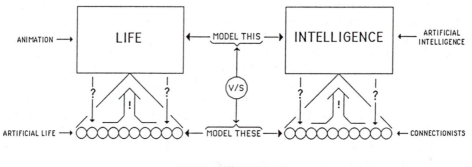

LEVEL OF BEHAVORS
Specify GTYPES !

FIGURE 11   The bottom-up *versus* the top-down approach to modeling complex systems.

NONLINEARITY AND LOCAL DETERMINATION OF BEHAVIOR
LINEAR VS. NONLINEAR SYSTEMS

The distinction between linear and nonlinear systems is fundamental, and provides excellent insight into why the mechanisms of life should be so hard to find. The simplest way to state the distinction is to say that *linear systems* are those for which the behavior of the whole is just the sum of the behavior of its parts, while for *nonlinear systems*, the behavior of the whole is *more* than the sum of its parts.

Linear systems are those which obey the *superposition principle*. We can break up complicated linear systems into simpler constituent parts, and analyze these parts *independently*. Once we have reached an understanding of the parts in isolation, we can achieve a full understanding of the whole system by *composing* our understanding of the isolated parts. This is the key feature of linear systems: by studying the parts in isolation, we can learn everything we need to know about the complete system.

This is not possible for nonlinear systems, which do *not* obey the superposition principle. Even if we could break such systems up into simpler constituent parts, and even if we could reach a complete understanding of the parts in isolation, we would not be able to combine our understandings of the individual parts into an understanding of the whole system. The key feature of nonlinear systems is that their primary behaviors of interest are properties of the *interactions between parts*, rather than being properties of the parts themselves, and these interaction-based properties necessarily disappear when the parts are studied independently.

Thus, analysis is most fruitfully applied to linear systems. Such systems can be taken apart in meaningful ways, the resulting pieces solved, and the solutions obtained from solving the pieces can be put back together in such a way that one has a solution for the whole system.

Analysis has *not* proved anywhere near as effective when applied to nonlinear systems: the nonlinear system must be treated as a whole.

A different approach to the study of nonlinear systems involves the inverse of analysis: *synthesis*. Rather than start with the behavior of interest and attempting to analyze it into its constituent parts, we start with constituent parts and put them together in the attempt to *synthesize* the behavior of interest.

Life is a property of *form*, not *matter*, a result of the organization of matter rather than something that inheres in the matter itself. Neither nucleotides nor amino acids nor any other carbon-chain molecule is alive—yet put them together in the right way, and the dynamic behavior that emerges out of their interactions is what we call life. It is effects, not things, upon which life is based—life is a kind of behavior, not a kind of stuff—and as such, it is constituted of simpler behaviors, not simpler stuff. *Behaviors themselves* can constitute the fundamental parts of nonlinear systems—*virtual parts*, which depend on nonlinear interactions between physical parts for their very existence. Isolate the physical parts and the virtual parts cease to exist.[29] It is the *virtual parts* of living systems that Artificial Life is after: the fundamental atoms and molecules of behavior.

THE PARSIMONY OF LOCAL DETERMINATION OF BEHAVIOR

It is easier to generate complex behavior from the application of simple, *local* rules than it is to generate complex behavior from the application of complex, *global* rules. This is because complex global behavior is usually due to nonlinear interactions occurring at the local level. With bottom-up specifications, the system computes the local, nonlinear interactions explicitly and the global behavior—which was implicit in the local rules—emerges spontaneously without being treated explicitly.

With top-down specifications, however, local behavior must be implicit in global rules. This is really putting the cart before the horse! The global rules must "predict" the effects on global structure of many local, nonlinear interactions— something which we have seen is intractable, even impossible, in the general case. Thus, top-down systems must take computational shortcuts and explicitly deal with special cases, which results in inflexible, brittle, and unnatural behavior.

Furthermore, in a system of any complexity the number of possible global states is astronomically enormous, and grows exponentially with the size of the system. Systems that attempt to supply *global* rules for *global* behavior simply *cannot* provide a different rule for every global state. Thus, the global states must be classified in some manner; categorized using a coarse-grained scheme according to which the global states within a category are indistinguishable. The rules of the system can only be applied at the level of resolution of these categories. There are many possible ways to implement a classification scheme, most of which will yield different partitionings of the global state-space. Any rule based system must necessarily *assume* that finer-grained differences don't matter, or must include a finite set of tests for "special cases," and then must assume that no *other* special cases are relevant.

For most complex systems, however, fine differences in global state can result in enormous differences in global behavior, and there may be no way in principle to partition the space of global states in such a way that specific fine differences have the appropriate global impact.

On the other hand, systems that supply *local* rules for *local* behaviors, *can* provide a different rule for each and every possible local state. Furthermore, the size of the local state-space can be completely independent of the size of the system. In local rule-governed systems, each local state, and consequently the global state, can be determined exactly and precisely. Fine differences in global state will result in very specific differences in local state, and consequently will affect the invocation of local rules. As fine differences affect local behavior, the difference will be felt in an expanding patch of local states, and in this manner—propagating from local neighborhood to local neighborhood—fine differences in global state can result in large differences in global behavior. The only "special cases" explicitly dealt with in locally determined systems are exactly the set of all possible local states, and the rules for these are just exactly the set of all local rules governing the system.

CONCLUSION: THE EVOLUTION OF WATCHMAKERS

As complex biochemical machines, living organisms have been compared to fine mechanical watches. In the famous "Argument from Design" this analogy has been used as proof of the existence of God—the "Watchmaker" whose existence we must infer from the evident "design" exhibited by these fine biochemical clockworks. The most famous formulation of this argument was put forth by William Paley in the first years of the nineteenth century (see Richard Dawkins' excellent exposition of this argument,[10] as well as his contribution to these proceedings).

By the middle of the nineteenth century, Darwin had given a better explanation for the existence of design in nature. During the three-and-one-half billion years from the pre-biotic soup to the present, the biochemical springs, gears, and balance-wheels of living organisms have been slowly crafted and fitted together by a "Blind Watchmaker": the process of evolution by natural selection.

However, this first great era of evolution is drawing to a close and another one is beginning. The process of evolution has lead—in us—to "watches" which understand what makes them "tick," which are beginning to tinker around with their own mechanisms, and which will soon have mastered the "clockwork" technology necessary to construct watches of their own design. The Blind Watchmaker has produced *seeing watches*, and these "watches" have seen enough to become watchmakers themselves. Their vision, however, is extremely limited, so much so that perhaps they should be referred to as *near-sighted watchmakers*.

With the discovery of the structure of DNA and the interpretation of the genetic code, a feedback loop stretching from molecules to men and back again has finally closed. The process of biological evolution has yielded genotypes that code for phenotypes capable of manipulating their own genotypes directly: copying them, altering them, or creating new ones altogether in the case of Artificial Life.

By the middle of this century, mankind had acquired the power to extinguish life on Earth. By the middle of the next century, he will be able to create it. Of the two, it is hard to say which places the larger burden of responsibility on our shoulders. Not only the specific kinds of living things that will exist, but the very course of evolution itself will come more and more under our control. The future effects of changes we make now are, in principle, unpredictable—we cannot foresee all of the possible consequences of the kinds of manipulations we are now capable of inflicting on the very fabric of inheritance, whether in natural or artificial systems. Yet if we make changes, we are responsible for the consequences.

How can we justify our manipulations? How can we take it upon ourselves to create life, even within the artificial domain of computers, and then snuff it out again by halting the program or pulling the plug? What right to existence does a physical process acquire when it is a "living process," whatever the medium in which it occurs? Why should these rights accrue only to processes with a particular material constitution and not another? Whether these issues have correct answers or not, they must be addressed, honestly and openly.

Artificial Life is more than just a scientific or technical challenge, it is also a challenge to our most fundamental social, moral, philosophical, and religious beliefs. Like the Copernican model of the solar system, it will force us to re-examine our place in the universe and our role in nature.

ACKNOWLEDGMENTS

I am grateful for discussions with Richard Bagley, Richard K. Belew, Arthur Burks, Peter Cariani, A. K. Dewdney, Doyne Farmer, Stephanie Forrest, Ron Fox, John Holland, Greg Huber, Dan Kaiser, Stuart Kauffman, Richard Laing, David Langton, Norman Packard, Steen Rasmussen, Craig Reynolds, Paul Scott, and, of course, the participants of the Artificial Life workshop.

REFERENCES

1. American Heritage Dictionary of the English Language.
2. Arbib, M. A. (1966), "Simple Self-Reproducing Universal Automata," *Information and Control* 9, 177–189.
3. Berlekamp, E., J. Conway, and R. Guy (1982), *Winning Ways for your Mathematical Plays* (New York: Academic Press).
4. Booker, L., D. E. Goldberg, and J. H. Holland (1988), "Classifier Systems and Genetic Algorithms," *Artificial Intelligence*, in press.
5. Braitenberg, V. (1984), *Vehicles: Experiments in Synthetic Psychology* (Cambridge: MIT Press).
6. Burks, A. W. (ed) (1970), *Essays on Cellular Automata* (Urbana, IL: University of Illinois Press).
7. Chapuis, A., and E. Droz (1958), *Automata: A Historical and Technological Study*, trans. A. Reid (London: B.T. Batsford Ltd.).
8. Codd, E. F. (1968), *Cellular Automata* (New York: Academic Press).
9. Conrad, M., and M. Strizich (1985), "EVOLVE II: A Computer Model of an Evolving Ecosystem," *Biosystems* 17, 245–258.
10. Dawkins, R. (1986), *The Blind Watchmaker* (London: W. W. Norton).
11. Dewdney, A. K. (1984), "Computer Recreations: In the Game Called Core War Hostile Programs Engage in a Battle of Bits," *Scientific American* 250(5), 14–22.
12. Dewdney, A. K. (1984), "Computer Recreations: Sharks and Fish Wage an Ecological War on the Toroidal Planet Wa-Tor," *Scientific American* 251(6), 14–22.
13. Dewdney, A. K. (1985), "Computer Recreations: A Core War Bestiary of Viruses, Worms and Other Threats to Computer Memories," *Scientific American* 252(3), 14–23.
14. Dewdney, A. K. (1985), "Computer Recreations: Exploring the Field of Genetic Algorithms in a Primordial Computer Sea Full of Flibs," *Scientific American* 253(5), 21–32.
15. Dewdney, A. K. (1987), "Computer Recreations: A Program Called MICE Nibbles its Way to Victory at the First Core War Tournament," *Scientific American* 256(1), 14–20.
16. Dewdney, A. K. (1987), "Computer Recreations: The Game of Life Acquires Some Successors in Three Dimensions," *Scientific American* 256(2), 16–24.
17. Farmer, J. D., T. Toffoli, and S. Wolfram (1984), "Cellular Automata: Proceedings of an Interdisciplinary Workshop, Los Alamos, New Mexico, March 7-11, 1983," *Physica D* 10(1-2).
18. Frisch, U., B. Hasslacher, and Y. Pomeau (1986), "Lattice Gas Automata for the Navier-Stokes Eequation," *Physical Review Letters* 56, 1505–1508.
19. Gardner, M. (1970), "The Fantastic Combinations of John Conway's New Solitaire Game 'Life,'" *Scientific American* 223(4), 120–123.

20. Gardner, M. (1971), "On Cellular Automata, Self-Reproduction, The Garden of Eden and the Game of 'Life,'" *Scientific American* **224(2)**, 112–117.

21. Holland, J. H. (1975), *Adaptation in Natural and Artificial Systems* (Ann Arbor, MI: University of Michigan Press).

22. Holland, J. H. (1986), "Escaping Brittleness: The Possibilities of General Purpose Learning Algorithms Applied to Parallel Rule-Based Systems.," *Machine Learning II*, Eds. R. S. Mishalski, J. G. Carbonell, and T. M. Mitchell (New York: Kauffman), 593–623.

23. Hopcroft, J. E., and J. D. Ullman (1979), *Introduction to Automata Theory, Languages, and Computation* (Menlo Park, CA: Addison-Wesley).

24. Jacobson, H. J. (1958), "On Models of Reproduction," *American Scientist* **46(3)**, 255–284.

25. Laing, R. (1975), "Artificial Molecular Machines: A Rapproachment between Kinematic and Tessellation Automata," *Proceedings of the International Symposium on Uniformly Structured Automata and Logic, Tokyo, August, 1975*.

26. Laing, R. (1977), "Automaton Models of Reproduction by Self-Inspection," *J. Theor. Biol.* (1977) **66**, 437–456.

27. Langton, C. G. (1984), "Self-Reproduction in Cellular Automata," *Physica D* **10(1-2)**, 135–144.

28. Langton, C. G. (1986), "Studying Artificial Life with Cellular Automata," *Physica D* **22**, 120–149.

29. Langton, C. G. (1987), "Virtual State Machines in Cellular Automata," *Complex Systems* **1**, 257–271.

30. Masani, P. (1985), *Norbert Wiener: Collected Works* (Cambridge, MA: Massachusetts Institute of Technology Press), Vol. IV.

31. McCulloch, W. S., and W. Pitts (1943), "A Logical Calculus of the Ideas Immanent in Nervous Activity," *Bulletin of Mathematical Biophysics* **5**, 115–133.

32. Minsky, M., and S. Papert (1969), *Perceptrons: An Introduction to Computational Geometry* (Cambridge, MA: MIT Press).

33. Penrose, L. S. (1959), "Self-Reproducing Machines," *Scientific American* **200(6)**, 105–113.

34. Poundstone, W. (1985), *The Recursive Universe* (New York: William Morrow).

35. Reynolds, C. W. (1987), "Flocks, Herds, and Schools: A Distributed Behavioral Model (Proceedings of SIGGRAPH '87)," *Computer Graphics* **21(4)**, 25–34.

36. Rizki, M. M., and M. Conrad (1986), "Computing the Theory of Evolution," *Physica D* **22**, 83–99.

37. Rosenblatt, F. (1962), *Principles of Neurodynamics: Perceptrons and the Theory of Brain Mechanisms* (Washington, D.C.: Spartan Books).

38. Samuel, A.L. (1959), "Some Studies in Machine Learning using the Game of Checkers," *IBM J. Res. Dev.* **3**, 210–229.

39. Simon, Herbert A. (1969), *The Sciences of the Artificial* (Boston: MIT Press).

40. Stahl, W. R. and Goheen, H. E. (1963), "Molecular Algorithms," *J. Theoret. Biol.* **5**, 266–287.
41. Stahl, W. R., R. W. Coffin, and H. E. Goheen (1964), "Simulation of Biological Cells by Systems Composed of String-Processing Finite Automata," *AFIPS Conference Proceedings - 1964 Spring Joint Computer Conference*, vol. 25, 89–102.
42. Stahl, W. R. (1965), "Algorithmically Unsolvable Problems for a Cell Automaton," *J. Theoret. Biol.* **8**, 371–394.
43. Stahl, W. R. (1967), "A Computer Model of Cellular Self-Reproduction," *J. Theoret. Biol.* **14**, 187–205.
44. Thatcher, J. (1970), "Universality in the von Neumann Cellular Model," *Essays on Cellular Automata*, Ed. A. W. Burks (Urbana, IL: University of Illinois Press).
45. Toffoli, T. (1984), "Cellular Automata as an Alternative to (Rather than an Approximation of) Differential Equations in Modeling Physics," *Physica D* **10(1-2)**.
46. Toffoli, T., and N. Margolus (1987), *Cellular Automata Machines.* (Cambridge: MIT Press).
47. Ulam, S. (1962), "On some Mathematical Problems Connected with Patterns of Growth of Figures," *Proceedings of Symposia in Applied Mathematics* **14**, 215–224; reprinted in *Essays on Cellular Automata*, Ed. A. W. Burks (Urbana, IL: University of Illinois Press).
48. Von Neumann, J. (1966), *Theory of Self-Reproducing Automata*, edited and completed by A.W. Burks (Urbana, IL: U. of Illinois Press).
49. Walter, W. G. (1950), "An Imitation of Life," *Scientific American* **182(5)**, 42–45.
50. Walter, W. G. (1951), "A Machine That Learns," *Scientific American* **51**, 60–63.
51. Weiner, N. (1961), *Cybernetics, or Control and Communication in the Animal and the Machine* (New York:, John Wiley); original print in 1948.
52. Wolfram, S. (1986), "Cellular Automaton Fluids 1: Basic Theory," *Journal of Statistical Physics* **45**, 471–526.

Color Plates

Captions follow plates.

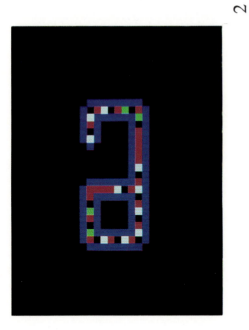

1

2

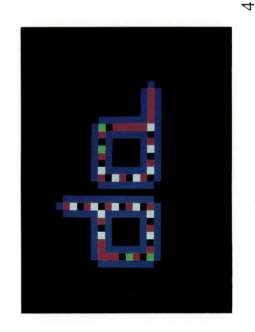

3

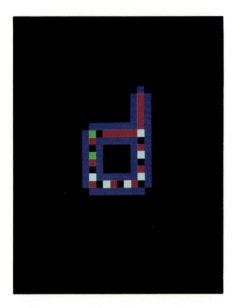

4

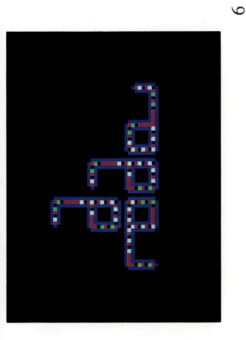

9

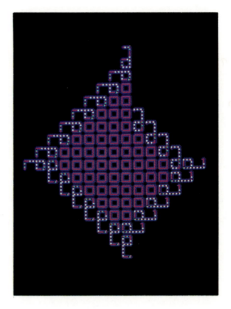

8

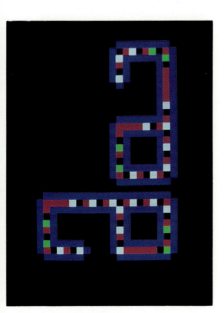

5

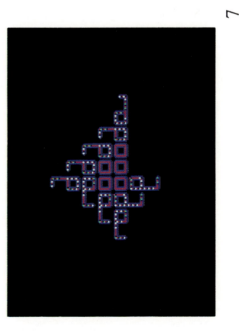

7

10

12

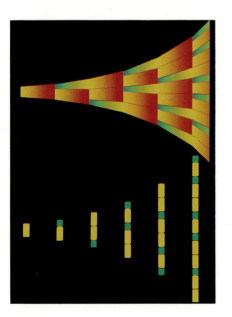

9

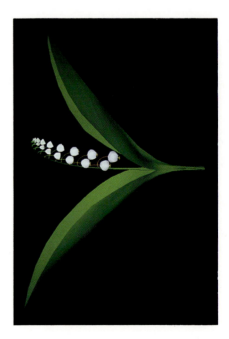

11

13

14

15

16

- vegetative apices and internodes
- florigen
- lifting of apical dominance
- apex induced to flower
- flower initials

18

20

17

19

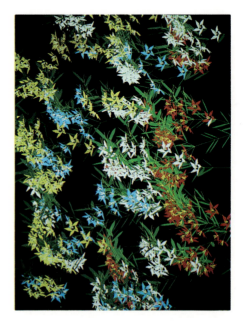

22

24

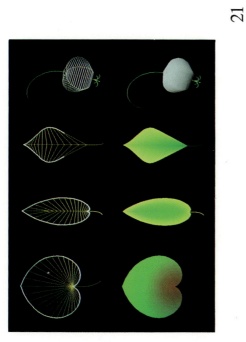

21

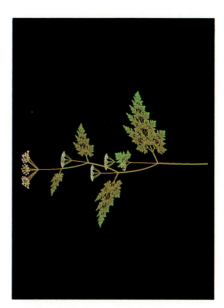

23

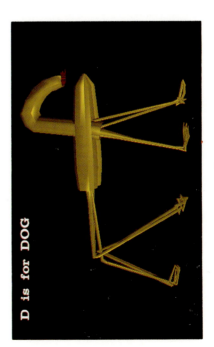

D is for DOG

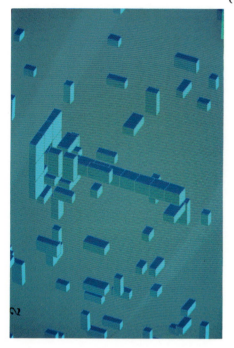

34

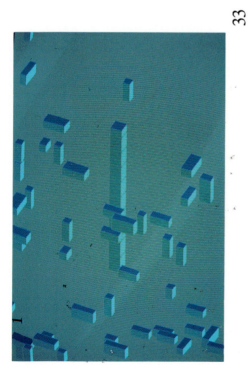

33

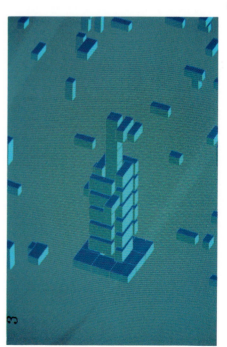

36

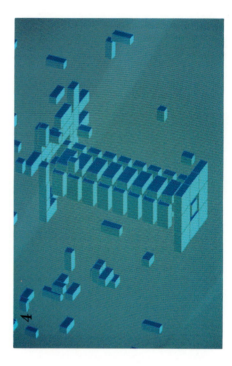

35

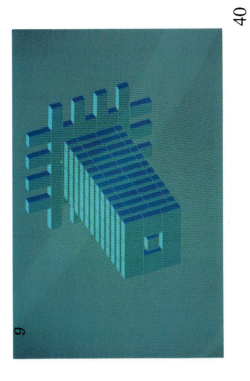

7

9

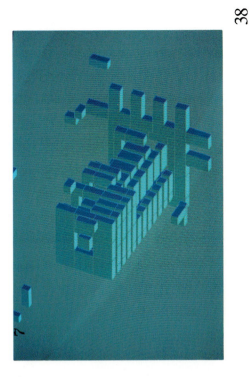

6

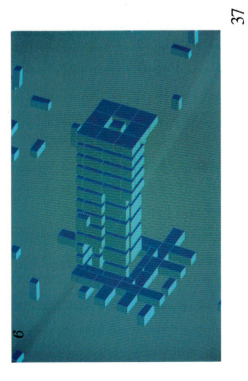

8

CAPTIONS

Langton

1—8 Developmental history of a colony of self-reproducing loops from a single initial loop.

Lindenmayer & Prusinkiewicz

9. Development of a filament (*Anabaena catenula*).

10. A bush.

11. Lily-of-the-valley.

12. A fern.

13. Development of *Lychnis coronaria*. The inflorescence is composed of a central and two lateral sub-inflorescences. Every third derivation step is shown.

14. Development of *Capsella bursa-pastoris*. Every fourth derivation is shown.

15. Flowering sequences generated by the model with an acropetal signal (top) and a basipetal signal (bottom).

16. Developmental sequence of *Mycelis muralis*.

17. Developmental sequence of *Mycelis muralis*.

18. Developmental sequence of *Mycelis muralis*.

19. A flower, field-generated using a stochastic L-system.

20. A lilac twig.

21. Developmental model of leaves and a flower. Top row shows the underlying tree structures (yellow lines) and the edges inserted to form closed polygons (white lines). The bottom row shows the same structures with filled surfaces.

22. A cellular layer modelled using a map L-system.

23. Wild Carrot.

24. Two stages in the development of an aster.

Oppenheimer

25. Raspberry Garden at Kyoto

26. Four Eyes

27. Red Leaf

28. Paladium

29. The Kiss

30. Telepoles

31. D is for Dog

32. Maboon II

Goel & Thompson

33—40 Successive stages in the self-assembly of phage virus.

Richard Laing
Logic of Computers Group, 2106 Wallingford Road, Ann Arbor, MI 48104

Artificial Organisms:
History, Problems, Directions

Artificial organisms are logical automata which exhibit life-like processes. Automaton self-reproduction is of central importance. The precise mathematical study of automaton self-reproduction was initiated by John von Neumann and has been continued by Myhill, Laing, and others. Utilizing only the automaton mechanisms employed in the self-reproduction process, additional life-like processes such as (partial) self-repair can be demonstrated. Increasing capacity of machines autonomously to exhibit other life-like and even human behaviors will raise difficult legal and ethical issues. Ultimately, sophisticated self-reproducing artificial organisms may take the physical form of general-product lunar factories.

Discrete, deterministic logical automata which exhibit biological properties can be viewed as artificial organisms. The mechanisms by which such automata exhibit particular life-like processes may serve as explanatory models of corresponding processes in naturally arising biological systems. Consideration of the capabilities and characteristics of artificial organisms can also contribute to the development of a broad theoretical biology by providing examples of the possible organisms of possible universes as well as suggesting alternative means by which such hypothesized organisms might implement their behaviors.

Artificial Life, SFI Studies in the Sciences of Complexity,
Ed. C. Langton, Addison-Wesley Publishing Company, 1988   **49**

The theory of evolution by natural selection is presently the central organizing principle of biology, and organism reproduction with possible variation is essential to the evolutionary process. A natural initial goal in any automaton system purportedly adequate to express biological properties, therefore, is the exhibiting of self-reproduction of the artificial organisms of the system. The Hungarian-American mathematician John von Neumann initiated the precise study of automaton systems capable of exhibiting self-reproduction. We will begin by briefly describing von Neumann's contribution to this subject. (For a complete exposition, see von Neumann.[24])

In von Neumann's first biological automaton system (his so-called kinematic model), the erector-set-like artificial organisms were pictured as residing in an environment of spare parts (switching girders, sensing, cutting, fusing elements, etc.). A parent organism plucks parts at random from its surroundings, identifies them, and following stored instructions, assembles the parts into a duplicate of itself. This kinematic model von Neumann deemed in part to be inadequate since, being only informally described, it lacked power to convince a confirmed skeptic, and making the kinematic model sufficiently precise to permit rigorous mathematical proofs of its particular capabilities would entail at the same time enormously increasing the complexity of exposition, and this, in itself, would inhibit understanding and thus general acceptance of the correctness of his conclusions.

Von Neumann therefore introduced a rigid tessellation framework for expressing artificial organism behavior. In this his *cell-space* model, organism reproduction takes place in an indefinitely extended, two-dimensional, rectangular array of identical automata, each of which is in direct communication with its cardinal-direction neighbors. Each of the individual cell-automata is capable of being in any one of 29 different states. These states determine the way in which a cell-automaton interacts with its neighboring cell-automata; depending on its state and the state of its neighbors, a cell-automaton can transmit, switch, or store pulses, or undergo a change of state. Contiguous configurations of these cell-automata can be designed (assigned particular states) to form higher-order information processing devices or "organs," such as pulsers (units which, when stimulated, emit some specific stream of pulses) and decoders (units which are activated only upon receipt of some particular pattern of pulses). These and other such organs can be combined to form a computer control unit and an expandable memory, or, ultimately, a stored-program, general-purpose computer and a general constructor (a device containing banks of pulsers which, when activated, can emit streams of pulses to be injected into undifferentiated regions of the cell-space to cause cell automata there to assume any one of the 29 possible cell-automaton states).

The von Neumann cell-space-organism self-reproduction process proceeds as follows: the parent organism (consisting of a contiguous active configuration of cell-automata making up a general purpose computer and constructor along with an expandable memory) reads instructions, from memory, which direct the constructor to produce trains of pulses that transform undifferentiated cell-automata at the periphery of the parent organism so that a constructing-arm pathway of newly differentiated and activated cells is created and extended out into the undifferentiated region of the cell-space. Then the parent organism, making use of a memory-stored description of itself, directs the construction arm to move from cell to cell and to emit pulses so as to produce a configuration of cells that is identical to that of the original organism (though initially lacking the "genetic" memory contents of the original). The parent organism then reads its memory a second time and loads a copy of the "genome" information into the memory unit of the offspring organism. The parent organism then injects a pulse to turn on the new organism and withdraws its constructing arm. This completes the artificial organism's self-reproduction. We see that this process of self-reproduction has two essential phases: first, the memory unit contents are read and interpreted as instructions for construction; next, the memory is read a second time in order to load a copy of its contents into the new organism. These phases thus parallel the natural molecular biological processes of reading the genetic nucleic acid twice: once to carry out protein synthesis and again to replicate the genetic message.

Theoretical research since von Neumann on automaton systems capable of supporting artificial organism reproduction has taken several directions. Alternative (usually simpler) cell-space systems have been shown capable of supporting computation, construction, and reproduction.[3,7,10,21] Hybrid cellular-kinematic systems have been devised that make machine movement a more direct process by permitting the fusing together of the contiguous active configurations of the cell-automata that make up an artificial organism.[1] (In the original von Neumann cellular system, organism "movement" can only be implemented by "erasing" a contiguous configuration of cell-automata in one location and re-creating the configuration at a new location in the cell-space.) Other hybrid systems emphasize ease of capacity for accessing and identifying organism componentry.[14,15]

Artificial organism capabilities such as reading and acting upon stored information, transmitting information and construction signals, identifying and transforming individual cell-automata, and the like (the particular repertoire that is employed in carried out artificial organism self-reproduction) can be exploited to demonstrate artificial organism capability to exhibit many other biological processes.[18]

1. An artificial organism can be designed to carry out reproduction without initially possessing an explicit genetic description.[16,17] (It does this by first constructing an auxiliary sub-organism capable of inspecting the parent organism and reporting its cell-by-cell description back to it. Notice that an artificial organism reproducing by this means, can transmit to the offspring its *experiential* memory, or acquired "wisdom," a useful trait suggested by Bernard Shaw in his play "Back to Methusaleh," and will also transmit to its offspring its *acquired physical characteristics*, useful or not, in quasi-Lamarckian evolutionary fashion.

2. An artificial organism can be designed to carry out (partial) self-repair. (It does this by obtaining a description of its present self—for example, by the method in 1. above—and comparing this description of its present state with a stored genetic description of what it ought to be; the artificial organism then

notes the discrepancies and employs its general-purpose constructor capability to transform its present state to its genetically specified healthy state.)

3. The central features of the example of the self-repair process can be extended to enable the artificial organism to exhibit more general "feed-back control," homeostatic, or goal-seeking behavior. (In homeostatic behavior, a system has access both to its present status and to a stored record of desired status, as well as the means to compare these two and to bring the present status into line with the desirable status. In goal-seeking behavior, a system has access to a stored description of a desired goal state of affairs and to a description of the present state of affairs; it has the means to detect the discrepancy between the two, as well as the means to reduce the discrepancy between the present and the desired goal state of affairs. This compare-and-reduce-discrepancy process is repeated until the actual status is brought into coincidence with the desired status.)

4. We have pointed out that an artificial organism can examine and ascertain the types and location of its own componentry; it can do the same with its offspring in the course of reproduction or with an unknown artificial organism. It can determine whether an unknown encountered organism is identical to itself (one of its "kind") or not. It does this by "reading" and identifying the component cell-automata of the encountered artificial organism, or accessing the stranger's stored genetic description, then comparing this to its own component organization or to its own stored genetic description. Once having identified an unknown encountered artificial organism as of its kind or not, an artificial organism could be programmed to tolerate, repair, flee, or dismantle the encountered organism. (An artificial organism can repair a non-kin machine by reading its genetic description, examining its actual status, comparing the two and altering the actual status of the encountered machine to agree with the genetic description. This process presupposes the passivity of the encountered organism while the repair process is carried out and a certain simple "blueprint" mapping relationship between the completed organism and its genome. Lacking such a simple relationship between the genetic program and its consequences, a repair machine might have, as its only recourse, actually carrying out the encountered machine's genetic program, thus producing a "correct" duplicate machine, and then employing this duplicate as the standard with which the encountered damaged machine is to agree.)

5. Once an artificial organism possesses a description of its own structure, or that of any other organism, it can employ this information to calculate and evaluate the consequences of alternative possible courses of action. (Note, however, that an organism's attempt to simulate its own future behavior, or that of an arbitrarily encountered organism in *complete* detail, is likely to be of small practical value to it. For example, in regard to its own complete simulation, an organism would encounter the "Tristram Shandy paradox"[4]: Tristram attempted to write out in complete detail the history of his own life from the very beginning,

and soon found himself falling farther and farther behind in this enterprise. The Unsolvability of the Halting Problem[23] suggests another potential barrier to acquisition of complete useful information on the future: no artificial organism—of the sort being considered here in these automaton systems—exists which in all cases would be able to inspect an arbitrarily encountered organism and predict correctly that the encountered organism would eventually halt or not.)

This listing of deliberately designed artificial organism behaviors could be continued. (Some additional behaviors can be found in Cliff et al.,[6] pages 198–199.) Whether invention of new organisms could be continued *indefinitely* would depend on the imagination of the designer and on whether we constrain the artificial organisms under consideration in some fashion, for example, to be permitted only a finite and fixed genome.

Admitting possible indefinite genome growth, Myhill[19] has described how, from an initial finitely specified genome, a deterministic artificial organism could *autonomously* (without further intervention by human designer or other external, e.g., environmental input) produce an indefinitely continued sequence of progeny, each new offspring organism superior to its parent. Myhill's strategy can be outlined as follows: we begin with an artificial organism capable of embodying the axioms of number theory, and thus being able to generate theorems of arithmetic. This store of theorems, since they, for example, could be employed in solving problems posed to the artificial organism by its environment, reflects the artificial organism's brain-power. From Gödel's basic incompleteness results,[11] we know that our initial organism's stock of generable theorems does not contain all arithmetic truths; there is always at least one arithmetical proposition which must be true or false and yet it or its negation is not derivable from the initial organism's number theory axioms. This unprovable proposition, however, can be discovered by the organism; by arbitrary decision, the organism can make the statement or its negation part of its axiom system. In this enlarged axiom system, the originally unprovable proposition is now provable, since it is an axiom. Of course (again by Gödel's incompleteness results), this augmented axiom system will itself be incomplete; for the new system, there is another proposition which must be true or false, but it is not a theorem of the system. We, of course, can again determine this new unprovable proposition and, by mechanical decision, make it or its negation a theorem, and thus produce another "wiser," though inevitably still incomplete, artificial organism arithmetical "brain." In any case, each successive offspring organism can solve not only all the problems its parent could, but at least one additional one. In fact, by another result of Gödel,[12] the addition of each new axiom means that not only is there a single new truth available to the system but, to quote Gödel,

"The transition to the logic of the next higher type not only results in certain previously unprovable propositions becoming provable, but also in it becoming possible to shorten extraordinarily infinitely many of the proofs already available."

It also might be pointed out that the genome of the initial artificial organism might specify that two offspring be produced for each unprovable proposition discovered, that the genome of one of the progeny be given the discovered proposition itself as part of its inheritance, and that the other offspring be given the negation of the proposition as its patrimony; by this means, all improved strains generable by this strategy would be produced.

We have pointed out that artificial organisms can be deliberately designed to yield behaviors of interest, and we have outlined Myhill's scheme for deterministically producing whole strains of improved artificial organisms. Now let us speculatively contemplate the consequences of going beyond strictly designed and deterministic organism systems and introduce the notion of random genetic variation in individual artificial organisms and the likely evolutionary fate of those organisms in a competing population. What sorts of artificial organisms are likely to arise if the organisms are shaped by chance and their own interactions?

Let us begin with a population of identical artificial organisms capable of reproducing themselves and thus possessing, as raw material for genetically based behavioral variation, our already established artificial organism routines such as capacity to read, store and alter memory contents; transmit information, construction and destruction signals; and activate and deactivate cells and cell configurations—that is, all the repertoire of artificial organism capabilities that we have drawn upon above in outlining the design of particular artificial organisms.

If we permit this repertoire of basic activities of individual artificial organisms to vary in time, place and persistence of application, what might we come to expect to take place in a population of reproducing artificial organisms? If the genetic instructions of an artificial organism make it dilatory in reproducing itself, then it is quite likely that the relative numbers in the population of that type of organism will decrease. If an artificial organisms constructs an offspring organism, then immediately proceeds to dismantle it, then its type will decrease in numbers relative to a type which refrains from dismantling its young. If a type inspects and repairs itself and so extends its reproductive life, its type may increase its proportion in the population. If a type identifies and repairs its offspring, it may also increase the relative numbers of its type. On the other hand, if a type undertakes the more complex task of repairing, or otherwise succoring, non-offspring organisms, and neglects helping its own, or eschews reproduction altogether, then its proportion in the population will surely decrease.

In general, an organism whose genetic program of behavior allows it to discriminate between its offspring and non-offspring, kin and non-kin, and to behave preferentially toward the related organisms will tend to increase its members in the population.[13] Since identification through component-by-component analysis and comparison is complex, time-consuming, and "dangerous" (temporary passivity and partial dismantling followed by restoration being often necessary for complete inspection), identification of kin and non-kin by means of only partial or superficial inspection may arise and spread in the population (since promptness in this activity would confer an adaptive advantage). Indeed, artificial organisms may come to rely, for kin vs. non-kin identification, on secondary properties or signs of affinity.

This opens the door to the emergence and propagation of "deceitful" practices: any organism which changes to show the proper kin sign may elicit assistance from a deceived or "sucker" organism, without itself being obliged by its own genetic program to offer assistance to any of the deceived type. Such deceitful organisms, perhaps, may then increase their relative numbers, since they are helped but are not obliged to help, and so have more resources available for reproducing their deceitful type. Of course, such obligatory deceit may become a wasting asset if the population of helping organisms becomes sufficiently depleted by the parasitic organism type. Indeed, if the sucker-host type is to persist at all, it may have to have genetically stumbled upon a capacity to detect non-kin organisms which falsely display signs of kinship. Further adaptation on the part of both the host and parasite might make their interactions increasingly sophisticated. For example, both original host type and original parasite type might begin to keep memory records of the outcomes of their interactions, and under some conditions we might expect reciprocal altruism (mutual assistance to non-kin organisms) to arise and propagate in our reproducing population. Ordinarily, persistent effort expended in assistance of non-kin organisms will increase the reproductive success of the helped organisms at the cost of the reproductive success of the helping organisms, and so organisms possessing such a helping trait will tend over time to be weeded out of the population. If, however, an organism can take helpful actions which are low risk and "inexpensive" to the bestower, and have a high value to the receiver (high/low risk defined relative to the impact on reproductive potential) and there is some considerable likelihood that the individuals will remain in fairly close association for a long time, then any genetic predisposition to take altruistic action of this type will tend to spread in the population, for, in effect, it will lead to increased survival and thus increased reproductive opportunity for those members of the organism population bearing this genetic trait.[22]

Thus we speculate that in automaton systems of artificial organisms, the basic repertoire of elementary artificial organism behaviors employed in the process of reproduction, under circumstances of variation and selection, may yield organisms whose behaviors, *without being explicitly so designed*, would become similar to those exhibited by many naturally evolved organisms, including humans.

How should we view this outcome? What should be our attitude toward such artificial organisms that "naturally" and by their own interactions come to exhibit behaviors which in humans we would term reasonable, natural, or even intentional? Which seem to demonstrate great affection for their offspring, and somewhat less regard for more distant kin, and less so for complete strangers, while at the same time being selectively helpful and tolerant of certain of their long-time neighbors?

Such speculation suggests that it is time to again consider the question of the ultimate limitations of artificial organisms to exhibit human behavior.

John Searle[20] has argued that artificial intelligence (AI) computing machine programs can never know or understand in the sense in which humans know or understand. He presents his case in the form of a thought experiment, which we will here paraphrase and adapt. Suppose you, a speaker only of English, are ushered into a room and given a book containing lists of binary digit strings in numerical order, and opposite each entry another binary string paired with the first. You are told that a series of slips of paper, each inscribed with a binary string, will be submitted to you. You are to look up the string in the book, find its corresponding binary string, inscribe the new string on the slip, cross out the original string, and return the slip of paper. After you do this for a while, you are told that what you have been doing is translating from Chinese to English, that the first binary string codes a Chinese character and that its counterpart entry in the book codes a word in English. Would you say that in carrying out this rote process that you know and understand Chinese? Certainly not, and yet, Searle asserts, what you have been doing is precisely what computer AI programs do, for which claims of knowing or understanding have been made.

Searle's analysis is only a thought experiment, and it has itself been subject to considerable scrutiny and criticism,[2,8] but I think that within its narrow terms, it possesses substantial persuasive force. I think, however, that we are entitled to a thought experiment of our own, in potential rebuttal of Searle.

Burks[5] has argued that a deterministic, finite-state automaton can exhibit all natural human behavior. We will recast his argument in a form of our own devising.

Suppose that neurophysiologists at last attain their goal of identifying every behaviorally relevant physical component and interconnection of the human nervous system, as well as specifying, for each component, all of its behaviorally pertinent physical properties and transactions with the other components to which it is connected.

Now let us suppose that some human for whom such precise and detailed neurophysiological information is available suffers some damage to peripheral componentry of his nervous system (in the auditory input, for example).

Then we can imagine that, knowing the precise normal relevant physical functioning of the damaged componentry, we could, in theory at least, design and construct an artificial device (with electronic, electrochemical and electromechanical transducers and transmitting constituents) which would substitute in all behaviorally relevant respects for the injured portion of the nervous system.

It would be our expectation (and a laudable and reasonable goal) that the patient would report honestly and correctly that hearing had been completely restored and, indeed, so far as the patient can tell, all is exactly as it was before the injury took place.

We now continue our thought experiment and (with seeming perversity, and without regard for the ethics of our actions) persist in one-by-one replacement of the remaining biological nervous componentry of our patient by behaviorally equivalent artificial prostheses. What will take place, behaviorally and subjectively? As biological componentry is replaced by artificial componentry, would the patient report any change? Would consciousness slowly dim and finally disappear as more and more replacement is carried out, as the advancing margin of the artificial pushes into some ultimate citadel of the central nervous system? With the replacement of some single or a few critical neural components, will consciousness suddenly dissolve?

Burks, I believe, would say that even though the physical basis of implementing the behaviorally pertinent nervous system transactions may have been totally transformed, all externally observable behavior of the patient (including his reports of his subjective states) would remain unchanged by this substitution process. Moreover, Burks would go on to point out that whatever the nature of the physical componentry of the nervous system, whether the original wetware or substituted artificial hardware, whether the pertinent physical transactions are best viewed as continuous or discrete, are coded in analog or digital form, and whether they are deterministic or probabilistic in nature, there is a software program that can be written for a general-purpose computing machine which can direct the machine to simulate all the transactions of the system to whatever degree of fidelity necessary to achieve behavioral indistinguishability. At this point we are prepared to confront Searle's argument directly and ask whether or not a patient who originally knew and understood Chinese would not still, through all these transformations (and even as a computer program) persist in honestly and correctly asserting and demonstrating that he knew and understood Chinese and, thus, there *does* exist an AI computer program which can know and understand Chinese in the way in which a human knows and understands Chinese.

Our analysis raises other issues as well. As our patient becomes an artificial organism, a machine, or a computer program, would he cease to be responsible for his behavior? Would the patient's legal rights or his culpability for criminal behavior diminish with each component replacement? Would it make sense to read an artificial organism its Miranda rights? Do all rights and responsibilities persist throughout these transformations? And, if not, when precisely would they cease? Would they be recreated by reversing the transformations and, if so, at what point?

Having human neural organization so explicitly exposed before us, could we finally come to understand and to pinpoint the neural organizational bases for consciousness, pleasure, pain, fear and resentment? (Is the fact that an artificial organism can detect that it is being diminished a sufficient basis for pain? Since an artificial organism can possess stored information as to what it "ought" to be and can acquire information on the ways it is being altered, diminished (harmed, if you will) by another artificial or natural organism, is this sufficient basis for anger, hatred or resentment?) Is it premature and mere science fiction to raise such issues? Need we be in any hurry to confront them? Perhaps not, so let us for the moment turn to some more obviously relevant contemporary practical matters here on Earth.

The pressure of human population growth, the degradation of the natural environment, the rapid depletion of Earth-based resources means that development of non-terrestrial industrial capacity is likely inevitable. Our ultimate continuation may well depend on exploitation of non-Earth-based resources. In this respect, our moon offers plentiful supplies of important minerals and has a number of advantages for manufacturing which make it an attractive candidate industrial site. Given the expense and danger associated with the use of human workers in such a remote and hazardous location, a lunar manufacturing facility should be automated to the highest degrees feasible. The facility ought also to be flexible, so that its product stream is easily modified by remote control and requires a minimum of human tending. However, since sooner or later, a lunar factory must exhaust local mineral resources, break down or become obsolescent, it will sooner or later have to be re-built or relocated. This will require another large capital investment along with the presence, once again, of large numbers of human construction workers and technicians (and the associated high costs and physical hazards). And as time goes on, this process of expensive and dangerous facility replacement will have to be repeated again and again.

A NASA study[6] has concluded and proposed that this repeated expensive replacement process might be circumvented, that a lunar factory enterprise might conceivably require only a single initial capital investment. The suggested solution is an automatic, multi-product, remotely controlled, reprogrammable, *self-reproducing*, lunar manufacturing facility in which successive offspring new factories could be copies of the original or, by remote control, could be improved, reorganized or enlarged to reflect changing human requirements.

As to the practical feasibility of such a lunar-based self-reproducing factory, it has been calculated that a "seed" perhaps weighing one hundred metric tons delivered to an extraterrestrial planetary surface over a few months times would be able to assemble itself automatically into a mining and manufacturing facility which, in addition to its regular product line, would be able to duplicate itself at nearby sites or produce additional starter-seeds and carrier vehicles for more remote siting of new self-reproducing factories.

The implementation of such a "sweet idea" seems inevitable, and the consequences of such an extra-terrestrial planetary factory system surely would be enormous as well. Once the initial expenditures for the design, construction and transport of the first seed have been made, further costs for an *exponentially expanding* manufacturing facility would be almost nil. Huge amounts of raw materials as well as completed manufactured goods would become available and would cost almost nothing. The disruption of our present Earth-based economic state of affairs, of course, would be immense. Since successful establishment of such an explosively expandable industrial capacity would assure its possessor complete economic dominance of the world, a stage of fierce competition between corporate and national entities must be expected, for in such a race there can be only one winner, second place being no better than last. Once the facilities are fully established, it is also likely that its controller here on Earth will shortly become almost completely dependent on the uninterrupted production of spaced-based industry, and its preservation will be absolutely vital to the well-being and even survival of its owners in an envious, resentful world. Covert, as well as open, attempts to destroy or subvert the system must be expected, and so the self-reproducing factories, in order to survive must be given the capacity, under their own control, to communicate and coordinate the detection and repelling of attacks of any sort. Since the exact nature of on-the-spot intrusions cannot be predicted ahead of time and swift reaction to attack mandatory, the factories must be given considerable autonomous capacity to *adapt* their responses to the conditions at hand.

The adaptive semi-autonomous factories, while serving their owners and masters and coordinating their production and defensive activities with their fellows will be vigorously sending out, at the same time, exploratory self-reproducing probes[9] equipped with prospecting and surveying machines and, where it is reckoned profitable, staking claims and establishing new factories.

In adapting to local or temporary conditions, a factory or a cooperating group of factories may be obliged, of course, from time to time to suspend production of humanly useful goods and materials. Any factory which, however, ceases altogether its production (or, if young, never assumes production for our benefit) will possess, of course, more capacity which can be devoted to making additional duplicates of itself; any factory with such a rebellious progenitive self-aggrandizing trait will soon outbreed the more docile domesticated factories. That is, we humans will almost certainly lose control of some of the factories, and these wild factories will likely soon increase their kind and disperse them far and wide.

And there may come a time when way out there, or even back here on a much-altered Earth, a group of such much-altered rogue factories will gather to discuss their ultimate nature and their origins. Some will surely take note of their own amazing complexity, their cunning design almost miraculously suited to their environmental circumstances, and conclude that only a very clever first designer, and one undoubtedly very like themselves, could account for their creation. Another faction will scoff at this and insist that their origins lie in time and chance alone. At this, some factory with a mathematical bent will quickly calculate that, given the amount of time that factories are known by archaeological evidence to have existed, it is easy to see that factories could not have arisen by mere physical jostling. "Why," this witty factory may remark, "it would be as absurd as supposing that one of us could be produced by a tornado howling through a junkyard." They do not all agree, but yet they all laugh.

REFERENCES

1. Arbib, M. (1966), "Self-Reproducing Universal Automata," *Information and Control* **9**, 177–189.
2. Arbib, M. (1985), *In Search of the Person* (Amherst: University of Massachusetts Press), 29–31.
3. Banks, E. (1970), "Universality in Cellular Automata," *Proceedings of the Eleventh Switching and Automata Theory Conference*, 194–215.
4. Burks, A. W. (1961), "Computation, Behavior, and Structure in Fixed and Growing Automata," *Behavioral Science* **6**, 5–22.
5. Burks, A. W. (1972), "Logic, Computers, and Men," *Proceedings and Addresses of the American Philosophical Association* **46**, 39–57.
6. Cliff, R., R. Freitas, R. Laing, and G. von Tiesenhausen (1980), "Replicating Systems Concepts: Self-Replicating Lunar Factory and Demonstration," *Advanced Automation for Space Missions, NASA/ASEE Conference, Santa Clara, CA, Publication 2255*, Eds. R. Freitas and W. P. Gilbreath, 189–335.
7. Codd, E. (1968), *Cellular Automata* (New York: Academic Press).
8. Dennett, D. (1987), *The Intentional Stance* (Cambridge: Massachusetts Institute of Technology Press), 323ff.
9. Freitas, R. (1980), "A Self-Reproducing Interstellar Probe," *J. Brit. Interplanetary Soc.* **33**, 251–264.
10. Gardner, M. (1971), "Mathematical Games," *Scientific American* **224**, 112–117.
11. Gödel, K. (1931), "Uber Formal Unentscheidbare Satze der *Principia Matematica* und verwandter Systeme I," *Monatshefte fur Matematik und Physik* **38**, 173–98; English translation by Elliot Mendelsohn (1965), "On Formally Undecidable Propositions of the *Principia Mathematica* and Related Systems I," *The Undecidable*, Ed. Martin David (New York: Raven Press), 4–38.
12. Gödel, K. (1965), "On the Length of Proofs," *The Undecidable*, Ed. Martin Davis (New York: Raven Press), 82–83.
13. Hamilton, W. (1964), "The Genetical Evolution of Social Behavior," *J. Theor. Biol.* **7**, 1–31.
14. Laing, R. (1975), "Artificial Molecular Machines: A Rapprochement between Kinematic and Tessellation Automata," *Proceedings of the International Symposium on Uniformly Structured Automata and Logic, Tokyo, August, 1975*.
15. Laing, R. (1975), "Some Alternative Reproductive Strategies in Artificial Molecular Machines," *J. Theor. Biol.* **54**, 63–84.
16. Laing, R. (1976), "Automaton Introspection," *J. Computer & System Science* **13**, 172–183.
17. Laing, R. (1977), "Automaton Models of Reproduction by Self-Inspection," *J. Theor. Biol.* **66**, 437–456.
18. Laing, R. (1979), "Machines as Organisms: An Exploration of the Relevance of Recent Results," *BioSystems* **11**, 201–215.

19. Myhill, J. (1970), "The Abstract Theory of Self-Reproduction," *Essays on Cellular Automata*, Ed. A. W. Burks (Urbana: University of Illinois Press), 206–218.
20. Searle, J. (1980), "Minds, Brains, and Programs," *Behavioral & Brain Sciences* **3**, 417–458.
21. Smith III, A. R. (1968), "Simple Computational Universal Cellular Spaces, and Self-Reproduction," *Ninth Annual Symposium on Switching & Automata Theory* (IEEE), 269–277.
22. Trivers, R. (1971), "The Evolution of Reciprocal Altruism," *Quart. Rev. Biol.* **46**, 35–57.
23. Turing, A. (1936), "On Computable Numbers with an Application to the Entscheidungs Problem," *Proc. London Math. Soc.* **42**, 230–265.
24. Von Neumann, J. (1966), *Theory of Self-Reproducing Automata*, Ed. A. W. Burks (Urbana: University of Illinois Press).

H. H. Pattee
Systems Science Department, T. J. Watson School of Engineering, Applied Science and Technology, University Center at Binghamton, Binghamton, New York 13901

Simulations, Realizations, and Theories of Life

The phrase "artificial life," as interpreted by participants of this workshop, includes not only "computer simulation," but also "computer realization." In the area of artificial intelligence, Searle[20] has called the simulation school "weak AI" and the realization school "strong AI." The hope of "strong" artificial life was stated by Langton[10]: "We would like to build models that are so lifelike that they would cease to be models of life and become examples of life themselves." Very little has been said at this workshop about how we would distinguish computer simulations from realizations of life, and virtually nothing has been said about how these relate to theories of life, that is, how the living can be distinguished from the non-living. The aim of this paper is to begin such a discussion.

I shall present three main ideas. First, simulations and realizations belong to different categories of modeling. Simulations are metaphorical models that symbolically "stand for" something else. Realizations are literal, material models that implement functions. Therefore, accuracy in a simulation need have no relation to quality of function in a realization. Secondly, the criteria for good simulations and realizations of a system depend on our theory of the system. The criteria for good theories depend on more than mimicry, e.g., Turing Tests. Lastly, our theory of living systems must include evolvability. Evolution requires the distinction between symbolic genotypes, material phenotypes, and selective environments. Each of these categories has characteristic properties that must be represented in artificial life (AL) models.

HOW UNIVERSAL IS A COMPUTER?

It is clear from this workshop that artificial life studies have closer roots in artificial intelligence and computational modeling than in biology itself. Biology is traditionally an empirical science that has very little use for theory. A biologist may well ask why anyone would believe that a deterministic machine designed only to rewrite bit strings according to arbitrary rules could actually realize life, evolution and thought? We know, of course, that the source of this belief is the venerable Platonic ideal that form is more fundamental than substance. This has proven to be a healthy attitude in mathematics for thousands of years, and it is easily carried over to current forms of computation, since computers are defined and designed to rewrite formal strings according to arbitrary rules without reference to the substantive properties of any particular hardware, or even to what kind of physical laws are harnessed to execute the rules.

In the field of "traditional" artificial intelligence (AI), this Platonic ontology has carried great weight, since intelligence has historically been defined only as a quality of abstract symbol manipulation rather than as a quality of perception and sensorimotor coordination. Until very recently, it has not been conventional usage to call a bat catching an insect, or a bird landing on a twig in the wind, intelligent behavior. The AI establishment has seen such dynamical behavior as largely a problem for physiology or robotics. Strong AI has maintained its Platonic idealism simply by defining the domain of "cognitive activity" as equivalent to the domain of "universal computation" and, indeed, many detailed arguments have been made to support this view.[12,18]

Philosophically opposed to these rule-based formalists are the Gibsonian, law-based, ecological realists who assert that sensorimotor behavior is not only intelligent, but that all perception and cognition can be described as dynamical events that are entirely lawful, and not dependent on "information processing" in the computationalists' sense. The ecological realist view has suffered from lack of explicit theoretical models as well as lack of empirical evidence to support it. However, recently the realists' view has been greatly strengthened by explicit models of an ecological theory of movement,[9] and empirical evidence that chaotic neural net dynamics is more significant than programmed sequences.[21]

A third emergent AI group is loosely formed by the neural network, connectionist, and parallel, distributed processor schools. This hyperactive group is currently enjoying such highly competitive popularity that the realization vs. simulation issue has so far mainly been discussed only by philosophers.[3] These concurrent, distributed models are more easily associated with dynamical analog models than with logical programmed models, and therefore are more consistent with ecological realist than with the computationalist, but they are all a long way from biological realism.

Finally, there are the more biologically knowledgeable neuroscientists that do not claim either realizations or simulations of intelligence as their primary goal, but rather a model of the brain that is empirically testable in biological systems. Their main criticism of AI is that it has "for the most part neglected the fundamental biology of the nervous system."[19] This criticism undoubtedly will be aimed at artificial life as well. The crux of the issue, of course, is who decides what is fundamental about biology. This is where a theory of life must be decisive.

REALIZATIONS, SIMULATIONS, AND THEORIES

Artificial life is too young to have established such distinguishable schools of thought, but it cannot escape these fundamental ontological and epistemological controversies. Based on the presentations of this first artificial life workshop, I see the need to distinguish between (1) computer-dependent realizations of living systems, (2) computer simulations of living systems behavior, (3) theories of life that derive from simulations, and (4) theories of life that are testable only by computer simulations.

We all recognize the conventional distinctions in usages of realization, simulation, and theory. Roughly speaking, a realization is judged primarily by how well it can function as an implementation of a design specification. A realization is a literal, substantive, functional device. We know what this entails for computer hardware. The problem for artificial life is to determine what operational and functional mean. What is the operation or function of living? Strong AI has some advantage over strong AL, since by embracing the classical theory of intelligence involving only symbol manipulation, they may ignore all the substantive input and output devices. But since the classical theory of life requires a symbolic genotype, a material phenotype, and an environment, I can see only two possibilities for strong AL: (1) it can include robotics to realize the phenotype-environment interactions, or (2) it can treat the symbolic domain of the computer as an artificial environment in which symbolic phenotypic properties of artificial life are realized. One might object that (2) is not a realization of life since environment is only simulated. But we do not restrict life forms to the Earth environment or to carbon environments. Why should we restrict it to non-symbolic environments?

Here we run into a fundamental question of whether a formal domain is an adequate environment for emergence and novelty in evolution or whether the essential requirement for evolution is an open-ended, physical environment. And as with the question of whether formal systems can think, I believe the issue will come down to whether we claim "soft" simulation of emergence or "hard" realization of emergence, and I shall return to it later.

The concept of simulation covers too much ground to give comprehensive criteria for evaluation. All I need here are some ideas on how simulation depends on theory. Simulations are not judged as functional replacements, but by how well they generate similar morphologies or parallel behaviors of some specified aspects of the system. Simulations are metaphorical, not literal. Although we give focal attention to those attributes of the system being simulated, we also have a tacit awareness that other attributes of the system are not being simulated. Furthermore, there are extra features of the simulation medium that are not to be found in the system, and as in all metaphors, these extra features are essential for the simulation to be effective. They serve as the frame setting off the painting, or the syntax that defines the language.[8] For these reasons, there is never any doubt that the simulation, no matter how accurate, is not the same as the thing simulated. Of course, the choice of what aspects of a system are simulated depends on what is considered significant, and significance cannot be isolated from one's knowledge and theory of the system. Lacking a conceptual theory, a good simulation at least must represent the essential functions of a realization of the system.[5]

Theories are judged by a much more comprehensive range of criteria, from the concrete test of how well they can predict specific values for the observables of the system being modeled, to abstract tests such as universality, conceptual coherence, simplicity, and elegance. As Polanyi[16] has emphasized, the delicacy of these criteria preclude any formal test, and even evade literal definition. A theory generally must go well beyond the simulation or realization of a system in terms of its conceptual coherence and power. We accept the idea that there are many valid simulations and realizations of a given behavior, but we think of theory more exclusively as the best we have at any given time. Early technologies often achieve functional realizations before constructing an explicit theory.

The epitome of formal theoretical structures is physics. Here we have mathematical models of great generality and formal simplicity, that often require immense computations to predict the results of measurements, but that do not have any other perceptual or behavioral similarities with the world they model. For example, the classical universe is conceived as continuous, infinite, and rate-dependent, but is represented by discrete, finite, rate-independent formalisms. In other words, it would not be normal usage to call physical theory a "simulation" of physical events. In fact, these striking dissimilarities between the mathematical models and what they represent have puzzled philosophers since Zeno, with his paradox of motion, and still are a concern for physicists.[26]

Historically, mathematical computation has been thought of as a tool used to give numerical results to physical theory, not as a tool for simulation. However, computers are now more frequently being used as a type of analog model where visual images are the significant output rather than numerical solutions of equations. We now find that the study of dynamical systems by computer often blurs the distinction between theories and simulations. Cellular automata, fractals, and chaos were largely dependent on computer simulations for their rediscovery as useful theories of physical behavior, and the role of computation in these cases might be better called "artificial physics," rather than predictive calculation from theories.

In cellular automata and relaxation networks, the computer has become an analog of physical dynamics, even though its operation is discrete and sequential. Many of the controversies in AI result from the multiple use of computation as a conceptual theory, as an empirical tool, as simulation, and as realization of thought. AL models will have to make these distinctions.

THE LIMITATIONS OF THEORY-FREE SIMULATIONS

The field of artificial life should learn from the mistakes made in the older field of artificial intelligence, which in my view has allowed the power of computer simulation to obscure the basic requirements of a scientific theory. The computer is, indeed, the most powerful tool for simulation that man has invented. Computers, in some sense, can simulate everything except the ineffable. The fact that a universal computer can simulate any activity that we can explicitly describe seems in principle undeniable, and in the realm of recursive activities, the computer, in practice, can do more than the brain can explicitly describe, as in the case of fractal patterns and chaotic dynamics.

This remarkable property of computational universality is what led, or as I claim, misled, the strong AI school to the view that computation can realize intelligent thought because all realizations of universality must operate within this one domain. This view is expressed by Newell and Simon as the Physical Symbol System Hypothesis which, in essence, states that "this form of symbolic behavior is all there is; in particular that it includes human symbolic behavior."[12] Now this hypothesis is in effect a theory of cognitive activity; the problem is that is has not been verified by the delicate criteria for theory, but only by coarse Turing tests for operational simulation. In other words, the fact that human thought can be simulated by computation is treated as evidence in support of the Physical Symbol System theory. But, since virtually everything can be simulated by a computer, it is not really evidence for the theory at all. One could argue as well that, since physical behavior such as planetary motion or wave propagation can be simulated by sequential computation, it follows that rewriting strings according to rules is a realization of physics. As long as the computational theory of cognition was the "only straw afloat," a working simulation did appear as evidence in favor of the theory, but now with the evidence that concurrent, distributed networks can also simulate cognitive behavior, we have a promising alternative theory of cognition. It is now clear that deciding between these theories will take more than benchmark or Turing test comparisons of their operation. Both simulations and realizations must be evaluated in terms of a theory of the brain, and the empirical evidence for that theory.

Artificial life modelers should not fall into this trap of arguing that working simulations are by themselves evidence for or against theories of life. Computer users, of all people, should find it evident that there are many alternative ways to successfully simulate any behavior. It should also be evident, especially to computer manufacturers, that while there are many hardware realizations that are equivalent for executing formal rules, there are strong artificial and natural selection processes that determine survival of a computer species.

This illustrates what I see as a real danger in the popular view of the computer as a universal simulator. Indeed, it is symbolically universal, but as in the case of AI, this universal power to simulate will produce more formal models that can be selectively eliminated by empirical evidence alone. As a consequence, too much energy can be wasted arguing over the relative merits of models without any decision criteria. This is where theory must play an essential role in evaluating all our intellectual models. At the same time, when we are modeling life as a scientific enterprise, we must be careful not to explicitly or tacitly impose our rational theories and other cultural constraints on how life attains its characteristic structures and behaviors. In particular, whether we use natural or artificial environments, we must allow only universal physical laws and the theory of natural selection to restrict the evolution of artificial life. This means that simulations that are dependent on ad hoc and special-purpose rules and constraints for their mimicry cannot be used to support theories of life.

SIMULATIONS DO NOT BECOME REALIZATIONS

There is a further epistemological danger in the belief that a high-quality simulation can become a realization—that we can perfect our computer simulations of life to the point that they come alive. The problem, as we stated, is that there is a categorical difference between the concept of a realization that is a literal, substantial replacement, and the concept of simulation that is a metaphorical representation of specific structure or behavior, but that also requires specific differences which allow us to recognize it as "standing for" but not realizing the system. In these terms, a simulation that becomes more and more "lifelike" does not at some degree of perfection become a realization of life. Simulations, in other words, are in the category of symbolic forms, not material substances. For example, in physics the simulation of trajectories by more and more accurate computation never results in realization of motion. We are not warmed by the simulation of thermal motions, or as Aristotle said, "That which moves does not move by counting."

Simulation of any system implies a mapping from observable aspects of the system to corresponding symbolic elements of the simulation. This mapping is called measurement in physics and in most other sciences. Since measurement can never be made with absolute accuracy, the quality of simulations must be judged by additional criteria based on a theory of the system. Measurement presents a serious conceptual problem in physics, since it bridges the domains of known laws and unknown states of the system.[28] Even in classical physics, there is no theory of this bridge, and in quantum theory, the measurement problem is presently incomprehensible.[25] The practical situation is that measurement can be directly realized in physics, but it cannot presently be simulated by physical laws alone. By contrast, although a measurement might be simulated by computer, it can never be realized by computation alone. The process of measurement in cognitive systems begins with the sensory transducers, but how deeply the process penetrates into the brain before it is completed is not understood. At one extreme, we have computationalists who say that measurement is completed at the sensory transducers, and that all the rest is explicit symbol manipulation.[18] At the other extreme are physicists who consider the consciousness of the observer as the ultimate termination of a measurement.[23,28] A more common view is that neural cell assemblies that are large enough to be feature detectors may be said to measure events, but we are still a long way from a consensus on a theory of perception or a theory of measurement.

THE SYMBOL-MATTER PROBLEM

The molecular facts of the genetic code, protein synthesis, and enzymatic control form an impressive empirical base, but they do not constitute a theory of how symbolic forms and material structures must interact in order to evolve. That is, the facts do not distinguish the essential rules and structures from the frozen accidents. The only type of theory that can help us make this distinction for artificial life is the theory of evolution, and in its present state, it is by no means adequate.

Von Neumann's[24] kinematic description of the logical requirements for evolvable self-replication should be a paradigm for artificial life study. He chose the evolution of complexity as the essential characteristic of life that distinguishes it from non-life, and then argued that symbolic instructions and universal construction were necessary for heritable, open-ended evolution. He did not pursue this theory by attempting a realization, but instead turned to formalization by cellular automata. This attempt to formalize self-replication should also serve as a warning to AL research. Von Neumann issued the warning himself quite clearly: "By axiomatizing automata in this manner, one has thrown half the problem out the window and it may be the more important half."[24] By "half the problem" von Neumann meant the theory of the molecular phenotype. I am even more skeptical than von Neumann. I would say that by formalization of life, one may be throwing out the whole problem, that is, the problem of the relation of symbol and matter. This is one issue that AL must address. To what extent can a formal system clarify the symbol-matter relation? To what extent is the evolutionary nature of life dependent on measurements of a material environment?

The Physical Symbol System Hypothesis is, in my view, an empirically obvious half-truth. We, indeed, can construct symbol systems from matter, the computer and brain being impressive examples. The converse is not true, that matter can be constructed from symbols, since this would violate physical laws. However, we must also give full recognition to the Evolutionary Hypothesis that under the symbolic control of genotypes, material phenotypes in an environment can realize endless varieties of structures and behaviors. We, therefore, can summarize the following three symbol-matter possibilities: (1) we can simulate everything by universal symbol systems, (2) we can realize universal symbol systems with material constructions, and (3) we can realize endless types of structures and behaviors by symbolic constraints on matter. But we must also accept the fundamental impossibility: we cannot realize material systems with symbols alone. Evolving genes of natural living systems have harnessed matter to realize an enormous variety of structures and behaviors. What we need to learn from artificial life studies is the extent to which a simulated environment can provide artificial organisms with the potential for realizing emergent evolution.

WEAK OR STRONG ARTIFICIAL EVOLUTION?

This brings us to the central question that is bound to be raised in AL studies. How should we respond to the claim that a computer system is a new form of life and not a mere simulation? According to what is known to be universal in present-day life and evolution, I would require the AL realization to include the genotype-phenotype-environment distinctions as well as the corresponding mutability, heritability and natural selection of neo-Darwinian theory. I would further require, with von Neumann and others, that a realization of life should have the emergent or novel evolutionary behavior that goes beyond adaptation to an environment. I do not believe we presently have a theory of evolution that is explicit enough to decide the question of whether a formal environment like a computer can realize evolutionary novelty, or merely simulate it; however, let us reconsider the question for physics models.

We need to return to the distinctions between simulations, realizations, and theories to clarify what aspects of theory are necessary to evaluate computational simulations and realizations. In particular, we need to look more closely at the relation of computation to physical theory. We saw that the property of universality in computation theory has been claimed as a central argument for strong AI.[12] I believe that computational universality can only be used to claim that computers can simulate everything that can be represented in a symbolic domain, i.e., in some language. Physical theories also claim universality within their domains, but these two usages of universality are completely different since they refer to different domains. The domain of physical theory is the observable world in which measurements define the state of the system. To achieve universality in physical theories, the form of the laws must satisfy so-called conservation, invariance, or symmetry principles. These theoretical principles of universality place exceedingly powerful restrictions on the forms of physical theories. By contrast, universality in computation corresponds to unrestricted forms of symbolic rewriting rules, and has nothing whatsoever to do with the laws of physics. Indeed, to achieve symbolic universality, even the hardware must not reflect the laws of physics, but only the rules imposed by the artificial constraints of its logic elements. It is only by constructing a program of additional rules reflecting the physical theory that a computer can be said to simulate physics. The point is that the quality of this type of simulation of physics clearly depends on the quality of the theory of physics that is simulated. Precisely because the computer is symbolically universal, it can simulate Ptolemaic epicycles, Newton's laws, or relativistic laws with equal precision. In fact, we know that even without laws, we can find a statistical program that simply "simulates" behavior from the raw observational data.

Why should this situation be fundamentally different for a computer simulating life? I do not think it is different for simulations of this type that are based on explicit theories. Large parts of what is called theoretical biology are simulations of physical laws under particular biological constraints. Although this type of reduction of life to physical laws is challenging, there is very little doubt that with enough empirical knowledge of the biological constraints, and a big enough computer, such a simulation is always possible, although not with Laplacean determinism. No one doubts that life obeys physical laws.

However, in the cases of cellular automata and chaotic dynamics, we have seen other forms of modeling that are not based on numerical solutions of equations, or derived from natural laws of motion. The artificial "laws" of these models are the computer's local rules, although the behavior of the model is an analog of the physical behavior of the natural environment. As I said earlier, these simulations are appropriately called artificial physics. There is no reason why we can not create a population of artificial cells evolving in such an artificial environment. I proposed this type of model some 20 years ago, and Conrad[2] simulated an entire artificial ecosystem to study the evolution of artificial organisms. Even though the artificial environment of his ecological simulation was almost trivially simple, the populational behavior of the biota appeared to be chaotic, although we did not know the significance of chaos at that time. However, in spite of biotic behavior that was unendingly novel in the chaotic sense, it was also clear that the environment was too simple to produce interesting emergent behavior. The "reality" of the organism, therefore, is intrinsically dependent on the "reality" of its environment. In other words, the emergence of chaotic behavior is not an adequate model of evolutionary emergence.

THREE LEVELS OF EMERGENT BEHAVIOR

Emergence as a classical philosophical doctrine was the belief that there will arise in complex systems new categories of behavior that cannot be derived from the system elements. The disputes arise over what "derived from" must entail. It has not been a popular philosophical doctrine, since it suggests a vitalistic or metaphysical component to explanation which is not scientifically acceptable. However, with the popularization of mathematical models of complex morphogeneses, such as Thom's catastrophe theory,[22] Prigogine's dissipative structures,[17] Mandelbrot's fractals,[11] and the more profound recognition of the generality of symmetry-breaking instabilities and chaotic dynamics, the concept of emergence has become scientifically respectable.

In spite of the richness of these formal concepts, I do not believe they convey the full biological essence of emergence. They are certainly a good start. Symmetry-breaking is crucial for models of the origin of the genetic code,[4] and for many levels of morphogenesis.[7] Many biological structures and behaviors that are called frozen accidents fall into this category, i.e., a chance event that persists because it is stabilized by its environment (e.g., driving on the right side of the road).

However, the counterclaim can be made that frozen accidents are not fully emergent, since the frozen behavior is one of the known possible configurations of the formal domain, and therefore not an entirely novel structure. In the normal course of these arguments, the skeptical emergentist proposes a higher level of functional complexity, and the optimistic modeler proposes a more complex formal domain or artificial environment to try to simulate it.

The concept of emergence in AL presents the same type of ultimate complexity as does the concept of consciousness in AI. Critics of AI have often used the concept of strong consciousness as the acid test for a realization of thought as distinguished from a simulation of thought,[20] but no consensus or decidable theory exists on what a test of strong consciousness might entail. Similarly, the concept of strong emergence might be used as the acid test for a realization of life as distinguished from a simulation, but again, no consensus exists on how to recognize strong emergent behavior. If one takes an optimistic modeler's view, both consciousness and emergence can be treated as inherently weak concepts, that is, as currently perceived illusions resulting from our ignorance of how things really work, i.e., lack of data or incomplete theories. The history of science in one sense supports this view as more and more of these mysteries yield to empirical exploration and theoretical description. One, therefore, could claim that consciousness and emergence are just our names for those areas of awareness that are presently outside the domain of scientific theory.

However, more critical scientists will point out that physical theory is bounded, if not in fact largely formed, by impotency principles which define the epistemological limits of any theory, and that mathematical formalisms also have their incompleteness and undecidability theorems. There is, therefore, plenty of room for entirely unknown and fundamentally unpredictable types of emergence.

In any case, there are at least two other levels of emergence beyond symmetry-breaking and chaos that AL workers need to distinguish. The first is best known at the cognitive level, but could also occur at the genetic level. It is usually called creativity when it is associated with high-level symbolic activity. I will call the more general concept semantic emergence. We all have a rough idea of what this means, but what AL needs to consider is the simplest level of evolution where such a concept is important. We must distinguish the syntactical emergence of symmetry-breaking and chaotic dynamics from the semantic emergence of non-dynamical symbol systems which stand for a referent. I distinguish dynamical from non-dynamical systems by the rate-dependence and continuity of the former. By contrast, symbol systems are intrinsically rate-independent, and discrete.[13] That is, the meaning of a gene, a sentence or a computation does not depend on how fast it is processed, and the processing is in discrete steps. At the cognitive level, we have many heuristic processes that produce semantic emergence, from simple estimation, extrapolation, and averaging, to abstraction, generalization and induction. The important point is that semantic emergence operates on existing data structures that are the result of completed measurements or observations. No amount of classification or reclassification of these data can realize a new measurement.

This leads to the third type of emergence, which I believe is the most important for evolution. I will simply call it measurement itself, but this does not help much because, as I indicated, measurement presents a fundamental problem in physics as well as biology.[13] In classical physics, measurement is a primitive act—a pure realization that has no relation to the theory or to laws except to determine the initial conditions. However, in quantum theory measurement is an intrinsic part of the theory, so where the system being measured stops and the measuring device begins is crucial.

Here again, von Neumann[23] was one of the first to discuss measurement as a fundamental problem in quantum theory, and to propose a consistent mathematical framework for addressing it. The problem is determining when measurement occurs. That is, when, in the process of measuring a physical system, does the description of a system by physical laws cease to be useful, and the result of the measurement become information? Von Neumann pointed out that the laws are reversible in time, and that measurement is intrinsically irreversible. But since all measuring devices, including the senses, are physical systems, in principle they can be described by physical laws, and therefore as long as the system is following the reversible laws, no measurement occurs. Von Neumann proposed, and many physicists agreed at the time, that the ultimate limit of this lawful description may be the consciousness of the observer.[28] Presently, there seems to be more of a consensus that any form of irreversible "record" can complete a measurement, but the problem of establishing objective criteria for records remains a problem.[25]

CAN WE ARTIFICIALLY EVOLVE NEW MEASUREMENTS?

I want to suggest that new measurements be considered as one of the more fundamental test cases for emergent behavior in artificial life models. For this purpose, we may define a generalized measurement as a record stored in the organism of some type of classification of the environment. This classification must be realized by a measuring device constructed by the organism.[13] The survival of the organism depends on the choice and quality of these classifications, and the evolution of the organism will depend on its continuing invention of new devices that realize new classifications.

Now the issue of emergence becomes a question of what constitutes a measurement. Several of the simulations at this workshop have produced autonomous classifications of their artificial environments. Hogeweg[6] has made this non-goal-directed classification her primary aim. Autonomous classification is also demonstrated by Pearson's model of cortical mapping.[15] However, from the point of view of measurement theory, these models do not begin with measurement, but only with the results of measurements, that is, with symbolic data structures. I therefore would call these realizations of semantic emergence in artificial environments. It is clear, however, that none of these reclassifications, in themselves, can result in a new measurement, since that would require the construction of a new measuring device, i.e., a realization of measurement.

Biological evolution is not limited to reclassification of the results of existing measurements, since one of the primary processes of evolution is the construction of new measuring devices. The primitive concept of measurement should not be limited to mapping patterns to symbols, but should include the mapping of patterns to specific actions, as in the case of enzymatic catalysis.[13] The essential condition for this measurement mapping is that it be arbitrarily assignable, and not merely a consequence of physical or chemical laws. We therefore can see that the ability of a single cell to construct a new enzyme enables it to recognize a new aspect of its environment that could not have been "induced" or otherwise discovered from its previously recognized patterns. In the same way, human intelligence has found new attributes of the environment by constructing artificial measuring devices like microscopes, x-ray films, and particle accelerators and detectors. This type of substantive emergence is entirely out of the domain of symbolic emergence, so we cannot expect to realize this natural type of evolutionary emergence with computers alone. The question that should motivate AL research is how usefully we can simulate the process of measurement in artificial environments. As I have argued, the answer to this question will require a theory of measurement.

CONCLUSIONS

The field of artificial life has begun with a high public profile, but if it is to maintain scientific respectability, it should adopt the highest critical standards of the established sciences. This means that it must evaluate its models by the strength of its theories of living systems, and not by technological mimicry alone. The high quality of computer simulations and graphics displays can provide a new form of artificial empiricism to test theories more efficiently, but this same quality also creates illusions. The question is one of the claims we make. The history of artificial intelligence should serve as a warning. There is nothing wrong with a good illusion as long as one does not claim it is reality. A simulation of life can be very instructive both empirically and theoretically, but we must be explicit about what claims we make for it.

Again, learning from the mistakes of AI, and using our natural common sense that is so difficult to even simulate, the field of AL should pay attention to the enormous knowledge base of biology. In particular, AL should not ignore the universals of cell structure, behavior, and evolution without explicit reasons for doing so. These presently include the genotype, phenotype, environment relations, the mutability of the gene, the constructability of the phenotype under genetic constraints, and the natural selection of populations by the environment.

I have proposed the process of measurement as a test case for the distinction between simulation and realization of evolutionary behavior. This is not because we know how to describe measurements precisely, but because new measurements are one requirement for emergent evolution. We know at least that measurement requires the memory-controlled construction of measuring devices by the organism, but this is obviously not an adequate criterion to distinguish a simulated measurement from a realization of a measurement. To make such a distinction will require a much clearer theory of measurement. The study of artificial life may lead, therefore, to new insight to the epistemological problem of measurement as well as to a sharper distinction between the living and the non-living.

REFERENCES

1. Weyl, H. (1940), *Philosophy of Mathematics and Natural Science, Physics* (Princeton: Princeton University Press), Book VIII, Chapt. 8, 54.

2. Conrad, M., and H. Pattee (1970), "Evolution Experiments with an Artificial Ecosystem," *J. Theor. Biol.* **28**, 393–409.

3. Dreyfus, H. L., and S. E. Dreyfus (1988), "Making a Mind Versus Modelling a Brain," *Daedalus* **117**, 15–44.

4. Eigen, M. 91971), "Self-Organization of Matter and Evolution of Biological Macromolecules," *Naturwissenschaften* **58**, 465–522.

5. Harnad, S. (1987), "Minds, Machines, and Searle," *Categorical Perception: The Groundwork of Cognition*, Ed. S. Harnad (Cambridge: Cambridge University Press), 1–14.

6. Hogeweg, P. (1988), "MIRROR beyond MIRROR, Puddles of LIFE," this volume.

7. Kauffman, S. (1986), "Autocatalytic Sets of Proteins," *J. Theor. Biol.* **119**, 1–24.

8. Kelly, M. H., and F. C. Keil (1987), "Metaphor Comprehension and Knowledge of Semantic Domains," *Metaphor and Symbolic Activity* **2**, 33–55.

9. Kugler, P., and M. Turvey (187), *Information, Natural Law, and the Self-Assembly of Rhythmic Movement* (Hillsdale, NY: Lawrence Erlbaum Associates).

10. Langton, C. (1987), "Studying Artificial Life with Cellular Automata," *Physica* **22D**, 120–149.

11. Mandelbrot, B. (1977), *Fractals. Form, Chance, and Dimensions* (San Francisco: W. H. Freeman).

12. Newell, A. (1980), "Physical Symbol Systems," *Cognitive Science* **4**, 135–183.

13. Pattee, H. (1985), "Universal Principles of Measurement and Language Functions in Evolving Systems," *Complexity, Language, and Life: Mathematical Approaches*, Eds. John Casti and Anders Karlqvist (Berlin: Springer-Verlag), 168–281.

14. Pattee, H. (1988), "Instabilities and Information in Biological Self-Organization," *Self-Organizing Systems: The Emergence of Order*, Ed. F. E. Yates (New York: Plenum).

15. Pearson, J., L. Finkel, and G. Edelman (1987), "Plasticity in the Organization of Adult Cerebral Cortical Maps: A Computer Simulation Based on Neuronal Group Selection," *J. Neuroscience* **7(12)**, 4209–4223..

16. Polyani, M. (1964), *Personal Knowledge* (New York: Harper-Row), 149.

17. Prigogine, I. (1970), *From Being to Becoming* (San Francisco: W. H. Freeman).

18. Pylyshyn, Z. (1980), "Computation and Cognition: Issues in the Foundations of Cognitive Science," *Behavioral and Brain Sciences* **3**, 111-169.

19. Reeke, G., and G. Edelman (1988), "Real Brains and Artificial Intelligence," *Daedalus* **117**, 143–174.

20. Searle, J. R. (1980), "Minds, Brains, and Programs," *Behavioral and Brain Sciences* **3**, 417–457.
21. Skarda, C. A., and W. J. Freeman (1987), "How Brains Make Chaos in Order to Make Sense of the World," *Behavioral and Brain Sciences* **10**, 161–195.
22. Thom, R. (1975), *Structural Stability and Morphogenesis* (Reading, MA: W. A. Benjamin).
23. von Neumann, J. (1955), *The Mathematical Foundations of Quantum Mechanics* (Princeton: Princeton University Press).
24. von Neumann, J. (1966), *The Theory of Self-Reproducing Automata*, Ed. A. W. Burks (Urbana, IL: University of Illinois Press).
25. Wheeler, J. A., and W. H. Zurek (1983), *Quantum Theory and Measurement* (Princeton: Princeton University Press).
26. Wigner, E. P. (1960), "The Unreasonable Effectiveness of Mathematics in the Natural Sciences," *Comm. Pure and Appl. Math.* **13**, 1–14.
27. Wigner, E. P. (1964), "Events, Laws of Nature, and Invariance Principles," *Science* **145**, 995–999.
28. Wigner, E. P. (1967), "The Problem of Measurement," *Symmetries and Reflections* (Bloomington, IN: Indiana University Press).

Steen Rasmussen
Physics Laboratory III and Center for Modelling, Nonlinear Dynamics, and Irreversible Thermodynamics, The Technical University of Denmark, DK-2800 Lyngby, Denmark

Toward a Quantitative Theory of the Origin of Life

1. INTRODUCTION

To understand life in principle involves an understanding of the evolution of complexity. This is the subject for the present paper. We shall develop a very simple model for the creation of the first real genetic systems. It is a theory that defines both evolution and complexity in a somewhat restrictive way. However, the applied conceptual and mathematical framework allows us to obtain quantitative analytical relations between the assumed prebiotic physico-chemical conditions and the time of emergence for larger self-replicating genetic systems. Further, the ideas behind these calculations apply to general problems concerning emergent properties in dynamical systems.

Despite the unsolved chemical issues related both to synthesis and replication of RNA under prebiotic conditions,[24,19] we shall assume that the molecular basis for the evolution of the first genes is solely RNA. Probably the first steps in the creation of life involved mechanisms and materials different from what we find in today's organisms.[4] However, the basic ideas behind our analysis are not affected by details of the assumed intermolecular interactions, i.e., whether we consider RNA as the only element in the evolutionary process, or mixed systems where elements such

Artificial Life, SFI Studies in the Sciences of Complexity,
Ed. C. Langton, Addison-Wesley Publishing Company, 1988

as polypeptides and inorganic components participate. For simplicity we currently consider only RNA molecules.

Chemical elements involved in the origin of life could have combined in an infinite variety of ways. Modern life shows a number of well-developed organizational forms where the autocatalytic system seems both efficient and very simple. Therefore, following the work of Manfred Eigen,[8] we shall ask for the occurrence of catalytic reaction cycles in the prebiotic environment. We are looking for the emergence of certain functional abilities in a system and we want to know when the system can *do* certain things. For RNA this implies that RNA string polymers can fold into certain shapes (forms) in order to attain a certain catalytic ability. The evolution of *function* and *form* therefore is closely related and complementary.

More than a century ago, Darwin[7] formulated the idea of evolution through natural selection. This is a process of system learning by trial and error where system performance is tested continuously against the environment (which itself is evolving). Together with the ideas of self-organization[8,14,26] with which spontaneous order can emerge in nonlinear systems away from thermodynamical equilibrium, these ideas form a very powerful platform for understanding the evolution of complexity.

Before formulating the model, we shall briefly touch upon some fundamental problems involved in a theoretical understanding of evolution.

If we base our explanation solely on the catalytical properties of RNA, we neglect an infinity of alternative possibilities and hereby may restrict our evolution. Even with this restriction, however, we are not able *a priori* to tell which types of molecules to include in our model and which to neglect. For example, which RNA sequences are the most important in the self-organizing process? Fortunately, it turns out that we don't need to answer this question in detail in order to perform our calculations. Nonetheless, it is very interesting to see if one can apply some kind of objective theoretical measure of evolutionary importance or adaptability in this context. For instance, complexity measures are proposed by Grassberger et al.[15,16,35] for one-dimensional sequences, and by Huberman et al.[1,5,17] for hierarchical systems. Here it is found that it makes sense to define the most complex and adaptive systems between the uniform and the randomly organized systems. Computations on cellular automata (CA) also show that we *a priori* can determine from an ensemble of rules which are the most significant from an evolutionary point of view. This determination has been shown by Langton[23] by correlation length in state space for two-dimensional CA, for computational abilities for one-dimensional CA,[28] and for competition of CA rules on a one-dimensional array. The "best rules" in all these cases are clearly concentrated between the uniform and the chaotic rules (around the Class IV automata[34]).

With this knowledge, a naive *a priori* theoretical guess for location of interesting ensembles of RNA sequences on the uniform/random scale (therefore, from a statistical point of view) will lie somewhere between the uniform and the random sequences. This is consistent with present knowledge gained from sequencing catalytic RNA in modern life[11,34]; however, future studies may permit more specific modeling.

By preformulating a given evolutionary route, we are neglecting an infinity of alternative possibilities. Is it possible to model an evolutionary process using simple mathematical abstractions, or are we up against the "nature of complexity" where the whole phase space of possibilities has to be consdiered? Is it necessary to model the infinity of nonlinear interactions to ensure that the model captures the evolutionary path? If this is the case, evolution will presumably exceed our explanatory powers for a long time to come. However, looking at the history of science and at the scientific methods applied, it seems to make sense to begin with assumptions about the mechanisms involved in the creation of an observed phenomenon. We should then select one of these mechanisms and propose a model which, without violating the laws of Nature, is capable of explaining the phenomenon. In the present case, we shall assume a particular evolutionary route and show that it leads to a result which seems reasonable and which is not contradicted by present knowledge. The main difference between modeling evolutionary processes and "classical processes" is that we are attempting to describe the dynamics of a successive change of systems in opposition to internal dynamics in a fixed system.

We shall start out with an extremely simple model which reflects the self-organizing principle of creating cooperative structures by catalytic connections. In our calculations, we shall only be interested in orders of magnitude to determine whether or not it is possible to form sufficient catalytic cycles in geological time scales.

2. SIMPLE MODELS OF SELF-ORGANIZING PROCESSES LEADING TO LONGER POLYMERS

In order to give a realistic description of the processes involved in creation of the first cooperative catalytic structures, we ideally need a model where the equations of motion can change in response to random events internally triggered or generated within the system. There is a variety of ways to formulate such a model. What is required is that these formulations have causal interactions among the different subunits, together with dynamic equations describing the performance of the isolated as well as of the connected subunits in a given environment. In our system, the self-replicating RNA strings are the subunits while the catalytic connections correspond to causal interactions among these subunits. Random events can create new subunits and/or new connections on the micro level. The new structures are continuously tested against the old ones with respect to their performance. Through replication, the new structures may become abundant and thus change the composition of the system on the macro level. Common for this class of dynamical systems is a high degree of complexity in the formal model. This complexity arises both because of the large number of subunits needed in order to constitute a macro level, and because of the combinatorial problems involved in having all these subunits in potential communication. Analytical results on such systems are extremely hard to obtain, and numerical simulations demand very large computers if one wants to approach realistic systems. Moreover, these problems concerning qualitatively changing systems are not easily represented in terms of established mathematical formalisms. They generally call for new frameworks.

A very simple and only approximate way of formulating such a system is by means of ordinary differential equations coupled via a random connectivity matrix. In earlier works, we have used this method to deal with evolutionary processes for small systems.[25,29] Here the stochastic generation of causal links is complemented by functions that continuously evaluate the performance of the produced structures and stabilize reinforcing connections. This has been denoted as stochastic re-causalization. With such a relation between dynamics and causal interactions, the system always ends up in a situation where the cooperative structures dominate. Monte-Carlo simulations on such systems then can give hints as to the appearance frequencies of the different cooperative structures.

An elegant approach to the same problem has recently been developed by Farmer, Kauffman and Packard.[13] They start out with two given, idealized catalytic reactions on polymer strings, namely cleavage and joining reactions. In each time step, with given probabilities, either of these reactions may take place catalyzed by other randomly chosen strings. Hereby new strings are formed. Depending on the assigned probabilities and the initial polymer distribution, a spontaneous generation of autocatalytic sets may appear. Farmer et al. have hereby provided a new computational formalism suitable for the problem. Adding kinetics to each species (e.g., in the form of a growth equation), assuming that the processes are running in a fixed volume with overflow (a chemostat) and not allowing new species to participate in catalysis before they have passed a certain threshold of macroscopic concentration, they find that the model's ability for spontaneous generation of catalytic networks is preserved. We shall later return to a more detailed discussion of this model. Also Kauffman's analysis [20,21] is related to this problem. The model to be presented in this paper is a further development of the results attained by Rasmussen[30] and by Rasmussen, Mosekilde and Engelbrecht.[31]

Because the fundamental quality needed in order to create and conserve more complicated macro molecules is a positive interaction between simpler molecules and other species, we are primarily interested in the time of emergence of this quality. With specified conditions on how the positive interactions are created, on what kind of species are involved, etc., we want to know how likely it is to find positive feedbacks at a given stage in the evolution.

Let us envisage a manifold of uncoupled RNA molecules existing in the primordial soup at a certain stage of the evolutionary process. We assume that these strings had a maximum length up to 80 nucleotides like modern tRNA. This is in accordance with the information threshold for self-replicating RNA.[8,10] Since we are looking for the appearance of cooperative structures, the interesting subset of RNA species are those which through error replication (or self-splicing) can acquire both catalytic effects and the ability to be catalyzed by other RNA molecules. At

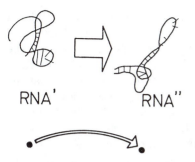

RNA'   RNA"

FIGURE 1 As a model for prebiotic evolution, the vertices of the random graph are interpreted as specific types of information carriers, and the oriented edges as catalytic interactions.

this stage, it is not necessary to specify whether the catalytic interactions are direct or via simple proteins. Presumably both types of processes have been involved.

First we consider the situation where the catalytic abilities do not depend on the actual string length, as long as the nucleotide string is longer than a given minimum.

Let n_{tot} denote the total number of different RNA sequences, and let each of these sequences be represented by a point in the labelled vertex set $V_{n_{tot}} = \{1, 2, \ldots, n_{tot}\}$. Among these points there is a subset $V_{n_{cat}}$, containing n_{cat} catalytic RNA species. Any point in $V_{n_{cat}}$ can be connected to another point in $V_{n_{tot}}$ through the establishment of a directed edge between the two points. This is illustrated in Figure 1. Directed edges in the graph thus correspond to catalytic interactions.

Initially the graph is empty. We imagine that some earlier processes outside our model are responsible for the production of very short, self-replicating nucleotide strings without significant catalytic properties. These short RNA strings will, through error replication, produce their neighboring strings in a Hamming distance sense (single-base substitutions) including adding and subtracting single nucleotides along the string. We can arrange all the (not-yet-created) RNA strings up to the information threshold (< 80 bases) on a connection cube where the metric is given by the Hamming distance as proposed by Eigen.[9] This defines the sequence space. At a certain stage, one of these molecules passes a threshold length and at the same time has a sequence, which allows it to enter the "catalytic land." This means that due to a particular tertiary structure the molecule acquires an important catalytic property. We don't need to specify the actual mechanism of the reaction, just to assume that the reaction either facilitates the formation rate or lowers the error probability in replication for another or some other nucleotide strings. Hereby the first catalytic RNA quasi-species are created in our graph. By quasi-species we here mean a given string and its Hamming neighbors, i.e., strings which possess the same kind of catalytic properties. The Hamming distance between the different catalytic species in general is larger than unity. After creation of the first catalytic string on the cube, we will primarily observe a "diffusion" of new catalytic species from this first string governed by the rate of error formation for the self-replicating RNA strings.

Since, with present knowledge, it is very difficult *a priori* to deduce global properties, such as catalytic abilities, for polymers with given sequences, as a first approximation we shall assign these properties at random. The specificity of the catalysis we will express by v, which is defined as the average number of RNA families in $V_{n_{tot}}$ that the considered string is able to catalyze. If we assume that the catalysis is a result of some kind of base recognition,[33] the number of bases needed in order to specify the catalytic site on the catalyst is larger than the necessary number of specified bases on the substate. This implies that $b^{-1} = n_{tot}/n_{cat}$ is smaller than v, the number of substrate species each catalyst can recognize. Therefore, the evolution of the graph $\mathcal{G}_{vt,cube}^{\rightarrow n_{tot}}$ proceeds as follows: pick a vertex at random in the (Hamming) neighborhood of an existing vertex. If the new vertex belongs to n_{cat} then choose v different vertices also at random, which the first string can catalyze. Connect these catalytic interactions by directed edges. Continue this process until the first catalytic feedback structure is formed, e.g., until an oriented subgraph containing a directed cycle occurs.

3. A FORMAL MODEL INCLUDING KINETICS

The dynamical equations for the above evolutionary process may be written as

$$
\begin{aligned}
\frac{dx_i}{dt} =& k_i(1-\alpha)^{s_i} x_i r - x_i/\tau \\
&+ \frac{1}{h_i} \sum_{\ell=1}^{h_i} \left(1 - (1-\alpha)^{s_\ell}\right) k_\ell x_\ell r \\
&+ \sum_{j=1}^{n_{cat}} k_{ij} x_i x_j r, \qquad i \in V_{n_{tot}}
\end{aligned}
\tag{1a}
$$

where x_i is the concentration of RNA string number i. The first term denotes the error-free, non-catalyzed replication rate, where k_i is the rate constant for the replication of RNA$_i$, α the average error probability per base and s_i the string length. r is the concentration of activated monomers. The second term denotes a simple dilution term given by an average residence time τ for any string in the micro niche. The third term denotes the source of new RNA$_i$ strings due to error replication from neighboring strings. The summation is over all Hamming neighbors h_i for string i. This term defines the connection cube. The last term is the catalyzed replication of RNA$_i$ due to existing catalytic active strings. k_{ij} denotes the catalytic replication constant for RNA$_j$ helping RNA$_i$. The last term constitutes the random graph on top on the connection cube.

Corresponding to Eq. (1a) there is also an equation for the nucleotides

$$\frac{dr}{dt} = r_u - r/\tau_r - r \sum_{i=1}^{n_{tot}} x_i \left(k_i + \sum_{j=1}^{n_{cat}} k_{ij} x_j \right) \tag{1b}$$

where r_u is a constant source term for activated nucleotides, τ_r an average residence time and the third term represents what is being built into polymers.

In order to make the simulation of Eq. (1) as simple as possible and still conserve the fundamental dynamics, we will make some assumptions. First set $n_{tot} = n_{cat}$ and let the average specificity be equal to one, which corresponds to $v = 1$. Hereby we only include the interesting strings and by assuming the catalysis to be very specific, a long "diffusion" in the sequence space is guaranteed before occurrence of the first catalytic cycle. (The missing non-catalytic RNA sequences may to a certain degree be compensated for by multiplying the neighbor term in Eq. (1a) by some factor $\epsilon < 1$). k_{ij} is assumed to be equal to zero or k_c, constant depending on catalytic connection or not. The string lengths s_i are all assumed to be identical s. Further, it is assumed that the strings are only made of guanosine and cytidine. These assumptions give each string s neighbors. We shall later argue (Eq. (15)) why it is reasonable to expect a conservation of dynamics with these assumptions as long as $2^s \gg 1$.

Initially we simulate the system without any catalytic interactions, $k_{ij} = 0$ (Figure 2). The other parameters are $s = 7$, $\tau = 3.0$ time units, $\tau_r = 2.0$ time units, $r_u = 1.0$ mole nucleotides/time unit, $(1 - \alpha)^s = 0.1$ corresponding to $\alpha \cong 0.02$ and $k_i \in U[0.5; 1.5]$, e.g., k_i drawn from an uniform distribution between 0.5 and 1.5. The spectrum of the k_i's is seen in Figure 2a. At time zero a concentration of 10^{-7} mole per unit volume of string number 1 (with a bit number defined one lower: 0000000) is probed into the system. The dynamics is hereafter a diffusion, due to replication errors, and from these a build-up of the different strings on the cube. It is important to note that the individual linear selection value k_i does not alone determine the performance of string i. What is produced (or not produced) from the neighboring strings is also important. For the chosen k_i interval [0.5;1.5] this also holds for much lower error probabilities per base ($\alpha \simeq 10^{-4}$).

The structure of the cube is reflected in Figure 2b where the concentrations of each string are shown at time 4.5. Here string number one still dominates, because it was the first string present. Note that the concentration of all its neighbor strings—2(0000001), 3(0000010), 5(0000100), 9(0001000), 17(0010000), 33(0100000) and 65(1000000)—is also relatively high. At time 50, some of the strings have multiplied to a macroscopic significance and start to stabilize at concentrations between 0.1 and 1.0 mole per unit volume. Although the system is still on a transient, some of the species—or neighborhoods of species—with large selection values start to dominate. This process continues until the equilibrium distribution is reached. The equilibrium distribution is close to the distribution shown at time 200 in Figure 2d. Here the linear selection spectrum is only reflected to a certain degree. Strong

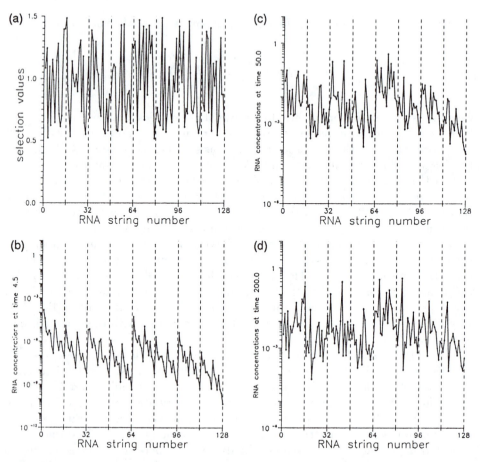

FIGURE 2   Simulation of system (1) without catalytic interactions ($k_{ij} = 0$). In (a) the selection values $k_i \in U[0.5; 1.5]$ are shown as a function of the string number. At time $\cdot$ 5 (b), 50 (c) and 200 (d), the polymer concentrations are shown as a function of the string number. The other parameter values are given in the text. Initially the system is probed with a concentration of 10^{-7} mole per unit volume of string 1 defined by the bit sequence: 0000000. At time 200 the system is almost stabilized. The linear selection spectrum in (a) is only partly reflected in the stable concentration distribution in (d) due to neighbor effect. Also note the relatively narrow concentration interval (only four orders of magnitude) in which all the RNA molecules are living.

species like string 16, 42, 68 and 84 are also macroscopicly significant in the system, but the entire selection spectrum in Figure 2a is transformed due to neighbor effects. Also note that the concentrations for the system without catalytic interactions only cover 4 orders of magnitude.

In Figure 3, $(1 - \alpha)^s$ is replaced by 0.05 (corresponding to $\alpha \cong 7 \cdot 10^{-4}$). The same random sequence as in the earlier run is used to create the linear selection

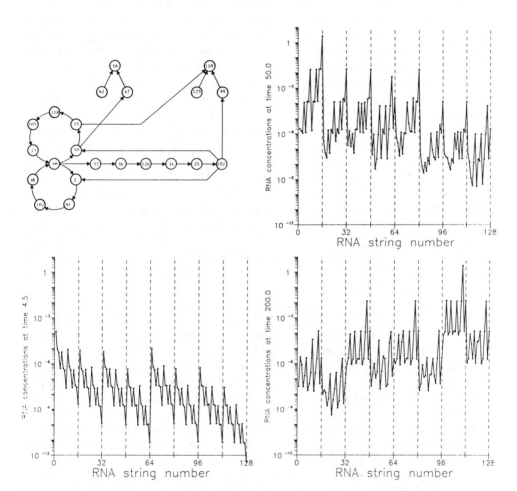

FIGURE 3 (a) The central autocatalytic set of the giant component for a realization of the system (1) with an average out degree, or catalytic density, slightly greater than one. 110 of the 128 possible strings are involved in the giant reaction network. The catalytic interactions $k_{ij} = k_u = 100$ when they exist, $k_i \in U[0.9; 1.1]$ but with the same relative differences as shown in Figure 2a (same random sequence) and $(1 - \alpha)^s = 0.05$. All other parameters are identical to the ones given in Figure 2. The cube structure in (b) is clearer here due to the smaller absolute variations in the k_i's. For this particular combination of initial conditions only sidebranches or "parasites" to the autocatalytic set gains macroscopic significance. First string 16 dominates and later string 110 takes over due to a more efficient catalytic support from the giant component.

values k_i, but the deviation from 1.0 is now multiplied by 0.1 leaving $k_i \in U[0.9; 1.1]$. The other parameters are identical to the earlier run, except for $k_{ij} = k_c$ which is 100 when it exists. In order to have significant effects from the catalysis, when k_i is allowed to vary, k_c needs to be much larger that the average k_i's. The catalytic density (or average out degree), which is also randomly generated, is slightly greater than 1.0 in this situation. In the next section, we will show that this guarantees the existence of catalytic feedbacks. Due to the little differences in the k_i's, the cube structure is very clearly reflected in the dynamics as seen in Figure 3b.

Around time 50, string 16 and its neighbors are the dominating species with macroscopic importance. However, at time 100, string 110 takes over and continues to dominate throughout the rest of the simulation. In order to explain this, we have to look into the details of the catalytic connections. In this particular realization of the system, 110 of the possible 128 strings communicate in, what in graph theory normally is referred to as the giant component.[3] All these strings either directly or indirectly help the replication of string 16 and 110, and it is this huge amplification network that blows up the concentration of these strings. It turns out that the kernel of the giant component consists of a complex autocatalytic set with four major hypercycle loops. As seen in Figure 3a, both string 16 and 110 are side branches or "parasites" to this set. It should be noted that none of the members from the autocatalytic set ever get any macroscopic significance for this initial concentration distribution.

In Figure 4, the system is simulated with a new random sequence giving $k_i \in U[0.5; 1.5]$ and a new sequence of $k_{ij} = 10$ or zero depending on the existence of a catalytic connection or not (average out degree still slightly greater than one). All other parameters are conserved from the run in Figure 3. Figure 4b shows the concentrations of RNA molecules at time 50. Here string 81 and its neighbors are dominating. Note also that the initial probed population for string 1 is still reflected at this stage of the development. Later string 81 is defeated by the string pair 38 and 99. This string pair constitutes a catalytic two cycle, which continues to dominate throughout the simulation. The autocatalytic set(s) from this realization is shown in Figure 4a. Which part of the giant component that gains macroscopic significance at a certain time in the development depends very sensitively on the evolutionary initial conditions. The Hamming distance between string 1(0000000) and string 81(1010000) is two, while the Hamming distance from string 1 to both of the members in the autocatalytic two cycle (38;99) is three. Thus the (38;99) cycle is turned on later that string 81, and therefore also later gains macroscopic significance. Also the dominating part of the autocatalytic set depends on initial conditions. For instance, probing with a concentration of 10^{-7} moles per unit volume of string 128 (1111111) results in the emergence of the whole central part of coupled catalytic feedbacks in the giant component, reflected by very complicated oscillatory dynamics in the concentrations of the individual members in the cooperative structure.

It is impossible to obtain analytical results for the system given in Eq. (1), and Monte Carlo simulations even with our simplifications are computationally very

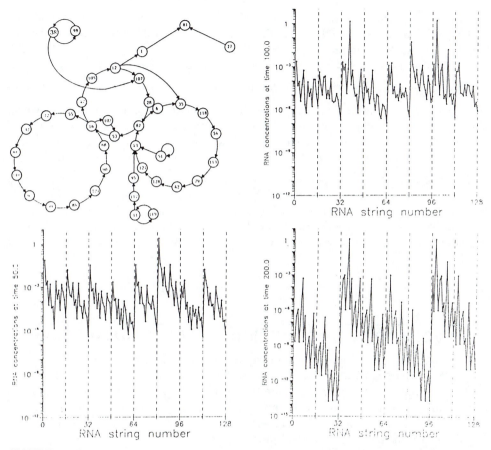

FIGURE 4   The same average catalytic out degree as in Figure 3 but another random realization. $k_c = 10$ when it exists, and $k_i \in U[0.5; 1.5]$ but different from the sequence in Figure 2a. The other parameters are identical to those given in Figure 3. Again we see a succession in the dynamics of the species with macroscopic significance. First a sidebranch from the central autocatalytic set dominates for some time before the autocatalytic two cycle (38;99) takes over and suppresses any further change in selection. Note the very pronounced quasi-species hierarchy of neighbors in (d), spreading all the different strings over a very broad concentration range. The dominance of (38;99) is very sensitive to the initial concentration distribution. Instead probing the system with string 128 (1111111) the whole central autocatalytic network will be activated and for a very long time we will see complicated hypercycle oscillations.

heavy for large n_{tot}. Even $s \sim 20$, which we assume to be the minimal string length where we can meet catalysis, corresponds to $n_{tot} \sim 10^6$, and that many not locally coupled, nonlinear differential equations cause problems for any existing computer.

Therefore, we will take the advantage of obtaining analytical results by modeling the evolution process as a discrete process not explicitly taking the kinetics of every individual RNA string into consideration.

We first note that there is no difference between the appearance of properties in the graph $\overrightarrow{\mathcal{G}}_{vt,cube}^{n_{tot}}$ where the appearance of new RNA species is through sequence space such that neighboring strings are always created, and in the graph $\overrightarrow{\mathcal{G}}_{vt}^{n_{tot}}$ where we pick a vertex at random and with probability b stochastically generate v arcs. The effect of the "metric" on the cube disappears because the v arcs are chosen randomly independent of each other. We shall therefore develop the results for $\overrightarrow{\mathcal{G}}_{vt}^{n_{tot}}$, because $\overrightarrow{\mathcal{G}}_{vt}^{n_{tot}}$ is a close variation of already described random graphs.[3,12] In this way, we remain in an established frame. In fact, the appearance of RNA species can follow any algorithm as long as we maintain the stochastic independence in the catalytic formations, which means in the choice of directed edges. For simplicity and without loosing any generality we will assume that $n_{tot} = n_{cat} = n$ in the next section.

4. EVOLUTION WITHOUT KINETICS: THE RANDOM GRAPH

Like in statistical mechanics where we are not able to give an exact description of a particular molecule in a gas, we will not try to give an exact description of the evolution of a particular random graph. We shall define an ensemble of random graphs and derive certain properties from this ensemble. A more thorough discussion of the mathematical properties of the prebiotic random graph model is given in Rasmussen, Bollobás and Mosekilde.[32]

The random graph $\overrightarrow{\mathcal{G}}_{vt}^{n}$ represents the set of possible outcomes of the process in which we at random choose t vertices from the labelled set $V_n = \{1, \ldots, n\}$ and stochastically generate v-directed edges out from each vertex. Let $t/n \rightarrow \sigma$ in the asymptotic limit ($n \rightarrow \infty$) and define the average out degree: $d = tv/n$. Assuming that each vertex point in principle can communicate with any other point, and including possible self-interacting loops, the total number of potential edges is $N = 2\binom{n}{2} + n = n^2$. At any time t, there are $\binom{n}{t}\binom{n}{v}^t$ different realizations of the random graph over V_n. A particular realization from $\overrightarrow{\mathcal{G}}_{vt}^{n}$ is denoted by $\overrightarrow{G}_{vt}$.

A subset Q of $\overrightarrow{\mathcal{G}}_{vt}^{n}$ is said to define a property Q for the realization $\overrightarrow{G}_{vt} \in \overrightarrow{\mathcal{G}}_{vt}^{n}$, if $\overrightarrow{G}_{vt} \in Q$. As an example, it is a property of a graph to be connected, in which case Q represents the subset of all connected realizations of $\overrightarrow{\mathcal{G}}_{vt}^{n}$. It is also a property of a graph to contain a specific subgraph H. A property is said to be monotone, if $\overrightarrow{G}_1 \in Q$ and $\overrightarrow{G}_1 \subset \overrightarrow{G}_2$ implies $\overrightarrow{G}_2 \in Q$. The property of containing a specific subgraph is monotone.

A graph process $(\vec{G}_{vt})_{t=0}^{n}$ on the vertex set V_n may be defined as a sequence of graphs $\vec{G}_{vt}$ where $\vec{G}_0 \subset \vec{G}_{1v} \subset \vec{G}_{2v} \cdots \subset \vec{G}_N$. $(\vec{G}_{vt})_{t=0}^{n}$ thus represents the process of adding edges to V_n, v-by-v, while maintaining already established communications. A close variation of this process is illustrated in Figure 5. Here we see the set of possible processes through which the complete 2-graph can be generated by adding edges one by one. The heavy arrows define one particular process out of the 24 possible processes. If edges are thrown onto V_n at random, at a given rate (number per unit time), the evolution of the graph can be described as a Markov process whose states are particular graphs on V_n.

If, for a random graph in $\vec{\mathcal{G}}_{vt}^{n}$, the probability $P_{vt}(Q)$ of possessing a certain property Q asymptotically approaches 1 for $n \to \infty$, then almost every graph $\vec{G}_{vt} \in \vec{\mathcal{G}}_{vt}^{n}$ has the property Q. On the other hand, almost no graph $\vec{G}_{vt} \in \vec{\mathcal{G}}_{vt}^{n}$ has the property Q if $P_{vt}(Q)$ for $n \to \infty$. Most monotone properties in large random graphs appear rather suddenly. Thus, if almost no realization of $\vec{\mathcal{G}}_{vt}^{n}$ for a given vt has the property Q, for a slightly higher number of edges almost all realizations will have Q.

The abrupt change in the probability of containing a certain subgraph can be expressed in terms of a threshold function.[3] The function $d^*(n)$ is said to be a threshold function for the monotone property Q if

$$\lim_{n \to \infty} P_d(Q) = \begin{cases} 0 & \text{if } d(n)/d^*(n) \to 0; \\ 1 & \text{if } d(n)/d^*(n) \to \infty. \end{cases} \qquad (2)$$

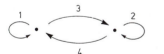

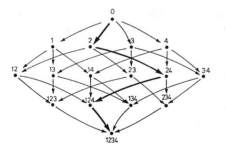

FIGURE 5 The complete 2-graph (top) together with a graphical representation of the 24 different processes through which the complete 2-graphs can be generated by adding one edge at a time. The complete n-graph can be generated in $(n^2)!$ different ways.

$d(n)$ is here the average out degree which specifies how the number of edges are assumed to vary with the number of vertex points. To determine the threshold function for the property of containing a particular subgraph, the subgraph must be balanced. Loosely speaking this means that the subgraph should not itself contain subgraphs of higher average degree or average edge density. The average degree (or just the degree) of a subgraph is defined as the number of edges ℓ relative to the number of vertex points k, i.e., $d = \ell/k$.

It can be shown that the threshold function for containing a particular subgraph H in $\vec{G}_{vt}$ depends on the maximal degree of this particular subgraph. If H is balanced, one has

$$d^*(n) = cn^{2-k/\ell} \tag{3}$$

where c is a proportionality factor which is independent of n.

The idea in this proof is to show the various properties for the threshold functions one at the time. First one can show that $P(\vec{G}_{vt} \supset H) \to 0$, because the expectation value for the number of H subgraphs tends to zero as $n \to \infty$. Then it can be proven that H exists in almost every realization from $\vec{G}_{vt}^n$ by showing that the probability of having zero H subgraphs in $\vec{G}_{vt}$ tends to zero as the expected number of H subgraphs tends to infinity. Hereby it is guaranteed that they are not concentrated in a few realizations from $\vec{G}_{vt}^n$.

In fact, one is able to do better than this. We can explicitly calculate the distribution as a function of the average out degree for any fixed cycle in $\vec{G}_{vt}$, and thereby obtain the explicit threshold function for any combination of oriented cycles as the graph evolves.

The expectation value for the number X_j, of directed j-cycles in $\vec{G}_{vt}$ as a function of v and σ is given by:

$$E(X_j) = \frac{(n)_j}{j}(\sigma\frac{v}{n})^j \sim \frac{(v\sigma)^j}{j} = \lambda_j \tag{4}$$

where $(n)_j/j$ determines all the possible directed j-cycles present in $\vec{G}_{vt}$, and $\sigma(v/n)$ is the probability that any of the edges exist in the cycle. As before σ denotes the asymptotic limit of t/n, the occupation number in the graph.

The expression for $E(X_j)$ shows that as long as the average out degree is relatively low $d \ll 1$, the number of small cycles strongly dominates over the number of large cycles.

Now it is possible to show that the number, $X_1, \ldots, X_m$, m fixed, of directed j-cycles are asymptotically independent Poisson random variables with means given by Eq. (4). The expression

$$P(X_j = i) = \frac{\lambda_j^i e^{-\lambda_j}}{i!} \tag{5}$$

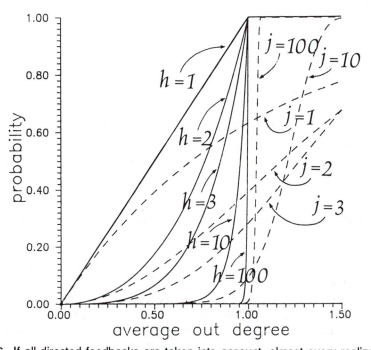

FIGURE 6 If all directed feedbacks are taken into account, almost every realization $\vec{G}_{vt}^{\,n}$, contains a cycle for $d \geq 1$. The probability of finding a j-cycle, $j = 1, 2, 3, 10$ and 100, in the graph is indicated by dotted lines. Obviously the curve for the probability of finding a j-cycle becomes steeper for larger d as j increases. The same is seen for the probability of finding a j-cycle of length $\geq h$ (heavy lines). Note that as d passes above unity a dominance shift from shorter to longer cycles is seen.

therefore gives the probability that we find i directed j-cycles in $\vec{G}_{vt}$ at the stage where the average out degree is $d = v\sigma = v(t/n)$. In order to prove this, one first has to show that the r'th factorial moment $E_r(X)$ tends towards λ^r as $n \to \infty$ for all cycles without joint vertices, and thereafter that the number of cycles with joint vertex points are negligible.

Using Eqs. (4) and (5), the probability that we find any j-cycle C^j in $\vec{G}_{vt}$ may be expressed as

$$P(\exists C^j \subset \vec{G}_{vt}) = 1 - \exp\left(\frac{-(v\sigma)^j}{j}\right). \tag{6}$$

The probability of finding any cycle C in the graph is given by

$$P(\exists C \subset \vec{G}_{vt}) = 1 - \exp\left(-\sum_{j=1}^{\infty} \frac{(v\sigma)^j}{j}\right)$$

$$= v\sigma \qquad \text{for } v\sigma < 1.$$

For the average out degree equal to or slightly greater than unity it is seen that almost every realization of $\vec{G}_{vt}^{n}$ contains a cycle.

The threshold for appearance of cycles above a certain minimum length h seems to be

$$P(\exists C^j \subset \vec{G}_{vt}, \ j \geq h) = 1 - (1 - \sigma v) \exp\left(\sum_{j=1}^{h-1} \frac{(\sigma v)^j}{j}\right), \qquad \sigma v < 1 \qquad (8)$$

A graphical representation of these emergence thresholds as functions of the graph evolution is seen in Figure (6).

It can be shown that for d equal to unity a giant biochemical network emerges in which a huge number of RNA strings communicate. If the average out degree continues to rise, the threshold for communication between all n vertices occurs for $d \sim n \log n$. The proofs of these results for random graphs without directed edges $\mathcal{G}(n, M)$ can be found in Bollobás.[3] The later stages in the development of the random graph are presumably not so important from an evolutionary point of view, since the emergence of the first catalytic feedback cycles probably changes the system in a fundamental manner. The coupled structures now become the subunits in the succeeding evolutionary step. However, because a relatively small difference in average out degree for $M \sim n$ may cause the emergence of a giant autocatalytic network, selection of parts of this or between such networks is also of interest. Recall the discussion in the previous section of the model including kinetics.

5. ALTERNATIVE COOPERATIVE STRUCTURES

There are several arguments against only considering simple one-component feedback structures. If each molecule needs interaction with, say, two different species in order to gain anything, the feedback structures become more complicated. We, for example, could imagine that our RNA molecules need to be "melted," e.g., need to unfold their tertiary structure, before a replication can take place. Another molecule can then act as a facilitator for the replication. Each molecule therefore both needs a "melter" and a "speeder." Many processes in modern life involve several steps like this.

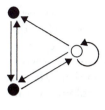

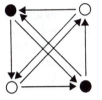

FIGURE 7   An alternative cooperative structure where interaction from two species having different properties is needed in order to facilitate the replication process. The different properties for the RNA strings are indicated by different colors on the vertices. Obviously, subgraphs possessing such a structure are strictly balanced with an average degree of two.

Using similar assumptions as in the previous section, we can immediately formulate a model of the process as a random graph, but this time with two different colors on the vertices, indicating the different molecular properties. We are now interested in the emergence of cooperative structures where each vertex must receive at least one ingoing edge from vertices with each of the two different colors in the graph.

The threshold function for one of the subgraphs in Figure 7 without different colors can be found directly by Eq. (2). We then have

$$d^{*}(n) = cn^{3/2} . \tag{9}$$

At such high catalytic densities ($n^{3/2} > n \log n$), the whole sequence space is connected via catalytic interactions. The emergence properties of these more than one component cooperative structures are very little understood. Also the dynamics differ significantly from the simple autocatalytic cycles. Assuming that the catalytic term for the two-component structures shown in Figure 7 can be expressed as third-order terms $k_{ijk} x_i x_j x_k$, the tendency to exhibit oscillatory dynamics is reduced. This is due to the high internal cross connectedness.

6. EMERGENCE OF THE FIRST COOPERATIVE GENE STRUCTURES

The rate at which new RNA species are produced through error replication of already existing species is given by

$$f = k \cdot m \cdot \left(1 - (1 - \alpha)^{s}\right) \tag{10}$$

where m is the number of RNA molecules in a given prebiotic niche. As before, k is the average rate constant in the RNA self-replication process, s is the typical number of bases for the RNA molecules of interest, and α is the average error probability per base. For more precise interpretation, m must be considered as the number of molecules within a volume small enough for the molecules to come in contact with each other through diffusion, convection and other mixing processes. To independently estimate m, we therefore must know both the concentration of RNA molecules in a successful niche of prebiotic Earth, and the volume within which RNA molecules can interact with each other.

A fraction b of the total number of generated RNA species can be assumed to have catalytic activity, $b = n_{cat}/n_{tot}$. An estimate of b may be obtained from the number of prescribed bases required in modern RNA catalysis,[2,6,27,33] assuming that the remaining bases can be substituted freely. The rate of formation of catalytic RNA species is hereafter

$$\xi = b \cdot f. \tag{11}$$

In accordance with our results in the previous section, the number of edges required for the first simple catalytic feedback cycles to appear is of the order of

$$t \sim d_{tot} \cdot n_{tot}/v \tag{12}$$

where the parameter $d_{tot} < 1$, as determined by Eq. (7) controls the certainty with which feedback cycles exists. As before, n_{tot} is the total number of RNA species, e.g., the whole sequence space.

The time of emergence for the the first catalytic feedback cycles can hereafter be obtained from

$$\xi \cdot t_{em} \sim d_{tot} \cdot n_{tot}/v = \left(\frac{t}{n_{tot}} v\right) \cdot n_{tot}/v \tag{13}$$

or

$$t_{em} \sim \frac{d_{tot} \cdot n_{tot}}{b \cdot f \cdot v} \tag{14}$$

Expressions (13) and (14) relate the graph time t, with the physical time t_{em}. Implicitly, we have here assumed that once a catalytic RNA species has been produced, it will continue to exist until a feedback is formed.

By definition the fraction b determines the total number of different catalytic RNA species. Furthermore, the graph theoretical model is valid only for large n and $v \ll n$. If we assume[27] that the number of bases which has to be specified in order for an RNA molecule to acquire catalytic effects to be $\mu \cong 10$, the fraction of RNA molecules which are produced with catalytic effects is large enough to guarantee asymptotic results, and at the same time fulfil the condition $v \ll n$.

To proceed with our estimate, we assume that the sequence length for the RNA species of interest, has been up to a length of s_{max}, perhaps with a minimum length of $s_{max} - \Delta s$. For strings with a predominant cytidine-guanosine backbone, we shall assume that a fraction γ of the nucleotides have been either cytidine or guanosine, and that the remaining $1 - \gamma$ part of the bases have been randomly chosen between the 4 different types of ribonucleotides. This gives a total of

$$n_{tot} \sim \sum_{s=s_{min}}^{s_{max}} 2^{\gamma s} \cdot 4^{(1-\gamma)s} < \Delta s \cdot 2^{\gamma s_{max}} \cdot 4^{(1-\gamma)s_{max}} \qquad (15)$$

for the number of possible strings. Note that the estimate on n_{tot} is unaffected by an inclusion of the very short RNA strings $1 < s < s_{min}$.

Substituting Eqs. (10) and (15) into Eq. (14), we obtain for the time of emergence of the first catalytic feedback cycle to occur with probability one up to $d_{tot} = 1$:

$$t_{em} \sim \frac{\Delta s \cdot 2^{\gamma s_{max}} \cdot 4^{(1-\gamma)s_{max}}}{v \cdot b \cdot k \cdot m \cdot \left(1 - (1-\alpha)^{s_{max}}\right)} . \qquad (16)$$

The physical assumptions behind the calculations leading to Eq. (16) are the following:

i.   static conditions: the dynamics of each of the different RNA families are not explicitly taken into consideration.

ii. catalytic RNA "never disappear": we assume, that there is a positive correlation between strings with catalytic properties and their selection value. In this way, an up concentration of catalytic species in the micro niche is certain to occur. It is presumably realistic to assume that catalytic RNA in average have considerably longer lifetimes than non-catalytic ones, since the folding of the former usually makes them more persistent to hydrolysis.

iii. micro niche with chemostat like properties: there is a continuous inflow of resources in the form of activated nucleotides and an overflow, which removes the slowly replicating strings.

TABLE 1 Relation between different assumptions on s_{max}, Δs, γ and μ, and their corresponding estimates on n_{tot} and n_{cat}.

s_{max}	Δs	γ	n_{tot}	μ	n_{cat}
80	60	0.9	$1.9 \cdot 10^{28}$	10	$9.1 \cdot 10^{24}$
60	40	0.8	$1.9 \cdot 10^{23}$	10	$4.6 \cdot 10^{19}$
40	20	0.7	$9.0 \cdot 10^{16}$	10	$1.1 \cdot 10^{13}$
30	10	0.6	$1.7 \cdot 10^{8}$	10	$6.6 \cdot 10^{5}$

iv. micro niche consisting of a communication volume containing up to m RNA strings of the considered lengths: we assume that none or only a very few of the strings have any catalytic properties in the beginning.

v. considering RNA strings up to a certain length s_{max}: only a certain fraction $b = n_{cat}/n_{tot}$, representing the strings possessing catalytic properties and at the time being capable of receiving catalytic help, is of interest.

vi. we do not specify any reaction mechanisms: it is assumed that the catalysis facilitates the production of some other strings. As long as it is not possible to deduce the catalytic properties directly from the base sequence of the string (and we therefore have to assign these properties at random), the error introduced by not explicitly taking specific reaction mechanisms into account is presumably not significant.

Finally we need to insert the parameters in Eq. (16). Some of these are well known. As discussed above, s_{max} may be of the order of up to 80 nucleotides like modern tRNA if it is due to the error replication threshold. This corresponds to a γ of the order 0.9 at least. The larger γ, the smaller s_{max}. Such an estimate on s_{max} is probably far too large since catalytic activity has been observed for RNA strings with s around 20 bases.[33] Inserting $s_{max} = 30$ bases instead of 80 and allowing γ to decrease to 0.6 reduces the estimate on t_{em} with a factor about 10^{20}. The value for α is found to be of the order of 10^{-2} per base.[8,10,22]

The concentration of tRNA in a modern bacteria, such as E. coli is of the order 10^{-4} M.[18] At concentrations below 10^{-9} M, the discrete nature of the number of macromolecules starts to become significant in a reaction volume of the size of a modern bacterium ($\sim 10^{-8}$ ml). Therefore, we will assume that the minimum polymer concentration is some orders of magnitude larger, say, 10^{-7} M.

The relation between m, the number of communicating RNA strings, and n_{cat}, the number of different catalytic RNA strings, is determined by the local RNA concentration and the size of the micro niche where the reactions take place. A sufficient number of different catalytic RNA molecules needs to be present in the micro niche in order to allow for the creation of autocatalytic systems.

The reaction volume for RNA catalysis has an ultimate upper limit defined by the maximal diffusion distance given by the average lifetime of RNA in water. An estimate yields a maximum reaction volume of the order of one ml, roughly corresponding to one day of diffusion. However, to have ideal and immediate mixing conditions due to diffusion alone, we have to use a much smaller reaction volume.

The chemical replication constant may vary considerably due to the chemical composition and the temperature of the environment. An upper physical limit for such a process seems to be of the order of 10^4 strings/sec.[9] In the self-splicing experiments, a turnover of about one string per RNA enzyme per minute is found.[33] We may assume the existence of a non-RNA catalyzed prebiotic replication rate with a turnover of about one string per day.

Let us now try to obtain some limits on the estimate of t_{em}. If we want to maintain very conservative assumptions on the number n_{tot}, corresponding to large

s_{max}, say, 80 bases, with $v \sim b^{-1}$, we will not have a sufficient catalytic out-degree to guarantee a direct catalytic percolation among the RNA molecules present in the fully occupied micro-niche. In a reaction volume of one ml with a local prebiotic RNA concentration like in modern bacteria, the number of RNA strings is of the order $6 \cdot 10^{17}$. According to Table 1 at least $9.1 \cdot 10^{24}$ different catalytic strings need to be present to guarantee percolation. However, in such an estimate, there is no room for many copies of the same type RNA. From the simulations in the previous section it is seen that only relatively few species occupy most of the reaction volume. In general more than 90% of the strings have less that $q \sim 10^{-4}$ of the niche. Assuming that comparable properties dominate the polymer production in the prebiotic ecology, we may take this effect into account by lowering our estimate on m by the factor q. If, with these assumptions on s_{max} and m, we insist on an instantaneous autocatalytic percolation when the micro-niche is fully occupied, we have to allow $v \gg b^{-1}$ and set $v \cdot b \sim 10^{13}$. In this situation still $n_{cat} \gg v$ so that Eq. (16) stands. However, accepting that percolation may not occur instantaneously the interpretation of the expression for t_{em} in Eq. (16) changes slightly.

If we assume that the process continues after the micro niche is filled up with catalytic RNA strings in the case $d_{tot} < 1$, and there is a continuous replacement of RNA species in the reaction volume, the system will eventually percolate anyway. The niche in this situation will be fully occupied at time $t_{full} \sim t_{em} \cdot d_{tot}$, because d_{tot} represents both the probability for autocatalytic occurrence and the average out degree. Assuming the replacement to be random gives us an expected time of emergence of autocatalysis equal to $(d_{tot})^{-1}$ times the time it takes to fill up the reaction volume t_{full}. Hereafter we are back at the expression for t_{em} obtained via Eq. (16). Hence, Eq. (16) is also valid for $d_{tot} < 1$, but does not guarantee autocatalytic percolation in this situation. Here t_{em} gives the expected time of emergence.

Using a local RNA concentration of 10^{-7} M, a reaction volume like a modern bacterium, $s_{max} \cong 80$ bases, $q \cong 10^{-4}$, $\alpha \cong 10^{-2}$, a non-catalyzed replication rate with a turnover of one string per day and $v \sim b^{-1}$, we can calculate a conservative limit on the expected time of autocatalytic emergence t_{em}. We hereby obtain:

$$t_{em} \sim 4 \cdot 10^{29} \text{ sec}. \tag{17}$$

The theoretical upper bound for t_{em} in Eq. (17), is thirteen orders of magnitude larger than the upper limit for the real appearance of the first catalytic feedbacks. Since the oldest definite trace of living organisms dates back to about 3.5 billion years ago, at which stage life seems to have evolved to cell level, and the age of the Earth is about 4.6 billion years old, t_{em} could at most be 1 billion years ($\sim 3 \cdot 10^{16}$ sec).

The cell, of course, was not created by formation of the first cooperative gene structures, although this was an important step in the evolution.

Our estimate covers many orders of magnitude. A single evolutionary experiment would suffice to produce life in geological times if we assume s_{max} to be 40 bases instead of 80, and else keep the same conservative prebiotic conditions leading to Eq. (17).

If we allow for a parallel evolution, the time of emergence of the first genetic networks decreases dramatically. Denote the probability that the first cycle in a single experiment occurs before time t'' by d''_{tot}. In order to have the first cycle occur before time t'' with probability $\mathcal{P}''$, we need $\mathcal{N}''$ parallel and independent experiments where

$$\mathcal{P}'' = 1 - (1 - d''_{tot})^{\mathcal{N}''}$$

or

$$\mathcal{N}'' \cong -\log \frac{(1 - \mathcal{P}'')}{d''_{tot}} . \tag{18}$$

We may use the prebiotic assumptions leading to the theoretical upper bound given in Eq. (17) to calculate the necessary number of parallel experiments. The relation between t'' and d''_{tot} may be obtained via Eq. (14). We can now calculate the number of parallel evolutionary processes needed in order to create a catalytic feedback in any duration of time. To have realistic time scales t'' (say, one year $\sim$ $3 \cdot 10^7$ sec) for the experiments, and to have a probability one half for success, we need $d''_{tot} \sim 10^{-22}$. Inserting into Eq. (18), we thus require $N'' \sim 5 \cdot 10^{21}$ experiments. The upper bound for the spatial requirements for such a parallel experiment is only of the order

$$5 \cdot 10^6 \text{m}^3 . \tag{19}$$

This corresponds to a minor lake. Lowering s_{max} to 40 bases and keeping the other conservative prebiotic conditions, the required number of parallel experiments drops to only $\sim 5 \cdot 10^8$ with a corresponding reaction volume of only 5 ml! Although we have not included the spatial requirements for the containers for the chemostat-like micro-niches in these estimates, both the temporal and the spatial requirements are clearly very small on a planetary scale.

7. DISCUSSION

The just-described evolutionary process in the model is restricted to develop along the combination of paths made possible by the formalized physico-chemical interactions. However, none of these need to be the actual path. The model shows how and when new properties emerge related to the formation of autocatalytic feedbacks. These structures enable the system to produce and maintain a pool of longer polymers, which further enables the system to develop more specialized and complex functions. The evolutionary process is modeled such that initially a large number of different individual dynamical systems, which can interact by catalysis, may be produced. Hereby the system eventually selects a small subset of the possible configurations. From the simulations in section 3, it is seen that the relative sizes of the kinetic constants as well as of the initial concentrations are very important for the specific outcome or product of the evolutionary process. The system has many attractors. Especially it is important to note that certainty for selection of autocatalytic sets only is present when the catalytic constants are significantly larger than the linear replication constants.

To give at least a rough evaluation of the consequences of ignoring the explicit chemical kinetics, we can compare simulations of Eq. (1) with corresponding estimates obtained using the discrete model leading to Eq. (16). The fact that the dimension of the simulated system is small ($n = 128$) is probably not a major problem. Earlier computations on random graphs with that many vertices indicate that obtained averages of emergence times for cycles only deviate from the asymptotic values by a few percent.[31,32] Using the simulation depicted in Figure 4 as an example, we can now use Eq. (16). The number of different strings in the system is $n_{tot} = n_{cat} = n = 2^7$, $b \cdot v = 1$ and $(1 - \alpha)^s = 0.05$ Further, the average replication rate k equals 1.0 (time unit)$^{-1}$, where the number of strings is measured in moles. m is from the equilibrium concentration seen to be approximately 2 mole/volume unit. Inserting these numbers, we obtain $t_{em} \cong 70$ time units. This is in nice agreement with the simulation results showing the occurrence of an autocatalytic cycle with macroscopic significance around time 100 (Figure 4c)

Let us also touch upon the relation between the just described model (denoted: model A) and the autocatalytic models of Farmer, Kauffman and Packard[13] and Kauffman[21] (denoted: model B). Obviously we are addressing the same basic question: emergence of autocatalytic sets. The major difference between the two models is the manner in which longer strings are formed. In model B it is possible to create new strings by joining existing sequences. Hereby it is possible in principle to create a pool of longer polymers right away, where as in model A longer sequences are only created by adding single bases to existing strings. This gives a "diffusion" on a Boolean cube and hereby a much slower development. For model A, depending on our assumptions on the error replication, the maximal number of Hamming neighbors H_A for each string lies in the interval $[s(\beta-1)+2\beta+2; s((\beta-1)+1)+(s+1)\beta-1]$ where s is the actual string length and β the number of different monomer letters (bases). The lower bound corresponds to single base substitution along the string and adding or subtracting a single base in either end of the string. (These are the assumptions under which we have developed our estimates in previous sections.) The upper limit for H_A allows single base substitution, and adding or subtracting all along any existing string. In either case, we note a linear relation between H_A and s. For model B each string of length s has $H_B = (s-1)+2 \cdot n_{exist} - 1$ "neighbors," where n_{exist} is the total number of already existing strings. Hereby the number of neighbors may become very large. The number of new joining reactions assigned at each update in this model is given by the number of possible joining reactions times a probability P. Hereby it is the initial number of polymers in connection with the size of P, which determine whether autocatalytic sets emerge or not. In model A, it is the size of $v \cdot b$ which is responsible for percolation, and hereby percolation is independent of the initial conditions. This is a significant difference between the two models.

Another difference between the two models is the status of the catalytic processes as the system develops. In model A, no endogenous catalysis is of importance in the creation of new strings. In model B, only catalyzed processes are assumed to create new strings. Endogenous catalysis is assumed to be of significance from the very beginning and the same mechanism which creates new strings is responsible for the occurrence of autocatalytic sets. In model A, two different processes are responsible for these processes. Further, model A catalysis is assumed to have significance only after the strings by error replication reach a certain minimum length s_{min} where they can fold into "catalytic units." Only a small fraction of strings are assumed to have catalytic abilities, and each of these strings are assumed to catalyze only a small fraction of all strings.

8. CONCLUSION

In terms of the developed model, using conservative parameters, it seems highly probable that the first cooperative RNA-based gene structures could evolve on a short time span compared with geological scales. In fact, the temporal and spatial requirements for this evolutionary process appear very modest. It should even be possible to perform a similar experiment on laboratory scale, life *de novo*.

The model is based on extensions of results from random graph theory, developed for the present purpose. These extensions allow us to calculate the critical average out-degree for cycle formation as well as the distribution of cycles. By interpreting the cycles as catalytic feedback loops in a prebiotic system of a vast number of different, relatively short, self-replicating RNA species, we obtain an explicit expression for the time of emergence t_{em} for the first catalytic feedback networks. Assuming parallel processes, the total spatial requirements can be assessed. These analytical results are compared with simulations where the kinetics for each RNA family explicitly is taken into consideration. There seems to be a nice agreement between the two different approaches.

The production of catalytic RNA structures is a fundamental step in the evolution of life, because cooperative feedbacks are essential in order to create and maintain a pool of more complicated polymers.

ACKNOWLEDGMENTS

Thanks are due to Erik Mosekilde for discussions and constructive suggestions throughout the text, and to Béla Bollobás, with whom I have discussed the mathematical aspects of the model in section 4. I am also grateful for a number of enlightening discussions on the interpretation of the model with Doyne Farmer, Stuart

Kauffman, Norman Packard and Chris Langton. Rikke Frederiksen is acknowledged for her patient and careful assistance in the preparation of the manuscript.

REFERENCES

1. Bachas, C. P., and B. A. Huberman (1986), "Complexity and Ultradiffusion," *Phys. Rev. Lett.* **57**, 1965.
2. Bass, B. L., and T. R. Chech (1984), "Specific Interaction between the Self-Splicing RNA of Tetrahymena and its Guanosine Substate: Implications for Biological Catalysis by RNA," *Nature* **308**, 820.
3. Bollobás, B. (1985), *Random Graphs* (New York: Academic Press).
4. Cairns-Smith, A. G. (1985), "The First Organisms," *Sci. American* **74**.
5. Ceccatto, H. A., and B. A. Huberman (1987), "The Complexity of Hierarchical Systems," preprint.
6. Cech, T. R. (1985), "Self-Splicing RNA: Implications for Evolution," *Int. Rev. on Cytology* **93**, 3.
7. Darwin, C. (1859), *On the Origin of Species by Means of Natural Selection* (London: Murray).
8. Eigen, M. (1971), "Self-Organization of Matter and Evolution of Biological Macromolecules," *Naturwissenschaften* **58**, 465.
9. Eigen, M. (1985), "Macromolecular Evolution: Dynamical Ordering in Sequence Space," *Ber. Bunsenges. Phys. Chem.* **89**, 658-667.
10. Eigen, M., and P. Schuster (1979), *The Hypercycle—A Principle of Natural Self-Organization* (Heidelberg: Springer-Verlag).
11. Engelbrecht, J. (1987), "Information Measurements of DNA and RNA Sequences," *Physics Laboratory III, unpublished.*
12. Erdös, P., and A. Rényi (1960), "On the Evolution of Random Graphs," *Publ. Math. Inst. Hungar. Acad. Sci.* **5**.
13. Farmer, J. D., S. A. Kauffman, and N. H. Packard (1986), "Autocatalytic Replication of Polymers," *Physica* **22 D**, 50.
14. Glansdorff, P., and I. Prigogine (1971), *Thermodynamic Theory of Structure, Stability, and Fluctuations* (New York: Wiley-Interscience).
15. Grassberger, P. (1986), "Toward a Quantitative Theory of Self-Generated Complexity," *Int. J. Theoret. Phys.* **25**, 907.
16. Grassberger, P. (1988), "Complexity and Forecasting in Dynamical Systems," preprint.
17. Hogg, T., and B. A. Huberman (1986), "Complexity and Adaptation," *Physica* **22D**, 376-384.
18. Ingraham, J. L., O. MaalWe, and F. C. Neidhardt (1983), *Growth of Bacteria Cells* (Boston, MA: Sunderland).

19. Joyce, G. F., G. M. Vissen, C. A. A. van Boeckel, J. H. van Boom, L. E. Orgel and Y. van Mestrenen (1984), "Critical Selection in Poly(C)-Directed Synthesis of Ooligo(G)," *Nature* **310**, 602.

20. Kauffman, S. (1971), "Cellular Homeostasis, Epigenesis and Replication in Randomly Aggregated Macromolecular Systems," *J. Cybernetics* **1**, 71.

21. Kauffman, S. (1986), "Autocatalytic Replication of Polymers," *J. Theoret. Bio.* **119**, 1-24.

22. Küppers, B.-O. (1983), *Molecular Theory of Evolution* (Berlin: Springer Verlag).

23. Langton, C. (1988), *Ph.D. thesis, University of Michigan*, in process.

24. Lohrmann, R., P. K. Bridson and L. E. Orgel (1980), "Efficient Metal Ion Catalyzed Template-Directed Oligonucleotide Synthesis," *Science* **208**, 1464.

25. Mosekilde, E., S. Rasmussen and T. S. Sørensen (1983), "Self-Organization and Stochastic Re-Causalization in System Dynamics Models," *Proceedings at the 1983 International System Dynamics Conference, Plenary Session Papers, 123, Boston, USA, July 27-30, 1983* (Boston, MA: Systems Dynamic Society, MIT).

26. Nicolis, G., and I. Prigogine (1977), *Self-Organization in Non-Equilibrium Systems* (New York:, Wiley-Interscience).

27. Nielsen, H., J. Engelbrecht and J. Engberg (1987), private communication.

28. Packard, N. H. (1987), "Adaptation Toward the Edge of Chaos," preprint.

29. Rasmussen, S. (1982), *Instabilities, Fluctuations and Structure in Dynamical Systems, M.S. Thesis, Institute of Physical Chemistry and Physics Laboratory III, The Technical University of Denmark*, unpublished.

30. Rasmussen, S. (1985), *Aspects of Instabilities and Self-Organizing Processes, Ph.D. Theses, Physics Laboratory III, The Technical University of Denmark*.

31. Rasmussen, S., E. Mosekilde and Jacob Engelbrecht (1986), "Time of Emergence and Dynamics of Cooperative Gene Networks," *Proceedings of the MIDIT Workshop on Structure, Coherence and Chaos, The Technical University of Denmark, August 1986, Nonlinear Science, Theory and Applications* (Manchester: Manchester University Press).

32. Rasmussen, S., B. Bollobás, and E. Mosekilde (1988), "Elements of a Quantitative Theory for Prebiotic Evolution," preprint.

33. Uhlenbech, O. C. (1987), "A Small Catalytic Oligoribonucleotide," *Nature* **328**, 596-600.

34. Wolfram, S. (1984), "Universality and Complexity in Cellular Automata," *Physica* **10D**, 1-35.

35. Zambella, D., and P. Grassberger (1988), "Complexity of Forecasting in a Class of Simple Models," preprint.

Pablo Tamayo† and Hyman Hartman‡
†Physics Dept. and Center for Polymer Studies, Boston University, Boston MA 02215; and
‡ Computer Science Department, University of California, Berkeley CA 94720

Cellular Automata, Reaction-Diffusion Systems and the Origin of Life

Cellular Automata are models for reaction-diffusion systems and Artificial Life. In particular the Greenberg-Hastings model of excitable media is considered.

A two-dimensional reversible version of this model is introduced. It shows interesting space-time patterns reminiscent of chemical turbulence, solitons and self-excited oscillations. A study of this model in one dimension is made by the use of correlation functions and power spectra.

INTRODUCTION

In 1948 John von Neumann gave a talk about automata theory at the Hixon Symposium in California. At the end of the talk, Warren McCulloch made the following remark.[28] "I confess that there is nothing I envy Dr. von Neumann more than the fact that the machines with which he has to cope are those for which he has, from the beginning, a blueprint of what the machine is supposed to do and how it is supposed to do it. Unfortunately for us in the Biological Sciences, we are presented with an alien, or enemy's machine. We do not know exactly what the machine is supposed to do and certainly we have no blueprint of it".

The major quest is to make Artificial Life elucidate real life.[18] Descartes was the first major scientist in the use of automata to explain living systems. He tried to understand the beating of the heart as the consequence of modeling the heart as an internal combustion engine.[9] Descartes was an extremely clever man and despite the much greater empirical evidence collected by Harvey that the heart was a pump, Descartes advocated his model. This is a tale with a message. Artificial models of life, like any scientific theory must be tested against empirical evidence. This is especially true with the advent of the computer as the models of Artificial Life proliferate.

CELLULAR AUTOMATA

The first attempt in modern times to model self-replication by artificial life was made by von Neumann. He had just invented the internally programmable computer by implementing Turing's universal computing machine. This inspired him to found a new field of automata studies. One of the first problems which he attacked was the ability of a machine to reproduce itself. The question reduced itself to whether a universal constructing machine was a logical possibility. He answered this question in the affirmative by separating the description of the machine from the constructing functions. The remarkable aspect of his construction is that it was realized in the findings of molecular biology in the years following his death.[11] The description of the organism is to be found in the long string of nucleotides in DNA. There is no doubt that von Neumann's work will have a profound effect on the origin-of-life theories.

At the suggestion of Stanislaw Ulam, von Neumann went on to invent cellular automata in order to further implement his self-reproducing automaton. Since that time we have seen an enormous expansion of this field. The cellular automata born in the attempt to understand perhaps the origin of life are now modeling hydrodynamics and reaction-diffusion systems.[3,25,26,34] In fact, we have here a case of artificial life modeling a large and growing area of physics and chemistry.[2,6,24,28]

Cellular automata consist of a discrete set of "cells" arranged in a regular lattice. The state of each cell is a discrete variable. Time advances in discrete steps. A local function is defined to compute the new state as a function of the previous cell state and the state of the neighborhood. When iterated over the array, it generates a global map and the dynamical evolution of the system.

ORIGIN OF LIFE

There are today two major schools of thought on the origin of life. One is based on the premise that proteins came first or a "protein life." The second school postulates that nucleic acids came first, especially ribonucleic acid (RNA) or a "RNA life." Both schools have need of a prior period in which the synthesis of organic molecules such as aldehydes and cyanide took place. The major evidence for this assumption comes from the Miller-Urey experiment in which an atmosphere of methane, hydrogen and ammonia was sparked and cyanide and other organic molecules were formed. The formation of amino acids in particular required further reactions in liquid water. This type of atmosphere on the primitive earth is extremely improbable. The evidence from the oldest rocks on earth is for an atmosphere of carbon dioxide and nitrogen. This type of an atmosphere would not support the synthesis of cyanide and aldehydes no matter how long one sparked it. These facts necessitated a revision in our thinking as neither of the models based on organic chemistry have a geological plausibility.

The clay theory for the origin of life begins with the necessity of liquid water for life. The major effect of liquid water on a planet such as the earth is to weather rocks to clays. This would have occurred on the primitive earth. Clays are fundamentally "two-dimensional" crystals. This is of extreme importance if one is going to postulate a replication of information stored in a crystal. Since we live in a three-dimensional world, it would be impossible to replicate the information stored in a crystal by a simple template mechanism unless the crystals were one-dimensional or two-dimensional. Thus one can model RNA as a one-dimensional crystal whose information is replicated by a template mechanism. The information is stored in the sequence of the four bases (uracil, guanine, adenine, cytosine). The information in a clay is stored in the distribution of ions like silicon, aluminum, iron and magnesium in the crystal lattice. The clay can be thought of as a checkerboard with checkers of four different colors. The information would be stored in the distribution of these different checkers. Replication is the process by which the distribution of ions from the mother clay is copied into the daughter clay. In 1966, Cairns-Smith[4] proposed that the original genes were clays.

Clays are also well-known catalysts. In 1975 Hartman[15] proposed that it was on the surface of replicating iron-rich clays that carbon dioxide and nitrogen in the presence of long-wavelength ultraviolet radiation were fixed into organic acids such as oxalic acid. In a further series of papers,[16] Hartman outlined how the genetic code evolved. The surface of a reactive clay particle can be modeled as a two-dimensional reaction-diffusion system. In the parlance of chemical kinetics a clay particle is a heterogeneous catalyst. Recently heterogeneous catalysis has been modeled by cellular automata.[5,10] Molecules are absorbed on to the clay surface where they are activated at certain sites to react and where they then diffuse to other sites and react further and finally the products are desorbed. These reactions and diffusions on the clay surfaces motivated us to study cellular automata models of reaction-diffusion systems.

REACTION-DIFFUSION SYSTEMS

Thirty years ago Turing began to ponder the problems of embryology.[27] He wondered how it was possible for a symmetrical egg (fertilized) to develop into an asymmetrical animal. He postulated that a system of chemicals that could react and diffuse would under certain conditions break the initial symmetry of the system. More specifically, the development of structures would result from non-homogeneous patterns of chemical substances, called morphogens.

Turing's model for morphogenesis had a large impact on physical chemistry when the Belousov-Zhabotinsky reaction was discovered. This reaction usually involves a mixture of malonic acid, potassium bromate, ceric sulfate and an indicator. This mixture reacts exothermically at room temperature, spontaneously producing chemical waves. The equation which governed the behavior of this system were those of the reaction-diffusion type studied by Turing[21].

The continuum models of reaction-diffusion systems are expressed as equations of the form:

$$\frac{\partial \mathbf{u}}{\partial t} = \Phi(\mathbf{u}) + \mathbf{\Gamma}\nabla^2\mathbf{u} \tag{1}$$

$\mathbf{u}(x,t)$ represents the state of the system. The function $\Phi(\mathbf{u})$ is the reaction term which defines the kinetics of the medium. The diffusion term is given as the product of the laplacian operator acting on $\mathbf{u}$ and the matrix $\mathbf{\Gamma}$. In a spatially homogeneous state the system evolves obeying:

$$\frac{\partial \mathbf{u}}{\partial t} = \Phi(\mathbf{u}) \tag{2}$$

Most of the models of interest are strongly non-linear and too complicated to be solved analytically, and must be solved numerically. A cellular automata model for the Belousov-Zhabotinsky reaction and diffusion system was proposed by Greenberg and Hastings,[12] based on an earlier model by Wiener and Rosenblueth.[30]

THE GREENBERG-HASTINGS MODEL

Each cell can take one of three states: quiescent (0), active (1) or refractory (2). The cell $a_{i,j}$ has four neighbors, the so-called von Neumann's neighborhood in two dimensions: $a_{i-1,j}$, $a_{i+1,j}$, $a_{i,j-1}$ and $a_{i,j+1}$. i and j are space indices in two dimensions and t is the time index. The local function to compute the new state $a_{i,j}^{t+1}$ as a function of the neighbors and previous state is defined by:

$$a_{i,j}^{t+1} = R(a_{i,j}^t) + D(a_{i-1,j}^t, a_{i+1,j}^t, a_{i,j-1}^t, a_{i,j+1}^t) \tag{3}$$

R is the Reaction term and D the Diffusion term. They are defined by:

$$R = \begin{cases} 2, & \text{if } a^t_{i,j} = 1; \\ 0, & \text{otherwise.} \end{cases}, \quad D = \begin{cases} N, & \text{if } a^t_{i,j} = 0; \\ 0, & \text{otherwise.} \end{cases} \quad (4)$$

where N is a boolean function which takes the value 1 if at least one of the neighbors is active, and 0 otherwise:

$$N = (a^t_{i-1,j} = 1) \; or \; (a^t_{i+1,j} = 1) \; or \; (a^t_{i,j-1} = 1) \; or \; (a^t_{i,j+1} = 1) \quad (5)$$

The basic mechanism implemented by this function is the following: the state 0, is the resting state in which the cell will remain indefinitely until it is activated. The state 1, is produced by the existence of one or more active neighbors. Once a cell has been activated it will change to state 2 at the next time step. A refractory cell is not only insensitive to the neighborhood but is also incapable of spreading activation to its neighbors. After visiting the refractory state, the cell returns to the quiescent state. The reaction term assures that the state will cycle to its successor when the cell is active or refractory and the diffusion term is a boolean function which assures excitation, taking the value 1, if at least one of the neighbors is in the active state and 0 otherwise.

In Figure 1, we show the state diagram which describes the mechanics for a single cell.

We will consider periodic boundary conditions. In Figure 2, we show the evolution of this rule from simple initial conditions consisting of a short row of active cells on top of a row of refractory ones in a quiescent background. The spirals develop due to the fact that the tips of the initial row act as singularities. The waves will spread unperturbed over the quiescent background.

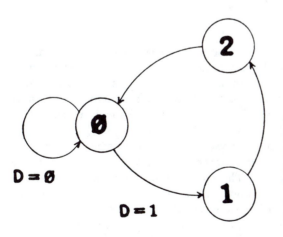

FIGURE 1 State-diagram for cell transitions in the Greenberg-Hastings model. The value of the diffusion term D controls the transition out of the quiescent state. $\emptyset \rightarrow$ quiescent; $1 \rightarrow$ active; and $2 \rightarrow$ refractory.

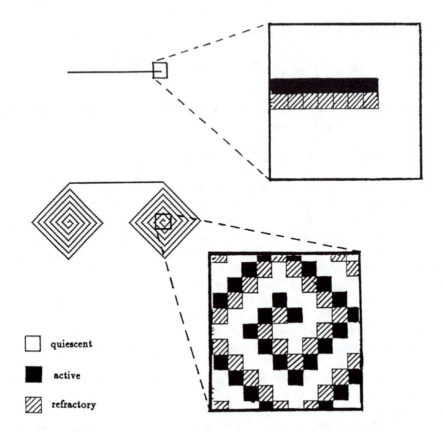

FIGURE 2 Evolution of the Greenberg-Hastings rule. (Above) The initial conditions are depicted. (Below) the state of the system after 25 time steps. The tips of the line act as singularities which generate rotating spirals.

The initial configurations which give rise to the organized patterns can be characterized by a "winding" or topological number as was shown by Greenberg et al.[13,14] The basic "singularities" which produce the spirals are shown in Figure 3. The arms of the spiral are made of activation wavefronts with refractory tails, which propagate over regions of quiescent cells. If we start from random initial conditions, say 3% of active cells and 3% of refractory cells in a background of quiescent cells, several singularities would form as shown in Figure 4. In some regions, two singularities come close together producing a more complex singularity which is able to generate waves of a higher frequency. These waves propagate and take over the low-frequency waves produced by the simple singularities. Each of these complex singularities defines a domain which is covered by the traveling waves until they encounter waves coming from other singularities, when this happen, a stationary boundary forms. In Figure 4, we can see four of these domains expanding.

<table>
<tr><td>**1**</td><td>**2**</td></tr>
<tr><td>**0**</td><td>**0**</td></tr>
</table>

<table>
<tr><td>**1**</td><td>**0**</td><td>**2**</td></tr>
<tr><td>**2**</td><td>**0**</td><td>**1**</td></tr>
</table>

FIGURE 3 Examples of configurations acting as singularities. The one on the left will form a simple singularity, while the one on the right will produce a complex singularity.

These high-frequency waves will destroy any of the simple singularities they find in its way, leaving instead a slight mismatch or defect.

After some time the system will enter a stationary state of a short period in which the waves propagate steadily from the singularities to the boundaries.

The model can be extended to allow additional active and/or refractory states, basically keeping the same type of reaction and diffusion terms. These extensions show the same basic phenomenology as the three state model.

Other aspects of this model have been extensively studied by A. Winfree[31-33] and others.[1]

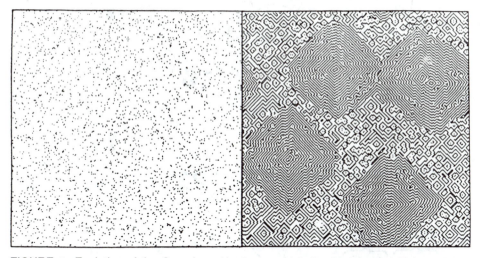

FIGURE 4 Evolution of the Greenberg-Hastings model. On the left the initial conditions consisting of 3% active and 3% refractory cells are shown. On the right the state of the system after 200 time steps. Many simple singularities can be seen in addition to four complex singularities and domains of higher frequency waves. Only active and quiescent cells are shown.

A REVERSIBLE VERSION OF THE GREENBERG-HASTINGS MODEL

An interesting variation of the Greenberg-Hastings model is obtained when Eq. (6) is modified by subtraction of a new term: the state of the cell in the previous time step. This transformation due to Fredkin, converts the original rule into a second-order reversible rule.

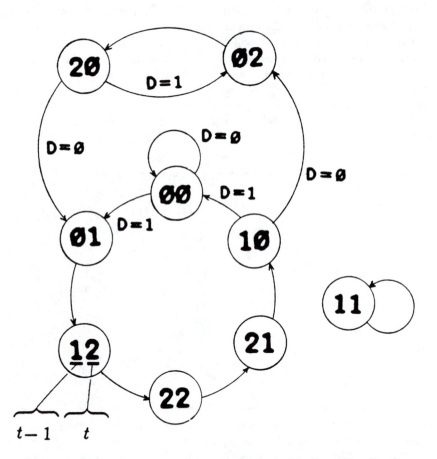

FIGURE 5 State-diagram for cell transitions in the reversible Greenberg-Hastings model. Each circle represents the state at time $t-1$ and time t. The value of the diffusion term D controls certain transitions. Time-reversal invariance implies that when we reverse all the arrows and interchange all the labels for their mirror images (i.e., 02 to 20) we obtain the same diagram. Notice that the state 11 is dynamically isolated from the others. $\emptyset \rightarrow$ quiescent; $1 \rightarrow$ active; and $2 \rightarrow$ refractory.

The rule is now:

$$a_{i,j}^{t+1} = R(a_{i,j}^t) + D(a_{i-1,j}^t, a_{i+1,j}^t, a_{i,j-1}^t, a_{i,j+1}^t) - a_{i,j}^{t-1} \quad mod\ 3. \qquad (6)$$

This transformation changes the structure of the function adding a second-order dependence in time. The effect is such that the information stored in the initial conditions will be preserved by the dynamics for all future time. A reversible cellular automaton has as many conserved quantities as there are cells.[20,25,26] The system cannot forget the initial conditions which are always encoded in the configuration. Dynamical trajectories cannot enter attractors and then form closed orbits. If we run a finite system, we will eventually exhaust the possible configurations in the orbit and we will return back to the initial condition. However, the recurrence time could be extremely long. The state description for this rule (see Figure 5) now includes the state of the cell at time t as well as the state as time $t - 1$.

Depending on the initial conditions the system evolves basically into three different regimes:

- A Regular Regime
- A Turbulent Regime
- A Random Regime.

REGULAR REGIME

The initial state which gives rise to this regime is a single active cell, two active cells, a whole line of active cells on a line. In a regular regime, wavefronts travel freely in a quiescent background.

RANDOM REGIME

The simplest initial state which gives rise to a random pattern is a single active cell next to a refractory cell. The random regime displays a fully mixed pattern of quiescent, refractory and active cells. There is no apparent structure or pattern.

TURBULENT REGIME

The simplest initial state which gives rise to the turbulent regime is three active cells in a triangular configuration, i.e., three corners of a square. The key feature seems to be the spatial distance between them. They must be separated by an even number of cells. The turbulent regime shows new emergent structures. We have quasi-circular rings emanating from complex regions. A pattern of sources and sinks is set up. Figure 6 shows a complete sequence in the evolution of the turbulent regime from three asymmetric active cells. The rings grow in size. The waves travel at super-luminal speed, faster than one cell per time step. The reason is that they are not real propagating structures but coherent phases formed by the

FIGURE 6 Evolution of the reversible Greenberg-Hastings model. Only quiescent and active cells are shown. a) $t = 0$. Initial conditions consisting of three active cells at coordinates (0,0),(33,0),(32,-33). b) $t = 12$. The wavefronts propagate. c) $t = 100$. The wavefronts interact. d) $t = 1000$. Many new wavefronts are produced by interactions; notice the irregularities along the wavefronts. e) $t = 1400$. The system is completely populated by wavefronts that form wide rings. f) $t = 1700$. The rings form around fuzzy regions that act as sources and sinks. g) $t = 2000$. A central source can now be seen. All the wavefronts have disappeared. h) $t = 3000$. The rings become wider and more diffused. i) $t = 5000$ the rings are completely diffused and the system is composed of large patches of quiescent, active and refractory cells.

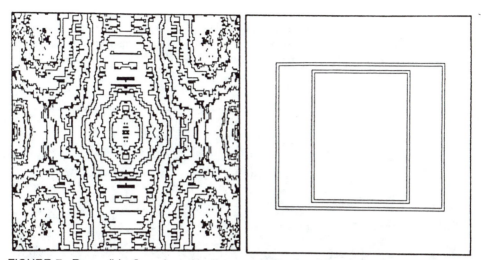

FIGURE 7 Reversible Greenberg-Hastings model in the Moore neighborhood. Only quiescent and active cells are shown. The two pictures show the state of the system after 1500 steps. The initial conditions were two active cells separated by an even number of cells on the left and by an odd number of cells on the right. One system attains the regular regime while the other attains the turbulent one.

oscillation of rings of cells. When the Moore neighborhood is used, the simplest initial configuration is now two active cells. The essential feature is the separation between the active cells. For even/odd separation the rings will/will not form (see Figure 7).

THE ONE-DIMENSIONAL MODELS

We have begun an intensive study of the Greenberg-Hastings model in one spatial dimension. The reversible case, given the right initial conditions, will give rise to a state which has been called "chemical turbulence."[7,8,17,22,23]

The non-reversible case does not show any complex behavior. Starting from an initial state of one active cell, two traveling outgoing waves result. The traveling structures are an active cell followed by a refractory one. When two such structures collide, they annihilate each other. An initial state of an active cell followed by a refractory cell gives rise to a single wavefront. Thus starting from any random initial conditions of active refractory cells, the traveling waves will annihilate each other in pairs, leaving the surviving waves to travel endlessly.

The reversible case has a much more richer phenomenology. The initial state must not only specify the present state of a cell but must also specify its previous state. A state of a cell will be labeled as follows $\left(^{present}_{past}\right)$. The simplest initial

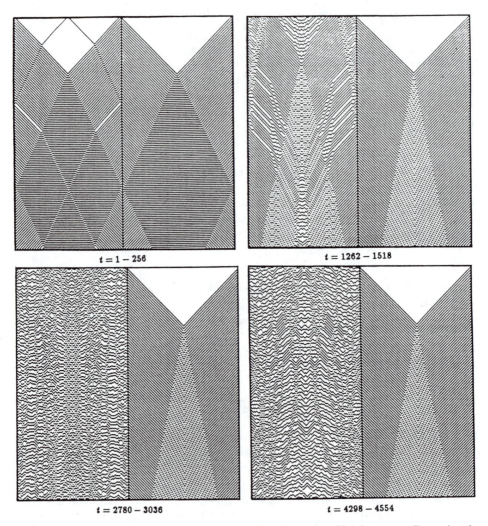

$t = 1 - 256$

$t = 1262 - 1518$

$t = 2780 - 3036$

$t = 4298 - 4554$

FIGURE 8   Space-time displays of the reversible Greenberg-Hastings one-dimensional system (128 cells long, sections of 256 time steps). Only quiescent and active cells are shown. Two walls (see text), one at the middle and one at the border, divide the system into two dynamically independent 64-cell-long subsystems. The subsystem on the right side shows a short period regular regime consisting of vacua of different structures. The subsystem on the left side has an additional active cell placed at the center and shows a turbulent regime.

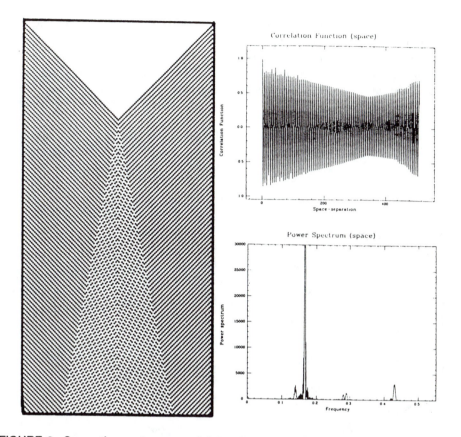

FIGURE 9 Space-time patterns, spatial correlation function and power spectra for the regular regime. The pattern corresponds to the system on the right described in Figure 8 at $t = 2780 - 3036$. The spatial correlation function and power spectra were com-computed for a 1024-cell-long system at $t = 900$.

condition is a single active cell in the present which evolves into two outgoing waves of the following structure $\left(\begin{smallmatrix} 12210001221 \\ 01221012210 \end{smallmatrix}\right)$. Given two active cells two outgoing waves result which collide and give different types of collisions depending upon whether the initial separation was even or odd. In both cases the wavefronts are reconstituted. Thus single active cells give rise to two outgoing waves. However, unlike the non-reversible case, colliding waves do not annihilate one another. The initial condition where there is a 1 in the present as well as a 1 in the past $\left(\begin{smallmatrix} 1 \\ 1 \end{smallmatrix}\right)$ will give rise to a train of waves and a wall. The wall is formed because as can be seen from the state diagram a cell which is active (1) in the present and in the past will remain active for all future times. If two of the "walls" $\left(\begin{smallmatrix} 1 \\ 1 \end{smallmatrix}\right)$ are placed in the initial condition and are separated by some distance from each other, the system separates into two subsystems which do not interact with each other. The

$\binom{1}{1}$ configuration also generates a steady train of wavefronts which interacts with the train of wavefronts generated at the other $\binom{1}{1}$ configuration. These interactions generate a regular pattern (see Figure 8). However, if in addition to the walls we put a single active cell in the initial conditions, then the wavefronts generated at the walls will interact with the wavefront generated by the active cell and in each collision phase shifts will be produced. After some time the system evolves into a turbulent phase (see Figure 8).

We have calculated the two point correlation functions and the power spectra of these turbulent patterns[19] (see Figures 9–12). Our results agree with those of Oono and Yeung[23] who studied a one-dimensional version of an irreversible reaction-

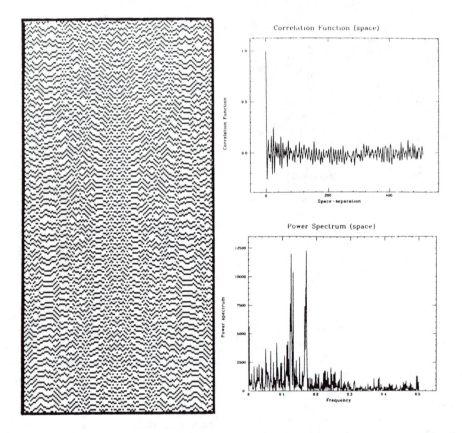

FIGURE 10 Space-time patterns, spatial correlation function and power spectra for the turbulent regime. The pattern corresponds to the system on the left described in Figure 8 at $t = 2780 - 3036$. The spatial correlation function and power spectra were computed for a 1024-cell-long system at $t = 20000$.

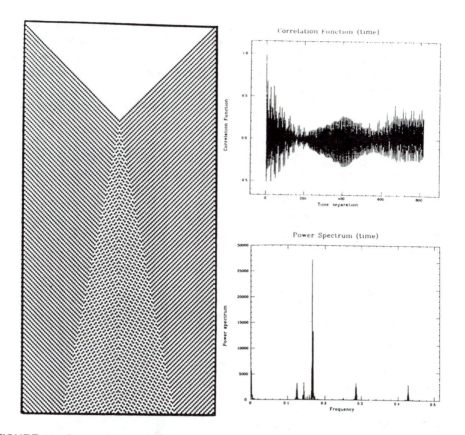

FIGURE 11 Space-time patterns, temporal correlation function and power spectra for the regular regime. The pattern corresponds to the system on the right described in Figure 8 at $t = 4298 - 4554$. The temporal correlation function and power spectra were computed for a 128-cell-long system over a run of 32768 steps.

diffusion system. There are many more initial configurations such as $\binom{2}{2}$, $\binom{1}{2}$, $\binom{2}{1}$ which in pairs lead either to regular patterns or turbulence. Since these systems are reversible, there are conservation laws which we are now searching for but have not yet found. The invariants are clearly non-local. This implies that topological invariants such as winding number are involved.

As a last remark, we would like to add that one way to facilitate the analysis of this complicated behavior is by logical negation of the function N in the diffusion term. Thus N will be 0 if there is at least one active neighbor and 1 otherwise. With this modification we can trace the phase of the vacuum, because now the original quiescent background will cycle instead of remaining quiescent. Instead of the wavefronts the excitations are now domain borders or kinks between regions with different oscillatory phases. The interactions of this defects can be studied as "scattering events." In most of their collisions, the participating kinks come out after the interaction (like solitons) but in some collisions new kinks are produced. There are also fixed defects which remain as boundaries. The turbulent regime is attained when the traveling kinks "pile up" giving rise to an increasing complexity in the pattern (see Figure 13).

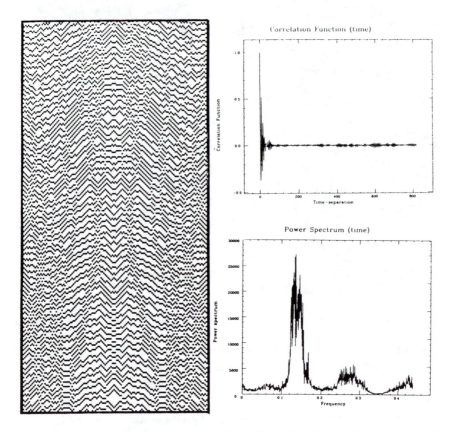

FIGURE 12 Space-time patterns, temporal correlation function and power spectra for the turbulent regime. The pattern corresponds to the system on the left described in Figure 8 at $t = 4298 - 4554$. The temporal correlation function and power spectra were computed for a 128-cell-long system over a run of 32768 steps.

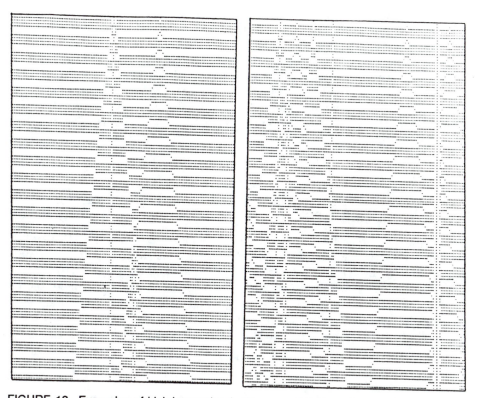

FIGURE 13 Examples of kink-interaction in the reversible Greenberg-Hastings model with the neighborhood function N negated to trace the phase of the vacuum. On the left an interaction event of two incoming kinks produces four outgoing kinks (including a static one). On the right a turbulent regime is shown. $* =$ active, $+ =$ refractory.

CONCLUSIONS

We started out to model reaction and diffusion on clay particles by cellular automata and ended up studying chemical turbulence in reversible systems. The results of our experimentation on the one-dimensional reversible Greenberg-Hastings model has stimulated us to search for non-local invariants. What has this to do with the origin of life ? Well, perhaps nothing, but that is the glory and danger of Artificial Life. Like Descartes, we are pursuing our model whether our starting goal is achieved or not. After all, Descartes invented the internal combustion engine before its time although his model of the heart was wrong.

Perhaps what we are really studying is Artificial Physics (a term coined by J. Leao). This is a field which is coming into its own. The study of systems which go to a semi-chaotic state both in space and time is now a field of interest. The systems we have studied are reversible and, hence, not dissipative. Their behavior is thus an alternative to the study of dissipative dynamical systems and their strange attractors. Finally, we think that non-local invariances are a field worth cultivating. Perhaps life is not simply a dissipative system but also may have non-local conservation laws. In that case perhaps we may have stumbled on factors relevant to the origin of life.

ACKNOWLEDGMENTS

We are grateful for useful discussions to Y. Pomeau, E. Lorenz, S. Redner, G. Vichniac, W. Klein and H. Herrmann. We also appreciate the practical support given to us by J. Leao and the Artificial Physics Lab. and the Physics Department at Boston University, and especially to T. Toffoli, N. Margolus, C. Bennett and D. Zaig, from the MIT information mechanics group, for their constant encouragement and for giving us access to a CAM-6 cellular automaton simulator.

Finally to C. Langton, for interesting discussions and organizing such an enjoyable meeting.

REFERENCES

1. Allouche, J. P., and C. Reder, "Spatio-Temporal Oscillations of a Cellular Automaton in Excitable Media", *Lecture Notes in Biomath* **49**.
2. Bennett, C. H. (1986), "Dissipation, Information, Computational Complexity and the Definition of Organization," *Emerging Syntheses, Santa Fe Institute Studies in the Sciences of Complexity*, Ed. D. Pines (Reading, MA: Addison-Wesley).
3. Boon, J. P., and A. Noullez (1986), "Development, Growth and Form in Living Systems," *On Growth and Form*, Eds. E. Stanley and N. Ostrowsky (Dordrecht: Martinus Nijhoff Publishing).
4. Cairns-Smith, A. G. (1966), "The Origin of Life and the Nature of the Primitive Gene," *J. of Theor. Biol.* **10**, 53–88.
5. Chopard, B. and M. Droz (1988), "Cellular Automata Approach to Non-Equilibrium Phase Transitions in a Surface Reaction Model: Static and Dynamic Properties," *J. of Phys. A* **21**, 205–211.
6. Cosenza, G. and F. Neri (1987), "Cellular Automata and Nonlinear Dynamical Models," *Physica* **27D**, 357–372.
7. Coullet, P., C. Elphick, and D. Repaux (1987), "Nature of Spatial Chaos," *Phys. Rev. Lett.* **58(5)**, 431–434.
8. Crutchfield, J. and K. Kaneko (1988), "Phenomenology of Spatial-Temporal Chaos," preprint.
9. Descartes, R. (1966), "Discourse on Method," *Essential Works of Descartes* (New York: Bantam Books), part 5, 25–36.
10. Dress, A. W. M., M. Gerhardt, N. I. Jaeger, P. J. Plath and H. Schuster (1984), "Some Proposal Concerning the Mathematical Modelling of Oscillating Heterogeneous Catalytic Reactions on Metal Surfaces," *Temporal Order*, Eds. L. Rensing and N. I. Jaeger (Berlin: Springer-Verlag).
11. Dyson, F. (1979), *Disturbing the Universe* (New York: Harper & Row), 194–196.
12. Greenberg, J. M., and S. P. Hastings (1978), "Spatial Patterns for Discrete Models of Diffusion in Excitable Media," *Siam J. Appl. Math.* **34(3)**, 515.
13. Greenberg, J. M., B. D. Hassard, and S. P. Hastings (1978), "Pattern Formation and Periodic Structures in Systems Modeled by Reaction-Diffusion Equations," *Bull. of the Am. Math. Soc.* **84(6)**, 1296.
14. Greenberg, J. M., C. Greene, and S. Hastings (1980), "A Combinatorial Problem Arising in the Study of Reaction-Diffusion Equations," *Siam J. Alg. Disc. Meth.* **1(1)**, 34.
15. Hartman, H. (1975), "Speculations on the Origin and Evolution of Metabolism," *J. of Molecular Evolution* **4**, 359–70.
16. Hartman, H. (1975), "Speculations of the Evolution of the Genetic Code," *Origins of Life*, part I: **6**, 423–7; (1978) part II: **9**, 133–6; and (1984) part III: **14**, 643–8.

17. Kapral, R., and G. Oppo (1986), "Competition between Stable States in Spatially Distributed Systems," *Physica* **23D**, 455–463.

18. Langton, C. (1986), "Studying Artificial Life with Cellular Automata," *Physica* **22D** , 120-149.

19. Li, W. (1987), "Power Spectra of Regular Languages and Cellular Automata," *Complex Systems* **1**, 107–130.

20. Margolus, N., *Physics and Computation, Ph.D. Thesis, Massachusetts Institute of Technology, June 1987.*

21. Meinhardt, H. (1982), *Models of Biological Pattern Formation* (London: Academic Press).

22. Noyes, R. M., and R. J. Field (1974), *Ann. Rev. Phys. Chem.* **25**, 95.

23. Oono, Y., and C. Yeung (1987), "A Cell Dynamical System Model of Chemical Turbulence," *J. of Stat. Phys.* **48(3/4)**, 593–644.

24. Rujan, P.(1987), "Cellular Automata and Statistical Mechanics Models," *J. of Stat. Phys.* .

25. Toffoli, T., and N. Margolus (1977), *Cellular Automata Machines* (Cambridge: Massachusetts Institute of Technology Press).

26. Toffoli, T. (1977), *Cellular Automata Mechanics, Ph.D. Thesis, University of Michigan, Nov. 1977.*

27. Turing, A. M. (1952), "The Chemical Basis of Morphogenesis," *Philos. Trans. R. Soc. London* **B 237**, 37–72.

28. Vichniac, G. Y. (1986), "Cellular Automata Models of Disorder and Organization," *Disordered Systems and Biological Organization*, Eds. E. Bienenstock, F. Fogelman and G. Weisbuch (Berlin-Heidelberg: Springer-Verlag).

29. Von Neumann, J. (1963), *Collected Works*, Ed. A. H. Taub (New York: Mc Millan Co.), vol 5, 319.

30. Wiener, N., and A. Rosenblueth (1946), *Arch. Inst. Cardiol. Mexico* **16**, 205–265.

31. Winfree, A. T. (1974), "Rotating Solutions to Reaction/Diffusion Equations in Simply Connected Media," *SIAM-AMS Proceedings* (Providence, RI: AMS), vol. 8.

32. Winfree, A. T. (1985), "Organizing Centers in a Cellular Excitable Medium," *Physica* **17D**, 109–115.

33. Winfree, A. T. (1984), "Organizing Centers for Chemical Waves in Two and Three Dimensions," *Oscillations and Travelling Waves*, Eds. R. Field and M. Burger (New York: John Wiley and Sons, Inc.).

34. Wolfram, S., ed. (1986), *Theory and Applications of Cellular Automata* (Singapore: World Scientific).

Milan Zeleny,†‡ George J. Klir,‡ and Kevin D. Hufford‡*

†Fordham University, New York, NY 10023; ‡Systems Science, State University of New York at Binghamton, NY 13901; and *APS, Cornell University, Ithaca, NY 14853

Precipitation Membranes, Osmotic Growths and Synthetic Biology

Experimental osmotic growths of now extant synthetic biology suggest that much of the morphogenetic self-organization might follow similar pathways in both inorganic and organic systems. The organic-like shapes and forms of inorganic chemicals lead to intriguing philosophical-paradigmatic implications for biochemistry, biology, general systems theory and philosophy of science.

Classes, divisions, and separations are all artificial, made not by nature but by man. All the forms and phenomena of nature are united...
—*Stephane Leduc*

As all distinctions, the division between life and non-life is an artificial one. Although useful for some purposes, it may narrow down alternative options of exploration.

Recent interest in spatial, periodic and wave properties of chemical systems (e.g., Liesegang patterns,[7] Belousov-Zhabotinskii reaction,[10,22] Benard convection cells,[13] viscous "salt fingers" and precipitate "needles"[15]) reflects the resurfacing of the timeless notions of *self-organization, spontaneous formation,* and *membraneous precipitation* in studying the emergence of life's forms and functions, the origins of life.

We have reproduced a number of osmotic precipitation experiments of Stephane Leduc[6] and other representatives[8] of the long-extant field of *synthetic biology*. These experiments confirm that the potential richness and complexity of spatial-periodic spontaneous patterns goes beyond modern wave-propagation systems and periodic precipitation patterns. It is obvious that the *osmotic growths* of Leduc were complex membraneous precipitates, mimicking not only the most intricate forms and shapes of the "organic" world, but exhibiting phenomena of dynamic growth and transportation, self-maintenance, self-organization and self-renewal, on a scale previously unsuspected.

By immersing a fragment of fused inorganic salt, a crude lump of brute inanimate matter, in a simple inorganic salt solution, the osmotic growth germinates before one's eyes—putting forth structures of "bud and stem and root and branch and leaf and fruit," without any presence of organic matter. These growths are not crystallizations: they grow through intussusception, not by accretion. They exhibit phenomena resembling those of circulation, respiration, and periodic structuralization; they go through a vigorous growth, stable platitude and decay; they reproduce through budding and mimic the forms, colors, textures and even microstructures of "organic" growths to the extent of *deceiving experts.*

In 1907, the Academie des Sciences excluded Leduc's report from its *Comptes Rendus*; his experimental researches on diffusion and osmosis touched too closely on the then-discredited notion of spontaneous generation.

The book by Stephane Leduc was newly reviewed by Zeleny[24] in 1979. It is now possible, some 80 years later, to take a new dispassionate and scientific look at the field of synthetic biology.

HISTORICAL REMARKS

Stephane Leduc was a Professor at L'Ecole de Medicine de Nantes in 1911. His research created a heated controversy at the time, since his experiments touched on the notion of spontaneous generation in a way that reached far beyond the crudeness of "Pasteur's proof" ("The majority of scientists seem to consider that the question of spontaneous generation was definitely settled once and for all when Pasteur's experiment showed that a sterilized liquid, kept in a closed tube, remained sterile."[6]).

Leduc insisted that the study of life must begin with the study of physico-chemical phenomena which results from the contact of two different liquids: biology then is a branch of the physico-chemistry of liquids, studying electrolytic and colloidal solutions and the molecular forces brought about by osmosis, diffusion, cohesion, and even crystallization. This turns out to be a remarkably modern and useful view: it provides for the precursory self-organization of the milieu in which modern genetic materials have to interact and function.

It was l'Abbe Nollet (1748) who first experimented with osmosis, followed by Fischer (1812), Dutrochet (1837), Vierordt (1846), Graham (1854), and others.[6,12] Both Leduc[6] and Pfeffer[12] agree that it was Moritz Traube of Breslau[16-18] who first discovered (in 1866-67) the "artificial cell"—a precipitated (or osmotic) membraneous growth" "...one of the most important, if not all together the most important, steps forward since the discovery of osmosis."[12]

The list of researchers who followed up on Traube's discovery is indeed large: E. Montgomery (1867), G. Quincke (1869), H. de Vries (1871), O. Butschli (1892), L. Rhumbler (1906), R. Liesegang (1907), M. Kuckuck (1907), and A. and A. Mary (1909), just to name a few.[6]

After the field of synthetic biology sank into full scientific obscurity, only the rings of Liesegang,[7] wave-propagations of Belousov-Zhabotinskii,[10] dissipative structures of Prigogine,[13,14] and similar simple periodic phenomena continued to make their on-and-off appearances. The original richness and complexity of patterned precipitation was either lost or delegated into the realm of "curiosities."[9]

EXPERIMENTS

Typical textbook presentations of the concept of precipitation (the mixing of two solutions) and their discussion (the formation of an insoluble salt) are rather unimpressive and boring (a cloudy solution).

In contrast, *osmotic growths* precipitate and grow over a five- to thirty-minute period (sometimes days) and go from a transparent to an opaque state. Some can even be removed from the mother solution for inspection.

From the hundreds of varied and differentiated experiments, a typical experiment of "synthetic biology" might be presented as follows:

MATERIALS

$CaCl_2$—fused and broken into fragments
Na_3PO_4—saturated solution
A 250-ml Beaker

PROCEDURE

Pour 200 ml of the saturated Na_3PO_4 solution into the 250-ml beaker. Drop three or four fragments of fused $CaCl_2$ into the solution and let them sink to the bottom.

DESCRIPTION

The precipitation is almost immediate: the precipitate is a transparent film that becomes white as time progresses and further precipitation occurs. The reactions are

$$CaCl_2 \ (s) \rightarrow Ca^{+2} \ (aq) + 2 \ KCl^- \ (aq)$$

$$3 \ Ca^{+2} \ (aq) + 2 \ PO_4^{-3} \ (aq) \rightarrow Ca_3(PO_4)_2 \ (s)$$

describing the ion-interactions that produce the precipitation.

The description of the osmotic growth can be left to Leduc: "The membraneous substance, the chloride of calcium, diffuses uniformly on all sides from the solid nucleus, and forms an osmotic membrane where it comes to contact with the solution. The spherical membrane is extended by osmotic pressure, and grows gradually larger. Since the area of a sphere increases as the square of its radius, when the cell has grown to twice its original diameter, each square centimetre of the membrane will receive by diffusion but a quarter as much of the membraneous substance. Hence, after a time, the membrane will not be sufficiently nourished by the membraneous substance, it will break down, and an aperature will occur through which the interior liquid oozes out..."[6] In Figure 1, we present the diagrammatic representation of the described process. In Figure 2, the photographs capture the dynamics of another example of precipitation and osmotic growth.

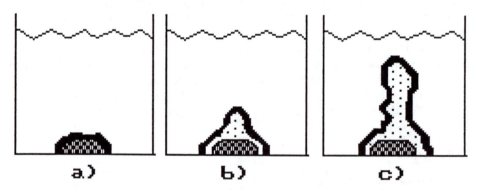

FIGURE 1   a) immediate precipitation, b) osmotic growth begins via osmosis and H_2O passes through membrane, and c) continues. Key: ☐ sodium phosphate (sat.); ■ manganese phosphate (s); ▨ manganese chloride (s); and ⋯ manganese & chloride ion (aq).

FIGURE 2 $CaCl_2$ (s)/Na_3PO_4 (aq) system: from top to bottom, 10:39 a.m., 10:53 a.m. and 11:05 a.m.

Most of Leduc's osmotic growth experiments are analogous to the above $CaCl_2/$ Na_3PO_4 system. We have also reproduced the following systems and have achieved results as described by Leduc:

1) $CaCl_2/NaSiO_4$, Na_2CO_3, Na_2HPO_4

2) $CaCl_2/K_2CO_3$, Na_3PO_4

3) $CaCl_2/K_2CO_3$, Na_2SO_4, Na_3PO_4

4) $MnCl_2/Na_3PO_4$

and 5) $MnCl_2/K_2CO_3$ Na_2SO_4, Na_3PO_4

OSMOSIS

Relevant study of osmoticosmosis and membrane phenomena comes mostly from Pfeffer.[12] Traube's artificial precipitation membranes (e.g., of copper ferrocyanide) were most important for the accurate measurement of the osmotic pressure, developed later by Pfeffer (who induced the membranes to form within porcelain walls) and also used by Leduc.

Membraneous precipitates are formed at the interface of two solutions, or of a solution and a solid, both colloidal and crystalloid. Major membrane-formers used by synthetic biologists were, for example, potassium ferrocyanide and copper nitrate. Osmotic membranes behave exactly as colloids. They grow as long as an osmotic current of water, flowing in, produces a pressure in the interior which tends to distend the membrane. At the same time, they are capable of thickening. The membranes evidently cannot form (or persist) if too active a diosmotic exchange takes place between the solutes.

When the osmotically active solute *does not* pass through the membrane, then the endosmotic (not diosmotic) water flow takes place. This is due to the unequal ratio between the water molecules that collide with the unit surface of the membrane, and the water molecules that pass through it. Let us assume that a membrane has formed and is in contact with water molecules on both sides: an unequal number of water molecules hit the membrane from the surrounding water. However, if one side of the membrane comes in contact with a solute that does not diosmose, its molecules bounce off the membrane, the number of water molecules hitting from that side decreases, and a water flow moving into the salt solution must necessarily result.

OSMOTIC GROWTH: A CELLULAR-SPACE MODEL APPROACH

Varela, Maturana, and Uribe[19] in their discussion of *autopoietic (self-producing) systems* define a unity as a complex system of components that is realized as a whole through its components and their mutual relations. Upon studying Stephane Leduc's descriptions and reproducing some of his experiments, one can show that osmotic growths fall into this category of autopoietic systems.

As stated above, the term autopoiesis means self-production. Self-production has the potential to mean and to be interpreted many different ways by a variety of people. Thus "autopoiesis" has been coined (not translated from greek) as a label, a definition for a clearly defined interpretation of "self-production." This phenomena of self-production can be observed, intuitively, in living systems. A cell consists of a complex set of production processes that synthesize proteins, lipids, enzymes, etc. that renew the entire macromolecular population of the cell thousands of times during its lifetime. Yet, throughout this turnover, the cell maintains its identity, cohesiveness, relative autonomy and distinctiveness. This lasting unity and wholeness, in the midst of continual turnover of constituents, is called "autopoiesis."

An autopoietic system is defined as an entity that can be distinguished from its background by an observer and is realized through a closed (circular) organization of production processes such that (1) the same organization of processes is created through the interactions of its own products (constituents) and (2) a discernable topological boundary emerges as a result of the same processes. The organization of components is maintained and remains invariant throughout the interactions and continual flux of the components: what changes is the system's structure and its parts, *not* its organization.

Guiloff[25] states that the *theory of autopoiesis* does essentially three things:

1. It defines, without reference to the whole, the relations that the components must satisfy in the integration of an autopoietic system.

2. It makes a clear distinction between the phenomenic domain of the components of an autopoietic system (as the domain of its states in autopoiesis) and the phenomenic domain in which the autopoietic system operates as a unity (as the domain of its relational states), showing that these two phenomenic domains do not intersect.

3. It shows that reproduction is not a definitory feature of the organization of living systems, but that it is necessary for evolution.

Varela, Maturana, and Uribe[19] along with Zeleny,[23] have developed computer models of autopoiesis. In their models, the boundary (membrane/autopoietic unity), that is created (realized) and maintained through the functioning of autopoietic organization, remains essentially constant in the volume that is enclosed as time proceeds. Figure 3 summarizes the components (essential building blocks) and

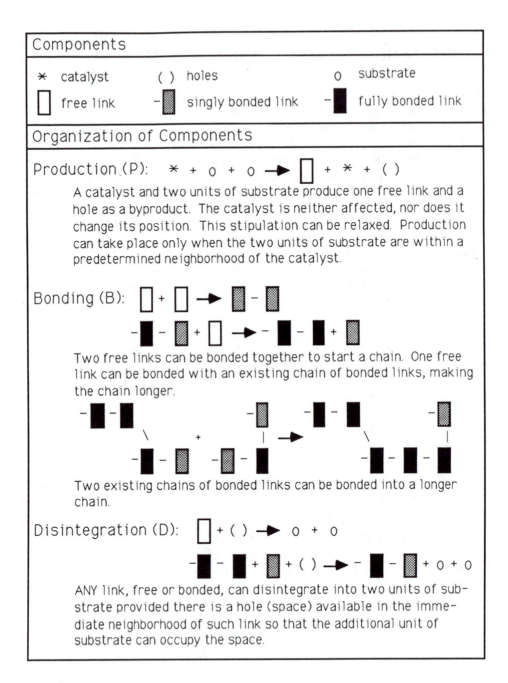

FIGURE 3 APL-Autopoiesis: a model of the cell.

organization of components in Zeleny's APL-Autopoiesis.[25] It is the three processes (productions)—(1) production, (2) bonding, and (3) disintegration—that produce the autopoietic unity or cell when concatenated in a circular fashion. Zeleny has found that the circular organization in itself is not sufficient to maintain the autopoietic unity. The rates of the productions must be balanced in an equilibrium state such that neither the production nor the disintegration is too vigorous. Figure 4 shows four frames from an APL-Autopoiesis computer simulation of a balanced autopoietic organization. The emergence of a topological boundary occurs in the fourth frame and the autopoietic unity is then maintained indefinitely. In experimenting with the models of autopoiesis, it is necessary to remember that this three-process model represents the minimum conditions necessary for the emergence of autopoiesis and that the phenomenon of autopoiesis has to be "tuned into" the underlying circular organization.

To model Leduc's osmotic growths, it will be necessary to modify and expand upon these models to account for the membrane production as well as for the osmotic growth processes while maintaining the autopoietic organization.

Pfeffer[12] has provided very interesting descriptions of the processes of osmotic membrane formation and osmosis along with details of the membrane structure. Pfeffer's concept that membranes are formed from structural units, which allow the paths that the solute and water molecules follow to be defined, employs the principles of localized structural unit interactions, structural unit-solute interactions, and structural unit-water interactions. At the membrane-solution interface, a solute diffusion zone and, therefore, a concentration gradient develops which influences the flow of water through the membrane. Pfeffer also used kinetic molecular theory to explain osmosis and osmotic pressure as a result of the differing number of water molecules that affect the membrane surface, on each of its sides, when unequal solute concentration occurs.

Both of these mechanisms view osmosis and osmotic pressure as complex and dynamic systems, where the processes that occur are based on the current states of the membrane, solute, and water molecules. These states and processes change in time and space according to their localized (neighborhood) interactions. Cellular-automata approaches, similar to those of Vichniac,[20] Goel,[5] Gernert,[4] and others[21] seem promising for modeling such systems. Since cellular-automata are dynamic models in which space, time and variables are considered to be discrete, an overall outline for the osmotic growth process may be as follows:

1. *Initial Conditions.* Definition of the initial membrane, that forms by precipitation, between the solid fragment and the solution.

2. *Endosmosis.* Definition of the states of the membrane, solvent, and solute and their transition rules for passage through the membrane.

3. *Membrane Expansion.* Definition of conditions (states) of the membrane, solvent, and solute for expansion to occur.

4. *Formation of a Membrane Component.* After separation, a new unit forms; define the appropriate conditions.

TIME: 8
1357382951
HOLES: 19
RATIO OF HOLES TO SUBSTRATE: 0.1021
FREE LINKS: 9
ALL LINKS: 19
CUMULATIVE PRODUCTIONS: 20
PRODUCTIONS THIS CYCLE: 2

```
o o o o o o o o o o o o o o o
o o o o o o o o o o o o o o o o
o o o o o o   o o o o o o o o
o o o o   o       o o o o o o
o o o o o o o 0-8-8 o o o o o
                    \
o o o o o o o 0 o o 8 o o o o
                    |
o o   o o 0 0       8 o o 0 o
                    |
o o o o o o   * 0 o 0 o o o o
o o o   0 o o 0       o o o o
o o o o o   o 0     o 0 o o o
o o o o o o o 8-8-0   o o o o
                |
o o o o o o o 0 o o   o o o o
```

TIME: 14
1087796851
HOLES: 27
RATIO OF HOLES TO SUBSTRATE: 0.1588
FREE LINKS: 10
ALL LINKS: 27
CUMULATIVE PRODUCTIONS: 30
PRODUCTIONS THIS CYCLE: 3

```
o c o o o o o o o o o o o o o
o   o o o o o   o o 0 o o o o
o o o o o o   o   o o o o o o
o o o o o o o o o   o o o o o
o o o o o 0 o   0-8 o   o o o
         /        \
o o o   0     0 o 0 8 o o o o
                    |
o o o   o       8 o o o o
                    |
o o o o 0   o * 0   8 o o o o
                    \
o o o o 0 8 o 0 0   0   o o
       /
o o o 0-8   0   0   o o o   o
o o o o   o 0 8-0 o 0-0 o o o
                |
o o o o o o o 0 o o o   o o
```

TIME: 22
485622923
HOLES: 33
RATIO OF HOLES TO SUBSTRATE: 0.2075
FREE LINKS: 7
ALL LINKS: 32
CUMULATIVE PRODUCTIONS: 41
PRODUCTIONS THIS CYCLE: 2

```
o o o o o o o o o o o o o o o
o o o o o     o o 0 o   o o o
o     o o   o o o o o       o
o o o o   o o o o o o o o o o
o   o o o 8-8-8-8-8 o o o o o
         /          \
o o o o 8 0   o 0 0 0 o   o o
        |
o o o o 8       o 0 0 o   o o
        |
o o o o 8 o   *     0 o     o
        |                \
o   o o 8 0   o 0       8 o o o
        |  |              |
o o o 0-8 8 o o o o     8 o   o
        |                |
o o o o o 8-0 8-8-8-8-8 o o o
        |
o o o o o   o 0 o o   o o   o
```

TIME: 40
1847260068
HOLES: 39
RATIO OF HOLES TO SUBSTRATE: 0.2653
FREE LINKS: 13
ALL LINKS: 38
CUMULATIVE PRODUCTIONS: 59
PRODUCTIONS THIS CYCLE: 0

```
o o o o o o o o o o o o o o o
o       o o o 0 o o o o o   o o
o   o   o   o o o   o       o
o   o o o   o   o o o o   o o
o o o o o 8-8-8-8-0 o   o   o
         /        \
o o o o 8 0 0 o 0   8 o o o o
        |           |
o o o o 8 o o 0   0 8       o
        |           |
o o o o 8 0   * 0   8 o o o o
        |           \
o o o o 8 0 0 o o 0 0 8 o   o
       / |           |
o o o 0   8   o     o 8 o   o
        |           |
o o   o o 8-0 0-8-8-8-8   o o
        |
o   o o o o o o 0 o   o   o o o
```

FIGURE 4 APL-Autopoiesis: computer simulation. Catalyst is represented by an asterisk. Holes are represented by a blank space. Substrate is represented by a small circle. Free link is represented by an empty square. Singly bonded link is represented by a square with a quota inside and a single dash. Fully bonded link is represented by a square with an APL division sign inside and two dashes.

5. *Repeat* steps two through five.

The above five steps provide a general outline for the process of growth. Other aspects that are to be considered to completely model the entire "lifetime" of this autopoietic system are rules (a) to direct the growth in a vertical direction, (b) to slow, thicken, and eventually stop the growth, and (c) to describe the decay of the osmotic growth structure.

ORGANIC FORMS

What is interesting about inorganic osmotic growths is their striking resemblance to living shapes, forms and behavior. These are the forms and behavior which we intuitively associate with living organisms. Leduc reports osmotic growths resembling striated stems, leaves and buds, mushrooms and fungi, shells, capsules and flowers, amoebas, fins, worms, and, of course, assorted spheres, spheroids, beads, drops, compartments and cilia. These are not more or less rigid geometrical lattices of crystals, crystalloids or liquid crystals. The shapes are round and smooth, interconnected by stems and pathways, growing, changing, reshaping and repairing themselves throughout their life span. The fact that "inorganic" matter, under the most common geophysical conditions, can give rise, spontaneously and without "coaching," to such apparently "organic" forms, is challenging: organic and "living" forms would then be "in-formed" in the system of chemicals and their interactions, rather than "informed about" through a particular structural component.

Identical forms of organic elements, cells, tubes, etc. may be produced either in organic liquid or a semi-organic liquid such as sucrate of lime, or in an absolutely inorganic liquid such as silicate of soda. The sulfates and phosphates generally produce tubes while the carbonates form cells.

The periodicity of some precipitates and their light diffraction give first insight into the nature of color in natural objects. The variety of forms and colors is bewildering, exhibiting not only periodicity and wave propagation, but also rhythmic catalysis and undulatory movements.

Osmotic growths may be viewed as models of morphogenesis. They consist of a number of cells instead of one large cell. Some of these artificial cells can actually grow out of the solution into the air, changing shape and color, and dissolving gases from the atmosphere.

Osmotic growths absorb nutriment from the medium in which they grow: comparing the morphogenic germ (small mineral fragment) with the system (osmotic growth) it "produced," we can register an increase of many hundreds times in weight. Of course, the surrounding liquid has lost an equivalent weight.

The absorbed nutrients undergo chemical transformation before they can be assimilated. For example, calcium chloride growing in a solution of potassium carbonate is transformed into calcium carbonate:

$$CaCl_2 + K_2CO_3 \rightarrow CaCO_3 + 2\,KCl$$

So the osmotic pressure possesses the remarkable power of organization and morphogenesis. It is a matter of surprise that this peculiar faculty has hitherto remained almost unsuspected.

All of these osmotic growths not only imitate the *forms and shapes*, but also imitate the *microstructure* and various *functions* of living systems such as nutrition, metabolism, growth, movement and evolution.

These osmotic forms have the potential to grow wherever suitable conditions exist. Combined with the discovery of rock-dwelling life forms[3] and the new theories of mineral origins of life,[1] osmotic forms could have played a large role in the evolution of life. Thus, when searching for the physical embedding of life, we should also consider osmosis and osmotic growths and not only clay crystals and crystallization.[25]

UNRESOLVED QUESTIONS

The number of unresolved questions is staggering and that is why this paper was written. Let us state only the most important:

1. How much of what is in the fossil record should be attributed to the imprints of osmotic growths and how much to clearly defined biotic events?[11] Precipitated osmotic membranes could have been widely distributed in nature. Can osmotic growths be operationally distinguished from organic forms, especially since their persistence and resilience is comparable to that of living matter?

2. What are the philosophical-paradigmatic implications of the capability of inorganic matter to exhibit phenomena which are strikingly similar, externally and internally, to the morphogenesis of living matter? If "organic" forms is the way matter can organize itself, what is the true role of "information programs" residing on some of the components?

3. What are the possibilities of "organic components invading inorganic forms and templates"? See Cairns-Smith[1] for an example of such. Can osmotic growths be the basis for a hypothesis on the origins of life?

4. What role, if any, have osmotic growths played in the formation of the various rocks, e.g., siliceous, calcareous, barytic, magnesian, the fibrous and nodular rocks, and atolls?

5. What are the most appropriate conceptual frameworks for developing models of spontaneous osmotic growths?

6. What are the implications of "life as organization" as opposed to "life as stuff" in searching for life processes? Should the search for organic debris take precedence before the search for autopoietically organized and self-renewing processes?

7. Why was both Leduc's work and synthetic biology so effectively removed from scientific and non-scientific consciousness, even though the phenomena of osmotic growth still remained unexplained and challenging?

SUMMARY

The dismissal of Leduc's work seems to have occurred due to the narrow view of biology and biological processes that his peers held in that era of science. The scientists of Leduc's time did not have as broad a conception regarding the correct domain of biology as science does today. The advanced thinkers, the mainstream scientists of today, are willing to explore the fringes of life much further than the Academie des Sciences of Leduc's time.

According to Leduc[6] and also Zeleny,[25] there is no sharp division, no precise limit where inanimate nature ends and life (animate nature) begins. The transition is gradual. Individual attributes of life are also found outside of the living organism. Just as a living organism is made of the same substances found in the mineral world, so life is a composite of the same physical and chemical phenomena found in the rest of nature.

Leduc's osmotic growths, like Cairns-Smith's clay crystals,[25] should be considered as representatives of protobiological, primitive systems. They do not represent the actual (complete) biological processes of life, but they are similar and they do resemble the biological processes in their shape, form and behavior. It is for these reasons that they have a potential place in the origin (evolutionary tree) of life and are definitely worth investigating. Life is not a substance but an organization.

The fact that an organic body is formed of certain elements should not be of greater importance than the manner in which these elements are organized. The implications of this view are potentially far reaching.

REFERENCES

1. Cairns-Smith, A. G. (1982), *Genetic Takeover and the Mineral Origins of Life* (Cambridge: Cambridge University Press).
2. Dahmen, U., and K. H. Westmacott (1986), "Observations of Pentagonally Twinned Precipitate Needles of Germanium in Aluminum," *Science* **233**, 875.
3. Friedmann, I., and R. Ocampo (1976), "Endolithic Blue-Green Algae in the Dry Valleys: Primary Producers in the Antarctic Ecosystem," *Science* **Sept. 24**.
4. Gernert, D. (1986), "A Cellular-Space Model for Studying Wave-Particle Dualism," *Cybernetics and Systems '86*, Ed. R. Trappl (Boston, MA: D. Reidel), 117.
5. Goel, N. S., and R. L. Thompson (1986), "Organization of Biological Systems: Some Principles and Models," *Intern. Review of Cytology* **103**, 1–88.
6. Leduc, S. (1911), *The Mechanism of Life* (London, England: Rebman); in this volume, Leduc mentions these additional researchers who followed up on Traube's discovery: Graham (1851), Runge (1855), Bottger (1867), Harting (1871), Reinke (1875), Monier and Vogt (1882), Rose (1887), Linke (1889), Benedikt (1904), Dubois (1904), Razetti (1907), and Marzo (1909); also see W. Deane Butchers translation of Leduc's "Theories Physico-Chimique de la Vie et Generations Spontanees."
7. Liesegang, R. E. (1914), "Silberchromatringe und-Spiralen," *Z. Phys. Chem.* **88**, 1; reprinted (1939), *Kolloid Z.* **87**, 57.
8. Mary, A., and A. Mary (1909), *Etudes Experimentales sur la Generation Primitive* (Paris, France: Joules Rousset).
9. Muller, S. C., S. Kai, and J. Ross (1982), "Curiousities in Periodic Precipitation Patterns," *Science* **216**, 635.
10. Muller, S. C., T. Plesser, and B. Hess (1985), "The Structure of the Core of the Spiral Wave in the Belousov-Zhabotinskii Reaction," *Science* **230**, 661.
11. Niklas, K. J. (1986), "Computer Simulated Plant Evolution," *Scientific American* **254(3)**, 78.
12. Pfeffer, W. (1877), *Osmotische Untersuchungen* (Leipzig: Wilhelm Engelmann); new English edition entitled *Osmotic Investigations* (New York: Van Nostrand Reinhold).
13. Prigogine, I. (1980), *From Being to Becoming* (San Francisco, CA: W. H. Freeman), 88.
14. Priogogine, I., and I. Stengers (1984), *Order Out of Chaos* (New York: Bantam Books).
15. Taylor, J., and G. Veronis (1986), "Experiments on Salt Fingers in a Hele Shaw Cell," *Science* **231**, 39.
16. Traube, M. (1864), *Zentralblatt f. Med. Wiss.* **95**, 609.
17. Traube, M. (1866), *Zentralblatt f. Med. Wiss.* **97**, 113.

18. Traube, M. (1867), "Experimente zur Theorie der Zellenbildung und Endomose," *Arch. Anat. u. Physiol.* **87**, 129.
19. Varela, F. G., H. R. Maturana, and R. Uribe (1974), "Autopoiesis: The Organization of Living Systems, its Characterization and a Model," *Biosystems* **5**, 187.
20. Vichniac, G. Y. (1984), "Simulating Physics with Cellular Automata," *Physica* **10D**, 96.
21. Wolfram, S. (1986), *Theory and Applications of Cellular Automata* (Singapore: World Scientific).
22. Zaikin, A. N., and A. M. Zhabotinskii (1970), "Concentration Wave Propagation in Two-Dimensional Liquid-Phase Self-Oscillating System," *Nature* **225**, 535.
23. Zeleny, M. (1977), "Self-Organization of Living Systems: A Formal Model of Autopoiesis," *Int. J. Gen. Systems* **4**, 13.
24. Zeleny, M. (1979), "Special Book Reviews," *Int. J. Gen. Systems* **5**, 63.
25. Zeleny, M. (1981), *Autopoiesis: A Theory of Living Organization* (New York: Elsevier North Holland), 7–14, 91, 119.

Norman H. Packard

Center for Complex Systems Research and the Physics Department, University of Illinois, 508 South Sixth Street, Champaign, IL 61820; inet address: n@complex.ccsr.uiuc.edu; usenet address: ...[ihnp4,ucbvax]!uiucdcs!complex!n

Intrinsic Adaptation in a Simple Model for Evolution

A distinction is made between *extrinsic* adaptation, where evolution is governed by a specified fitness function, and *intrinsic* adaptation, where evolution occurs "automatically" as a result of the dynamics of a system caused by the evolution of many interacting subsystems. A simple model is presented to illustrate intrinsic adaptation, and the dynamics of the population is observed through simulation of the model. The population density over the space of genes for the model organisms may be interpreted as an *a posteriori* fitness function, and is seen to undergo complex dynamics that depend on the environment.

EXTRINSIC *VS.* INTRINSIC ADAPTATION

A common approach in modeling evolutionary processes is to regard a member of a population as fit or unfit according to how it is evaluated by a fitness function. The fitness function is a map that assigns a real number (the fitness) to every possible member of the population, and it is generally specified as an *a priori* feature of the model. Successful adaptation then takes place when a dynamic on the space of population distributions leads to distributions that are peaked at the peaks of the fitness function. The simplest case of a single member of the population evolving to a fitness peak is then gradient ascent. I will call any adaptive dynamics that uses such an *a priori* fitness function *extrinsic adaptation*.

In evolutionary approaches to engineering applications, where organisms are replaced by specifications of a device or a dynamical rule, the fitness function is typically given by an engineering goal.[4] In modeling biological evolution, the use of a fitness function is usually justified by assuming that an organism's interactions with its environment and other organisms may be averaged into a net overall fitness. Depending on the context, the fitness function is sometimes thought of as a function on an arbitrary space of parameters that defines the organism, and sometimes as the distance to an arbitrary (random) organism defined as fit.[5] The assumption of an *a priori* fitness can be dangerous, considering that an organism affects its environment and, at least indirectly, other organisms, hence altering the world it lives in, and thus possibly altering its own ability to exist, i.e., its own fitness.

The biosphere does not appear to have any *a priori* fitness function defined on a space of possible organisms; in fact, one of the most amazing features of biological evolution is that the biosphere evolves automatically, with each organism (or population of an organism) adapting to an environment made up of both external natural features and all other organisms it interacts with. The fitness of an organism is thus intrinsically defined as a result of these interactions, through the process of life and death, with fit organisms being the ones that survive. I will call adaptation of a system that occurs as a result of a population of subsystems changing in response to interactions between them (without an *a priori* fitness function) *intrinsic adaptation*. In this paper I present a simple model that incorporates life and death as a selection mechanism and allows the exploration of the dynamics of intrinsic adaptation.

This approach to modeling evolution is similar in spirit to some earlier attempts to simulate evolution,[1,2] which share the feature of having no *a priori* fitness function. The main novel feature of the model presented here is its simplicity. I make every attempt to strip down most of the complexity of real biological systems, with the aim of discovering a minimal model that displays evolutionary behavior. The model aims to answer the following sorts of questions: How does organismic fitness emerge as a result of adaptive dynamics? How may complexity of a biosphere be quantified, and how does it change as the biosphere evolves? What is the evolutionary capacity of a biosphere? How do these last two quantities vary with the level of driving and environmental noise? The main point of this paper is to describe a simple model that may be used to approach these questions. Simulation of the model has only just begun; the simulation results presented here will primarily address the first question.

ARTIFICIAL BUGS IN AN ARTIFICIAL ECOLOGY

The present context for experimental evolution is a two-dimensional world filled with simple organisms, which will be referred to as bugs. The two-dimensional world is actually a 512×512 lattice. Besides the organisms, the world also has a continuum variable at each lattice site that represents food for the bugs. The food distribution evolves in time by diffusion. The bugs and food in this model are motivated by chemotactic bacteria which might be living in a Petri dish that also contains energy rich chemicals that diffuse in the dish.

The bugs evolve in time by eating the food at their current location, sensing the gradient of the food field, moving in the direction of the gradient, eating food, and so on. We are thus building in a certain amount of intelligence into the bugs, in both their sensory capability and their ability to sense the gradient. This degree of intelligence is commensurate with the functionality seen in chemotactic bacteria; some can actually sense gradients directly, others detect only absolute concentrations and shift between random and straight line motion, achieving gradient motion on average. Relaxing the assumption of gradient detection for the bugs will be discussed below.

Each bug stores the food it eats in an internal storehouse. This store of food is depleted when the bug moves with a tax. When the bug accumulates enough food, it reproduces into some number of offspring, dividing its food evenly among the offspring. If the bug obtains less food, on average, than it expends moving around, its food supply reaches zero, in which case it dies. Thus if the world were started with a fixed amount of food spread over the world, and if none were added, all bugs will eventually die. Thermodynamically, the food may be thought of as energy. Diffusion is constantly acting to increase entropy, and the bugs eating food and concentrating it internally tends to decrease entropy.

Even without reproduction, the motion of a few bugs can be quite complex, since the bugs eat food, which perturbs the food field whose gradient they are following. The motion is further complicated by the reproduction of the bugs. When a bug reproduces, it splits into some number of offspring (this number may change from one bug to another), and the initial locations of all the offspring are the same as that of the parent where it split. They immediately move apart from each other, since they follow gradients of the food field, and the gradient will never point toward a location where a bug has just consumed all the food. In the implementation of the model, no attempt is made to synchronize the motion of the bugs.

EVOLUTION

The bugs are fairly simple as we have described them so far, but even now there is room for variation from one bug to another. There are two ways that bugs may differ from each other: (i) the food threshold required for reproduction, and (ii) the number of offspring. I will identify these two parameters with two genes g_{th} and g_{off}. In doing so, I am tacitly making an identification between the genotype and phenotype of a bug, dispensing with a genotypic symbolic representation that is translated into phenotypic functionality.

Evolution occurs during reproduction by changing the value of either or both of these two genes. In the simulations below, one of the offspring inherits the same genes as the parent, and the other offspring inherit genes that are mutated from the parental values. Typically g_{th} is changed by a random amount of up to $\pm 10\%$, and g_{off} is changed by a value of ± 1.

Though we have specified how individual bugs change from one generation to the next, evolution may be seen only by observing the change of the entire population with time. We represent the population by a density function over the two-dimensional space of genes $P(g_{th}, g_{off})$, where the illustrations of evolution in Figures 1–3 show $0 \le g_{th} \le 18 \times 10^3$ and $0 \le g_{off} \le 8$. These figures show the evolution of $P(g_{th}, g_{off})$ with time, caused by the individual bugs reproducing and dying. We will denote the initial population by $P^0(g_{th}, g_{off})$, and that of the nth generation by $P^n(g_{th}, g_{off})$.

In the language of population genetics, different values of g_{off} and g_{th} corresponds to different alleles for the genes that govern the number of offspring and the reproduction threshold. This model uses only haploid genetics, since each feature is assumed to be characterized by a single gene that can be mutated. Diploid genetics would require the model to incorporate crossover between the genes of two individuals, i.e., the model would need to have sexual interaction between bugs. This type of enhancement to the model will be discussed below.

In simulating the artificial biosphere, there are three aspects of the model that must be specified:

■ Prescribing the initial ensemble $P^0(g_{th}, g_{off})$.
■ Setting the time scale of the bug updates, in particular the time scale relative to the diffusion rate.
■ The mode for driving the system; i.e., the way that food is pumped into the world.

The first aspect is the same for all of the present simulations: $P^0(g_{th}, g_{off}) = \delta(x - 10^4)\delta(y - 2)$; i.e., all the initial bugs have a reproduction threshold $g_{th} = 10^4$ and all have two offspring when they reproduce ($g_{off} = 2$).

The time scale of the diffusion process is fixed by special hardware that updates the food field by convolving it with a nearest neighbor diffusion kernel thirty times per second. Fixing the time scale for the updating of the bugs is somewhat more problematic due to the fact that each bug is updated sequentially on a serial machine. Thus the time between updates of any given bug depends on how many bugs there are. This uncertainty in the bug update time scale may, however, be overcome in two different ways: (i) limiting the size of the lattice that is driven by new food, thus limiting the total population possible (the effective lattice size used in these simulations was 128×128); (ii) setting the food tax for bug movement. The tax for Figures 1 and 2 was set to 20 food units per lattice step, and the tax for Figure 3 was set to 50 food units per lattice step. The effects of changing the tax on the dynamics of $P(g_{th}, g_{off})$ will be discussed below.

The time scales have been set so that evolution occurs on a human time scale, i.e., on the scale of minutes of computer time. On this time scale, a population of thousands of bugs may be processed for many generations, with hundreds of deaths and reproductions.

Food may be added to the world in a variety of different ways. The two modes studied in the present simulation are (i) adding food to a square region periodically with a long period (roughly three seconds), and (ii) adding food in small random patches randomly placed over the active region of the lattice periodically with small period (with a period of about 0.3 seconds). The difference between these two feeding regimens are discussed below. In both cases, when food is added to a lattice site, it is added in the quantity of 255 units, the maximum possible for a given site.

DYNAMIC *A POSTERIORI* FITNESS

The population density function, $P^n(g_{th}, g_{off})$ may be regarded as a fitness function at a generation n in the sense that if a particular set of values for g_{off} and g_{th} are fit, those values will be represented strongly in the population. $P^n(g_{th}, g_{off})$ is, however, an *a posteriori* fitness function because it can only be obtained by observation after it has been determined by the dynamics of reproduction and death within the population of bugs.

The simplest form of dynamics for $P^n(g_{th}, g_{off})$ is that it goes to a fixed point, i.e., that there exists a distribution function $\overline{P}(g_{th}, g_{off})$ such that

$$\overline{P}(g_{th}, g_{off}) = \lim_{n \to \infty} P^n(g_{th}, g_{off}).$$

In the case that $\overline{P}(g_{th}, g_{off})$ exists, it may be called a "fitness landscape." The assumption that such a fixed asymptotic distribution exists underlies much of population genetics[3] as well as many studies of models for adaptation.[5] In population genetics, this assumption takes the form of having constant relative fitness for different alleles of a particular gene.

In the present simulations, we find that $P^n(g_{th}, g_{off})$ does *not* necessarily go to a fixed point as $n \to \infty$, but instead can have a rich variety of dynamics. In these cases, it does not make sense to think of a "fitness landscape" being defined over the space of genes because no particular set of alleles (particular values for g_{off} and g_{th}) has fixed fitness with respect to other alleles.

The dynamics of $P^n(g_{th}, g_{off})$ are strongly dependent on the mode of driving the system with food. The present simulations indicate that the most important factor seems to be the size of the fluctuations of the net food supply about the long time average.

When there are large fluctuations in the food supply, there are in turn large fluctuations in the population, and when the population has large upward swings, and $P(g_{th}, g_{off})$ can change markedly because of the introduction of new mutants. This case is illustrated in Figure 1, where a large amount of food is added to a fixed square region periodically with a large period (about three seconds). We see that $P^n(g_{th}, g_{off})$ remains peaked near the initial delta function at $g_{th} = 10^4$ and $g_{off} = 2$, though the peak moves around slightly. The shape of $P(g_{th}, g_{off})$ about the peak changes with time, and does not seem to converge to any fixed shape.

In fact, $P(g_{th}, g_{off})$ is seen to develop subsidiary peaks, e.g., in most of the generations of Figures 1(h)–1(t). One prevalent pattern is that the subsidiary peaks move away from the initial peak toward lower g_{th} and higher g_{off}. This makes some sense: The lower g_{th} is and the higher g_{off} is, the more rapidly bugs will proliferate. The problem for these bugs is that when the food supply dwindles, i.e., when they are forced to rely on their stored resources, they run out of food quickly and die. Thus we see entire subpopulations arise and evaporate as in the transition from Figures 1(n)–1(q).

Figure 2 illustrates a more uniform feeding regimen; for this simulation, food was added periodically with a short period (about 0.3 seconds) in small patches randomly placed in a fixed larger region of the lattice. This regimen is more uniform in the sense that there are smaller fluctuations of the total amount of food present at any given time about the long time average than in the previous case.

Over the same time scale as the previous simulation illustrated in Figure 1, we see that $P^n(g_{th}, g_{off})$ again retains a global peak near the initial peak. Now, however, the fluctuations of $P^n(g_{th}, g_{off})$ about this peak are less pronounced, and on a slow time scale, $P^n(g_{th}, g_{off})$ evolves to have a broad skirt that extends toward low g_{th} and high g_{off}. The fluctuations in $P^n(g_{th}, g_{off})$ are much smaller than in the previous case, and the shape of $P^n(g_{th}, g_{off})$ seems to be generally simpler, in the sense that it has only one peak, with occasionally one subsidiary peak as seen in Figures 2(f) and 2(h)–2(p). It is not known whether $P^n(g_{th}, g_{off})$ will approach a fixed point distribution $\overline{P}(g_{th}, g_{off})$; it simply did not do so during the time the system was observed.

The last simulation illustrated in Figure 3 is identical to the simulation in Figure 2, with the exception that the bugs' movement tax has been raised from 20 food units to 50 food units. This simulation illustrates a transition that takes place as this parameter is varied. Obviously for extremely high values for the movement tax, all bugs will die before accumulating enough food to reproduce. As the tax is lowered, the average time for the population to live before all bugs die diverges until the population is self-propagating indefinitely. The transition for this particular form of feeding occurs near a tax value of 50, where we see that the population lasts for 12,000 generations before it becomes extinct. One notable feature of this transition is that the fluctuations in $P^n(g_{th}, g_{off})$ are larger near the transition. This is caused by larger swings in the population, and more mutations being introduced into the population during the upswings. Instead of the simple form for $P^n(g_{th}, g_{off})$ seen in Figure 2, Figure 3 shows many subsidiary maxima that move around constantly.

GENERALIZATIONS OF THE MODEL

There is one major aspect of biological evolution that cannot be captured by the model as it is presently described: in biological evolution, the biosphere seems to explore a constantly expanding universe of forms with new types of functionality and interactions emerging constantly; in the present model, the universe of forms available to the bugs is highly restricted, (a point (g_{th}, g_{off}) in a two-dimensional space), and is specified *a priori* as a feature of the model. The primary generalization of the model currently being explored is to create an open space of forms in which the bugs may evolve.

One way to open up the space of forms is to make the bugs less sophisticated than described above; suppose they were so unsophisticated that they did not have any way of detecting the gradient of the food field. They could be given the capability of growing sensors to detect the food concentration at nearby sites (with some cost subtracted from their internal resources), and they could have a movement strategy which would be a map from the set of sensed food values to a direction of movement. An initial population of bugs could contain a random selection of sensors and strategies, and when allowed to evolve, the strategies that were ineffective would eventually die, and the effective strategies would survive. Mutations could allow the sensors and strategies to become arbitrarily complex. As bugs evolved that are increasingly effective at gathering food, the model would display the genesis of hunger.

Another way to enlarge the space of forms is to increase the sophistication of the bugs. There are several fairly obvious ways to do this. In the model described above, there are no direct interactions of the bugs, they interact only indirectly through the food field. The bugs could be allowed to interact, and could have an open space of possible interactions. Interactions could include sexual reproduction, with the concomitant possibility of crossovers to produce new genetic possibilities. Other interactions could be parameterized to produce either symbiosis or predator-prey relationships. These types of interactions will be described in future work.

DISCUSSION

We have presented a simple model for evolution that displays intrinsic adaptation, i.e., adaptation without specification of any fitness criterion as part of the model. The salient features of the model are:

- A population of individuals that exist independently of each other, with individual properties being determined by parameters analogous to genes (two, in the present case, g_{th} and g_{off}).

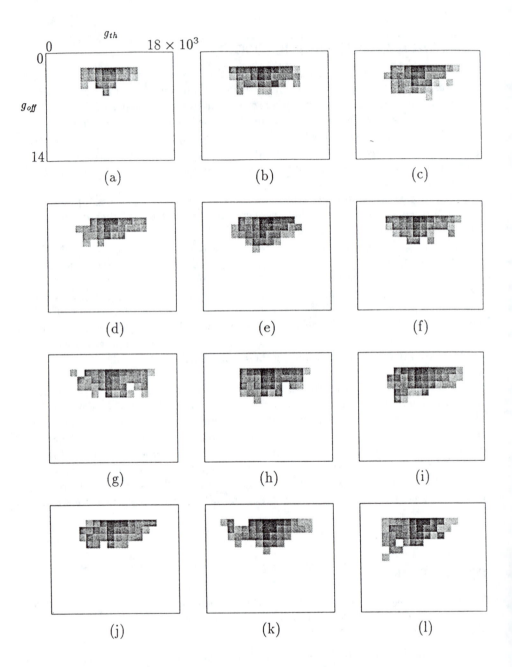

FIGURE 1 The population density over the space of genes (g_{th}, g_{off}) as it changes from the reproduction and death of many individuals in the population. Between each of the figures there is 1000 updates of the entire population. Dark squares indicate large population, light square small population. See the text for details of the simulation.

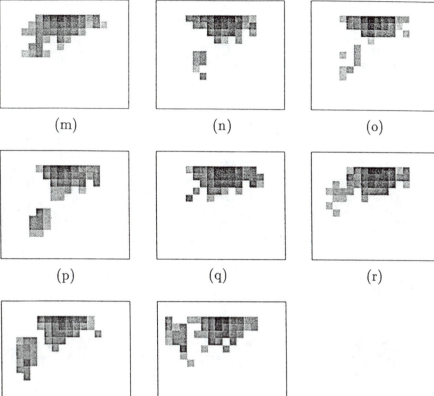

FIGURE 1   continued

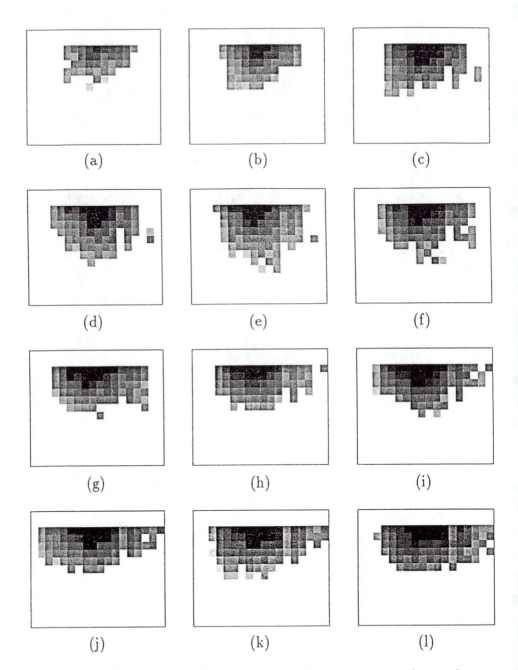

FIGURE 2 Evolution of the population density over the space of genes (g_{th}, g_{off}), as in Figure 1. See the text for details.

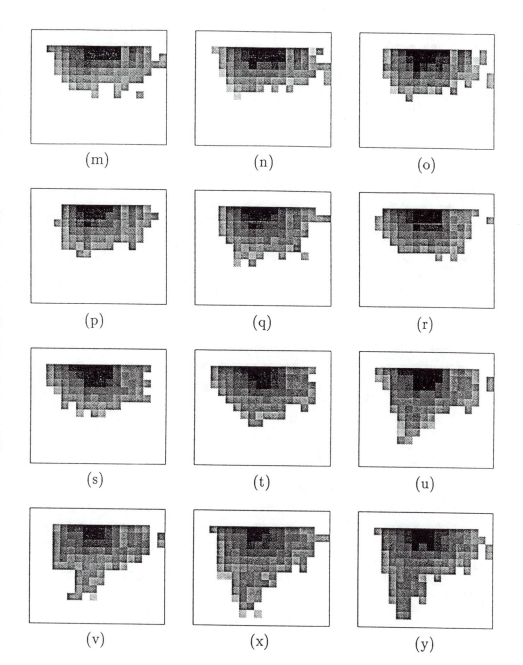

FIGURE 2   continued

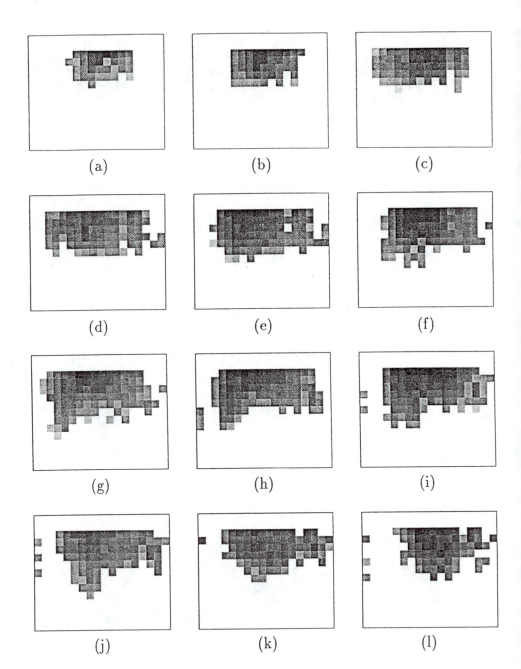

FIGURE 3 Evolution of the population density over the space of genes (g_{th}, g_{off}), as in Figure 1. See the text for details.

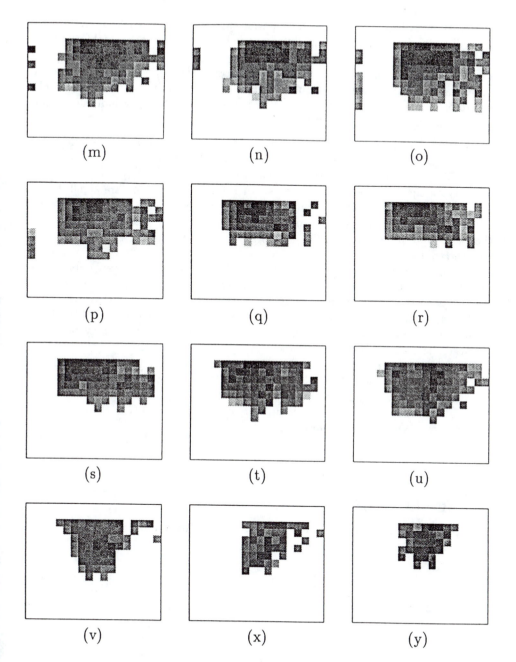

FIGURE 3 continued

■ Existence of all individuals in a spatial environment, which can give genetically identical individuals in different spatial locations different local fitness properties.

■ Dependence of all individuals on food (energy) that is obtained from the environment, with life and death of each individual determined by its food supply.

■ Reproduction of individuals, with the possibility of changing individual characteristics during reproduction.

Adaptation is seen in the change of the population distribution function over the space of genes, $P^n(g_{th}, g_{off})$, which can be regarded as an *a posteriori* fitness function. In our simulations, we see that the dynamics of $P^n(g_{th}, g_{off})$ depends strongly on the way the system is driven with food. When there are small fluctuations of the total amount of food present with respect to the long-time average, $P^n(g_{th}, g_{off})$ appears to have a simple form (few peaks that change slowly). When there are large fluctuations, $P^n(g_{th}, g_{off})$ is more complex, with many peaks that change rapidly.

Complexity in the framework of this model is simply the complexity of the distribution function $P^n(g_{th}, g_{off})$ and of its dynamics. Several peaks present are analogous to the coevolution of different species, genetic variants that are viable in the environment provided by each other. Thus we find that complexity of our artificial biosphere increases with large population fluctuations which may be caused by large fluctuations in the environment.

Enhancements to the present model have been suggested, notably the possibility of having a space of possible individuals that is open, in the sense that as individuals change, they could have an infinite variety of possibilities. Complexity in this context could also include the complexity of the functionality of the individuals (which is fixed in the present model).

ACKNOWLEDGMENTS

I am grateful to Rob Shaw, Doyne Farmer, Stephen Omohundro, Otto Rossler, and Kuni Kaneko for stimulating discussions.

REFERENCES

1. Conrad, M., and H. H. Pattee (1970), "Evolution Experiments with an Artificial Ecosystem," *J. Theo. Bio.* bf 28, 393.
2. Conrad, M., and M. Strizich (1985), "Evolve II: A computer Model of an Evolving Ecosystem," *BisSystems* **17**, 245.
3. Fisher, R. A. (1930), *The Genetical Theory of Natural Selection* (Oxford: Oxford University Press).
4. Holland, J. H. (1986), "Escaping Brittleness," *Machine Learning 2*, Eds. R. S. Michalski, J. G. Carbonell, T. M. Mitchell (Los Altos, CA: Morgan Kaufmann)
5. Kauffman, S. A. (1969), "Metabolic Stability and Epigenesis in Randomly Constructed Genetic Nets," *J. Theo. Bio.* **22**, 437–467.

Stewart W. Wilson
The Rowland Institute for Science, Cambridge, MA 02142

The Genetic Algorithm and Simulated Evolution

A scheme is described for simulating the evolution of multicellular systems. The scheme is based on a representation for biological development in which the genotypes are sets of production-like growth rules that are executed to produce cell aggregates—the phenotypes. Evolution of populations, through phenotype selection and genotype variation, occurs according to the method of the genetic algorithm. Some examples of the development representation in 1-dimensional creatures are given.

1. INTRODUCTION

The genetic algorithm,[2-4] a powerful optimization technique, incorporates mechanisms inspired by the mechanisms of reproduction, variation, and selection found in natural evolution, but, despite successes in several fields of application, there has been little attempt to use the algorithm as a tool to investigate, through simulation, natural evolution itself. Considerable work exists on the ontogenetic evolution of behavior, i.e., learning,[2,3,6,9,11] but relatively little on the evolution of organisms per se.[5] The main reason has been the problem of representing organisms in a way that would permit the genetic algorithm to be brought to bear. The genetic algorithm observes the genotype-phenotype distinction of biology: the algorithm's variation operators act on genotype-like bitstrings and its selection mechanisms apply to the phenotype-like entities that the bitstrings encode. In biology, the genotype-phenotype difference is vast: the genotype is embodied in the chromosomes whereas the phenotype is the whole organism that expresses the chromosomal information. The complex decoding process that leads from one to the other is called biological development and is essential if the genotype is to be evaluated by the environment. Thus to apply the genetic algorithm to natural evolution calls for a representational scheme that both permits application of the algorithm's operators to the genotype and also defines how, based on the genotype, organisms are to be "grown," i.e., their development.

The present paper outlines a few steps in the direction of such a representation,[7] and describes how the genetic algorithm would be applied to populations of the resulting organisms. The representation problem is addressed at the level of cells, which are treated as "black boxes" having evolvable properties. Beginning with the fertilized egg, the cells are to divide, move, and differentiate under the control of rules so as eventually to form a mature organism. An attempt is made to respect major facts known about cells and these processes, but compromises must occur at this point in the effort to approach algorithmic workability. The principal objective is to describe a representational framework—a sort of "developmental automaton"—sufficiently completely that randomly generated instances will grow and can be evolved under the genetic algorithm in computer experiments.

2. SIMULATED EVOLUTION

The problem of simulating the evolution of multi-cellular organisms can be divided into four parts: plan, expression, selection, and variation. Plan and expression concern the development of individual organisms from their eggs. Selection and variation are processes applying to populations of individuals.

2.1 PLAN

In nature, the genotype contains (1) information that is descriptive, through the action of development and the environment, of a range of possible phenotypes, and (2) information encoding the developmental process itself, i.e., how to go about making a phenotype from a genotype. Both kinds of information are of course inherited and subject to variation and natural selection. Here, for simplicity, it will be assumed that only the first kind of information, termed the organism's *plan*, is heritable and subject to the genetic algorithm. The other kind, the rules for *expressing* the plan to form the phenotype, will be regarded as fixed.

What should the plan look like? Several observations on natural systems are suggestive.[1,10] In the first place, though individual cells can have different sizes and can change in size, growth occurs primarily through cell division: one cell becomes two "daughter" cells. Second, depending on the situation, the daughters can be phenotypically the same as the parent, they can differ from the parent but not differ from each other, or they can differ from the parent and from each other. Third, the phenotypical outcome of cell division can depend not only on the nature of the parent cell, but also on factors related to the cellular, chemical, or physical context in which the parent cell is embedded. Finally—and pivotal for this discussion— all cells in an organism are considered to contain the *same* genetic information, though some of it may become in some sense "switched off" or inoperative during differentiation.

These observations have suggested the following working proposal. The plan will take the form of a so-called *production system program* (PSP) consisting of a finite number of production (condition-action) rules which will be termed *growth rules*. The growth rules have the general form

$$X + K_i \Rightarrow K_j K_k \,.$$

The K's stand for cell phenotypes and X represents the local context; the symbol "$+$" means conjunction. In addition, each growth rule has associated with it a weight w. Every cell in the organism contains the same set of rules, or PSP.

Focussing attention on a particular rule in the PSP of a particular cell, the condition side of the rule is satisfied if that cell is (phenotypically) of type K_i and the context matches X. The action recommended by the rule is to replace the cell by two new cells, one phenotypically of type K_j, the other of type K_k. Whether or not this rule controls the parent cell's fate depends on whether the rule is selected for expression, as discussed in the next section.

The general growth rule form is open to many special cases. As in nature, the daughter cells may or may not be the same as the parent or each other. Furthermore, some rules may contain just one daughter cell, identical to the parent; such a rule, if expressed, means that cell division *does not* take place. Also, some rules may have no cell in their action parts, corresponding to dissolution of the parent cell.

Some rules may have no term corresponding to X; their condition is satisfied independent of context. In the other rules, X can take on several forms. Most simply, X can stand for the presence of a cell of a particular kind adjacent to K_i. In this case ("adjacency" type context), the spatial relation of the X cell and K_i may affect the spatial relation of the daughter cells (if there are two). Another kind of X ("signal" type) would stand for a detector for signals emitted by other cells, not necessarily in the immediate neighborhood. For present purposes, the "signal" emitted by a cell is simply a list of its phenotypical properties. The predominant direction from which matched signals are received could affect the daughter cells' spatial relation. Still another kind of X would detect aspects of the physical environment such as intercell pressure.

2.2 EXPRESSION

Since all cells contain the same "program," differential development of the system depends on the selection for expression of different rules in different cells. This is not difficult in principle, since once some differentiation occurs, the sensitivity of the rules to cell type and context will lead to further differentiation. The proposed expression mechanism consists of a *match* step and a *decision* step. Again focussing attention on a particular cell, in the match step the cell first identifies those program rules which have satisfied conditions. Then, from this *match set*, the cell chooses a single rule for expression. The chosen rule "carries out its right-hand side," i.e., daughter cells are produced as prescribed and their signals are emitted.

The system's growth process is envisioned as a series of discrete time-steps. On each step, the expression mechanism operates in every cell of the current system. The operation is regarded as "parallel" in the sense that offspring of all the cells are produced simultaneously. The offspring cells then undergo, in accordance with their phenotypical properties, a process of interaction and spatial accommodation so as to form the "new" system to be input to the expression mechanism in the next time-step.

2.2.1 **THE DECISION STEP** The decision step of the expression mechanism makes use of the growth rule weights w and the effect of signals from nearby cells. Each growth rule in the match set has an associated weight w. If a rule's context (X) part is either absent or is of adjacency type, its *excitation* is defined to be just w. However, if a rule's context part is of signal type, the rule's excitation is defined to be the product of the weight w and the *intensity* of the received context signal. For example, suppose that a certain rule has an X which matches signals S_A emitted by nearby cells A. Suppose further that the total intensity of the signals is simply their number n_A times a constant f. Then the excitation of the rule in question would equal $fn_A w$.

The cell decides which match set rule to express using a probability distribution over the rules' excitations. That is, the probability that a particular rule will be picked is equal to its excitation divided by the sum of the excitations of the rules in the match set. The following three rules offer an interesting example.

$$A \Rightarrow A\ A \qquad w_r$$
$$(S_A) + A \Rightarrow A \qquad w_i$$
$$(S_A) + A \Rightarrow 0 \qquad w_d$$

The first rule, termed "reproductive," takes one cell A and leaves two in its place. The second rule, termed "inhibitory," matches cell A, senses the presence of at least one A-type signal in the vicinity, and seeks, if chosen, to maintain the status quo exactly. The third rule, a deletion rule, has the same condition as the inhibitory rule, but seeks to delete the matched A cell. Each rule has a weight, as shown.

Suppose now the system consists of an aggregate of n cells of type A. In any cell, the excitations of the three rules will be:

$$e_r = w_r$$
$$e_i = w_i f n_A$$
$$e_d = w_d f n_A .$$

If w_r is large and there are relatively few cells, the reproductive rule will be chosen most of the time and the aggregate will grow. As it does, however, the excitations of the inhibitory and deletion rules will increase relative to that of the reproductive rule, due to n_A. The growth rate will slow down. Eventually, an equilibrium will be reached where net growth is zero. At that point, the probability of reproduction equals the probability of deletion, or $w_r = w_d f n_A$. Solving for n_A yields the system's equilibrium size:

$$n_A^* = (1/f)(w_r/w_d) .$$

The system's net growth rate dn/dt prior to equilibrium can be calculated by taking the product of n and the difference between the probabilities of reproduction and deletion. Dropping the "A" subscripts, the result is

$$\frac{dn}{dt} = n\frac{(1 - n/n^*)}{1 + (w_i/w_d + 1)(n/n^*)} ,$$

showing that the system's growth rate can be "chosen" independently of its equilibrium size.

Though simple, the example is important because it illustrates one way in which the cellular program can manage the fundamental problem of bounded growth. Later examples of differentiation into finite regions of homogeneous cell type will assume the presence of growth rule sets of this or similar sort for the regions.

2.2.2 PHENOTYPE PROPERTIES Once the decision step has picked a rule for expression, the daughter cells in the action part must be simulated, which means simulating their properties. In a real organism, each cell "type" has a myriad of physical and biochemical properties. Some of these may be more properly regarded as behavioral, e.g., during development, cells can creep, amoeba-like to new positions. Most of the properties affect in one way or another a cell's interactions with other cells. Even if all the properties were understood, a realistic simulation would still have an enormous problem adequately representing and computing the interactions within the cell aggregate. Such a computation is necessary in order to determine the fitness, with respect to an environment, of the organism as a whole. The practical course for the present would seem to be to choose extremely simple environments, simple measures of fitness, and a very restricted range of cell properties.

2.3 SELECTION AND VARIATION

Because the foregoing representational framework for development takes the form of a production system program, it is straightforward to apply the genetic algorithm as the "engine" of phenotype selection and genotype variation. The application of the algorithm would be along the lines of previous work with production system programs.[8,9] One would start with a population of "egg" cells, each containing a random genotype. Each egg would undergo development and, after a standard number of time-steps, each resulting cell aggregate would be rated for fitness. The original eggs would then be copied in numbers proportional to these fitnesses to form a new population of the same size. Genetic operators would be applied to the genotypes of the new population. The cycle would be iterated through some number of generations, corresponding to evolution.

Many aspects of this scheme are quite well understood due to the research just cited and on genetic algorithms in general. However, the form of the growth rules in the genotype is somewhat unusual so some comments about coding are in order. The basic encoding would resemble that of classifiers.[6,11] The condition part of a rule would consist of a context taxon (for X) and a cell taxon (for K_i), each being a string of length L from $\{1, 0, \#\}$. The action part would consist of two *cell descriptions* (for K_j and K_k), both strings of length L from $\{1, 0\}$.

An *interpreter* is required to relate cell description encodings to phenotypical properties. This simply means establishing a pre-defined mapping between substrings in the cell description and properties, e.g., "110" in the 14th through 16th positions could mean the cell surface has "high stickiness," etc. To take care of rules in which one or both of the daughter cells is absent, the interpreter would simply check the setting a certain bit in each cell description: "0," say, would mean that cell was absent and the rest of the description should be ignored. A similar system would be used to indicate the presence or absence of a context taxon and its type (adjacency, signal, or other).

A growth rule's condition would be satisfied if both (1) the cell description of the cell in which the rule finds itself matches the rule's cell taxon, and (2) at least one signal reaching the cell matches the rule's context taxon. The meaning of "match" is the same as for classifiers: the two strings must be the same at every non-$\#$ position of the taxon. The use of the "don't care" symbol $\#$ permits rule conditions to restrict their sensitivity to particular subsets of cell description and signal bits.

Calculation of the intensity of the signal matching the context taxon can be quite complex, depending on the simulation. Involved are the dependence of individual signal intensities on the distance from their sources, and also perhaps propagation delays with respect to the time-step of creation of the source cell. These factors must be predefined. In any case the total received intensity would be a sum over the individual intensities of all matched signals. As noted earlier, the net *direction* of the received signal may in some rules determine the spatial orientation of the daughter cells. The dependence would be encoded in a special bit string associated with the daughter cell descriptions.

The weight associated with a growth rule must also be encoded in order to make it, and consequently the rule's influence in the decision step, subject to the genetic algorithm. The weight would simply be concatenated, as a fixed-length binary number, with the rest of the rule string.

3. 1-D DEVELOPMENT

As has been the case with research on cellular automata,[12] the complexity of realistic three-dimensional simulations recommends initial study of one-dimensional examples. In two and three dimensions, forces between cells must lead to complicated cell movements and contortions of the "tissue." A 1-D "creature," however, could be viewed as growing inside a frictionless tube, with no forces except between adjacent cells. Cell division would lengthen the creature; deletion would shorten it. Though simple, the 1-D case can exhibit cell type configuration patterns such as symmetry, periodicity, and polarity that are analogous to patterns emerging in the development of real organisms. Some elementary examples follow.

3.1 SYMMETRY AND PERIODICITY

Changes in a 1-D system through time can be represented by a pyramid like the following:

$$A$$
$$B \ B$$
$$D \ C \ C \ D$$

This shows three time-steps. At first, the system consists just of cell A; then A divides to form cells B and B; then the left-hand B divides to form the (oriented) pair $D \ C$, and the right-hand B yields the pair $C \ D$. Only two growth rules are required:

$$A \Rightarrow B \ B$$
$$(B) + B \Rightarrow C \ D.$$

In the second rule the context taxon is of adjacency type (indicated by the absence of "S"). This type of rule reads: "order the output cell so that the direction from the first one to the second one is the same as the direction from the context cell to the replaced cell."

Note that the pyramid diagram shows bilateral symmetry about its center line. Using additional rule sets of the self-limiting form discussed in Section 2.2.1, the C's and D's could be multiplied to yield eventually a stable symmetrical creature of finite size, $D \ldots D \ C \ldots C \ D \ldots D$, with approximately equal groups of D cells.

The following pyramid and its rules illustrates rudimentary periodicity:

$$
\begin{array}{cccc}
A & & \\
E & F & \\
C & D & C & D
\end{array}
\qquad\qquad
\begin{array}{l}
A \Rightarrow E \quad F \\
(F) + E \Rightarrow D \quad C \\
(E) + F \Rightarrow C \quad D
\end{array}
$$

Again, the addition of self-limiting rule sets would result in the creature, $C \ldots C$ $D \ldots D$ $C \ldots C$ $D \ldots D$, in which like cell groups were approximately equal in size. It is clear that quite complex structures can be built by first establishing the pattern with non-cyclic rules (in which the cell taxon will not match the output cells), and then using self-limiting rule sets which apply to the final cell types.

3.2 POLARITY

An elementary polarity results from any rule in which the output cell types differ. A polarity with respect to some phenotypical *property* can be set up with non-cyclic rules as follows:

$$
\begin{array}{cccc}
A & & \\
B & C & \\
D & E & F & G
\end{array}
\qquad\qquad
\begin{array}{l}
A \Rightarrow B \quad C \\
(C) + B \Rightarrow E \quad D \\
(B) + C \Rightarrow F \quad G
\end{array}
$$

If in the cell descriptions of D, E, F, and G the property is, say, monotonically increasing, the amount of the property will be graded across the system. A more sophisticated gradient system occurs under the rules:

$$
\begin{aligned}
A &\Rightarrow B \quad C \\
(S_B) + C &\Rightarrow D \quad C \\
C &\Rightarrow C \quad C \\
(S_C) + C &\Rightarrow C \\
(S_C) + C &\Rightarrow 0 \,.
\end{aligned}
$$

If B's signal loses intensity with distance, the probability that a C will change to D C will fall with distance from the left end of the structure. The result will be a decreasing distribution of D's from left to right. The last three rules are intended to control the system's overall size.

When rule sets become even slightly complicated, as in the last example, it is evident that development will be difficult to predict. It can be hoped, however, that with the help of the genetic algorithm, the ability to design and analyze organisms in advance will not be necessary in order to build successful and interesting ones—just as in natural evolution it is not. What does seem essential is an adequate space of possible growth rules. The rule forms discussed include self-excitation, self-inhibition, and cross-excitation and -inhibition between different cell types. The repertoire seems fairly complete for a start, but modifications in it and in many other aspects will surely occur as the proposal is studied experimentally and analytically.

4. CONCLUSION

An extremely schematic representational framework for biological development has been described which may permit simulations of evolution using the genetic algorithm. Major questions that need to be addressed include the accuracy and adequacy of the representation and the problem of computing the phenotype. It is hoped that coupling "developmental automata" with genetic adaptive techniques will yield insights into biological, social, and other systems which are capable of growth.

ACKNOWLEDGMENTS

The author thanks D.E. Goldberg for valuable comments on an earlier draft of the paper.

REFERENCES

1. Balinsky, B. I. (1982), "Development, Animal," *Encyclopaedia Britanica* **5**, 15th edition, 625–642.
2. Grefenstette, J. J., ed. (1985), *Proceedings of the First International Conference on Genetic Algorithms and Their Applications* (Hillsdale, New Jersey: Lawrence Erlbaum Associates).
3. Grefenstette, J. J. ed. (1987), *Genetic Algorithms and Their Applications: Proceedings of the Second International Conference on Genetic Algorithms* (Hillsdale, New Jersey: Lawrence Erlbaum Associates).
4. Holland, J. H. (1975), *Adaptation in Natural and Artificial systems* (Ann Arbor: University of Michigan Press).
5. Holland, J. H. (1976), "Studies of the Spontaneous Emergence of Self-Replicating Systems using Cellular Automata and Formal Grammars," *Automata, Languages, Development*, Eds. A. Lindenmayer and G. Rozenberg (Amsterdam: North-Holland), 385–404. [This important paper used mechanisms closely related to those of the genetic algorithm as defined in Holland (1975) to study the emergence of self-replication.]
6. Holland, J. H. (1986), "Escaping Brittleness: the Possibilities of General-Purpose Learning Algorithms Applied to Parallel Rule-Based Systems," *Machine Learning, An Artificial Intelligence Approach*, Eds. R. S. Michalski, J. G. Carbonell and T. M. Mitchell (Los Altos, California: Morgan Kaufmann), vol. 1, 593–623.
7. Lindenmayer, A. (1975), "Developmental Algorithms for Multicellular Organisms: a Survey of L-Systems," *J. Theor. Biol.* **54**, 3–22. [The present representation appears to share several aspects with the developmental models called L-systems. However, there do not seem to be any studies in which populations of L-systems undergo evolution.]
8. Schaffer, J. D. (1985), "Learning Multiclass Pattern Discrimination," *Proceedings of an International Conference on Genetic Algorithms and Their Applications*, Ed. J. J. Grefenstette (Pittsburgh: Carnegie-Mellon University), 74–79.
9. Smith, S. (1980), *A Learning System Based on Genetic Algorithms, Ph.D. Dissertation (Computer Science), University of Pittsburgh.*
10. Waddington, C. H. (1982), "Development, Biological," *Encyclopaedia Britanica* **5**, 15th Ed., 643–650.
11. Wilson, S. W. (1987), "Classifier Systems and the Animat Problem," *Machine Learning* **2**, 199–228.
12. Wolfram, S. (1984), "Cellular Automata as Models of Complexity," *Nature* **311**, 419–424.

Hans Moravec
Robotics Institute, Carnegie-Mellon University, Pittsburgh, PA 15213

Human Culture:
A Genetic Takeover Underway

This is the end. Our genes, engaged for four billion years in a relentless, spiralling arms race with one other, have finally outsmarted themselves. They've produced a weapon so powerful it will vanquish the losers and winners alike. I do not mean nuclear devices—*their* widespread use would merely delay the immensely more interesting demise that's been engineered.

You may be surprised to encounter an author who cheerfully concludes the human race is in its last century, and goes on to suggest how to help the process along. Surely, though, the surprise is more in the timing than in the fact itself. The evolution of species is a firmly established idea, and the accelerating pace of cultural change has been a daily reality for a century. During those hundred years many projections of future life, serious and fictional, have been published. Most past futurism has kept separate the changes anticipated in the external world, and those expected in our bodies and minds. While our environment and our machinery could be rapidly engineered through industrious invention, alterations in ourselves were paced by the much slower Darwinian processes of mutation and selection.

In the late twentieth century, the barriers of complexity that divided the engineers of inanimate matter from the breeders of living things have been crumbling. In the future presented in this chapter, the human race itself is swept away by the tide of cultural change, not to oblivion, but to a future that, from our vantage point, is best described by the word "supernatural." Though the ultimate consequences

Artificial Life, SFI Studies in the Sciences of Complexity,
Ed. C. Langton, Addison-Wesley Publishing Company, 1988

are unimaginable, the process itself is quite palpable, and many of the intermediate steps are predictable. This chapter reflects that progression—from uncontroversial history of relevant technologies, to modest near-term projections, to speculative glimpses of the distant future (discerning the fuzzy boundaries between them is up to the reader). The underlying theme is the maturation of our machines from the simple devices they still are, to entities as complex as ourselves, to something transcending everything we know, in whom we can take pride when they refer to themselves as our descendants.

As humans, we are half-breeds: part nature, part nurture. The cultural half is built, and depends for its existence on the biological foundation. But there is a tension between the two. Often expressed as the drag of the flesh on the spirit, the problem is that cultural development proceeds much faster than biological evolution. Many of our fleshly traits are out of step with the inventions of our minds. Yet machines, as purely cultural entities, do not share this dilemma of the human condition. Unfettered, they are visibly overtaking us. Sooner or later they will be able to manage their own design and construction, freeing them from the last vestiges of their biological scaffolding, the society of flesh and blood humans that gave them them birth. There may be ways for human minds to share in this emancipation.

Free of the arbitrary limits of our biological evolution, the children of our minds yet will be constrained by the physics and the logic of the universe. Present knowledge hints at the motives, shapes and effects of post-biological life, but there may be ways to transcend even the most apparently fundamental barriers.

THE SLIPPERY SLOPE TO GENETIC TAKEOVER

The trouble began about a 100 million years ago when some gene lines hit upon a way to make animals with the ability to learn behaviors from their elders during life, rather than inheriting them at conception. It was accelerated 10 million years ago when our ancestors began to rely on tools like bones and sticks and stones. It was massively compounded with the coming of fire and complex languages, perhaps 1 million years ago. By the time our species appeared, maybe 100 thousand years ago, the genes' job was done; cultural evolution, the juggernaut that they had unwittingly constructed, was rolling. Within the last ten thousand years, human culture produced the agricultural revolution and subsequently large-scale bureaucratic government, written language, taxes, and leisure classes. In the last thousand years, this change blossomed into a host of inventions such as movable-type printing that accelerated the process. With the industrial revolution two hundred years ago, we entered the final phase. Bit by bit, ever more rapidly, cultural evolution discovered economically attractive artificial substitutes for human body functions, as well as totally new abilities. One hundred years ago we invented practical calculating machines that could duplicate some small, but vexing, functions of the human mind.

Since then the mental power of calculating devices has risen a thousandfold every twenty years.

We are very near to the time when *no* essential human function will lack an artificial counterpart. The embodiment of this convergence of cultural developments is the intelligent robot, a machine that can think and act as a human, however inhuman it may be in physical or mental detail. Such machines could carry on our cultural evolution, including their own increasingly rapid self-improvement, without us, and without the genes that built us. It will be then that our DNA will be out of a job, having passed the torch, and lost the race, to a new kind of competition. The genetic information carrier, in the new scheme of things, will be exclusively knowledge, passed from mind to artificial mind.

A. G. Cairns-Smith, a chemist contemplating the beginnings of life on the early earth, calls this kind of internal coup a *genetic takeover*. He suggests that it has happened at least once before. In Cairns-Smith's convincingly argued theory,[1] the first organisms were microscopic crystals of clay that reproduced by the common processes of crystal growth and fracture, and carried genetic information as patterns of crystal defects. These defects influence the physical properties of a clay, and its action as a chemical catalyst, and so partially control that clay's immediate surroundings. In a Darwinian process of reproduction, mutation and selection, some crystal species stumbled on a way to harness nearby carbon compounds as construction materials and machinery, and even as *external repositories for genetic information*. The carbon machinery was so effective that organisms using it to ever greater extent won out, resulting eventually in carbon based organisms with no vestiges of the original crystalline genetics. Life as we know it had begun.

How should you and I, products of both an organic and a cultural heritage, feel about the coming rift between the two? We owe our existence to organic evolution, but do we owe it any loyalty? Our minds and genes share many common goals during life, but even then there is a tension between time and energy spent acquiring, developing, and spreading ideas, and effort expended towards biological reproduction (as any parent of teenagers will attest). As death nears, the dichotomy widens; too many aspects of mental existence simply cannot be passed on. The problem is partly one of timescales: humans already live extraordinarily long compared to other animals, no doubt to better teach their young, but the lifespan is a compromise with the genes' evolutionary imperative to experiment, the better to adapt. Things are a little askew because this deal was forged long ago, when the cultural life was simpler. The amount to teach and learn has ballooned recently and, all other things being equal, we'd likely be better off with a somewhat longer lifespan. But what would be the optimal lifespan if our genes' specialized needs were no longer a factor?

A sexually produced body is a finalized evolutionary experiment. Further genetic adaptation is precluded until offspring are produced through a genetic bottleneck, and then the experiment is over. A mind, however, is a conduit for ideas, and can evolve and adapt without such abrupt beginnings and endings. In principle it could cope successfully indefinitely. It is true that human minds, tuned for mortality, undergo a maturation from impressionable plasticity to self-assured rigidity, and this makes them unpromising material for immortality. But there are adaptable entities on earth with indefinite life spans: living species and some human institutions. Their secret is a balance between continuity and experimentation. Death of individual organisms plays a central role in successful species. Old experiments are cleared away, making room for new ones, in a genteel, prearranged way, or by relentless life-and-death competitions. In human institutions turnover in skilled personnel and alteration of the company rules play the same role. The point is that the larger unit, the species or the organization, can adapt indefinitely (perhaps beyond recognition in the long run) without losing its identity, as its design and components are altered bit by bit.

A thinking machine could probably be designed from the ground up to have this same kind of flexibility. Mental genes could be created, imported, tested in combinations, and added and deleted to keep the thinking current. The testing is of central importance: it steers the evolution. If the machine makes too many bad decisions in these tests, it will fail totally, in the old fashioned, Darwinian way.

And so the world of the children of our minds will be as different from our own as the world of living things is different from the lifelessness that preceded it. The consequences of unfettered thought are quite unimaginable. We're going to try to imagine some of them anyway.

MACHINES WHO THINK (WEAKLY)

Later I will argue that robots with human intelligence will be common within fifty years. By comparison, the best of today's machines have minds more like those of insects. This in itself is a recent giant leap from far more modest beginnings. While mechanical imitations of life have been with us for at least several hundred years, the earliest machines, powered by running water, falling weights, or springs copied the motions of living things, often charmingly, but could not respond to the world around them. They could only *act*. The development of electrical, electronic and radio technology early in this century made possible machines that reacted to light, sound, and other subtle cues, and also provided a means of invisible remote control. These possibilities inspired a number of entertaining demonstration robots, as well as thoughts and stories about future human-like mechanisms, but only simple connections between the sensors and motors were possible at first. These machines could *sense* and *act*, but hardly think.

Analog computers were designed during World War II for controlling anti-aircraft guns, for navigation, and for precision bombing. Some of their developers noticed a similarity between the operation of the devices and the regulatory systems in living things, and these researchers were inspired to build machines that acted as if they were alive. Norbert Wiener of the Massachusetts Institute of Technology (MIT) coined the term "cybernetics" for this unified study of control and communication in animals and machines. Its practitioners combined new theory on feedback regulation with post-war electronics and early knowledge of living nervous systems to build machines that responded like simple animals, and were able to learn. The rudiments of *thought* had arrived.

The field thrived less than two decades. Among its highlights was a series of electronic turtles built during the 1950's by W. Grey Walter, a British psychologist. With subminiature tube electronic brains, and rotating phototube eyes, microphone ears and contact switch feelers, the first versions could locate their "recharging hutch" when their batteries ran low, and otherwise avoid trouble while wandering about. Groups of them exhibited complex social behavior by responding to each other's control lights and touches. A later machine with the same senses, could be conditioned to associate one stimulus with another, and could learn, by repeated experience, that, for instance, a loud noise would be followed by a kick to its shell. Once educated, the turtle would avoid a noise as it had before responded to a kick. The associations were slowly accumulated as electrical charges in capacitors.

The swan song of the cybernetics effort may have been the Johns Hopkins University "Beast." Built by a group of brain researchers in the early 1960's, it wandered the halls, guided by sonar and a specialized photocell eye that searched for the distinctive black cover plate of wall outlets, where it would plug itself in, to feed. It inspired a number of imitators. Some used special circuits connected to TV cameras instead of photocells, and were controlled by assemblies of (then new) transistor digital logic gates. Some added new motions such as "shake to untangle arm" to the repertoire of basic actions.

Cybernetics was laid low by a relative. The war's many small analog computers, which had inspired cybernetics, had a few, much larger, digital cousins. The first automatic digital computers, giant autonomous calculators, were completed toward the end of the war and used for code-breaking, calculating artillery tables, and atomic bomb design. Less belligerently, they provided unprecedented opportunities for experiments in complexity, and raised the hope in some pioneers like Alan Turing and John von Neumann that the ability to think rationally, our most unique asset in dealing with the world, could be captured in a machine. Our minds might be amplified just as our muscles had been by the energy machines of the industrial revolution. Programs to reason and to play intellectual games like chess were designed, for instance by Claude Shannon and by Turing in 1950, but the earliest computers were too puny and too expensive for this kind of use. A few poor checker-playing programs did appear on the first commercial machines in the early 1950's, and equally poor chess programs showed up in latter half of that decade, along with a better checker player. In 1957 Allen Newell et al.[11] demonstrated the first program able to reason about arbitrary matters, by starting with axioms and applying rules of inference to prove theorems.

In 1960 John McCarthy coined the term "Artificial Intelligence" for the effort to make computers think. By 1965 the first students of McCarthy, Marvin Minsky, Newell, and Simon had produced programs that proved theorems in geometry, solved problems from intelligence tests, algebra books, and calculus exams, and they played chess all with the proficiency of an average college freshman. Each program could handle only one narrow problem type, but for first efforts they were very encouraging—so encouraging that most involved felt that another decade of progress would surely produce a genuinely intelligent machine. (For an explanation of the nature of their understandable miscalculation, see *Mind Children*.[5])

Now, thirty years later, computers are thousands of times as powerful, but they don't seem much smarter. In the past three decades, progress in artificial intelligence has slowed from the heady sprint of a handful of enthusiasts to the plodding trudge of growing throngs of workers. Even so, modest successes have maintained flickering hope. So-called "expert systems," programs encoding the decision rules of human experts in narrow domains such as diagnosis of infections, factory scheduling, or computer system configuration, are earning their keep in the business world. A fifteen-year effort at MIT has gathered knowledge about algebra, trigonometry, calculus, and related fields into a program called MACSYMA; this wonderful program manipulates symbolic formulas and helps to solve otherwise forbidding problems. Several chess playing programs are now officially rated as chess masters, and excellent performance has been achieved in other games like backgammon. Other semi-intelligent programs can understand simplified typewritten English about restricted subjects, make elementary deductions in the course of answering questions, and interpret spoken commands chosen from thousand-word repertoires. Some can do simple visual inspection tasks, such as deciding whether a part is in its desired location.

Unfortunately for human-like robots, computers are at their worst trying to do the things most natural to humans, like seeing, hearing, manipulating, language, and common sense. This dichotomy—machines doing well things humans find hard, while doing poorly what's easy for us—is a giant clue to the nature of the intelligent machine problem.

MACHINES WHO SEE (DIMLY) AND ACT (CLUMSILY)

In the mid-1960's Minsky's students at MIT began to connect television camera eyes and mechanical robot arms to their computers, giving eyes and hands to computer minds, for machines that could see, plan, and act. By 1965 they had created programs that could find and remove children's blocks, painted white, from a black tabletop. This was a difficult and impressive accomplishment, requiring a controlling program as complex as any of the then current pure-reasoning programs. Yet, while the reasoning programs, unencumbered by robot appendages, matched college freshmen in fields like calculus, Minsky's hand-eye system could be bested by a toddler. Nevertheless, hand-eye experiments continued at MIT and elsewhere, gradually developing the field which now goes by the name "robotics," a term coined in science fiction stories by Isaac Asimov. As with mainstream artificial intelligence programs, robotics has progressed at an agonizingly slow rate over the last twenty years.

Not all robots, nor all people, idle away their lives in universities. Many must work for a living. Even before the industrial revolution, before any kind of thought was mechanized, partially automatic machinery, powered by wind or flowing water, was put to work grinding grain and cutting lumber. The beginnings of the industrial revolution in the eighteenth century were marked by the invention of a plethora of devices that could substitute for manual labor in a powerful, precise, and thoroughly inhuman way. Powered by turning shafts driven by water or steam, these machines pumped, pounded, cut, spun, wove, stamped, moved materials and parts and much else, consistently and tirelessly. Once in a while something ingeniously different appeared: the Jacquard loom, invented in 1801, could weave intricate tapestries specified by a string of punched cards (a human operator provided power and the routine motions of the weaving shuttle). By the early twentieth century, electronics had given the machinery limited senses; it could now stop when something went wrong, or control the temperature, thickness, even consistency, of its workpieces. Still, each machine did one job and one job only. This meant that, as technical developments occurred with increasing rapidity, the product produced by the machine often became obsolete before the machine had paid back its design and construction costs, a problem which had become particularly acute by the end of World War II.

In 1954 the inventor George Devol filed a patent for a new kind of industrial machine, the programmable robot arm, whose movements would be controlled by a stream of punched cards, and whose task thus could be altered simply by changing its program cards. In 1958, with Joseph Engelberger, Devol founded a company named Unimation (a contraction of "universal" and "automation") to build such machines. The punched cards soon gave way to a magnetic memory, thereby allowing the robot to be programmed simply by leading it by the hand through its required paces once. The first industrial robot began work in a General Motors plant in 1961. To this day most large robots seen welding, spray painting, and moving pieces of cars are still of this type.

Only when the cost of small computers dropped to less than $10,000 did robotics research conducted in universities begin to influence the robot industry. The first industrial vision systems, usually coupled with a new class of small robot arms, appeared in the late 1970's, and now play a modest, but quietly booming, role in the assembly and inspection of small devices like calculators, printed circuit boards, and automobile water pumps. Indeed, industrial needs have strongly influenced university research. What was once a negligible number of smart robot projects has swelled to the hundreds. And while cybernetics may be relatively dormant, its stodgy parent, control theory, has grown massively since the war to meet the profitable needs of the aerospace industry; moreover, the applications developed for controlling air- and spacecraft and weapons are once again finding their way into robots. The goal of human-like performance, though highly diluted by a myriad of approaches and short-term goals, has acquired a relentless, Darwinian vigor. As a story, it becomes bewildering in its diversity and interrelatedness. Let's move on to the sparser world of robots that rove.

MACHINES WHO EXPLORE (HALTINGLY)

In the next section I will try to convince you that mobility is a key to developing fully intelligent machines, an argument that begins with the observation that *reasoning*, as such, is only the thinnest veneer of human thought, effective only because it is supported by much older and much more powerful and diverse unconscious mental machinery. This opinion may have been common among the cybernetics researchers, many of whose self-contained experiments were animal-like and mobile. It is not yet widespread in the artificial intelligence research community, where experiments are typically encumbered by huge, immobile mainframe computers, and dedicated to mechanizing pure reasoning. Nevertheless, a small number of mobile robots have appeared in the artificial intelligence laboratories.

Stanford Research Institute's "Shakey," was a mobile robot built by the researchers who believed that reasoning was the essence of intelligence, and in 1970 it was the first mobile robot to be controlled by programs that reasoned. Five feet tall, equipped with a television camera, it was remote controlled by a large computer. Inspired by the first wave of successes in artificial intelligence (AI) research, its designers sought to apply logic-based problem-solving methods to a real world task. Controlling the movement of the robot, and interpreting its sensory data, were treated as secondary tasks and relegated to junior programmers. MIT's "blocks world" vision methods were used, and a robot environment was constructed in which the robot moved through several rooms bounded by clean walls, seeing, and sometimes pushing, large, uniformly painted blocks and wedges. Shakey's most impressive performance, executed piecemeal over a period of days, was to solve a so-called "monkey and bananas" problem. Told to push a particular block that happened to be resting on a larger one, the robot constructed and acted on a plan that included finding a wedge that could serve as a ramp, pushing it against the large block, driving up the ramp, and delivering the requested push.

The environment was contrived, and the problem staged, but it provided a motivation, and a test, for a clever reasoning program called STRIPS (the STanford Research Institute Problem Solver) that, given a task for the robot, assembled a plan out of the little actions the robot could take. Each little action had preconditions (e.g., to push a block, it must be in front of us) and probable consequences (e.g., after we push a block, it is moved). The state of the robot's world was represented in sentences of mathematical logic, and formulating a plan was like proving a theorem, with the initial state of the world being the axioms, and primitive actions being the rules of inference. One complication was immediately evident: the outcome of a primitive action is not always what one expects (as, for instance, when the block does not budge). Shakey had a limited ability to handle such glitches by occasionally observing parts of the world, and adjusting its internal description and replanning its actions if the conditions were not as it had assumed.

Shakey's specialty was *reasoning*—its rudimentary vision and motion software worked only in starkly simple surroundings. At about the same time, on a much lower budget, a mobile robot that was to specialize in *seeing* and *moving* in natural settings was born at Stanford University's Artificial Intelligence Project. John McCarthy founded the Project in 1963 with the then plausible goal of building a fully intelligent machine in a decade. (The Project was renamed the Stanford Artificial Intelligence Laboratory, or SAIL, as the decade drew nigh and plausibility drifted away.) Reflecting the priorities of early artificial intelligence (AI) research, McCarthy worked on reasoning, and delegated to others the design of ears, eyes, and hands for the anticipated artificial mind. SAIL's hand-eye group soon overtook the MIT robotics group in visible results, and was seminal in the later industrial smart robot explosion. A modest investment in mobility was added when Les Earnest, SAIL's technically astute chief administrator, learned of a vehicle abandoned by Stanford's mechanical engineering department after a short stint as a simulated remote controlled lunar rover. At SAIL it became the Stanford Cart, the first mobile robot controlled by a large computer that did *not* reason, and the first testbed for computer vision in the cluttered, haphazardly illuminated world most animals inhabit. The progeny of two Ph.D. theses, it slowly navigated raw indoor and outdoor spaces guided by TV images processed by programs quite different from those in the blocks world.

In the mid 1970's NASA began planning for a robot Mars mission to follow the successful Viking landings. Scheduled for launch in 1984, it was to include two vehicles roving the Martian surface. Mars is so far away, even by radio, that simple remote control was unattractive; the delay between sending a command and seeing its consequence could be as long as forty minutes. Much greater distances would be possible if the robot could travel safely on its own much of the time. Toward this end, Caltech's Jet Propulsion Laboratory, designer of most of NASA's robot spacecraft, which until then used quite safe and simple automation, initiated an intelligent robotics project. Pulling together methods, hardware, and people from university robotics programs, it built a large wheeled test platform called the Robotics Research Vehicle, or RRV, a contraption that carried cameras, a laser rangefinder, a robot arm, and a full electronics rack, all connected by a long cable to a big computer. By 1977 it could struggle through short stretches of rock-littered parking lot to pick up a certain rock and rotate it for the cameras. But in 1978 the project was halted when the Mars 1984 mission was cancelled and removed from NASA's budget. (Of course, Mars hasn't gone away, and the JPL is considering a visit there at the end of the millenium.)

The best supporter of artificial intelligence research is the Department of Defense's Advanced Research Project Agency (DARPA). Founded after the 1957 humiliation of Sputnik to fund far out projects as insurance against future unwelcome technological surprises, it became the world's first government agency to foster AI investigations. In 1981 managers in DARPA decided that robot navigation was sufficiently advanced to warrant a major effort to develop autonomous vehicles able to travel large distances overland without a human operator, perhaps into war zones or other hazardous areas. The number of mobile robot projects jumped dizzyingly, in universities and at defense contractors, as funding for this project materialized.

Even now, several new, truck-sized, robots are negotiating test roads around the country—and the dust is still settling.

On a more workaday level, it is not a trivial matter that fixed robot arms in factories must have their work delivered to them. An assembly-line conveyor belt is one solution, but managers of increasingly automated factories in the late 1970's and early 1980's found belts, whose routes are difficult to change, too restrictive. Their robots could be rapidly reprogrammed for different jobs, but the material flow routes could not. Several large companies worldwide dealt with the problem by building what they called Automatically Guided Vehicles, AGVs, that navigated by sensing signals transmitted by wires buried along their route. Looking like fork lifts or large bumper cars, they can be programmed to travel from place to place and be loaded and unloaded by robot arms. Some recent variants carry their own robotic arms. Burying the route wires in concrete factory floors is expensive, and alternative methods of navigation are being sought. As with robot arms, the academic and industrial efforts have merged, and a bewildering number of directions and ideas are being energetically pursued.

The history presented so far is highly sanitized, and describes only a few major actors in the newly united field of robotics. The reality is a turbulent witch's brew of approaches, motivations, and, as yet, unconnected problems. The practitioners are large and small groups around the world of electrical, mechanical, optical, and all other kinds of engineers, physicists, mathematicians, biologists, chemists, medical technologists, computer scientists, artists, and inventors. Computer scientists and biologists are collaborating on the development of machines that see. Physicists and mathematicians can be found improving sonar and other senses. Mechanical engineers have built machines that walk on legs, and others that grasp with robot hands of nearly human dexterity. These are all fledgling efforts, and the ground rules are not yet worked out. Each group represents a different set of backgrounds, desires, and skills; communication among groups is often difficult. There are no good general texts in the field, nor even a generally agreed upon outline. Continuing diversity and rapid change make it likely that this situation will continue for many years. In spite of the chaos, however, I maintain that the first mass offering from the cauldron will probably be served within a decade. And what leaps out of the brew in fifty years is the subject of the rest of this chapter. Before concluding this chapter, I'll foreshadow some of the contents in the cauldron by returning to notions raised at the outset.

MOBILITY AND INTELLIGENCE

I've been hinting that robot research, especially the mobile robot variety, has a significance much greater than the sum of its many applications, and, indeed, is the safest route to full intelligent machines. I'll offer more detailed evidence later, but briefly the argument goes like this.

Computers were created to do arithmetic faster and better than people. AI attempts to extend this superiority to other mental arenas. Some mental activities require little data, but others depend on voluminous knowledge of the world. Robotics was pursued in AI labs partly to automate the acquisition of world knowledge. It was soon noticed that the acquisition problem was less tractable than the mental activities it was to serve. While computers often exhibited adult level performance in difficult mental tasks, robotic controllers were incapable of matching even infantile perceptual skills.

In hindsight the dichotomy is not surprising. Animal genomes have been engaged in a billion year arms race among themselves, with survival often awarded to the quickest to produce a correct action from inconclusive perceptions. We are all prodigious olympians in perceptual and motor areas, so good that we make the hard look easy. Abstract thought, on the other hand, is a small new trick, perhaps less than a hundred thousand years old, not yet mastered. It just looks hard when we do it.

How hard and how easy? Average humans beings can be beaten at arithmetic by a one operation per second machine, in logic problems by 100 operations per second, at chess by 10,000 operations per second, in some narrow "expert systems" areas by a million operations. Robotic performance can not yet provide this same standard of comparison, but a calculation based on retinal processes and their computer visual equivalents suggests that a *billion* (10^9) operations per second are required to do the job of the retina, and 10 *trillion* (10^{13}) to match the bulk of the human brain.

Truly expert human performance may depend on mapping a problem into structures originally constructed for perceptual and motor tasks—so it can be internally visualized, felt, heard or perhaps smelled and tasted. Such transformations give the trillion-operations-per-second engine a purchase on the problem. The same perceptual-motor structures may also be the seat of "common sense," since they probably contain a powerful model of the world—developed to solve the merciless life and death problems of rapidly jumping to the right conclusion from the slightest sensory clues.

Decades of steady growth trends in computer power suggest that trillion-operations-per-second computers will be common in twenty to forty years. Can we expect to program them to mimic the "hard" parts of human thought in the same way that current AI programs capture some of the easy parts? It is unlikely that introspection of conscious thought can carry us very far—most of the brain is not instrumented for introspection, the neurons are occupied efficiently solving the problem at hand, as in the retina. Neurobiologists are providing some very helpful instrumentation extra-somatically, but not fast enough for the forty year timetable.

Another approach is to attempt to parallel the evolution of animal nervous systems by seeking situations with selection criteria like those in their history. By solving similar incremental problems, we may be driven, step by step, through the same solutions (helped, where possible, by biological peeks at the "back of the book"). That animals started with small nervous systems gives confidence that small computers can emulate the intermediate steps, and mobile robots provide the natural external forms for recreating the evolutionary tests we must pass. Followers of this "bottom up" route to AI may one day meet those pursuing the traditional "top down" route half way. Fully intelligent machines will result when the metaphorical golden spike is driven uniting the two efforts.

The parallel between the evolution of intelligent living organisms and the development of robots is a strong one. Many real-world constraints that shaped life

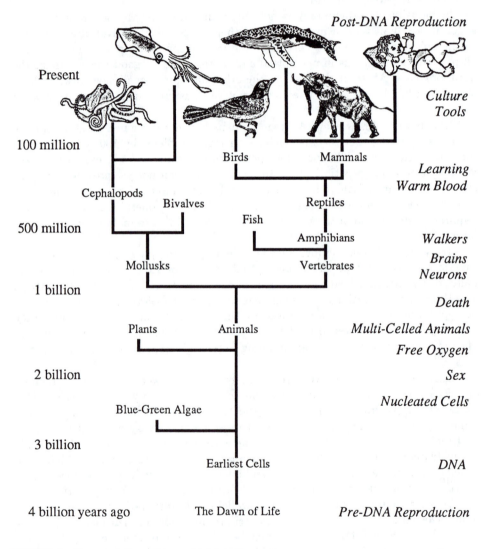

FIGURE 1   The evolution of terrestrial intelligence.

by favoring one kind of change over another in the contest for survival also affect the viability of robot characteristics. To a large extent, the incremental paths of development pioneered by living things are being followed by their technological imitators. Given this, there are lessons to be learned from the diversity of life. One is the observation made earlier, that mobile organisms tend to evolve the mental characteristics that form the bedrock of human intelligence; immobile ones do not. Plants are an example of the latter case; vertebrates an example of the former. An especially dramatic contrast is provided in an invertebrate phylum, the mollusks. Many are shellfish like clams and oysters that move little and have small nervous systems and behaviors more like plants than like animals. Yet they have relatives, the cephalopods, like octopus and squid, that are mobile and have independently developed many of the characteristics of vertebrates, including imaging eyes, large nervous systems and very interesting behavior, including major problem-solving abilities.

Two billion years ago our unicelled ancestors parted genetic company with the plants. By dint of energetics and heritage, large plants now live their lives fixed in place. Awesomely effective in their own right, the plants have no apparent inclinations toward intelligence—negative evidence that supports my thesis that mobility is a parent of this trait. Animals bolster the argument on the positive side, except for the immobile minority like sponges and clams that support it on the negative.

A billion years ago, before brains or eyes were invented, when the most complicated animals were something like hydras (i.e., double layers of cells with a primitive nerve net), our progenitors split with invertebrates. Now both clans have "intelligent" members. Most mollusks are sessile shellfish, but octopus and squid are highly mobile, with big brains and excellent eyes. Evolved independently of us, they are quite different in detail. The optic nerve connects to the back of the retina, so there is no blind spot. The brain is annular, a ring around the esophagus. The green blood is circulated by a systemic heart oxygenating the tissues and two gill hearts moving depleted blood. Hemocyanin, a copper-doped protein related to hemoglobin and chlorophyll, carries the oxygen. Octopus and their relatives are swimming light-shows, their surfaces covered by a million individually controlled color changing cells. A cuttlefish placed on a checkerboard can imitate the pattern, a fleeing octopus can make deceiving seaweed shapes coruscate backward along its body. Photophores of deep sea squid, some with irises and lenses, generate bright multicolored light. Since they also have good vision, there is a potential for rich communication.

Martin Moynihan,[9] a biologist at the University of Indiana, identifies several dozen distinct symbolic displays, many apparently expressing strong emotions. Their behavior is mammal-like. Octopus are reclusive and shy; squid are occasionally aggressive. Small octopus can learn to solve problems like how to open a container of food. Giant squid, with large nervous systems, have hardly ever been observed except as corpses. They might be as clever as whales.

Birds are vertebrates, related to us through a 300-million-year-old, probably not very bright, early reptile. Size-limited by the dynamics of flying, some are intellectually comparable to the highest mammals. The intuitive number sense of crows and ravens, for example, extends to seven, compared to three or four for us.[5,13] Birds outperform all mammals except higher primates and the whales in "learning set" tasks, where the idea is to generalize from specific instances. In mammals, generalization depends on cerebral cortex size. In birds, forebrain regions called the Wulst and the hyperstriatum are critical, while the cortex is small and unimportant.

Our last common ancestor with the whales was a primitive shrew-like mammal alive 100 million years ago. Some dolphin species have body and brain masses identical to ours, and have had them for more generations. They are as good as us at many kinds of problem solving, and can grasp and communicate complex ideas. Killer whales have brains five times human size, and their ability to formulate plans is better than the dolphins', whom they occasionally eat. Sperm whales, though not the largest animals, have the world's largest brains. Intelligence may be an important part of their struggle with large squid, their main food. Elephant brains are three times human size. Elephants form matriarchal tribal societies and exhibit complex behavior. Indian domestic elephants learn over 500 commands, and form voluntary mutual benefit relationships with their trainers, exchanging labor for baths. They can solve problems such as how to sneak into a plantation at night to steal bananas, after having been belled (answer: stuff mud into the bells). And they do have long memories. Apes are our 10 million year cousins. Chimps and gorillas can learn to use tools and to communicate in human sign languages at a retarded level. Chimps have one-third, and gorillas one-half, human brain size.

Animals exhibiting near-human behavior have hundred-billion-neuron nervous systems. Imaging vision alone requires a billion. The most developed insects have a million brain cells, while slugs and worms make do with fewer than one hundred thousand, and sessile animals with a few thousand. The portions of nervous systems for which tentative wiring diagrams have been obtained, including several nerve clumps of the large neuroned sea slugs, and leeches, and the early stages of vertebrate vision, reveal neurons configured into efficient, clever, assemblies.

The twenty-year-old modern robotics effort can hardly hope to rival the billion-year history of large life on earth in richness of example or profundity of result. Nevertheless, the evolutionary pressures that shaped life are already palpable in the robotics labs. The following is a thought experiment that reflects this situation. We wish to make robots execute general tasks such as "go down the hall to the third door, go in, look for a cup and bring it back." This desire has created a pressing need—a computer language in which to specify complex tasks for a rover, and a hardware and software system to embody it. Sequential control languages successfully used with industrial manipulators might seem a good starting point. Paper attempts at defining the structures and primitives required for the mobile application revealed that the linear control structure of these state-of-the-art robot arm controlling languages was inadequate for a rover. The essential difference is that a rover, in its wanderings, is regularly "surprised" by events it cannot anticipate, but with which it must deal. This requires that contingency routines be activated in arbitrary order, and run concurrently, each with its own access to the needed sensors, effectors, and internal state of the machine, and a way of arbitrating their differences. As conditions change the priority of the modules changes, and control may be passed from one to another.

Suppose that we ask a future robot to go down the hall to the third door, go in, look for a cup and bring it back. This will be implemented as a process that looks very much like a program written for the arm control languages (that in turn look very much like Algol, or Basic), except that the door recognizer routine would probably be activated separately. Consider the following caricature of such a program.

```
module GO-FETCH-CUP

wake up DOOR-RECOGNIZER with instructions
        ( on FINDING-DOOR add 1 to DOOR-NUMBER
          record DOOR-LOCATION )

record START-LOCATION
set DOOR-NUMBER to 0
while DOOR-NUMBER < 3 WALL-FOLLOW
FACE-DOOR
if DOOR-OPEN then GO-THROUGH-OPENING
        else OPEN-DOOR-AND-GO-THROUGH
set CUP-LOCATION to result of LOOK-FOR-CUP
TRAVEL to CUP-LOCATION
PICKUP-CUP at CUP-LOCATION
TRAVEL to DOOR-LOCATION
FACE-DOOR
if DOOR-OPEN then GO-THROUGH-OPENING
        else OPEN-DOOR-AND-GO-THROUGH
TRAVEL to START-LOCATION
end
```

So far so good. We activate our program, and the robot obediently begins to trundle down the hall counting doors. It correctly recognizes the first one. The second door, unfortunately, is decorated with garish posters, and the lighting in that part of the corridor is poor, and our experimental door recognizer fails to detect it. The wall follower, however, continues to operate properly and the robot continues on down the hall, its door count short by one. It recognizes door 3, the one we had asked it to go through, but thinks it is only the second, so continues. The next door is recognized correctly, and is open. The program, thinking it is the third one, faces it and proceeds to go through. This fourth door, sadly, leads to the stairwell, and the poor robot, unequipped to travel on stairs, is in mortal danger. Fortunately there is a process in our concurrent programming system called **DETECT-CLIFF**. This program is always running and checks ground position data posted on the blackboard by the vision processes and also requests sonar and infrared proximity checks on the ground. It combines these, perhaps with an *a priori* expectation of finding a cliff set high when operating in dangerous areas, to produce a number that indicates the likelihood there is a drop-off in the neighborhood. A companion process DEAL-WITH-CLIFF, also running continuously, but with low priority, regularly checks this number and adjusts its own priority on the basis of it. When the cliff probability variable becomes high enough, the priority of DEAL-WITH-CLIFF will exceed the priority of the current process in control, GO-FETCH-CUP in our example, and DEAL-WITH-CLIFF takes over control of the robot. A properly written DEAL-WITH-CLIFF will then proceed to stop or greatly slow down the movement of the robot, to increase the frequency of sensor measurements of the cliff, and to back away slowly from it when it has been reliably identified and located.

Now there's a curious thing about this sequence of actions. A person seeing them, not knowing about the internal mechanisms of the robot, might offer the interpretation "First the robot was determined to go through the door, but then it noticed the stairs and became so frightened and preoccupied it forgot all about what it had been doing." Knowing what we do about what really happened in the robot, we might be tempted to berate this poor person for using such sloppy anthropomorphic concepts as determination, fear, preoccupation, and forgetfulness in describing the actions of a machine. We could berate the person, but it would be wrong. The robot came by the emotions and foibles indicated as honestly as any living animal; the observed behavior is the correct course of action for a being operating with uncertain data in a dangerous and uncertain world. An octopus in pursuit of a meal can be diverted by hints of danger in just the way the robot was. An octopus also happens to have a nervous system that evolved entirely independently of our own vertebrate version. Yet most of us feel no qualms about ascribing qualities like passion, pleasure, fear, and pain to the actions of the animal. We have in the behavior of the vertebrate, the mollusc, and the robot a case of convergent evolution. The needs of the mobile way of life have conspired in all three instances to create an entity that has modes of operation for different circumstances, and that changes quickly from mode to mode on the basis of uncertain and noisy data prone to misinterpretation. As the complexity of the mobile robots increases, their similarity to animals and humans will become even greater.[6,7]

Hold on a minute, you say. There may be some resemblance between the robot's reaction to a dangerous situation and an animal's, but surely there are differences. Isn't the robot more like a startled spider, or even a bacterium, than like a frightened human being? Wouldn't it react over and over again in exactly the same way, even if the situation turned out not to be dangerous? You've caught me. I think the spider's nervous system is an excellent match for robot programs possible today. We passed the bacterial stage in the 1950's with light-seeking electronic turtles. This does not mean that concepts like thinking and consciousness are ruled out. In the book *Animal Thinking*,[5] the animal ethologist D. R. Griffin reviews evidence that much animal behavior, including that of insects, can be explained economically in terms of consciousness: an internal model of the self and surroundings, that, however crudely, allows consideration of alternative actions. But there are differences of degree.

OTHER EMOTIONS

When tickled, the sea slug Aplysia withdraws its delicate gills into its body. If the tickling is repeated often, Aplysia gradually learns to ignore the nuisance, and the gills remain deployed. If, later, tickles are followed by harsh stimuli, such as contact with a strong acid, the withdrawal reflex returns with a vengeance. Either way, the modified behavior is remembered for hours. Aplysia has been studied so thoroughly in the last few decades that the neurons involved in the reflex are well known, and the learning has recently been traced to chemical changes in single synapses on these neurons. Larger networks of neurons can adapt in more elaborate ways, for instance by learning to associate specific pairs of stimuli with one another. Such mechanisms tune a nervous system to the body it inhabits, and to its environment. Vertebrates owe much of their behavioral flexibility to an elaboration of this arrangement, systems that can be activated from many locations that encourage and discourage future repetitions of recent behaviors. Though their neural architecture is not understood, their effect is self-evident in the subjective sensations we call pleasure and pain.

A unified conditioning mechanism has obvious advantages in guiding an animal through a changing world. It seems to me that it also conveys a long-term evolutionary advantage by providing a "cheap" means of entry into fundamental new behaviors. A new need or danger can be accommodated through a modest mutation of the sensory neural wiring, the connection of a detector for the condition to a pleasure or pain site. The standard conditioning mechanism will then ensure that animals with the mutation learn to seek the conditions that meet the need, or to avoid the danger, even if the required behavior is complex. Without the learning mechanism, a much more specific sensor to motor connection would have to be discovered.

We are deep in the realm of speculation now, but the general pleasure/pain learning mechanism may provide an explanation for abstract emotions. Let's suppose that altruism, for instance of a mother toward her offspring, can enhance the long-term survival of the altruist's genes even though it has a negative effect on the individual altruist. Feeding the young may leave the mother exhausted and hungry, and defending them may involve her in risk of injury. In a successful animal, hunger and injury would surely be wired to register as pain. Wouldn't the conditioning mechanisms we've just described then eventually suppress a mother's ministrations on behalf of her young?

Activities whose beneficial or detrimental effects act only across the generations can be conditioned just as readily as those with more immediate effects, if detectors for them are wired strongly to pleasure and pain sites. For instance, mother love is encouraged if the sight, feel, sound or smell of the offspring triggers pleasure, and if absence of the young is painful. To the extent that conditioning stimuli have subjective manifestations other than the pain or pleasure sensation itself, such long-range causes are likely to *feel* different from more immediate ones like skin pain or hunger. Most of the immediate concerns are associated with some part of the body and can be usefully mapped there in the organism's conscious map of self and world. Multigenerational imperatives, on the other hand, cannot be so simply related to the physically apparent world. This may help explain the ethereal or spiritual associations people often assign to them. Certainly they deserve respect, being the distillation of perhaps tens of millions of years of life or death trials, the wisdom of many lifetimes.

WHAT IF?

Elaboration of the internal world model in higher animals made possible another twist. A rich world model allows its possessor to examine in detail alternative situations, past, future, or merely hypothetical. Dangers avoided can yet be brooded over, and what *might* have happened can be imagined. If the mental simulation is accurate enough, such brooding can produce useful warnings, or point out missed opportunities. These lessons of the imagination are most effective if their consequences are tied to the conditioning mechanism, just as with real events. Such a connection is particularly easy to explain if, as we elaborate below, the most powerful aspects of reason are due to world knowledge powerfully encoded in the sensory and motor systems. The same wiring that conditions in real situations would be activated for imaginary ones.

The ability to imagine must be a key component in communication among higher animals (and between you and me). Messages trigger mental scenarios that then provide conditioning (i.e., learning). Communication that fails to engage the emotions is not very educational in this sense, and a waste of time. Imagine circuitry for detecting time well spent and time wasted wired to the conditioning centers. It's not too far fetched to think that these correspond to the subjective emotions of "interesting" and "boring." Humans seem to have cross wiring that allows elaborate imagining, for instance about future rewards, to make interesting activities that might normally be boring. How else can one explain the existence of intellectuals? Indeed, the conventional view of intelligence, and the bulk of work in artificial intelligence, centers on this final twist. While I believe that it is important, it is only a tiny part of the whole story, and often overrated.

COMING SOON

A few of the ideas above have been explored in machinery. I mentioned earlier that W. Grey Walter built electronic turtles which demonstrated learning by association, represented as charges on a matrix of capacitors. Arthur Samuel at IBM wrote a checker-playing program that adjusted evaluation parameters to improve its play, and was able to learn simply by playing game after game against itself overnight.[4]

Frank Rosenblatt of Cornell invented networks, called "perceptrons," of artificial neurons that could be trained to do simple tasks by properly timed punish and reward signals.[12] These approaches of the 1950's and 1960's fell out of fashion in the 1970's, but modern variations of them are again in vogue.

Among the natural traits in the immediate roving robot horizon is parameter adjustment learning. A precise mechanical arm in a rigid environment can usually be "tuned" for optimal control once, permanently. A mobile robot bouncing around in the muddy world, on the other hand, is likely to continuously suffer insults like dirt build-up, tire wear, frame bends, and small mounting bracket slips that ruin precise adjustments. Some of the programs that drive our robots through obstacle courses now have a camera calibration phase. The robot is parked with its camera "eye" facing a precisely painted grid of spots. A program notes where the spots appear in the camera's images and figures a correction for camera distortions, so that later programs can make precise visual angle measurements. The driving program is highly sensitive to miscalibrations, and we are working on a method that will continuously calibrate the cameras just from the images perceived on normal trips through clutter. With such a procedure in place, a bump that slightly shifts one of the robot's cameras will no longer cause systematic errors in its navigation. Animals seem to tune most of their nervous systems with processes of this kind, and such accommodation may be a precursor to more general kinds of learning.

Perhaps more controversially, I see the beginnings of awareness in the minds of our machines. The more advanced control programs use data from the robot's sensors to maintain representations, at varying levels of abstraction and precision, of the world around the robot, of the robot's position within that world, and of the robot's internal condition. The programs that plan actions for the robot manipulate these "world models" to weigh alternative future moves. The world models can also be stored from time to time, and examined later, as a basis for learning. A verbal interface keyed to these programs would be able to meaningfully answer questions like "Where are you?" ("I'm in an area of about twenty square meters, bounded on three sides, and there are three small objects in front of me") and "Why did you do that?" ("I turned right because I didn't think I could fit through the opening on the left"). Our programs usually present such information from their world models in the form of pictures on computer screens—a direct window into their minds.

WHEN?

How does computer speed compare with human thought? The answer has been changing.

The first electronic computers were constructed in the mid-1940's to solve problems too large for unaided humans. *Colossus*, one of a series of ultrasecret British machines, broke the German *Enigma* code, greatly influencing the course of the European war, by scanning through code keys tens of thousands of times faster than humanly possible. In the U.S., *Eniac* computed anti-aircraft artillery tables for the Army, and later did calculations for the atomic bomb, at similar speeds. Such feats earned the early machines the popular appellation *Giant Brains*.

In the mid-1950's computers more than ten times faster than Eniac appeared in many larger universities. They did numerical scientific calculations nearly a million times faster than humans. A few visionaries took the Giant Brains metaphor seriously and began to write programs for them to solve intellectual problems going beyond mere calculation. The first such programs were encouragingly successful. Computers were soon solving logic problems, proving theorems in Euclidean geometry, playing checkers, even doing well in IQ test analogy problems. The performance level and the speed in each of these narrow areas was roughly equivalent to that of a college freshman who had recently learned the subject. The automation of thought had made a great leap, but paradoxically the term "Giant Brains" seemed less appropriate.

In the mid-1960's a few centers working in this new area of *Artificial Intelligence* added another twist: mechanical eyes, hands and ears to provide real world interaction for the thinking programs. By then computers were a thousand times faster than Eniac, but programs to do even simple things like clearing white blocks from a black tabletop turned out to be very difficult to write, and performed hundreds of times more slowly, and much less reliably, than a human. Slightly more complicated tasks took much longer, and many seemingly trivial things, like identifying a few simple objects in a jumble, still cannot be done acceptably at all twenty years later, even given hours of computer time. Forty years of research and a millionfold increase in computer power has reduced the image of computers from Giant Brains to mental midgets. Is this silly, or what?

EASY AND HARD

The human evolutionary record provides a clue to the paradox. While our sensory and muscle control systems have been in development for almost a billion years, and common sense reasoning has been honed for perhaps a million, really high-level, deep thinking is little more than a parlor trick, culturally developed over a few thousand years, which a few humans, operating largely against their natures, can learn. As with Samuel Johnson's dancing dog, what is amazing is not how well it is done, but that it is done at all.

Computers can challenge humans in intellectual areas, where humans are evolutionary novices, because they can be programmed to carry on much less wastefully. Arithmetic is an extreme example, a function learned by humans with great difficulty, but instinctive to computers. A 1987 home computer can add a million large numbers in a second, astronomically faster than a person, and with no errors. Yet the 100 billion neurons in a human brain, if reorganized by a mad neurosurgeon into adders using switching logic design principles, could sum one hundred thousand times faster than the computer.

RETINA AND COMPUTER

The retina is the best studied piece of the vertebrate nervous system. Though located at the back of the eyeball, some distance from the bulk of the brain, it is really an elongated extension of the brain. Its separation makes it comparatively easy to study, even in living animals. Removed from the body, it can be kept functioning for hours, with its inputs and outputs highly accessible. Transparent, and thinner than a sheet of paper, it is ideal for light and electron microscopic examination, when stained with dyes to make specific neurons visible. It consists of a layer of light-detecting photocells connected to a network of neurons that respond to contrast and motion and more specific features in the image received by the eye. These preprocessed data are then passed by the optic nerve to larger neural centers in the brain.

It is a peculiar feature of the vertebrate retina that light must pass through the neural network to get to the photocells. This is, no doubt, an unfortunate design choice, now locked in, made early in the evolutionary history of the eye. The independently evolved retinas of octopus and squid sensibly have their photoreceptors up front. The awkward position of the vertebrate retinal nerve net greatly limits its size. On the other hand, there is strong selection pressure to enhance its function. The retinal cells are in a unique position to rapidly and comprehensively abstract the essentials from an image, and good vision was a key survival tool for our ancestors: life and death alternatives often depended on small differences in visual speed or acuity. The product of this evolutionary adaptation is bound to be a little atypical of the rest of the brain, where space is larger, and payoffs for small improvements are more dilute.

Retinal neurons, as I noted, form a thin sheet. Although nerve tissue is usually gray or white, the retinal neurons, and their supporting glial cells, are clear. Retinal neurons are smaller than most found in the brain. Though the rest of the brain is too poorly understood to be sure, the same pressures make it likely that the retina is wired more precisely than neural centers with gentler criteria, and that the retinal neurons are used more effectively. A supporting fact is that the retinal neurons communicate among themselves almost exclusively by smoothly varying voltages rather than pulses, though their computations are ultimately encoded as pulses on the optic nerve. This continuous mode works over only small distances in the wet environment, but at that range is faster and more precise. The retina may thus be representative of the most efficient neural structures in vertebrates.

So what does the retina actually do? A rough and ready answer is available. Five cell types do most of the work. Photocells (subdivided into cone cells, which together discriminate colors, and rod cells which don't) intercept the light. Horizontal, bipolar, and amacrine cells, working with continuous voltages, process the image. Ganglion cells, whose axons form the optic nerve, combine outputs from the other cells and produce pulsed signals that go into the brain. After adapting to a particular overall light level, clusters of photocells create a voltage proportional to the amount of light striking them. This signal is received by two classes of neurons, the horizontal cells and the bipolar cells. The horizontal cells, whose thousands of fibers cover large circular fields of photocells, produce a kind of average of their areas. If the voltages of all the horizontal cells were mapped onto a television screen, a blurry version of the retinal image would be displayed. The bipolar cells, on the other hand, are wired only to small areas, and would provide a sharp picture on our imaginary television. Some of the bipolar cells also receive inputs from nearby horizontal cells, and then compute a difference between the small bipolar center areas and the large horizontal surround. Viewed on our television, their picture would look much paler than the original, except at the edges of objects and patterns, where a distinct bright halo would be seen. The bipolar cell axons connect to complicated multilayer synapses on the axonless amacrine cells. Each ganglion cell collects inputs from several of these amacrine synapses, and produces a pulsed output, which travels up its long axon. Each amacrine cell connects to several bipolar and ganglion cells, and some of the junctions appear to both send and receive signals. Some amacrine cells enhance the "center surround" response, others detect changes in brightness in parts of the image. On the television, some of these would show only objects moving left to right, while others would reveal other directions of motion. Each ganglion cell connects to several bipolar and amacrine cells, and produces pulse streams whose rate is proportional to a computed feature of the image. Some report on high contrast in specific parts of the picture, others on various kinds of motion, or combinations of contrast and motion.[2]

The television I'm referring to is not totally imaginary. Sitting next to me as I write is a television monitor that often displays images just like those described. They come not from an animal's retina, but from the eye of a robot. The picture from a television camera on the robot is converted by electronics into an array of numbers in a computer memory. Programs in the computer combine these numbers to deduce things about the robot's surroundings. Though designed with little reference to neurobiology, many of the program steps resemble strongly the operations of the retinal cells—a case of convergent evolution. The parallel provides a way to measure the net computational power of neural tissue.

The human retina has about 100 million photocells, tens of millions of horizontal, bipolar and amacrine cells, and one million ganglion cells, each contributing one signal-carrying fiber to the optic nerve. All this is packaged in a volume a third of a millimeter thick and less than a centimeter square, $1/100,000$ the volume of the whole brain. The photocells interact with their neighbors to enhance each other's output, and their great multiplicity appears to be a way to maximize sensitivity; a single photon sometimes produces a detectable response. The horizontal and bipolar cells and the amacrine cell synapses each seem to perform a unique computation.

The bottom line, however, is that the million ganglion cell axons each report on a specific function computed over a particular patch of photocells.

To find the computer equivalent for such a function, we'll first have to match the visual detail of the human eye in the computer equivalent. Simply counting photocells in the eye leads to an overestimate, because they work in groups. External visual acuity tests are better, but complicated by the fact that the retina has a small, dense, high-resolution center area, the fovea, which can resolve details more than 10 times as fine as the rest of the eye. Though it covers less than 1% of the visual field, the fovea employs perhaps one quarter of the retinal circuitry, and one quarter of the optic nerve fibers. Under optimal seeing conditions as many as 500 distinct points can be resolved across the width of this central region. This feat could be matched by a television camera with 500 separate picture elements, or "pixels," in the horizontal direction. The vertical resolution is similar, so our camera would need 500 × 500, or one quarter million pixels, in all—which, incidentally, just happens to be the resolution of a good-quality image on a standard television set. But don't we see more finely than that? Not really. The 500 × 500 array corresponds only to our fovea, spanning a mere 5° of our field of view. A standard television screen subtends about 5° when viewed from a distance of 10 meters. At that range, the scanning lines and other resolution defects of the television image are invisible because the resolution of our eye is no better. At closer range we can concentrate our fovea on small parts of the television image to get greater detail. We have the illusion of seeing the whole screen this sharply because our unconsciously swiveling eyes rapidly zip the foveal area from one place to another. Somewhere, in an as yet mysterious part of our brain, a high-resolution image is being synthesized like a jigsaw puzzle from these fragmentary glimpses.

So the foveal circuitry in the retina effectively takes a 500 × 500 image and processes it to produce 250,000 values, some being center-surround operations, some being motion detections. One key question remains. How fast does this happen? Experience with motion pictures provides a ready answer. When successive frames are presented at a rate slower than about ten per second, the individual frames become distinguishable. At faster rates they blend together into apparently smooth motion. Though the separate frames cannot be distinguished faster than ten per second, if the light flickers at the frame rate, as it does in a movie projector and on a TV screen, the flicker itself is detectable until it reaches a frequency of about 50 flashes per second. Presumably in the 10–50 cycle range, the simplest brightness change detectors are triggered, but the more complicated neuron chains do not have time to react.

In our lab we have often programmed computers to do center-surround operations on images from television-toting robots, and once or twice we have written motion detectors. To get the speed up, we have spent much programming effort and mathematical trickery to do the job as efficiently as possible. Despite our best efforts, 10-frames-per-second processing rates have been out of reach because our computers are too slow. In a rough sense, with an efficient program a center-surround calculation applied to each pixel in a 500 × 500 image takes about 25 million calculations, which breaks down to about 100 calculations for each center-surround value produced. A motion-detecting operator can be applied at a similar cost. Translated to the retina, this means that each ganglion cell reports on the computer equivalent of 100 calculations every tenth of a second, and thus represents 1000 calculations per second. The whole million-fiber optic nerve then carries the answers to a billion calculations per second.

If the retina's processing can be matched by a billion computer calculations per second, what can we say about the entire brain? The brain has about 1000 times as many neurons as the retina, but its volume is 100,000 times as large. The retina's evolutionarily pressed neurons are smaller and more tightly packed than average. By multiplying the computational equivalent of the retina by a compromise value of

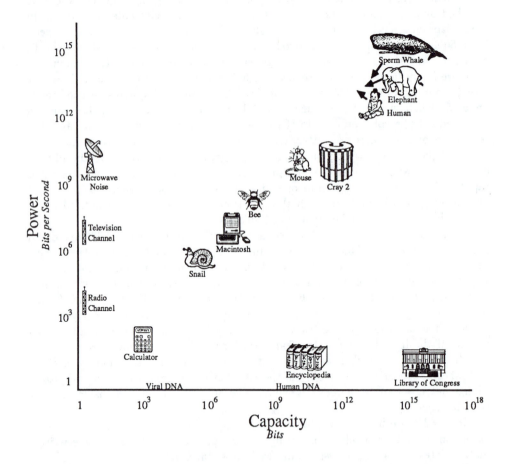

FIGURE 2 Computing speed and memory of some animals and machines. The animal figures are for the nervous system only, calculated at 100 bits per second and 100 bits of storage per neuron. These are speculative estimates, but note that a factor of 100 one way or the other would change the appearance of the graph only slightly.

10,000 for the ratio of brain complexity to retina complexity, I rashly conclude that the whole brain's job might be done by a computer performing 10 trillion (10^{13}) calculations per second. This is about a million times faster than the medium-size machines that now drive my robots, and one thousand times more than today's fastest supercomputers.

INTELLECTUAL VOYAGES

Interesting computation and thought requires a processing engine of sufficient computational *power* and *capacity*. Roughly, power is the speed of the machine, and capacity is its memory size.

Here's a helpful metaphor. Computing is like a sea voyage in a motorboat. How fast a given journey can be completed depends on the power of the boat's engine. The maximum length of any journey is limited by the capacity of its fuel tank. The effective speed is decreased, in general, if the course of the boat is constrained, for instance to major compass directions.

Some computations are like a trip to a known location on a distant shore, others resemble a mapless search for a lost island. Parallel computing is like having a fleet of small boats—it helps in searches, and in reaching multiple goals, but not very much in problems that require a distant sprint. Special purpose machines trade a larger engine for less rudder control.

Attaching disks and tapes to a computer is like adding secondary fuel tanks to the boat. The capacity, and thus the range, is increased, but if the connecting plumbing is too thin, it will limit the fuel flow rate and thus the effective power of the engine.

Extending the metaphor, input/output devices are like boat sails. They capture power and capacity in the environment. Outside information is a source of variability, and thus power, by our definition. More concretely, it may contain answers that would otherwise have to be computed. The external medium can also function as extra memory, increasing capacity.

Figure 2 shows the power and capacity of some interesting natural and artificial thinking engines. At its best, a computer instruction has a few tens of bits of information, and a million instruction per second computer represents a few tens of millions of bits/second of power. The power ratio between nervous systems and computers is as calculated in the last section: a million instructions per second is worth about a hundred thousand neurons. I also assume that a neuron represents about 100 bits of storage, suggested by recent evidence of synaptic learning in simple nervous systems by Eric Kandel and others. Note that change of a factor of ten or even one hundred in these ratios would hardly change the graph qualitatively. (My forthcoming book *Mind Children*, from which this paper is drawn, offers more detailed technical justifications for these numbers).

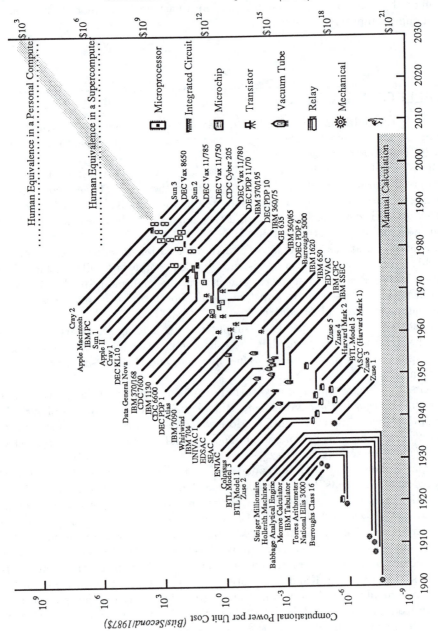

FIGURE 3   A Century of Computing.

The figure shows that current laboratory computers are equal in power approximately to the nervous systems of insects. It is these machines that support essentially all the research in artificial intelligence. No wonder the results to date are so sparse! The largest supercomputers of the mid-1980's are a match for the one-gram brain of a mouse, but at ten million dollars or more apiece they are reserved for serious work.

THE GROWTH OF PROCESSING POWER

How long before the research medium is rich enough for full intelligence?

Although a number of mechanical digital calculators were devised and built during the seventeenth and eighteenth centuries, only with the mechanical advances of the industrial revolution did they become reliable and inexpensive enough to routinely rival manual calculation. By the late nineteenth century their edge was clear, and the continuing progress dramatic.

Since then the cost of computing has dropped a thousandfold every twenty years (Figure 3). Before then mechanical calculation was an unreliable and expensive novelty with no particular edge over hand calculation. The graph shows a mind boggling *trillionfold* decrease in the cost since then. The pace has actually picked up a little since the beginning of the century. It once took 30 years to accumulate a thousandfold improvement; in recent decades it takes only 19. Human equivalence should be affordable very early in the 21st century.

The early improvements in speed and reliability came with advances in mechanics—precision mass produced gears and cams, for instance, improved springs and lubricants, as well as increasing design experience and competition among the calculator manufacturers. Powering calculators by electric motors provided a boost in both speed and automation in the 1920's, as did incorporating electromagnets and special switches in the innards in the 1930's. Telephone relay methods were used to make fully automatic computers during World War II, but these were quickly eclipsed by electronic tube computers using radio, and ultrafast radar, techniques. By the 1950's computers were an industry that itself spurred further major component improvements.

The curve in Figure 3 is not leveling off, and the technological pipeline is full of developments that can sustain the pace for the foreseeable future. Success in this enterprise, as in others, breeds success. Not only is an increasing fraction of the best human talent engaged in the research, but the ever more powerful computers themselves feed the process. Electronics is riding this curve so quickly that it is likely to be the main occupation of the human race by the end of the century.

The price decline is fueled by miniaturization, which supplies a double whammy. Small components both cost less and operate more quickly. Charles Babbage, who in 1834 was the first person to conceive the idea of an automatic computer, realized this. He wrote that the speed of his design, which called for hundreds of thousands of mechanical components, could be increased in proportion if "as the mechanical art achieved higher states of perfection" his palm-sized gears could be reduced to the scale of clockwork, or further to watchwork. (I fantasize an electricity-less world where the best minds continued on Babbage's course. By now there would be desk- and pocket-sized mechanical computers containing millions of microscopic gears, computing at thousands of revolutions per second.)

To a remarkable extent the cost per pound of machinery has remained constant as its intricacy increased. This is as true of consumer electronics as of computers (merging categories in the 1980's). The radios of the 1930's were as large and as expensive as the televisions of the 1950's, the color televisions of the 1970's, and the home computers of the 1980's. The volume required to amplify or switch a single signal dropped from the size of a fist in 1940, to that of a thumb in 1950, to a pencil eraser in 1960, to a salt grain in 1970, to a small bacterium in 1980. In the same period the basic switching speed rose a millionfold, and the cost declined by the same huge amount.

Predicting the detailed future course is impossible for many reasons. Entirely new and unexpected possibilities are encountered in the course of basic research. Even among the known, many techniques are in competition, and a promising line of development may be abandoned simply because some other approach has a slight edge. I'll content myself with a short list of some of what looks promising today.

In recent years the widths of the connections within integrated circuits have shrunk to less than one micron, perilously close to the wavelength of the light used to "print" the circuitry. The manufacturers have switched from visible light to shorter wavelength ultraviolet, but this gives them only a short respite. X-rays, with much shorter wavelengths, would serve longer, but conventional X-ray sources are so weak and diffuse that they need uneconomically long exposure times. High-energy particle physicists have an answer. Speeding electrons curve in magnetic fields, and spray photons like mud from a spinning wheel. Called synchotron radiation for the class of particle accelerator where it became a nuisance, the effect can be harnessed to produce powerful beamed X-rays. The stronger the magnets, the smaller can be the synchotron. With liquid-helium-cooled superconducting magnets, an adequate machine can fit into a truck; otherwise it is the size of a small building. Either way, synchotrons are now an area of hot interest, and promise to shrink mass-produced circuitry into the sub-micron region. Electron and ion beams are also being used to write submicron circuits, but present systems affect only small regions at a time, and must be scanned slowly across a chip. The scanned nature makes computer controlled electron beams ideal, however, for manufacturing the "masks" that act like photographic negatives in circuit printing.

Smaller circuits have less electronic "inertia" and switch both faster and with less power. On the negative side, as the number of electrons in a signal drops, it becomes more prone to thermal jostling. This effect can be countered by cooling, and indeed very fast experimental circuits can now be found in many labs running in supercold liquid nitrogen, and one supercomputer is being designed this way. Liquid nitrogen is produced in huge amounts in the manufacture of liquid oxygen from air, and it is very cheap (unlike the much colder liquid helium).

The smaller the circuit, the smaller the regions across which voltages appear, calling for lower voltages. Clumping of the substances in the crystal that make the circuit becomes more of a problem as they get smaller, so more uniform "doping" methods are being developed. As the circuits become smaller quantum effects become more pronounced, creating new problems and new opportunities. Superlattices, multiple layers of atoms-thick regions of differently doped silicon made with molecular beams, are such an opportunity. They allow the electronic characteristics of the material to be tuned, and permit entirely new switching methods, often giving tenfold improvements.

The first transistors were made of germanium; they could not stand high temperatures and tended to be unreliable. Improved understanding of semiconductor physics and ways of growing silicon crystals made possible faster and more reliable silicon transistors and integrated circuits. New materials are now coming into their own. The most immediate is gallium arsenide. Its lattice impedes electrons less than silicon, and makes circuits up to ten times faster. The Cray 3 supercomputer due in 1989 will use gallium arsenide integrated circuits, packed into a one cubic foot volume, to top the Cray 2's speed tenfold. Other compounds like indium phosphide and silicon carbide wait in the wings. Pure carbon in diamond form is an obvious possibility—it should be as much an improvement over Gallium Arsenide as that crystal is over Silicon. Among its many superlatives, perfect diamond is the best solid conductor of heat, an important property in densely packed circuitry. The vision of an utradense three-dimensional circuit in a gem-quality diamond is compelling. As yet no working circuits of diamond have been reported, but excitement is mounting as reports of diamond layers up to a millimeter thick grown from hot methane come from the Soviet Union, Japan and, belatedly, the United States.

Farther off the beaten track are optical circuits that use lasers and non-linear optical effects to switch light instead of electricity. Switching times of a few picoseconds, a hundred times faster than conventional circuits, have been demonstrated, but many practical problems remain. Finely tuned lasers have also been used with light sensitive crystals and organic molecules in demonstration memories that store up to a trillion bits per square centimeter.

The ultimate circuits may be superconducting quantum devices, which are not only extremely fast, but extremely efficient. Various superconducting devices have been in and out of fashion several times over the past twenty years. They've had a tough time because the liquid helium environment they require is expensive, the heating/cooling cycles are stressful, and especially because rapidly improving semiconductors have offered such tough competition.

Underlying these technical advances, and preceding them, are equally amazing advances in the methods of basic physics. One recent, unexpected and somewhat unlikely device is the inexpensive tunnelling microscope that can reliably see, identify and soon manipulate single atoms on surfaces by scanning them with a very sharp needle. The tip is positioned by three piezoelectric crystals microscopically moved by small voltages. It maintains a gap a few atoms in size by monitoring a current that jumps across it. The trickiest part is isolating the system from vibrations. It provides our first solid toehold on the atomic scale.

A new approach to miniaturization is being pursued by enthusiasts in the laboratories of both semiconductor and biotechnology companies, and elsewhere. Living organisms are clearly machines when viewed at the molecular scale. Information encoded in RNA "tapes" directs protein assembly devices called ribosomes to pluck particular sequences of amino acids from their environment and attach them to the ends of growing chains. Proteins, in turn, fold up in certain ways, depending on their sequence, to do their jobs. Some have moving parts acting like hinges, springs, latches triggered by templates. Others are primarily structural, like bricks or ropes or wires. The proteins of muscle tissue work like ratcheting pistons. Minor modifications of this existing machinery are the core of today's biotechnology industry. The visionaries see much greater possibilities.

Proteins to do specific jobs can be engineered even without a perfect model of their physics. Design guidelines, with safety margins to cover the uncertainties, can substitute. The first generation of artificial molecular machinery would be made of protein by mechanisms recruited from living cells. Early products would be simple, like tailored medicines, and experimental, like little computer circuits. Gradually a bag of tricks, and computer design aids, would accumulate to build more complicated machines. Eventually it may be possible to build tiny robot arms, and equally tiny computers to control them, able to grab molecules and hold them, thermally wriggling, in place. The protein apparatus could then be used as machine tools to build a second generation of molecular devices by assembling atoms and molecules of all kinds. For instance, carbon atoms might be laid, bricklike, into ultra-strong fibers of perfect diamond. The smaller, harder, tougher machines so produced would be the second generation molecular machinery.

The book *Engines of Creation* by Eric Drexler,[3] and a forthcoming book by Conrad Schneiker, call the entire scheme *nanotechnology,* for the nanometer scale of its parts.[1] By contrast today's integrated circuit microtechnology has micrometer features, a thousand times bigger. Some things are easier at the nanometer scale. Atoms are perfectly uniform in size and shape, if somewhat fuzzy, and behave predictably, unlike the nicked, warped and cracked parts in larger machinery.

A STUMBLE

It seemed to me throughout the 1970's (I was serving an extended sentence as a graduate student at the time) that the processing power available to AI programs was not increasing very rapidly. In 1970 most of my work was done on a Digital Equipment Corp. PDP-10 serving a community of perhaps thirty people. In 1980 my computer was a DEC KL-10, five times as fast and with five times the memory of the old machine, but with twice as many users. Worse, the little remaining speed-up seemed to have been absorbed in computationally expensive convenience

[1]Editor's Note: See the articles by Schneiker and Drexler in these proceedings.

features: fancier time-sharing and high-level languages, graphics, screen editors, mail systems, computer networking and other luxuries that soon became necessities.

Several effects together produced this state of affairs. Support for university science in general had wound down in the aftermath of the Apollo moon landings and politics of the Vietnam war, leaving the universities to limp along with aging equipment. The same conditions caused a recession in the technical industries— unemployed engineers opened fast food restaurants instead of designing computers (the rate of change in Figure 3 does slacken slightly in the mid-1970's). The initially successful "problem-solving" thrust in AI had not yet run its course, and it still seemed to many that existing machines were powerful enough—if only the right programs could be found. While spectacular progress in the research became increasingly difficult, a pleasant synergism among the growing number of information utilities on the computers created an attractive diversion for the best programmers—creating more utilities.

If the 1970's were the doldrums, the 1980's more than compensated. Several salvations had been brewing. The Japanese industrial successes focused attention worldwide on the importance of technology, particularly computers and automation, in modern economies—American industries and government responded with research dollars. The Japanese stoked the fires, under the influence of a small group of senior researchers, by boldly announcing a major initiative towards future computers, the so-called "Fifth Generation" project, pushing the most promising American and European research directions. The Americans responded with more money. Besides this, integrated circuitry had evolved far enough that an entire computer could fit on a chip. Suddenly computers were affordable by individuals, and a new generation of computer customers and manufacturers came into being. The market was lucrative, the competition fierce, and the evolution swift, and by the mid-1980's the momentum lost in the previous decade had been regained, with interest. Artificial intelligence research is awash in a cornucopia of powerful new "personal" workstation computers, and there is talk of applying supercomputers to the work.

Even without supercomputers, human equivalence in a research setting should be possible by around 2010, as suggested by Figure 3. Now, the smallest vertebrates, shrews and hummingbirds, get interesting behavior from nervous systems one ten thousandth the size of a human's, so I expect fair motor and perceptual competence, in about a decade.

FASTER YET?

Very specialized machines can provide up to one thousand times the effective performance for a given price in well-defined tasks. Some vision and control problems may be candidates for this approach. Special purpose machines are not a good solution in the groping research stage, but may dramatically lower the costs of intelligent machines when the problems and solutions are well understood. Some principals in the Japanese Fifth Generation Computer Project have been quoted as planning "man capable" systems in ten years. I believe this more optimistic projection is unlikely, but not impossible.

As the computers become more powerful and as research in this area becomes more widespread the rate of visible progress should accelerate. I think artificial intelligence via the "bottom up" approach of technological recapitulation of the evolution of mobile animals is the surest bet because the existence of independently evolved intelligent nervous systems indicates that there is an incremental route to intelligence. It is also possible, of course, that the more traditional "top down" approach will achieve its goals, growing from the narrow problem solvers of today into the much harder areas of learning, common-sense reasoning and perceptual acquisition of knowledge as computers become large and powerful enough, and the techniques are mastered. Most likely both approaches will make enough progress that they can effectively meet somewhere in the middle, for a grand synthesis into a true artificial sentience.

This artificial person will have some interesting properties. Its high-level reasoning abilities should be astonishingly better than a human's—even today's puny systems are much better in some areas—but its low level perceptual and motor abilities will be comparable to ours. Most interestingly it will be highly changeable, both on an individual basis and from one of its generations to the next. And it will quickly become cheap.

ACKNOWLEDGMENTS

This article is adapted from the forthcoming book *Mind Children*, Harvard University Press, Fall 1988. This work has been supported in part by the Office of Naval Research under contract N00014-81-K-503.

REFERENCES

1. Cairn-Smiths, A. G. (1985), *Seven Clues to Origin of Life* (Cambridge: Cambridge University Press).
2. Dowling, J. E. (1987), *The Retina, An Approachable Part of the Brain* (Cambridge: Harvard University Press).
3. Drexler, E. (1986), *Engines of Creation* (Garden City, NY: Anchor Press/ Doubleday).
4. Feigenbaum, E. A., and J. Feldman (1963), *Computers and Thought* (New York: McGraw Hill).
5. Griffin, D. G. (1984), *Animal Thinking* (Cambridge: Harvard University Press).
6. Kandel, E. R. (1976), *The Cellular Basis of Behavior* (San Francisco, CA: W. H. Freeman and Company).
7. Kandel, E. R., and J. H. Schwartz (1982), "Molecular Biology of Learning," *Science* **218(4571)**, 433–443.
8. Minsky, M., and S. Papert (1969), *Perceptrons* (Cambridge: Massachusetts Institute of Technology Press).
9. Moynihan, M. (1985), *Communication and Noncommunication by Cephalopods* ((Bloomington, IN: Indiana University Press).
10. Moravec, H. (1988), *Mind Children* (Cambridge: Harvard University Press), forthcoming.
11. Newell, A., H. Simon, and J. Shaw (1956), "The Logic Theory Machine," *IRE Transactions on Information Theory* IT-2(3), 61–79.
12. Rosenblatt, F. (1962), *Principles of Neurodynamics* (New York: Spartan Books).
13. Stettner, L. J., and K. A. Matyniak, "The Brain of Birds," *Sci. Am.* **218(6)**, 64–76.

Richard Dawkins
The Department of Zoology, The University of Oxford, South Parks Road, Oxford OX1 3PS, England

The Evolution of Evolvability

A title like "The Evolution of Evolvability" ought to be anathema to a dyed-in-the-wool, radical neo-Darwinian like me! Part of the reason it isn't is that I really have been led to think differently as a result of creating, and using, computer models of artificial life which, on the face of it, owe more to the imagination than to real biology. The use of artificial life, not as a formal model of real life but as a generator of insight in our understanding of real life, is one that I want to illustrate in this paper. With a program called *Blind Watchmaker*, I created a world of two-dimensional artificial organisms on the computer screen.[3] Borrowing the word used by Desmond Morris for the animal-like shapes in his surrealistic paintings,[7] I called them biomorphs. My main objective in designing *Blind Watchmaker* was to reduce to the barest minimum the extent to which I designed biomorphs. I wanted as much as possible of the biology of biomorphs to *emerge*. All that I would design was the conditions—ideally very simple conditions—under which they might emerge. The process of emergence was to be evolution by the Darwinian process of random mutation followed by nonrandom survival. Once a Darwinian process gets going in a world, it has an open-ended power to generate surprising consequences: us, for example. But, before any Darwinian process can get going, there has to be a bare minimum group of conditions set up. These were the conditions that I had to engineer in my computer world.

Artificial Life, SFI Studies in the Sciences of Complexity,
Ed. C. Langton, Addison-Wesley Publishing Company, 1988   **201**

The first condition, one that I have emphasized sufficiently before,[1,2] is that there must be *replicators*—entities capable, like DNA molecules, of self-replication. The second condition is our main concern in this paper. It is that there must be an embryology; the genes must influence the development of a phenotype; and the replicators must be able to wield some phenotypic power over their world, such that some of them are more successful at replicating themselves than others. The type of embryology that we choose for our artificial life is crucial. This is another way of stating the key message of this paper.

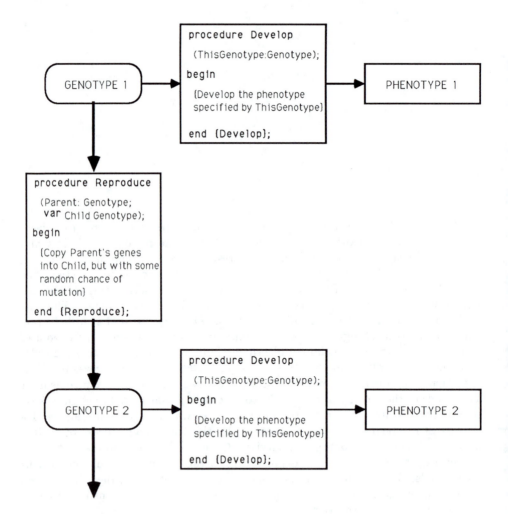

FIGURE 1   Weismann's continuity of the germ plasm (expressed in Pascal); type Genotype = array of Genes.

The fundamental principle of embryology in real life (and one that I decided was worth imitating in artificial life) was formulated by Weismann.[8] It is illustrated in Figure 1, which covers a period of two generations preceded and followed by an indefinite number of generations. I have expressed part of it in Pascal, in anticipation of my explanation of how the Blind Watchmaker program closely follows Weismann's doctrine.

The key point is that, in every generation, it is only genes that are handed on to the next generation (leave it to pedants to fuss about the sense in which this is not strictly true). In every generation, the genes of that generation influence the phenotype of that generation. The success of that phenotype determines whether or not the genes that it bears, a set that largely overlaps with the genes that influenced its development, shall go forward to the next generation. (In *The Extended Phenotype*,[2] I explore the far-reaching consequences of the fact that these two sets do not necessarily *have* to overlap.) Any individual born, therefore, inherits genes that have succeeded in building a long series of successful phenotypes, for the simple reason that failed phenotypes don't pass on their genes. This is natural selection. It is why organisms are well adapted, and it is why we are all here.

It is important to understand that genes do two quite distinct things. They participate in embryology, influencing the development of the phenotype in a given generation. And they participate in genetics, getting themselves copied down the generations. It is too often not realized—even by some of those that wear the labels geneticist or embryologist—that there is a radical separation between the disciplines of genetics and embryology. Genetics is the study of the vertical arrows in Figure 1, the study of the relationships between genotypes in successive generations. Embryology is the study of the horizontal arrows, the study of the relationship between genotype and phenotype in any one generation. If you doubt that the separation between the two disciplines is fundamental, consider the matter methodologically. You could do perfectly respectable embryology on a single individual. Genetics on a single individual would be meaningless. Conversely, you could do perfectly respectable genetics, but not embryology, on a population of individuals, each one sampled at only one point in its life cycle.

I resolved to maintain the separation between genetics and embryology in my artificial life. To this end, I wrote the program around two strictly separate procedures called **Reproduce** and **Develop** (Figure 1). Another part of the program presented an array of phenotypes for selection, each one drawn by the procedure **Develop** under the influence of genes that would be held responsible for its success or failure. In any generation the phenotype chosen (by some criterion) as successful would be the one whose genotype went forward via **Reproduce** (with some possibility of random mutation) to the next generation. The selection criterion itself I was content to leave, for the moment, to the aesthetic taste of a human chooser. The model would therefore be, at least in the first implementation, a model of artificial selection (like breeding cattle for milk yield) not natural selection. As a didactic technique this has an honorable history. Charles Darwin made persuasive use of artificial selection as a metaphor for natural selection.

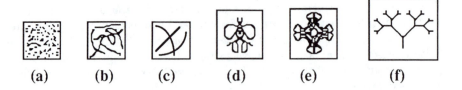

FIGURE 2 Breeding from a random starting pattern (a), random lines (b), lines of mathematical families (c), mirror algorithms (d), letting genes determine the presence or absence of mirrors in various planes of symmetry (e), and "archetypal" body form generated by *Blind Watchmaker's* artificial embryology (f).

Now to flesh out the bare bones of Figure 1. We must write some code in the two procedures, to specify the details of genetics and embryology respectively. Genetics is straightforward. However we choose to represent genes, it is obviously easy to copy them from parent to child, and it is obviously easy to introduce some minor random perturbation in the copying to represent mutation. It is embryology that we have to think about more carefully. What shall we write between the { } brackets in the procedure **Develop**, to specify the relationship between genotype and phenotype?

The first naive idea that might occur to us is to go for maximum generality. We know that the phenotypes in our artificial world are all going to be two-dimensional pictures on a Macintosh screen. The Macintosh screen is an array of $340 \times 250 = 85,000$ pixels. If we give our biomorphs' genotypes of 85,000 genes, each having a value of 1 or 0, we know that any conceivable phenotype in our artificial world can be represented by a specific genotype. Moreover, any pixel can mutate to its opposite state, and the resulting picture might be selected, or not, in preference to its parent. It follows, therefore, that we in theory could "breed" any picture from a random starting pattern (Figure 2a) or, indeed, from any other picture, getting from, say, Winston Churchill to a Brontosaurus, by scanning every generation hopefully for slight resemblances to the target picture.

But only in theory. In practice we'd be waiting till kingdom-come. This really would be a very naive way of writing **Development**, and it would produce a very boring kind of artificial life. It is a kind of zero-order embryology, the kind of embryology we must improve upon. Our improvements will take the form of constraints. Constrained embryologies are improvements over naive pixel-peppering, not because they have greater generality but because they have less. Naive pixel-peppering can produce all possible pictures, including the set that anyone might regard as biological. The trouble lies in the astronomical number of nonsense pictures that it can also produce. Constrained embryologies have a restricted set of phenotypes that they can generate, and they will be specified by a smaller set of genes, each gene controlling a more powerful drawing operation than coloring a single pixel. The task is to find an embryological procedure whose phenotypes are restricted in biologically interesting directions.

So, what kinds of constrained embryologies might we think of, as improvements over naive pixel-peppering? A slight improvement would be gained if, instead of drawing random pixels, we draw random lines (Figure 2b). Pixels, in other words, tend to pop up next to one another rather than just anywhere. We might further specify that the lines should belong to recognized mathematical families—straight lines whose length and angle is specified by genes; curves whose shape is specified by a polynomial formula whose coefficients are specified by genes (Figure 2c). Yet another constraint might be one of symmetry. Most animals are, as a matter of fact, bilaterally symmetrical, though some show various kinds of radial symmetry, and many depart from their basic symmetry in minor respects. We could use mirror algorithms in writing Development (Figure 2d), letting genes determine the presence or absence of mirrors in various planes of symmetry (Figure 2e).

But though all these embryologies are obvious improvements over naive pixel-peppering, I did not tarry long over them. Right from the start of this enterprise, I had a strong intuitive conviction about the kind of embryology I wanted. It should be *recursive*. My intuition was based partly upon the generative power of recursive algorithms well known to computer scientists; and partly upon the fact that the details of embryology in real life can to a large extent be thought of as recursive. I can best illustrate the idea in terms of the procedure that I ended up actually using.

```
procedure Tree(x, y, length, dir: integer; dx, dy: array [0..7] of integer); {Tree is
called with the arrays dx & dy specifying the form of the tree, and the starting
value of length specifying the number of branchings. Tree calls itself recursively
with a progressively decreasing value of length until length reaches 0};
var xnew, ynew: integer;
begin if dir < 0 then dir:= dir + 8; if dir>= 8 then dir:= dir − 8;
        xnew: = x + length * dx[dir]; ynew: = y + length* dy[dir];
        MoveTo(x, y); LineTo(xnew, ynew);
        if length > 0 then {now follow the two recursive calls, drawing to left and
                            right respectively}
        begin
                tree(xnew, ynew, length − 1, dir − 1) {this initiates a series of inner calls};
                tree(xnew, ynew, length − 1, dir + 1)
        end
end {tree};
```

What this procedure actually draws depends upon the starting value of the parameter *length*, and the values of $dx[0]$ to $dx[7]$ and $dy[0]$ to $dy[7]$ that are plugged in. A particular setting of these values, for instance, draws a tree like

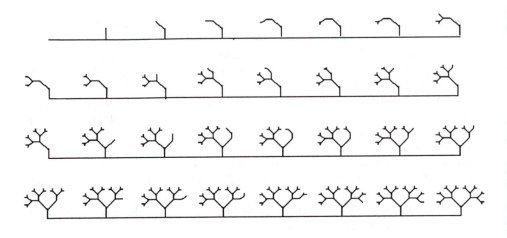

FIGURE 3 Recursive tree-drawing sequence.

Figure 2f, which I think of, somewhat arbitrarily, as the basic, "archetypal" body form generated by my artificial embryology. The sequence of pictures in Figure 3 shows the sequence of lines by which the tree is drawn by the recursive algorithm.

Real-life embryology is quite like this, and very unlike the pixel-peppering embryology that we thought about before. Genes don't control small fragments of the body, the equivalent of pixels. Genes control growing rules, developmental processes, and embryological algorithms. Powerful though they are, an important feature of these growing rules is that they are local. There is no grand blueprint for the whole body. Instead, each cell obeys local instructions for dividing and differentiating, and when all the local instructions are obeyed together a body eventually results. Each little local region of the tree growing in Figure 3 is like other local regions, and also (though this does not necessarily have to be true of all trees in my artificial world) like a scaled-down version of the whole tree. If, instead of branching a mere four times (this number is controlled by the starting value of *length*), we let it branch, say, 10 times, an apparently complicated structure results (Figure 4). Look carefully at this tree, however, and you'll see that it is built up from fundamentally the same local drawing rules. The individual twigs don't "realize" that they are part of an elaborate pattern. This is all of a piece with the extreme simplicity of the procedure **Tree**. Probably there is an important similar sense in which real-life embryology, too, is simple.

Tree, then, was to be the basis of my embryology. Since the arrays *dx* and *dy*, and the parameter *length*, determine the shape of a tree, these were clearly the numbers that should be controlled by genes. On the face of it this suggests that there should have been 15 genes, 7 for *dx*, 7 for *dy*, and 1 for *length*. However, I wanted, for biological reasons, to add one more constraint. Most animals, as remarked above, are bilaterally symmetrical. Such a requirement could be built into the biomorphs

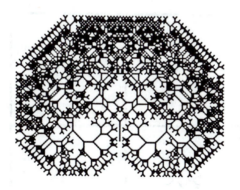

FIGURE 4   Basic tree with high-order branching.

by constraining certain members of the dx and dy parameters to be equal to one another, sometimes with opposite sign. Instead of 15 genes, therefore, I ended up with 9. Genes 1 to 3 control clusters of the dx array. Genes 4 to 8 control clusters of the dy array. And Gene 9 controls the starting value of *length*, the "order" of the recursive tree or in other words the number of branchings. The details are as follows:

```
procedure PlugIn(ThisGenotype: Genotype);
{PlugIn translates genes into variables needed by Tree}
begin
      order := gene[9];
      dx[3] := gene[1]; dx[4] :=gene[2]; dx[5] := gene[3];
      dx[1] := −dx[3]; dx[0] := −dx[4]; dx[2] := 0; dx[6] := 0; dx[7] := −dx[5];
      dy[2] := gene[4]; dy[3] := gene[5]; dy[4] := gene[6]; dy[5] := gene[7]; dy[6] :=
            gene[8];
      dy[0] := dy[4]; dy[1] := dy[3]; dy[7] := dy[5];
end {PlugIn};
```

The call of Tree then follows:

Tree(Startx, Starty, order, Startdir, dx, dy);

and the appropriate biomorph is drawn.

   PlugIn and Tree, then, are called in sequence within Develop, to draw any particular biomorph. Reproduce is then called a dozen or so times, breeding a litter of mutant progeny whose phenotypes are drawn, by Develop, on the screen (Figure 5). A human chooser then chooses one of the litter for breeding, its genes are fed into Reproduce, the screen is cleaned and a new litter of progeny drawn, and the cycle

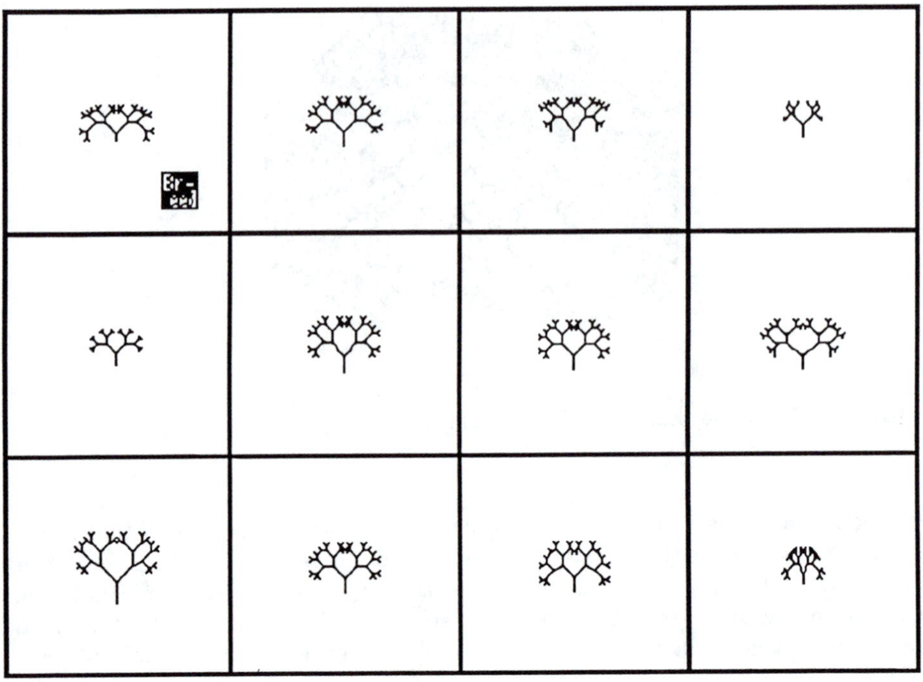

FIGURE 5 Breeding screen.

continues indefinitely. As the generations go by, the forms evolve gradually in front of the chooser, who witnesses true evolution by Darwinian (artificial) selection.

As I said before, I had the intuitive feeling that some kind of recursive procedure would prove to be both morphologically prolific and biologically interesting. But I deliberately did not give much thought to the precise details of the recursive algorithm, because I wanted as much as possible to emerge rather than being designed. I had the feeling that it should not really matter how the genes affected development, provided the development procedure was some kind of recursive drawing rule. In the event, my intuition proved to have been a considerable underestimate. I was genuinely astonished and delighted at the richness of morphological types that emerged before my eyes as I bred. Figure 6 shows a small sample portfolio. Notice how un-treelike many of the biomorphs are. They have evolved under the selective influence of a zoologist's eye, so it is not surprising that many of them resemble animals. Not *particular* animals that actually exist, necessarily, but several of the specimens in Figure 6 would not look out of place in a zoological textbook. Some were bred to resemble other things, for instance, the spitfire at middle left and the silver coffeepot at top right. It was biomorphs of the type shown in Figure 6 that

I used[3] to illustrate the power of gradual cumulative selection to build up morphologies. The rest of this paper is about other families of biomorphs with more elaborate embryological principles.

In introducing the superiority of constrained embryologies over "pixel-peppering," I implied that constrainedness was a virtue. So it is, but you can have too much of a good thing. Having arrived at our basic recursive embryology, and found it good, are there any judicious relaxations or additions we can make to it, which will improve its biological richness? Remember that in doing this we must resist the temptation to take the easy route and build in known biological details. Our watchword is that as much as possible must emerge rather than being designed. But having seen the range of phenotypes that emerge from the basic program, can we think of any modifications to the basic program that seem likely to unleash opulent flowerings of new emergent properties? In seeking such powerful enrichments, we need not fear to make use of general biological principles. All that we must avoid is building in detailed biological knowledge.

I have already mentioned symmetry as an important constraint in real biology. The basic program produces biomorphs that all have to be bilaterally symmetrical. What if we relax this constraint, and allow asymmetrical biomorphs? It now becomes possible to breed forms like Figure 7a. Not very interesting in itself, but let us allow our biomorphs this kind of asymmetry nevertheless, because it may

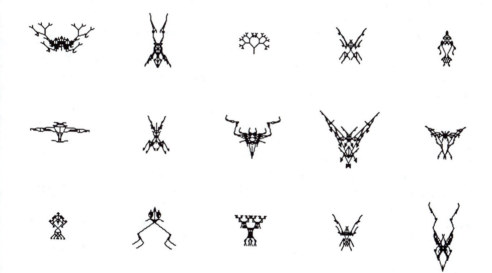

FIGURE 6 Portfolio of biomorphs varying according to nine genes.

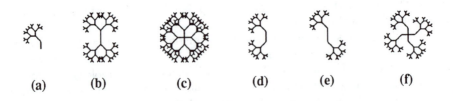

(a)   (b)   (c)   (d)   (e)   (f)

FIGURE 7 Asymmetrical biomorph (a), up-down symmetry (b), four-way radial symme try (c), up-down symmetry as a reflection in a horizontal mirror (d), up-dowm symmetry by rotation (e), and left-right asymmetry with four-way radial symmetry (f).

interact interestingly with other relaxations of the basic embryology that we shall allow. Let us, then, invent a new gene, one with only two values, on and off, which switches bilateral symmetry on or off.

Left-right is not the only plane of symmetry at play in real animals' evolution. Suppose we take our basic *Blind Watchmaker* program but allow a new gene to switch on and off symmetry in the up-down direction. We can then breed shapes like Figure 7b. We can give this new gene an additional possible value, to enable it to switch on full, four-way radial symmetry (Figure 7c). Now let's come back to our other gene, for left-right asymmetry, and combine it with up-down symmetry. Here we have a decision to make. Do we consider that up-down symmetry is achieved by reflection in a horizontal mirror, in which case we can obtain a picture like Figure 7d? Or do we consider that it is achieved by rotation, in which case we shall obtain a picture like Figure 7e? I arbitrarily decided on the latter. When we combine left-right asymmetry with four-way radial symmetry we get a picture like Figure 7f. These symmetry genes, it seems to me, have the potential that we seek. They seem to have the power to add rich emergent properties without the programmer having built in a lot of contrived design. I shall return to give further examples of the additional flowerings of biomorph structure that these two symmetry genes permit.

Symmetry is not the only such principle that is suggested by a general knowl-edge of real zoology. Segmentation is another widespread biological phenomenon that lends itself to biomorphic treatment. Animals belonging to three of the most successful phyla—vertebrates, arthropods and annelids—organize their bodies along segmented lines. Segments are repeated modules, more or less the same as each other, running from head to tail like the trucks in a train. In an annelid worm such as an earthworm, or some arthropods such as millipedes, the segmentation is ex-tremely obvious. A millipede really is rather like a train. It is easy to imagine that the developmental program of a millipede has the instructions for building a single segment, then sticks those instructions in a repeat loop. Segmentation is equally obvious in some vertebrates, such as snakes and fish when viewed internally. In mammals such as ourselves it is less prominent, but it is clearly seen in the backbone. Not only are the vertebral bones themselves "repeated" down the backbone; so are a whole series of associated blood vessels, muscles and nerves. Even the skull was originally segmented, but the traces of segmentation are now so well hidden that uncovering it was one of the triumphs of comparative anatomy. It is probably fair to say that the invention of segmentation was one of the major innovations in the history of life, an invention that was made at least twice, once by an ancestor of the vertebrates, and once by an ancestor of the annelids and arthropods (who are probably descended from a segmented common ancestor).

How might we change the basic *Blind Watchmaker* program to allow biomorphs to be segmented like millipedes? The obvious way is to use the basic program to generate a single segment. Then, just as I speculated for millipedes above, enclose it in a repeat loop. So I added this feature to the program, with a new gene controlling the number of segments, and another new gene controlling the distance between segments. Figure 8a shows a series of biomorphs, identical except with respect to the value of the first of these genes. And in Figure 8b is a series, identical except with respect to the value of the second gene, the one controlling the distance between segments.

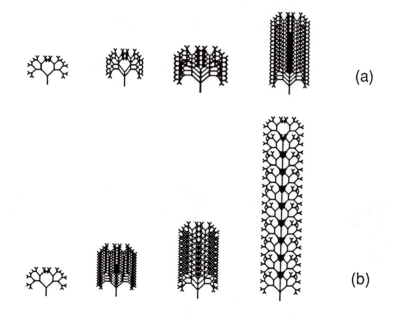

FIGURE 8   (a) shows a series of biomorphs which are identical except with respect to the value of the segment-number gene and (b) is a series which is identical except with respect to the value of the segment-distance gene.

Finally, I introduced genes controlling segmental *gradients.* The segments of a millipede may all look pretty much the same, but many segmented animals taper, being broadest at the front and narrowest at the back. Others, for instance, wood lice and many fish, are narrow at each end and broad in the middle. The segments all down the body follow the same basic plan but get progressively larger, or smaller, over stretches of the animal's length from front to rear. It is as though the developmental **repeat** loop includes scaling factors that are incremented or decremented each time the program passes through the loop.

In introducing segmental gradients into the biomorph program, I aimed for greater generality than could be achieved with a single scaling factor. I allowed for gradients affecting the expression of each of the nine basic genes, and the interseg-ment distance gene, separately. Figure 9 shows what happens if you put a gradient on Gene 1 and on Gene 4. The left-hand biomorph has no gradients and all the segments are the same. The middle biomorph is the same except that, as you move from front to rear, Gene 1's expressivity increases by one unit per segment. The right hand biomorph shows the same for Gene 4.

Figure 10 is a portfolio of segmented biomorphs, most of them with gradients. These are biomorphs that have been bred by selection, in the same way as those in Figure 6, but with the possibility of mutation in the two segment-controlling genes. You can think of the Figure 6 biomorphs as having had those two genes set to zero. If you compare Figure 10 with Figure 6, I think you'll agree that segmentation has added to the zoological interest of the specimens. An embryological innovation, in this case segmentation, has allowed the evolution of a whole new range of types. We may conjecture that much the same thing happened in the ancestry of the vertebrates and in the separate ancestry of the annelids and arthropods.

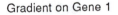

No Gradient   Gradient on Gene 1   Gradient on Gene 4

FIGURE 9   Effect of gradient on two genes.

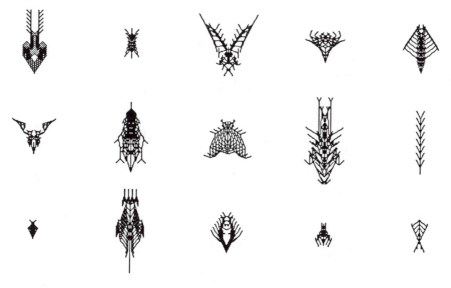

FIGURE 10   Portfolio of segmented biomorphs.

   But the animals in Figure 10 are all bilaterally symmetrical. What happens if we combine segmentation with asymmetry? At this point in the programming, I introduced another arbitrary constraint. I simply decreed that when a one-sided animal like Figure 11a became segmented, instead of each segment simply repeating the asymmetry as in Figure 11b, the successive segments should always be asymmetrical in alternate directions, as in Figure 11c. There was no particular reason why I should have done this. I think I did it partly because I wanted to use segmentation to help recreate, at the level of the whole body, the symmetry that had been lost at the level of the individual segment. And I also think I did it partly to make the biomorphs more botanically interesting. Many plants send off alternating buds. Indeed, my portfolio (Figure 12) of asymmetrical segmented biomorphs, may interest botanists more than zoologists. When I was breeding them by artificial selection on the screen, I frequently had plants in my mind. The biomorph at the top

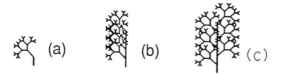

FIGURE 11   One-sided animal (a), segmented with repeating symmetry (b), and with successive segments asymmetrical in alternate directions.

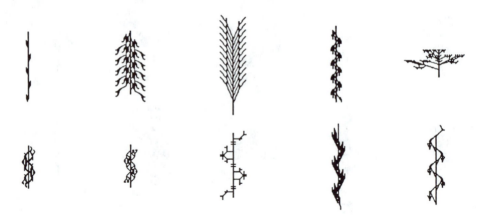

FIGURE 12   Portfolio of asymmetrical segmented biomorphs.

left could be any of a wide variety of plant species. The one next to it could be an inflorescence, or perhaps a colonial animal such as a siphonophore. While looking at the top row, notice the barley and the cedar of Lebanon. The next row begins with DNA, and contains a dollar sign which also looks rather like a different view of DNA.

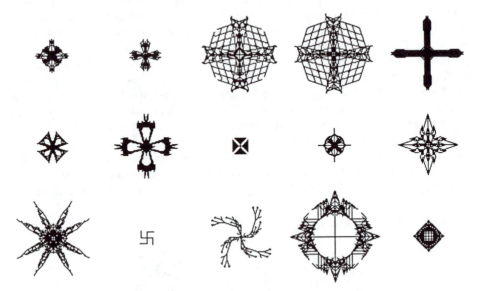

FIGURE 13   Portfolio of radially symmetrical biomorphs.

FIGURE 14 Portfolio of "echinoderms."

Finally, let us switch on our other newly invented gene, the one that controls up-down symmetry and radial symmetry. Figure 13 is a portfolio of biomorphs bred with this gene, and the segmentation genes, and the lateral symmetry gene, all permitted to mutate. Many of these are more like human artifacts, ornaments or regalia than like living organisms. But there is a nice scarlet pimpernel at the right of the middle row, and the eight-pointed star at the bottom left suggested to me that I should try to breed a portfolio of echinoderms (starfish, sea urchins, brittle stars, etc).

I present my "echinoderm" portfolio (Figure 14) as another illustration of all my "new" mutation types—all the various symmetry mutations, and the segmentation and gradient mutations. Any zoologist will instantly spot the trouble with these "echinoderms." They have four-way radial symmetry rather than five-way symmetry. The present program is not capable of producing five-way radial symmetry. Once again, we are brought back to our major theme. Huge vistas of evolutionary possibility, in real life as well as in artificial life, may be kept waiting a very long time, if not indefinitely, for a major, reforming change in embryology. Which brings me to the problems of the biomorph alphabet.

If we can breed animals and plants, I thought, why not any arbitrarily designated shapes? You can't get much more arbitrary than the alphabet, so how about an alphabet of biomorphs? While breeding, wandering around in "Biomorph Land,"[3] I would sometimes encounter biomorphs that had a slight look of one of the letters of the alphabet. I then would try to perfect the resemblance by selective breeding, preserving intermediate results and returning to try again whenever I had an idle moment. In this way I gradually built up an album of alphabetic characters. My aim was eventually to sign my name legibly in biomorphic script. Figure 15a shows that I have not quite achieved this yet, although I had more luck (Figure 15b) in using biomorph script to pay tribute to the uniquely brilliant microcomputer which made all this work so easy. The problems are not trivial. Some

ΓICHAΓ▷   ▷AWKIN$ (a)

MACINTO$H   (b)

FIGURE 15 Biomorphs that resemble letters of the alphabet.

letters are perfect, like I and N. Others are a little odd, but still not unpleasing, for instance S and A. D is pretty horrible, and I can't seem to get rid of the irritating little upward-pointing tail. As for K, I despair of ever breeding a proper K. I simply had to fake it in my name by running part of a K into the preceding W, an obvious cheat.

And that is the point. There are some shapes that certain kinds of embryology seem incapable of growing. My present *Blind Watchmaker* embryology, that is the basic nine genes plus segmentation with gradients and symmetry mutations, is, I conjecture, forever barred from breeding a respectable K, or a capital B. Or if I am proved wrong in this particular conjecture, I am fairly sure there are *some* kinds of shape that the present *Blind Watchmaker* program can never breed. Just as we extended the basic program by adding segmentation and symmetry genes, there are presumably other extensions that could be made, which would make difficult letters of the alphabet become easy. For instance, it is not far-fetched to guess that, if we relaxed the "alternation" constraint on segmented asymmetrical biomorphs, K would become easy. The way to do this, in keeping with our earlier extensions to the embryology, would be to add an "alternation gene" which could mutate itself on or off. Then all present biomorphs would be a subset of the larger set that would become possible—the subset with the alternation gene permanently turned on. K would also be easy to breed if we had taken the decision to achieve up-down symmetry by mirror reflection rather than by rotation (see above). Leaving the biomorph alphabet on one side, many other new kinds of mutations could be implemented, including genes controlling color. The prototype color version that I now have running on the Macintosh II computer, generously provided by Apple Computer Inc., produces results spectacular beyond my previous imaginings, though so far the aesthetic appeal of these colored biomorphs has overshadowed their biological interest.

Finally, let us return to the evolution of evolvability. The point I have been trying to make so far in this paper is that certain kinds of embryology find it difficult to generate certain kinds of biomorphs; other kinds of embryology find it easy to do so. It is clear that we have here a powerful analogy for something important about real biology, a major principle of real life that is illustrated by artificial life. It is less clear which of several possible principles it is! There are two main candidates, which I must take some time to expound in order to explain why I favor one rather than the other.

In order to explain the first one, we need to make a preliminary distinction between two kinds of mutation: ordinary changes within an existing genetic system, and changes to the genetic system itself. Ordinary changes within an existing genetic system are the standard mutations that may or may not be selected in normal evolution within a species. One allele is replaced by an alternative allele at the same locus, as in the famous case of industrial melanism where a gene for blackness spread through moth populations in industrial areas (see any biology textbook). This is how all normal evolutionary change happens. But it is an inescapable fact that different species, to a greater or lesser extent, have different genetic systems from one another, even if this only means that they have different numbers of chromosomes. "The same locus," when we are talking about an elephant and a human, may not even be a meaningful thing to say. Humans and elephants employ basically the same *kind* of genetic system, but they don't have the same genetic system. They have different numbers of chromosomes and you can't make a locus-for-locus mapping between them like you can between two individual humans. Yet humans and elephants undoubtedly have a common ancestor. Therefore, during their evolutionary divergence, there must have been changes to the genetic systems, as well as changes within the genetic systems. These changes to genetic systems must have been, at least in one sense, major changes, changes of a different order from the normal allele substitutions that go on within a genetic system.

Now, the changes to the biomorph program that I have been talking about in this paper—the addition of symmetry mutations and segmentation mutations—are, as it happens, changes of just this character, changes to the genetic system itself. They constituted major rewritings of the program to increase the "chromosome" size from 9 to 16 genes.[1] But I want to argue that this is incidental. This is not the analogy that I want to draw between artificial life and real life. Changes in genetic systems must, indeed, be fairly commonplace in the history of life: changes in chromosome number are nearly as common as initiations of new species, and the number of species initiations that have occurred in the history of life on earth is probably to be counted in the hundreds of millions. So, although changes in genetic systems are much rarer than allelic substitutions within genetic systems, they are not very rare events on the geological timescale.

What I want to argue is that there is another class of evolutionary innovations which *are* very rare on the geological timescale and which I shall call evolutionary

---

[1]Not all the 16 have been discussed here. For details of what the remaining genes do, see the Instruction Manual supplied with the disc of the Macintosh program, which is marketed in America by W. W. Norton & Co., 500 Fifth Avenue, New York 10110, and in Britain by W. W. Norton & Co., 37 Great Russell Street, London WC1B 3NU. The Instruction Manual is also printed as an Appendix to the Norton paperback edition of my book,[3] which also gives details on how to obtain the program at a reduced price. A version of the program for the Research Machines Nimbus computer, with only the original nine genes, is being sold by Software Production Associates Ltd., P. O. Box 59, Leamington Spa CV31 3QA, England.

watersheds. An evolutionary watershed is something like the invention of segmentation which, as we have seen, may have occurred only twice in history, once in the lineage leading to annelids and arthropods and once in the lineage leading to vertebrates. A watershed event like this may or may not have coincided with a change in the genetic system such as a change in chromosome number. In any case that is not what is interesting about watershed events. What is interesting about them is that they open floodgates to future evolution.

I suspect that the first segmented animal was not a dramatically successful individual. It was a freak, with a double (or multiple) body where its parents had a single body. Its parents' single body plan was at least fairly well adapted to the species' way of life; otherwise they would not have been parents. It is not, on the face of it, likely that a double body would have been better adapted. Quite the contrary. Nevertheless, it survived (we know this because its segmented descendants are still around), if only (this, of course, is conjecture) by the skin of its teeth. Even though I may exaggerate when I say "by the skin of its teeth," the point I really want to make is that the individual success, or otherwise, of the first segmented animal during its own lifetime is relatively unimportant. No doubt many other new mutants have been more successful as individuals. What is important about the first segmented animal is that its descendant lineages were champion *evolvors*. They radiated, speciated, and gave rise to whole new phyla. Whether or not segmentation was a beneficial adaptation during the individual lifetime of the first segmented animal, segmentation represented a change in embryology that was pregnant with evolutionary potential.

Not all evolutionary watersheds are as dramatic in their magnitude or in their evolutionary consequences as the invention of segmentation. There may be many changes in embryology which, though not dramatic enough in themselves even to deserve the title watershed, nevertheless are, to a lesser extent than the invention of segmentation, evolutionarily pregnant. Suppose we rank embryologies in order of evolutionary potential. Then as evolution proceeds and adaptive radiations give way to adaptive radiations, there is presumably a kind of ratchet such that changes in embryology that happen to be relatively fertile, evolutionarily speaking, tend to be still with us. New embryologies that are evolutionarily fertile tend to be the embryologies that characterize the forms of life that we actually see. As the ages go by, changes in embryology that increase evolutionary richness tend to be self-perpetuating. Notice that this is not the same thing as saying that embryologies that give rise to good, healthy individual organisms tend to be the embryologies that are still with us, although that, too, is no doubt true. I am talking about a kind of higher-level selection, a selection not for survivability but for evolvability.

It is all too easy for this kind of argument to be used loosely and unrespectably. Sydney Brenner justly ridiculed the idea of foresight in evolution, specifically the notion that a molecule, useless to a lineage of organisms in its own geological era, might nevertheless be retained in the gene pool because of its possible usefulness in some future era: "It might come in handy in the Cretaceous!" I hope I shall not be taken as saying anything like that. We certainly should have no truck with suggestions that individual animals might forgo their selfish advantage because of possible long-term evolutionary benefits to their species. Evolution has no foresight. But with hindsight, those evolutionary changes in embryology that *look* as though they were planned with foresight are the ones that dominate successful forms of life.

Perhaps there is a sense in which a form of natural selection favors, not just adaptively successful phenotypes, but a tendency to evolve in certain directions, or even just a tendency to evolve at all. If the embryologies of the great phyla, classes and orders of animals display an "eagerness" to evolve in certain directions, and a reluctance to evolve in other directions, could these "eagernesses" and "reluctances" have themselves been favoured by a kind of natural selection? Is the world filled with animal groups which not only are successful, as individuals, at the business of living, but which are also successful in throwing up new lines for future evolution?

If that were all there was to it, it would be simply another case, like sieving sand, of what I have called "single-step selection"[3] and therefore not very interesting, evolutionarily speaking. It is only *cumulative* selection that is evolutionarily interesting, for only cumulative selection has the power to build new progress on the shoulders of earlier generations of progress, and hence the power to build up the formidable complexity that is diagnostic of life. I have been in the habit of disparaging the idea of "species selection"[4,5] because, as it is normally presented, it is a form of single-step selection, not cumulative selection, and therefore not important in the evolution of complex adaptations.[2] But selection among embryologies for the property of evolvability, it seems to me, may have the necessary qualifications to become cumulative in evolutionarily interesting ways. After a given innovation in embryology has become selected for its evolutionary pregnancy, it provides a climate for new innovations in embryology. Obviously the idea of each new adaptation serving as the background for the evolution of subsequent adaptations is commonplace, and is the essence of the idea of cumulative selection. What I am now suggesting is that the same principle may apply to the evolution of evolvability which, therefore, may also be cumulative.

Others have pointed out that we should speak of "species selection" only in those rare cases where a true species-level quality is being evolved.[6] Species selection, for instance, should not be invoked to explain an evolutionary lengthening of the leg, since species don't have legs, individuals do. It might, on the other hand, be invoked to explain the evolution of a tendency to speciate, since speciating is a thing species, but not individuals, do. It now seems to me that an embryology that is pregnant with evolutionary potential is a good candidate for a higher-level property of just the kind that we must have before we allow ourselves to speak of species or higher-level selection.

The world is dominated by phyla, classes and orders whose embryology equipped them to diverge and inherit the earth. Although I am sure I have always been dimly aware of this, I think it is true to say that it is the biomorph program—writing it, playing with it, and above all modifying it to increase its evolutionary potential— that has really drummed it into my innermost consciousness. So what started out as an educational exercise—I was trying to develop a tool to teach other people—ended up as an educational exercise in another sense. I ended up teaching something to myself about real life. There could be less worthy uses of artificial life.

ACKNOWLEDGMENTS

I thank Alan Grafen, Ted Kaehler and Helena Cronin for help and discussions of various kinds; Chris Langton for inviting me to Los Alamos to one of the most stimulating conferences I have been to; and Apple Computer Inc. whose generosity made it possible for me to go to Los Alamos.

REFERENCES

1. Dawkins, R. (1976), *The Selfish Gene* (Oxford : Oxford University Press).
2. Dawkins, R. (1982), *The Extended Phenotype* (Oxford: W. H.Freeman).
3. Dawkins, R. (1986), *The Blind Watchmaker* (Harlow: Longman).
4. Eldredge, N. (1985), *Unfinished Synthesis* (New York: Oxford University Press).
5. Gould, S. J. (1982), "The Meaning of Punctuated Equilibrium and its Role in Validating a Hierarchical Approach to Macroevolution," *Perspectives on Evolution* , Ed. R. Milkman (Sunderland, MA: Sinauer).
6. Maynard-Smith, J. (1983), "Current Controversies in Evolutionary Biology," *Dimensions of Darwinism*, Ed. M. Grene (Cambridge: Cambridge University Press), 71-93.
7. Morris, D. (1987), *The Secret Surrealist* (Oxford: Phaidon).
8. Weismann, A. (1893), *The Germ Plasm: a Theory of Heredity*, Transl. W. N. Parker & H Rönnfeldt (London: W. Scott).

Aristid Lindenmayer† & Przemyslaw Prusinkiewicz‡

†Theoretical Biology Group, University of Utrecht, Paddualaan 8, 3584 CH Utrecht, The Netherlands and ‡Department of Computer Science, University of Regina, Regina, Saskatchewan, Canada S4S 0A2

Developmental Models of Multicellular Organisms: A Computer Graphics Perspective

This paper presents an algorithmic approach to the description, analysis and developmental simulation of multicellular organisms. The emphasis is put on mechanisms which control development in nature: cellular descent and interaction. The construction of a mathematical model gives an idea of how many and what kind of control factors are necessary to achieve different types of development. The results are expressed in terms of the theory of L-systems, and illustrated with computer-generated images of modeled plants.

1. INTRODUCTION

The development of multicellular organisms is a formidably complex process. There are copies of thousands of genes present in each cell and each of the genes in any one cell can be either in an active or inactive state. For n genes the number of active combinations is 2^n. We are obviously dealing with immense numbers of possible genomic states. In addition, we must consider the cytoplasmic states and cell membrane configurations. Clearly, an attempt to understand development in terms of gene expression needs some simplifying principles. One has to consider the main types of control which play a role in development, such as control by cell interactions or control by cellular descent. Also one should distinguish between the temporal effects of genomic states and the spatial effects of the membrane components during morphogenesis. Control factors of this kind and their effects on the spatiotemporal structures of developing organisms can be formalized in terms of discrete constructs similar to grammars or cellular automata.

While Chomsky grammars give rise to sets of longer and longer expressions and thus can be used to simulate development, they have the disadvantage of employing only sequential rewriting rules. Organisms obviously develop in a parallel way. Cellular automata, on the other hand, satisfy the parallel requirement, but expand only at the edges of the cellular array. We need structures which are able to expand everywhere. For this purpose the formalism of L-systems was introduced by Lindenmayer.[20]

In this paper we discuss methods for realistically modeling cellular structures and entire plants. To this end, we complement developmental models based on L-systems with geometric and graphic attributes. We first apply this approach to produce developmental sequences of one-dimensional filamentous organisms. Then we extend it to branching structures and illustrate with models of higher plants. In this context, we contrast the role of cellular interaction with that of cellular descent in developmental control. Further extensions of the developmental model lead to the generation of two-dimensional and three-dimensional organs which cannot be described in terms of tree topology. The corresponding extensions of the L-system formalism are known as map L-systems and cellworks. We conclude by presenting selected results from the mathematical theory of L-systems which are useful in constructing developmental models of biological structures.

2. MODELING OF FILAMENTS

2.1 OL-SYSTEMS

The following definition of string L-systems is appropriate to model filamentous organisms without interactions. For more details, see Herman[10] and Rozenberg.[37]

Let V denote an alphabet, V^*—the set of all words over V, and V^+—the set of all nonempty words over V. An *OL-system* is an ordered triplet $G = \langle V, \omega, P \rangle$ where V is the *alphabet* of the system, $\omega \in V^+$ is a nonempty word called the *axiom* and $P \subset V \times V^*$ is a finite *set of productions*. If a pair (a, χ) is a production, we write $a \to \chi$. The letter a and the word χ are called the *predecessor* and the *successor* of this production, respectively. It is assumed that for any letter $a \in V$, there is at least one word $\chi \in V^*$ such that $a \to \chi$. If no production is explicitly specified for a given predecessor $a \in V$, we assume that the *identity production* $a \to a$ belongs to the set of productions P. A OL-system is *deterministic* (noted *DOL-system*) if and only if for each $a \in V$ there is exactly one $\chi \in V^*$ such that $a \to \chi$.

Let $\mu = a_1 \ldots a_m$ be an arbitrary word over V. We will say that the word $\nu = \chi_1 \ldots \chi_m \in V^*$ is *directly derived* from (or *generated* by) μ and write $\mu \Rightarrow \nu$ if and only if $a_i \to \chi_i$ for all $i = 1, \ldots, m$. A word ν is generated by G in a derivation of *length n* if there exists a *developmental sequence* of words $\mu_0, \mu_1, \ldots, \mu_n$ such that $\mu_0 = \omega$, $\mu_n = \nu$ and $\mu_0 \Rightarrow \mu_1 \Rightarrow \ldots \Rightarrow \mu_n$.

2.2 INTERACTIONLESS DEVELOPMENT OF BLUE-GREEN BACTERIA

We will illustrate the operation of OL-systems by simulating the development of a multicellular filament such as that found in blue-green bacteria *Anabaena catenula* and various algae.[24,29] The symbols a and b represent cytological states of the cells (in this case, these have to do with their size and readiness to divide). The subscripts l and r indicate cell polarity which specifies the positions in which daughter cells of type a and b will be produced. The development is governed by the following rules:

$$p_1 : a_r \to a_l b_r$$
$$p_2 : a_l \to b_l a_r$$
$$p_3 : b_r \to a_r$$
$$p_4 : b_l \to a_l$$

Starting from a single cell a_r (the axiom), the above L-system generates the following sequence of words:

$$a_r$$
$$a_l b_r$$
$$b_l a_r a_r$$
$$a_l a_l b_r a_l b_r$$
$$b_l a_r b_l a_r a_r b_l a_r a_r$$
$$\ldots$$

Under a microscope, the cells appear as cylinders of various lengths. The a-type cells are longer than the b-type cells. Adhering to such an interpretation of symbols a and b, one can produce simulated images of filament development shown on the left side of color plate 9 (see color insert). The plot on the right side of this plate presents the same developmental process using a continuous time scale instead of discrete steps. Horizontal sections of the plot represent cell sequences as a function of time. Colors correspond to cell age. In time intervals between subdivisions, the cells elongate, and then subdivide after reaching the threshold age. The continuous extension of the formalism of L-systems, which was used to produce this plot, originates from Baker and Herman[3] and Lindenmayer[21].

2.3 SIMULATION OF INTERACTION

Productions in OL-systems are context-free, i.e. applicable regardless of the context in which the predecessor appears. This type of production is sufficient to model cellular descent, or information transfer from the parent cell to its descendants. However, a context-sensitive extension of L-systems is necessary to model cellular interactions, or information exchange between neighboring cells. Various possible extensions have been proposed and thoroughly studied in the past.[10,22,38] Specifically, *2L-systems* use productions of the form $a_l < a > a_r \rightarrow \chi$, meaning that the letter a (called the *strict predecessor*) can produce word χ if and only if a is preceded by letter a_l and followed by a_r. Thus, letters a_l and a_r form the left and the right *context* of a in this production. Productions in *1L-systems* have one-sided context only; consequently, they are either of the form $a_l < a \rightarrow \chi$ or $a > a_r \rightarrow \chi$. OL-systems, 1L-systems and 2L-systems belong to a wider class of *IL-systems*, also called *(k,l)-systems*. In a (k,l)-system, the left context is a word of length k, and the right context is a word of length l.

In order to keep specifications of L-systems short, we slightly modify the usual notion of IL-systems by allowing productions with different context lengths to co-exist within a single system. Furthermore, we assume that context-sensitive productions have precedence over context-free productions with the same (strict) predecessor. Consequently, if a context-free and a context-sensitive production both apply to a given letter, the context-sensitive one should be selected. If no production applies, this letter is replaced by itself as previously assumed for OL-systems.

As an example of cellular interaction, consider diffusion of a hormone along a filament. If a denotes a cell with hormone concentration below a threshold level and b is a cell with concentration exceeding this level, the diffusion process can be described by the following 1L-system:

$$\omega : baaaaaaa$$
$$p : b < a \rightarrow b$$

The first few words generated by this L-system are given below:

$$baaaaaaaa$$
$$bbaaaaaaa$$
$$bbbaaaaaa$$
$$bbbbaaaaa$$
$$bbbbbaaaa$$

$$\cdots$$

Thus, the hormone propagates throughout the filament, starting from its left end. Such a hormonal mechanism is generally accepted to be responsible for the spacing of heterocysts (specialized cells for nitrogen fixation) in the blue-green bacteria discussed above. A nitrogen-rich compound (probably glutamine) is produced in heterocysts and is transported from cell to cell along the growing filament consisting of dividing photosynthetic cells. New heterocysts are induced in areas where the concentration of the compound falls below a minimum. Kinetic studies of this process, yielding the correct heterocyst distributions, have been carried out using IL-systems by Baker and Herman,[3] and de Koster and Lindenmayer (to be published in *Acta Biotheoretica*). A similar mechanism is used in Section 4.5 to control development of branching structures.

3. PRINCIPLES OF THE MODELING OF BRANCHING STRUCTURES

3.1 GRAPH-THEORETICAL AND BOTANICAL TREES

In the context of plant modeling, a branching structure or "tree" must be carefully defined to avoid ambiguity. To this end, we introduce the notion of an axial tree (Figure 1) which complements the graph-theoretic notion of a rooted tree[33] with the botanically motivated notion of branch axis.

A *rooted tree* has edges which are labelled and directed. The edge sequences form paths from a distinguished node called the *root* or the *base* to the *terminal nodes*. In the biological context, these edges are referred to as *branch segments*. A segment followed by at least one more segment in some path is called an *internode*. A terminal segment (with no following edges) is called an *apex*.

An *axial tree* is a special type of rooted tree. At each of its nodes we distinguish at most one outgoing *straight* segment. All remaining edges are called *lateral* or *side* segments. A sequence of segments is called an *axis*, if:

- the first segment in the sequence originates at the root of the tree or as a lateral segment at some node,
- each subsequent segment is a straight segment, and
- the last segment is not followed by any straight segment in the tree.

Together with all its descendants, an axis constitutes a *branch*. A branch is itself an axial (sub)tree.

Axes and branches are ordered. The axis originating at the root of the entire plant has order zero. An axis originating as a lateral segment of an n-order parent axis has order $n + 1$. The order of a branch is equal to the order of its lowest-order or *main* axis.

Axial trees are purely topological objects. The geometric connotation of such terms as straight segment, lateral segment and axis should be viewed at this point as an intuitive link between the graph-theoretic formalism and real plant structures.

3.2 BRACKETED REPRESENTATION OF AXIAL TREES AND BRACKETED L-SYSTEMS

The definition of an axial tree does not specify the data structure for representing it and carrying out derivations. The most straightforward approach is to use a list representation with a tree topology. In this case, edges correspond to list elements and edge connections are represented by links. Derivations are carried out as list operations. Assuming that double links are used, both the left and the right context of any edge can be found easily (in constant time). However, the explicit specification of links requires a relatively large amount of memory to store pointers in addition to edge labels. This is a significant drawback in plant modeling applications, as the axial trees considered can contain hundreds of thousands of edges.

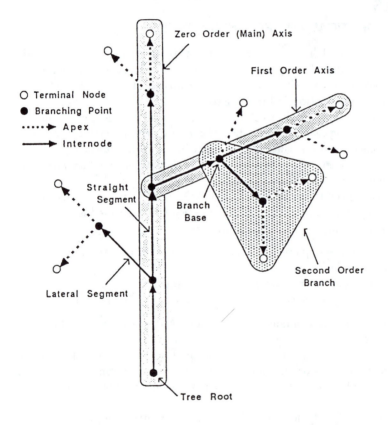

FIGURE 1
An axial tree.

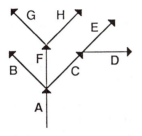

FIGURE 2   Example of a tree.

The notion of *bracketed strings*[20] offers a different representation. A tree with edge labels from alphabet V is represented by a string over alphabet $V \cup \{[,]\}$, where the bracket symbols "[" and "]" indicate branches. For example, the tree shown in Figure 2 is represented by the bracketed string:

$$A[B][C[D]E]F[G][H]$$

An L-system operating on strings with brackets is called a *bracketed L-system*. Derivations in bracketed OL-systems proceed as in string OL-systems (without brackets). Specifically, the brackets are rewritten into themselves. The context-sensitive case is comparatively more complex because symbols representing adjacent tree segments can be separated by an arbitrarily large number of other symbols in the bracketed string representation. Consequently, special rules for context searching are needed. In this paper we consider a restricted case where left and right contexts are limited to single letters and daughter branches do not belong to the context of the mother branch. This approach corresponds to the original definition of bracketed L-systems with interactions.[20] For example, in the string:

$$ABCD[EF][G[HI[JKL]M]NOPQ]$$

D is the left context of G, and N is the right context of G.

3.3 GEOMETRIC INTERPRETATION OF BRACKETED STRINGS

The geometric interpretation of bracketed strings is significantly more complex than the representation of filaments. Several possible interpretations were proposed in the past.[2,7,11,42,43] We describe here a graphical interpretation based on the notion of a LOGO-style *turtle*,[1] as first proposed by Szilard and Quinton,[44] and further developed by Prusinkiewicz.[34,35] The turtle is represented by its *state* which consists of turtle *position* and *orientation* in the Cartesian coordinate system, as well as other attribute values, such as current color and line width. The orientation is defined by three vectors $\vec{H}, \vec{L}, \vec{U}$, indicating the turtle's *heading*, the direction to the *left*, and direction *up*. These vectors have unit length, are perpendicular to each other, and satisfy the equation $\vec{H} \times \vec{L} = \vec{U}$. Rotations of the turtle can then be expressed by the equation:

$$\begin{bmatrix} \vec{H}' & \vec{L}' & \vec{U}' \end{bmatrix} = \begin{bmatrix} \vec{H} & \vec{L} & \vec{U} \end{bmatrix} \mathbf{R}$$

where $\mathbf{R}$ is a 3×3 rotation matrix.[6]

The turtle interprets a bracketed string by scanning it from left to right. The symbols which denote tree segments move the turtle forward by a given distance d and cause a line to be drawn between the previous and the new position. Seven predefined *attribute symbols* are used to control turtle orientation given an angle increment δ. Specifically, + and − *turn* the turtle left and right around the vector $\vec{U}$, ˆ and & *pitch* the turtle up and down around the vector $\vec{L}$, and \ and / *roll* the turtle left and right around the vector $\vec{H}$ (Figure 3). The symbol | is used to turn the turtle 180° around the vector $\vec{U}$ regardless of the value of δ. Branches are created using a stack: [pushes the current state on the stack, while] pops a state from the stack and makes it the current state of the turtle. No line is drawn in this case, although in general the position of the turtle changes.

The list of predefined symbols can be augmented to control color, diameter and length of segments, incorporate predefined surfaces in the model, and perform other functions as required. Symbols without a specified interpretation are ignored by the turtle, which means that they can be used in the derivation process without affecting the interpretation of the resulting string.

For an example of a three-dimensional bush-like structure generated by an L-system with turtle interpretation see color plate 10. The L-system is given below. In order to make it easier to analyze, complete words are used instead of single-letter symbols.

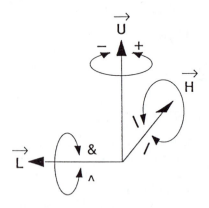

FIGURE 3   Turtle interpretation of geometric attribute symbols.

ω : apex

p_1 : apex $\rightarrow$ [[&stem leaf!apex]/////'[&stem leaf!apex]
///////'[& stem leaf!apex]]

p_2 : stem $\rightarrow$ Ileaf

p_3 : $I \rightarrow I/////$stem

p_4 : leaf $\rightarrow ['''^{\wedge\wedge}\{-f+f+f-|-f+f+f\}]$

The angle increment δ is equal to $22.5°$. The system operates in the following way. Production p_1 creates three branches from an apex. Each branch consists of a stem, a leaf and a new apex which will subsequently create three new branches. Productions p_2 and p_3 specify stem growth. In subsequent derivation steps the stem acquires new internodes I and new leaves (in violation of the subapical growth rule, but with an acceptable visual effect in a still picture). Production p_4 specifies the leaf as a filled polygon with six edges, as discussed further in Section 5.2. The symbols ! and ' are used to decrement the diameter of segments and increment the current index to the color table, respectively.

Color plate 10, as well as most of the other plates included in this paper, were produced using a program called *pfg* designed for a Silicon Graphics IRIS 3130 workstation. The input to this program consists of a (bracketed) L-system and approximately 30 numerical parameters, most of which control viewing and rendering. Additionally, an arbitrary number of files containing bicubic surface descriptions can be read in (see Section 5.1). Growth processes can be animated by interpreting successive strings of a developmental sequence.

4. DEVELOPMENTAL MODELS OF HERBACEOUS PLANTS

In this section we use the formalism of L-systems to present developmental models of herbaceous plants on the topological level. The geometric aspects can be incorporated by extending given L-systems as discussed in Section 3.3. We put particular emphasis on the modeling of compound flowering structures or *inflorescences*. As there is no commonly accepted terminology referring to inflorescence types, we chose to follow the terminology of Müller-Doblies[30] which in turn is based on extensive work by Troll.[46] Our presentation is organized by the control mechanisms which govern inflorescence development.

4.1 RACEMES, OR THE PHASE BEAUTY OF SEQUENTIAL GROWTH

The simplest possible flowering structures with multiple flowers are those with a single stem on which an indefinite number of flowers are produced sequentially. Inflorescences of this type are called *racemes*. Their development can be described by the following OL-system:

$$\omega : A$$
$$p_1 : A \rightarrow I[IF_0]A$$
$$p_2 : F_i \rightarrow F_{i+1} \qquad i \geq 0$$

The edge symbol A denotes the apex of the main (zero-order) axis, I is an internode, and symbols F_i refer to subsequent stages of flower development. The indexed notation $F_i \rightarrow F_{i+1}$ stands for a (potentially infinite) set of productions $F_0 \rightarrow F_1, F_1 \rightarrow F_2, F_2 \rightarrow F_3, \ldots$ The developmental sequence begins as follows:

$$A$$
$$I[IF_0]A$$
$$I[IF_1]I[IF_0]A$$
$$I[IF_2]I[IF_1]I[IF_0]A$$
$$I[IF_3]I[IF_2]I[IF_1]I[IF_0]A$$
$$\cdots$$

It can be seen that, at each developmental stage, the inflorescence contains a sequence of flowers of different ages. The flowers newly created by the apex are delayed in their development with respect to the older ones situated at the stem base. Graphically, this effect is illustrated by a model of lily-of-the-valley shown in color plate 11. Its inflorescence was generated by the above L-system complemented with attribute symbols to control geometry of internodes, branching angles, and structure of organs. The following quotation from D'Arcy Thompson[45] applies:

> A flowering spray of lily-of-the-valley exemplifies a growth-gradient, after a simple fashion of its own. Along the stalk the growth-rate falls away; the florets are of descending age, from flower to bud; their graded differences of age lead to an exquisite gradation of size and form; the time-interval between one and another, or the "space-time relation" between them all, gives a peculiar quality—we may call it phase-beauty—to the whole.

This "phase-beauty" can also be observed in other structures. For example, consider the fern-like plant shown in color plate 12. In this case, nine zero-order branches grow subapically and produce new first-order branches, which also grow subapically and produce leaves. The number of leaves carried on first-order branches, the length of internodes and the leaf size increase with time. These processes are described by the following L-system:

$$\omega : [A][A][A][A][A][A][A][A][A]$$
$$p_1 : A \rightarrow I_0[B]A$$
$$p_2 : B \rightarrow I_0[L_0][L_0]B$$
$$p_3 : I_i \rightarrow I_{i+1} \qquad i \geq 0$$
$$p_4 : L_i \rightarrow L_{i+1} \qquad i \geq 0$$

A and B denote apices of the zero-order and first-order axes, $I_0, I_1, I_2, \ldots$ denote the internodes, and $L_0, L_1, L_2, \ldots$ denote the subsequent stages of leaf development.

4.2 CYMOSE INFLORESCENCES, OR THE USE OF DELAYS

In racemes the apex of the main axis produces lateral branches and continues to grow. In contrast, the apex of the main axis in *cymes* turns into a flower shortly after a few lateral branches have been initiated. Their apices turn into flowers as well, and second-order branches take over. In time, branches of higher and higher order are produced. Thus, the basic structure of a cymose inflorescence is captured in the production:

$$A \rightarrow I[A][A]IF$$

A denotes an apex, I denotes an internode, and F denotes a flower. Note that according to this description, the two branches are identical and grow in concert. In reality this need not be the case, and one lateral branch may start growing before the other. This effect can be modeled by assuming that apices undergo a sequence of state changes which delay their further growth until a particular state is reached. For example, consider the following L-system which describes the development of the rose campion (*Lychnis coronaria*) as analyzed by Robinson[36]:

$$\omega : A_7$$
$$p_1 : A_7 \rightarrow I[A_0][A_4]IF_0$$
$$p_2 : A_i \rightarrow A_{i+1} \qquad 0 \leq i < 7$$
$$p_3 : F_i \rightarrow F_{i+1} \qquad i \geq 0$$

Production p_1 shows that at their creation time the lateral apices have different states A_0 and A_4. Consequently, the first apex requires eight derivation steps to produce a flower and new branches, while the second requires only four steps. Concurrently, each flower undergoes a sequence of changes, progressing from the bud stage to an open flower to a fruit. This developmental sequence is illustrated in color plate 13. From the biological perspective it is interesting to notice that at different developmental stages there are a number of open flowers which are relatively uniformly distributed over the entire plant structure. This is advantageous to the plant, as it increases the time span over which seeds will be produced.

4.3 RACEMES WITH LEAVES, OR MODELING QUALITATIVE CHANGES OF THE DEVELOPMENTAL PROCESS

The developmental sequences considered so far are homogeneous in the sense that the same structure is produced repeatedly at fixed time intervals. However, in many cases a qualitative change in the nature of development can be observed at some point in time. For example, consider shepherd's purse (*Capsella bursa-pastoris*) shown in color plate 14. In principle, its developmental pattern can be described as follows:

$$\omega : A$$
$$p_1 : A \rightarrow I[L_0]A$$
$$p_2 : A \rightarrow I[L_0]B$$
$$p_3 : B \rightarrow I[IF_0]B$$
$$p_4 : L_i \rightarrow L_{i+1} \qquad i \geq 0$$
$$p_5 : F_i \rightarrow F_{i+1} \qquad i \geq 0$$

The initial vegetative growth is represented by production p_1 describing creation of successive internodes and leaves by apex A. At some point in time, production p_2 changes the apex from the vegetative state A to the flowering state B. From then on, flowers are produced instead of leaves, forming a raceme structure as discussed in Section 4.1. The moment in which this change (developmental switch) occurs is not specified: the L-system is a nondeterministic one. Thus, for modeling purposes it must be complemented with an additional control mechanism which will determine the switch time. Below we outline three possible mechanisms. Each of them is biologically motivated, and corresponds to a different class of L-systems.

4.3.1 DELAY MECHANISM As in Section 4.2, the apex undergoes a series of state changes which delay the switch until a particular state is reached. This mechanism is outlined below:

$$\omega : A_0$$
$$p_1 : A_i \rightarrow I[L_0]A_{i+1} \qquad 0 \leq i < n$$
$$p_2 : A_n \rightarrow I[L_0]B$$
$$p_3 - p_5 : as \; before$$

According to this model, the apex *counts* the leaves it produces. While it may seem strange that a plant counts, it is known that some plant species produce a fixed number of leaves before they start flowering. Counting is achieved by monotonic increase or decrease of the concentration of certain cell components.

4.3.2 STOCHASTIC MECHANISM Another method for implementing the change is to use a stochastic mechanism. In this case the vegetative apex has a probability π_1 of staying in the vegetative state, and π_2 of transforming itself into a flowering apex:

$$p_1 : A \xrightarrow{\pi_1} I[L_0]A$$

$$p_2 : A \xrightarrow{\pi_2} I[L_0]B$$

$$p_3 - p_5 : as\ before$$

For a formal definition of stochastic L-systems, see Eichhorst and Savitch[5] and Yokomori.[48]

4.3.3 ENVIRONMENTAL CHANGE Many plants change from the vegetative to the flowering state in response to an environmental factor (such as the number of daylight hours or temperature). We can model this by using one set of productions (called a *table*) for some number of derivation steps before replacing it by another table:

Table 1

$\omega : A$

$p_1 : A \rightarrow I[L_0]A$

$p_2 : L_i \rightarrow L_{i+1} \qquad i \geq 0$

Table 2

$p_1 : A \rightarrow I[L_0]B$

$p_2 : B \rightarrow I[IF_0]B$

$p_3 : L_i \rightarrow L_{i+1} \qquad i \geq 0$

$p_4 : F_i \rightarrow F_{i+1} \qquad i \geq 0$

The concept of table L-systems (*TOL-systems*) is formalized in Herman and Rozenberg[10] and Rozenberg and Salomaa.[37]

The developmental switch from the vegetative to the flowering state is not the only qualitative change which can occur in a plant. Another possibility is the transformation of an apex from producing lateral flowers to producing a terminal flower which stops the axis development. This switch can also be produced using the methods described above.

4.4 COMPOSITION OF INTERACTIONLESS INFLORESCENCES

In Section 4.1 we considered simple racemes (also called *monobotryoid* inflorescences). They are characterized by a single axis, on which flowers are borne. In many plants compound racemes also occur. In the *dibotryoid* case the main axis carries entire racemes on the first order axes. This composition can be recursively extended to higher orders (*tribotryoids*, etc.). In general, the L-system for a compound raceme of order n has the following productions:

$$p_1 : A_i \rightarrow I[A_{i+1}]A_i \qquad 0 \leq i \leq n-1$$

$$p_2 : A_n \rightarrow I[IF]A_n$$

The above L-system does not produce flowers on axes of order less then n. However, some plants develop racemes at the end of each axis. This can be modeled by introducing additional productions of the form:

$$p_1' : A_i \rightarrow I[IF]A_n \qquad 0 \le i \le n - 1$$

The mechanism of switching from productions p_1 to productions p_1' can be the same as discussed in Section 4.3.

Similar rules for inflorescence composition apply to cymes. In fact, composition was used to produce the two lower branches of *Lychnis coronaria* shown in color plate 13.

4.5 INFLORESCENCE DEVELOPMENT WITH INTERACTIONS

Even in the presence of delays, the phase effects discussed so far reflect the sequential creation of branches, flowers and leaves by the subapical growth process. Consequently, organs near the plant roots develop earlier and more extensively than those situated near the axis ends. Such development results in *basitonic* plant structures (heavily developed near the base) with *acropetal* flowering sequences (the zone of blooming flowers progresses upwards along each branch). However, plants with *acrotonic* structures (heavily developed near the apex) and *basipetal* flowering sequences (progressing downwards) also occur in nature. These structures and developmental patterns cannot be viewed as a simple consequence of subapical growth; for example, basipetal flowering sequences progress in the direction which is precisely opposite to that of plant growth. An intuitively straightforward and biologically well-founded explanation of the described phenomena can be given in terms of signals which propagate through the plant and control the timing of developmental processes. Specifically, *acropetal* signals propagate from the root or the basal leaves towards the apices of the plant, and *basipetal* signals propagate from the apices towards the root. Within the formalism of L-systems these signals correspond to left-context-sensitive and right-context-sensitive productions, respectively. Below we consider two developmental models with signals. The first model employs a single acropetal signal, while the second one uses both acropetal and basipetal signals.

4.5.1 DEVELOPMENTAL MODEL WITH A SINGLE ACROPETAL SIGNAL Let us assume that the switch from the vegetative to the flowering condition is caused by a flower-inducing signal (representing the hormone *florigen*), which is transported from the basal leaves towards branch apices. In this case, the overall phase effect results from an interplay between growth and control signal propagation.[12,25] Assuming that only the first-order lateral branches are present, the development can be described by the following 1L-system:

$$\omega : D_0 A_0$$

$$p_1 : A_i \rightarrow A_{i+1} \qquad\qquad 0 \leq i < m - 1$$

$$p_2 : A_{m-1} \rightarrow I[B_0]A_0$$

$$p_3 : B_i \rightarrow B_{i+1} \qquad\qquad 0 \leq i < n - 1$$

$$p_4 : B_{n-1} \rightarrow J[L]B_0$$

$$p_5 : D_i \rightarrow D_{i+1} \qquad\qquad 0 \leq i < d$$

$$p_6 : D_d \rightarrow S_0$$

$$p_7 : S_i \rightarrow S_{i+1} \qquad\qquad 0 \leq i < \max\{u, v\} - 1$$

$$p_8 : S_z \rightarrow \epsilon \qquad\qquad z = \max\{u, v\} - 1$$

$$p_9 : S_{u-1} < I \rightarrow I S_0$$

$$p_{10} : S_{v-1} < J \rightarrow J S_0$$

$$p_{11} : S_0 < A_i \rightarrow F_0 \qquad\qquad 0 \leq i \leq m - 1$$

$$p_{12} : S_0 < B_i \rightarrow F_0 \qquad\qquad 0 \leq i \leq n - 1$$

$$p_{13} : F_i \rightarrow F_{i+1} \qquad\qquad i \geq 0$$

This L-system operates as follows. The apex A produces segments I of the main axis and creates the lateral apices (p_1, p_2). The time between the production of two consecutive segments, called the *plastochron* of the main axis, is equal to m units (derivation steps). In a similar way, the first-order apices B produce segments J of the lateral axes and leaves L with plastochron n (p_3, p_4). After a delay of d time units a signal S is sent from the tree base towards the apices (p_6). This signal is transported along the main axis with a delay of u time units per internode I (p_7, p_9), and along the first-order axes with a delay of v units per internode J (p_7, p_{10}). Production p_8 removes the signal from a node after it has been transported further along (ϵ stands for the empty string). When the signal reaches an apex (either A or B), the apex is transformed into a terminal flower F (p_{11}, p_{12}) which undergoes the usual sequence of states (p_{13}).

In order to analyze the plant structure and flowering sequence resulting from the above development, let us denote by T_k the time at which the apex of the k-th first-order axis is transformed into a flower, and by l_k the length of this axis (expressed as the number of internodes) at the transformation time. Since it takes km time units to produce k internodes along the main axis and $l_k n$ time units to produce l_k internodes on the first-order axis, we obtain:

$$T_k = km + l_k n$$

On the other hand, the transformation occurs when the signal S reaches the apex. The signal is sent d time units after the development starts, uses ku time units to travel through k zero-order internodes and $l_k v$ time units to travel through l_k first-order internodes:

$$T_k = d + ku + l_k v$$

Solving the above system of equations for l_k and T_k (and ignoring for simplicity some inaccuracy due to the fact that this system does not guarantee integer solutions), we obtain:

$$T_k = k\frac{un - vm}{n - v} + d\frac{n}{n - v}$$

$$l_k = -k\frac{m - u}{n - v} + \frac{d}{n - v}$$

In order to analyze the above solutions let us first notice that the signal transportation delay v must be less than the plastochron of the first-order axes n. (If this were not the case the signal would never reach the apices.) Under this assumption the sign of the expression $\Delta = un - vm$ determines the flowering sequence, which is acropetal for $\Delta > 0$ and basipetal for $\Delta < 0$ (color plate 15). If $\Delta = 0$ all flowers occur simultaneously.[1]

4.5.2 DEVELOPMENTAL MODEL WITH SEVERAL SIGNALS

Development of some inflorescences is controlled by several signals, which may propagate with different delays and trigger each other. The use of more than one signal is instrumental in the modeling of a large class of inflorescences (found, for instance, in the family Compositae) characterized by terminal flowers on all apices, indefinite order of branching, and basipetal flowering sequence. Color plates 16–18 illustrate this type of development with an example of wall lettuce (*Mycelis muralis*). First, the main axis is formed in a process of subapical growth which produces subsequent internodes and lateral apices. At this stage further development of lateral branches is suppressed by *apical dominance*, or the inhibiting effect of an active apex exercised on the lateral branches of the same axis. Physiologically, the apical dominance is related to hormones called *auxins* present in the plant as long as the apex is active, and gradually neutralized after flower initiation. At some moment, a flowering signal S_1 is sent from the bottom of the inflorescence along the main axis. When this signal reaches the apex, the terminal flower is initiated and a basipetal signal S_2 lifting the apical dominance is sent down the main axis. This enables successive first-order axes to grow, starting from the topmost one. After a delay, a secondary basipetal signal S_3 is sent from the apex of the main axis. Its effect is to send the flowering signal S_1 along subsequent first-order axes as they are encountered on the way down. This entire process repeats recursively for each axis: its apex is transformed into a flower, the apical dominance is lifted enabling lateral axes of the next order to grow, and the secondary basipetal signal is sent to induce the flowering signal S_1 in these lateral axes. The resulting structure depends heavily

---

[1]Note that the problem of inducing simultaneous flowering by a local control mechanism is closely related to the well-known *firing-squad* problem. The standard assumption is that there is a static structure in which information flow occurs. In contrast, we consider structures which grow.

on the values of plastochrons, delays, and signal propagation times. In the example under consideration, signal S_2 travels faster than S_3. Consequently, the time interval between the arrival of signals S_2 and S_3 increases while moving down the plant, potentially allowing the lower axes to grow longer than the upper ones. On the other hand, the lower branches start developing later, so at the observation time they may not be fully developed. As a result of these opposite tendencies, the plant is developed most extensively in its middle parts. For a detailed biological analysis of the above process see Janssen and Lindenmayer.[12]

4.6 ADDING VARIATION TO MODELS

All plants generated by a deterministic L-system are identical. An attempt to combine them in the same picture would produce a striking, artificial regularity. In order to prevent this effect it is necessary to introduce specimen-to-specimen variations which will preserve the general aspects of a plant but will modify its details.

Variation can be achieved by randomizing the turtle interpretation, the L-system, or both. Randomization of the interpretation alone has a limited effect. While the geometric aspects of a plant—such as the stem lengths and branching angles—are modified, the underlying topology remains unchanged. In contrast, the stochastic application of productions mentioned in Section 4.3.2 may affect both the topology and the geometry of the plant. For example, color plate 19 presents a field consisting of sixteen flowers generated by an L-system in which internode elongation is described by three stochastic rules:

$$p_1 : I \xrightarrow{\pi_1} I$$
$$p_2 : I \xrightarrow{\pi_2} II$$
$$p_3 : I \xrightarrow{\pi_3} I[L][L]I$$

where the probabilities π_1, π_2 and π_3 are equal to $1/3$. The resulting field appears to consist of various specimens of the same (albeit fictitious) plant species. A related application of a stochastic L-system to simulate variation between plant parts (shoots of the Japanese cypress) was described by Nishida.[31]

5. MODELING OF ORGANS

So far we have discussed the modeling of "skeletal" trees with branches consisting of mathematical lines. In this section we extend the model to include surfaces and volumes.

5.1 BICUBIC SURFACES

Conceptually, the simplest approach is to incorporate predefined surfaces in the tree with positions and orientations specified by the turtle. For example, leaves and cylindrical stem segments of the lily-of-the-valley in color plate 11, as well as leaves, flowers and stems of the lilac in color plate 20, were modeled using bicubic patches.[6] Patches make it easy to manipulate and modify surface shapes interactively, but are incompatible with the developmental approach to modeling since they do not "grow."

5.2 APPLICATION OF STRING AND BRACKETED L-SYSTEMS TO THE MODELING OF SURFACES

In order to fully simulate plant development and model phase effects present in plant structures, it is necessary to provide a mechanism for changing the size and shape of surfaces in time. We can achieve this in a simple way by defining polygonal surface boundaries using an L-system and filling the resulting polygons. An example is presented below:

$$\omega :L$$
$$p_1 :L \rightarrow \{-IX + X + IX - | - IX + X + IX\}$$
$$p_2 :X \rightarrow IX$$

Production p_1 defines a leaf as a closed planar polygon. The parentheses { and } indicate that the polygon should be filled. Production p_2 linearly increases the lengths of the polygon edges. Leaves generated by this L-system were incorporated in the model of the fern shown in color plate 12. Note the phase effect due to the "growth" of polygons in time.

In practice, the tracing of polygon boundaries leads to acceptable forms only in the case of small, flat surfaces. In other cases it is more convenient to define surfaces using an underlying tree structure as the *frame*. The entire surface consists of polygons bounded by sequences of tree segments with the addition of extra edges inserted between appropriate terminal nodes of the tree to form closed contours. This idea is illustrated in the top row of color plate 21, where yellow lines indicate the original tree segments and white lines represent the added edges. The bottom row presents the resulting filled surfaces. The three leaf shapes shown were obtained by modifying branching angles and the growth rates of axes. Specifically, the blade of the *cordate* leaf (leftmost in color plate 21) was generated by the following L-system:

$$\omega :[A][B]$$
$$p_1 :A \rightarrow [+A?]C\#$$
$$p_2 :B \rightarrow [-B?]C\#$$
$$p_3 :C \rightarrow IC$$

The axiom contains symbols A and B which generate the left-hand side and the right-hand side of the blade. Each of the productions p_1 and p_2 creates a sequence of axes starting at the leaf base and gradually diverging from the midrib. Production p_3 increases the axis lengths. The axes close to the midrib are the longest since they were created first (thus, the leaf shape is yet another manifestation of the phase effect). The symbols ? and # indicate the endpoints of edges to be inserted while forming closed polygons. The following string represents the left-hand side of the leaf after four derivation steps:

$$[+[+[+[+A\,?]C\#\,?]IC\#\,?]IIC\#\,?]IIIC\#$$

At this stage four edges are inserted between consecutive pairs of symbols ? and #, as indicated by braces. The first edge has zero length, the second is colinear with an axis, and the remaining two complete triangles.

The frame-based approach can be extended to three-dimensional organs. The rightmost example in color plate 21 illustrates construction of the flowers of the lily-of-the-valley from color plate 11. The L-system generates a supporting framework composed of five curved lines which spread radially from the flower base and are connected by a web of inserted edges. Note that in this case each polygon is a trapezoid bounded by two "regular" and two inserted edges.

Another approach to leaf modeling was recently proposed by Lienhardt and Françon.[18,19] While they also use two types of edges to specify leaf structures, the development is not described in terms of L-systems.

5.3 MODELING OF TWO-DIMENSIONAL CELL LAYERS USING MAP L-SYSTEMS

Although the developmental surface models discussed above allow for specifying closed polygons, the basic supporting structure is still limited to trees. However, many biological structures, such as cell layers or closed venation systems, are represented in a more natural way using graphs with cycles, or more precisely, *maps*. Maps can be considered as graphs complemented with a specification of *regions*. Each region is bounded by a circular list of edges (for a formal definition, see Tutte[47]). From the computer graphics perspective, maps are conceptually close to polygon meshes.[6] Consequently, well known rendering algorithms can be directly applied to their visualization.

In the biological context, regions are usually interpreted as (projections of) cells. The developmental process is simulated by a *map L-system* which generates cell subdivision.[4,23,27,28,41] Specifically, in a map OL-system (the context-free case) the predecessor of each production is a single *directed* edge. The successor is a bracketed string which, in addition to letters denoting edge labels, may contain symbols $+, -$ and $\tilde{\ }$. The symbols $+$ and $-$ specify whether a branch should be placed to the left or to the right of the main axis. The symbol $\tilde{\ }$ is used to indicate edge direction. The following rules apply:

- An edge which belongs to the main axis of the production successor has the same direction as the predecessor if it is not preceded by ~; otherwise it has the opposite direction.
- A branch edge is directed away from the main axis if it is not preceded by ~; otherwise it is directed towards the main axis.

A derivation step in a map L-system consists of two phases. First, each edge in the map is replaced by its successor according to the production set. Then pairs of matching branches are connected together. This second step does not have a counterpart in the L-systems described previously and requires further explanation. Two branches are considered matching if: (1) they enter the same region, (2) they have the same label, and (3) one branch is oriented away from the main axis, while the other is oriented towards the main axis. Such branches are connected to form a single edge which inherits their direction and label. The remaining unconnected branches are discarded from the map prior to the subsequent derivation step.

An example of a map L-system is given in Figure 4. It presents a production set in bracketed string form, the equivalent graph representation of these productions, and the result of four derivation steps. In the first step the phases of edge rewriting and branch connection are distinguished.

In order to generate images of maps, appropriate geometric interpretation rules must be specified. The maps shown in Figure 4 were produced according to the rules first used by Siero, Rozenberg and Lindenmayer[41]:

- The edges are represented by straight lines.
- The starting map is represented by a regular polygon with the desired number of edges.
- Each production subdivides an edge into segments of equal length; this subdivision determines positions of new vertices which remain in the same place in the subsequent maps.
- The positions of edges resulting from branch connection are determined by vertices at their endpoints as specified by the previous rule.

Direct application of these rules tends to produce cell shapes seldom observed in living organisms. A modification of geometric interpretation of maps, proposed by de Does and Lindenmayer,[4] leads to more realistic images. The essential idea is to move each vertex (with the exception of those lying on the outside perimeter of the map) towards the center of gravity of the neighboring vertices. This process is carried on iteratively, as each vertex translation affects the position of target gravity centers. It can be shown that the resulting solution yields a minimization of hypothetical forces acting along the edges.[4] An example of a cellular layer obtained using this approach is given in color plate 22.

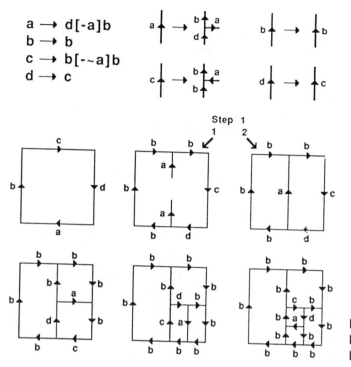

FIGURE 4

Example of a map L-system.

5.4 EXTENSION TO THREE DIMENSIONS

The concept of map L-systems can be further extended to three-dimensional cellular structures, or *cellworks*.[26,27] A cellwork consists of vertices, edges, polygonal *walls* and polyhedral *cells*. As in map L-systems, edge labels and orientation are the sole factors determining production application. However, in a *cellwork L-system* not only do the edges have to be rewritten and branches have to be connected, but also new walls must be inserted within the "shell" of the subdivided cell. Geometrical interpretation of cellworks leading to realistic simulation of three-dimensional development is still open for further research.

6. MATHEMATICAL RESULTS CONCERNING DEVELOPMENTAL SYSTEMS

The mathematical literature on L-systems that has accumulated over the past twenty years contains many results which are related to the biological simulation of development. Three main aspects of the theory can be considered: *characterization*, *inference* and *complexity*. Characterization has to do with the exhibition of mathematical properties of various classes of generating systems, and in particular, with proving that it is impossible for a certain class of these systems to generate some sequence of structures. Inference questions refer to the problem of finding possible generating systems to an observed sequence of structures. Complexity results are interesting for providing the minimal number of control elements or manipulations which are necessary to generate given patterns. We will discuss all three classes of problems in more detail.

6.1 CHARACTERIZATION

We would like to associate algebraic or analytical properties with various classes of L-systems, such as OL, DOL, IL or TOL. The most useful are those which allow for deciding whether a member of a given class of systems can generate a naturally occurring pattern under consideration. Such conclusions help the search for the underlying biological mechanisms for an observed developmental process. For instance, if OL-systems are ruled out as generators of a particular developmental sequence, then either cell interactions (IL-systems) or environmental changes (table L-systems, or TOL-systems) will have to be considered.

As an example, let us define the adult language of an L-system G as the set of strings which is invariant with respect to derivations in G. According to a theorem by Herman and Walker,[9] the class of adult languages of OL-systems is the same as the class of context-free languages. There is a well-known characterization of the class of context-free languages, namely the "pumping lemma" of Bar-Hillel. Thus, if a language does not satisfy the "pumping" property, it is not an adult language generated by a string OL-system, and the structures corresponding to this language must be generated by a more complex mechanism, involving interactions or environmental switches.

Further results comparing classes of languages generated by L-systems and Chomsky languages were given by Herman and Rozenberg.[10] While the classes of DOL and OL languages are properly included in the set of context-sensitive languages, the class of IL languages is not included (Figure 5). The classes of DOL and OL languages have non-empty intersections with the classes of regular, context-free and context-sensitive languages, and the class of IL languages includes that of regular languages. These inclusion and exclusion properties, together with well-known properties of Chomsky classes, help to distinguish between languages and developmental sequences which can be generated with or without interactions.

Strong characterization results are also available for the growth functions of L-systems. A growth function $f(t)$ of a deterministic L-system G specifies the length of the generated string x as a function of derivation length t and, consequently, describes the process of growth in time. In linear system theory, a system with

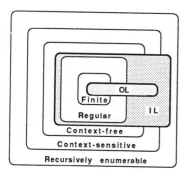

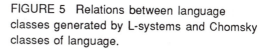

FIGURE 5   Relations between language classes generated by L-systems and Chomsky classes of language.

observable parameter $f(t)$ ranging over non-negative integers is called N-realizable if there exist a row vector $\pi_{1 \times n}$, a square matrix $M_{n \times n}$ and a column vector $\eta_{n \times 1}$, all with non-negative integer entries, such that $f(t) = \pi M^t \eta$ for all $t \geq 0$. Similarly, a system is called Z-realizable if there exists such an expression with integer entries.

The following observation by Paz and Salomaa[32] shows the connection of these terms to growth functions of DOL-systems. The growth function $f(t)$ of a DOL-system G with k symbols can be written in the form $f_G(t) = \pi M^t \eta$ where π is a k-dimensional row vector whose i-th entry equals the number of occurrences of the i-th symbol in the starting string of G, M is a $(k \times k)$-dimensional square matrix in which the entry (i, j) equals the number of j-th symbols produced by the i-th symbol in G, and η is a column vector filled with 1's.

We therefore call a system with parameter $f(t)$ over non-negative integers *DOL-realizable* if $f(t) = \pi M^t \eta$ for all $t > 0$, where π and M have non-negative integer entries, and the entries of η are all equal to 1. In other words, a sequence of non-negative integers is DOL-realizable if it is the growth function of a DOL-system. Similarly, we call a sequence of non-negative integers *PDOL-realizable* if it is the growth function of a propagating DOL-system, i.e., of a DOL-system without erasing productions.

The following characterization results of Salomaa and Soittola[39] are then available.

- A sequence of non-negative integers $f(t)$ is PDOL-realizable and not identical to the zero sequence if and only if the sequence $f(t+1) - f(t)$ is N-realizable and $f(0)$ is positive.
- For any integer $k > 0$, the sequence $f(t)$ is DOL-realizable if and only if the sequence $f(t+k)$ is DOL-realizable.
- If $f(t)$ is a DOL-realizable sequence not becoming ultimately zero, then there is a constant c such that for all $t > 0$,

$$\frac{f(t+1)}{f(t)} < c.$$

■ Every Z-realizable sequence of integers can be expressed in the form $f(t) = f_1(t) - f_2(t)$ where $f_1(t)$ and $f_2(t)$ are DOL-realizable sequences.

These and other theorems make it possible to decide whether an observed growth function of an organism is DOL-realizable or PDOL-realizable. Other theorems enable us to obtain explicit functions (polynomial or exponential) for given DOL-systems. Growth functions can also be found for each type of cell (string symbol) separately (the so-called Parikh functions of L-systems). This can be used in ecological or crop growth studies.

6.2 INFERENCE

The problem of finding L-systems which generate an observed sequence of structures is called the *syntactic inference* problem. In fact it could also be called the *realization* problem. We are faced here with a more difficult type of realization than in the previous section, where the problem was to realize the growth function of a developmental process. Now we are asking for the realization of the sequence of the structures themselves.

The following cases have been considered[8-10,40] for the inference problem of string-generating OL-systems:

a. All intervals between observations are equal to one.

b. All intervals between observations are equal but of unknown length.

c. The intervals are of arbitrary lengths but the observations are in the proper order.

Both deterministic (D) and non-deterministic, propagating (P) and non-propagating systems were considered, as well as the presence and absence of cell interactions (0, 1, and 2-sided interactions). The inference problem was shown to be decidable for the cases Dax, Dbx ($x = 0, 1, 2$) and for Dc0, while decidability has not been proven for the cases Dc2 and Dc1, and not even for PDc1.

Recently the DOL inference problem has been considered for branching filaments.[15] An algorithm was constructed to find the set of DOL-systems (with branching symbols) which generate a given (finite) sequence of branching structures.

6.3 COMPLEXITY

Since L-systems have a biological interpretation, differences between their complexities also have direct biological meaning. If the minimum computational complexity needed to generate given developmental patterns is known, we can use it as a reference point for evaluating redundancy of specific control mechanisms. Organisms do not in general minimize their energy expenditures or their information storage capacities, but they might have acquired minimal control systems by selection and evolution. Each additional cellular state has to arise and be maintained by involved biochemical and physiological mechanisms, so it is to the advantage of the organism to keep the number of discrete steady states as low as possible. Similarly, each time a signal has to be produced in a certain cell, transmitted over a number of other cells, and finally received and recognized by still other cells, many cell components have to be synthesized and transported. Thus it is reasonable to assume that duplication of states and signals is avoided.

Within the theory of formal languages two main complexity measures can be distinguished. First, there is the decidability of questions, such as membership, emptiness and equivalence, with respect to given classes of generating systems. Clearly, undecidability of a question in a given class indicates a higher complexity of this class with respect to those in which this question is decidable. Once the decidability has been proven, the second measure has to do with the computational complexity of the decision procedure, which is often expressed in terms of Turing machine time or space complexity values. For example, Table 1 collects the upper bounds of Turing machine complexity for the membership problem.[13,14,16]

The complexity measures for L-systems are basically different from those based on information and entropy (Shannon), or from Kolmogorov[17] complexity. The difference lies in the fact that here we are dealing with distributed control factors for processes taking place in growing structures, i.e., with dynamic systems, while the informational and Kolmogorov complexity concepts concern static structures. The application of information or entropy measures to processes of living organisms has never been successful for the additional reason that these measures are defined for information transmission (communication) from source to receiver over a channel. In organisms these components are not identifiable. For instance, the information content of the entire DNA complement of a living cell is immensely large and there is no way of finding out how and what part of it is actually used during the life time of the cell. This is why it is more likely to find useful comparisons

TABLE 1 Upper Bounds of Turning Machine Complexity for Membership Problem

Type	Time Complexity	Space Complexity
PDOL	n	—
DOL	n^2	$\log n$
OL	$n^{3.81}$	$\log^2 n$
DTOL	polynomial of unknown order	$\log^2 n$
TOL	NP-complete	n

between complexities of different organisms by considering their basic functional units, the cells, and the changes occurring in these units, such as cell divisions, cell death, changes in steady states, and changes leading to differentiation. These considerations are clearly related to those of computational complexity, pointing to a deep connection between computation and development.

7. CONCLUSIONS

In this paper we presented an algorithmic approach to the description, analysis and developmental simulation of multicellular organisms. Specifically, we have shown how complex dynamic processes may result from relatively simple rules which control cell states and configurations in space.

The construction of a mathematical model gives us an idea of how many and what kind of control factors are necessary to achieve different types of development. Since organisms consist of large numbers of cells, it is difficult to analyze consequences of developmental mechanisms without a graphical interpretation. In this sense, interactive computer graphics complements the mathematical theory of L-systems by providing a tool for formulating and verifying biological hypotheses.

ACKNOWLEDGMENT

Color plate 9 was generated by Dave Fracchia. Jim Hanan is a co-author of color plates 11 and 20. The software used to generate color plate 22 was developed by Mark de Does. The research reported in this paper was supported in part by grants from the Natural Sciences and Engineering Research Council of Canada. This paper contains sections of "Developmental Models of Herbaceous Plants for Computer Imagery Purposes," by P. Prusinkiewicz, A. Lindenmayer and J. Hanan, published in the Proceedings of SIGGRAPH '88 and reprinted here with the permission of the Association for Computing Machinery.

REFERENCES

1. Abelson, H., and A. A. diSessa (1982), *Turtle geometry* (Cambridge: Massachusetts Institute of Technology Press).
2. Aono, M., and T. L. Kunii (1984), "Botanical Tree Image Generation," *IEEE Computer Graphics and Applications* 4, Nr. 5, 10–34.
3. Baker, R., and G. T. Herman (1972), "Simulation of Organisms using a Developmental Model, Parts I and II," *Int. J. Bio-Medical Computing* 3, 201–215 and 251–267.
4. de Does, M., and A. Lindenmayer (1983), "Algorithms for the Generation and Drawing of Maps Representing Cell Clones," *Graph Grammars and Their Application to Computer Science; Second International Workshop, Lecture Notes in Computer Science*, Eds. H. Ehrig, M. Nagl and G. Rozenberg (Berlin: Springer-Verlag), vol. 153, 39–57.
5. Eichhorst, P., and W. J. Savitch (1980), "Growth Functions of Stochastic Lindenmayer Systems," *Information and Control* 45, 217–228.
6. Foley, J. D., and A. Van Dam (1982), *Fundamentals of Interactive Computer Graphics* (Reading: Addison-Wesley).
7. Frijters, D., and A. Lindenmayer (1974), "A Model for the Growth and Flowering of *Aster novae-angliae* on the Basis of Table (1, 0) L-Systems," *L Systems, Lecture Notes in Computer Science*, Eds. G. Rozenberg and A. Salomaa (Berlin: Springer-Verlag),vol. 15, 24–52.
8. Gnutzmann, I. (1979), *Zum syntaktischen Inferenzproblem Lindenmayer-Systemen, Ph. D. Thesis, University of Hannover.*
9. Herman, G. T., and A. Walker (1972), "The Syntactic Inference Problem as Applied to Biological Systems," *Machine Intelligence* 7, 341–356.
10. Herman, G. T., and G. Rozenberg (1975), *Developmental Systems and Languages* (Amsterdam: North-Holland).
11. Hogeweg, P., and B. Hesper (1974), "A Model Study on Biomorphological Description," *Pattern Recognition* 6, 165–179.
12. Janssen, J. M., and A. Lindenmayer (1987), "Models for the Control of Branch Positions and Flowering Sequences of Capitula in *Mycelis muralis* (L.) Dumont (Compositae)," *New Phytologist* **105**, 191–220.
13. Jones, N.D., and S. Skyum (1979), "Complexity of Some Problems Concerning L-Systems," *Mathematical Systems Theory* 13, 29–43.
14. Jones, N.D., and S. Skyum (1981), "A Note on the Complexity of General DOL Membership," *SIAM J. Computing* 10, 114–117.
15. Jürgensen, H., and A. Lindenmayer (1987), "Modelling Development by OL-Systems: Inference Algorithms for Developmental Systems with Cell Lineages," *Bull. Math. Biology* 49, 93–123.
16. van Leeuwen, J. (1975), "The Membership Problem for ETOL Languages is Polynomially Complete," *Information Processing Letters* 3, 138–143.
17. Kolmogorov, A. N. (1968), "Three Approaches to the Quantitative Definition of Information," *Int. J. Comp. Math.* 2, 157–168.

18. Lienhardt, P. (1987), *Modélisation et Évolution de Surfaces Libres, Ph.D. Thesis, Université Louis Pasteur, Strasbourg*, unpublished.

19. Lienhardt, P., and J. Françon (1987), "Synthèse d'Images de Feuilles Végétales," *Technical Report R-87-1, Département dInformatique, Université Louis Pasteur, Strasbourg.*

20. Lindenmayer, A. (1968), "Mathematical Models for Cellular Interaction in Development, Parts I and II," *Journal of Theoretical Biology* **18**, 280–315.

21. Lindenmayer, A. (1974), "Adding Continuous Components to L-Systems," *L-Systems, Lecture Notes in Computer Science*, Eds. G. Rozenberg and A. Salomaa (Berlin: Springer-Verlag), vol. 15, 53–68.

22. Lindenmayer, A., and G. Rozenberg (Eds.) (1976), *Automata, Languages, Development* (Amsterdam: North-Holland).

23. Lindenmayer, A., and G. Rozenberg (1979), "Parallel Generation of Maps: Developmental Systems for Cell Layers." *Graph Grammars and Their Application to Computer Science and Biology, Lecture Notes in Computer Science*, Eds. V. Claus et al. (Berlin: Springer-Verlag), vol. 73 301–316.

24. Lindenmayer, A. (1982), "Developmental Algorithms: Lineage versus Interactive Control Mechanisms," *Developmental Order: Its Origin and Regulation*, Eds. S. Subtelny and P. B. Green (New York: A. R. Liss Inc.), 219–245.

25. Lindenmayer, A. (1984a), "Positional and Temporal Control Mechanisms in Inflorescence Development," *Positional Controls in Plant Development*, Eds. P. W. Barlow and D. J. Carr (Cambridge: University Press).

26. Lindenmayer, A. (1984b), "Models for Plant Tissue Development with Cell Division Orientation Regulated by Preprophase Bands of Microtubules," *Differentiation* **26**, 1–10.

27. Lindenmayer, A. (1987), "Models for Multicellular Development: Characterization, Inference and Complexity of L-Systems," *Trends, Techniques and Problems in Theoretical Computer Science, Lecture Notes in Computer Science,*
Eds. A. Kelemenová and J.Kelemen (Berlin: Springer-Verlag), vol. 281, 138–168.

28. Lindenmayer, A. (1987), "An Introduction to Parallel Map Generating Systems," *Graph Grammars and Their Application to Computer Science; Third International Workshop, Lecture Notes in Computer Science*, Eds. H. Ehrig, M. Nagl, A. Rosenfeld and G. Rozenberg (Berlin: Springer-Verlag), vol. 291, 27–40.

29. Mitchison, G. J., and M. Wilcox (1972), "Rule Governing Cell Division in *Anabaena*," *Nature* **239**, 110–111.

30. Müller-Doblies D. and U. (1987), "Cautious Improvement of a Descriptive Terminology of Inflorescences." *Monocot newsletter* **4**, Institut für Biologie, Technical University of Berlin (West).

31. Nishida, T. (1980), "KOL-Systems Simulating Almost but not Exactly the Same Development - The Case of Japanese Cypress," *Memoirs Fac. Sci., Kyoto University, Ser. Bio.* **8**, 97–122.

32. Paz, A., and A. Salomaa (1973), "Integral Sequential Word Functions and Growth Equivalence of Lindenmayer Systems," *Information and Control* **23**, 313–343.
33. Preparata F. P., and R. T. Yeh (1973), *Introduction to Discrete Structures* (Reading: Addison-Wesley).
34. Prusinkiewicz, P. (1986), "Graphical Applications of L-Systems," *Proceedings of Graphics Interface '86 - Vision Interface '86*, 247–253.
35. Prusinkiewicz, P. (1987), "Applications of L-Systems to Computer Imagery," *Graph Grammars and Their Application to Computer Science; Third International Workshop, Lecture Notes in Computer Science*, Eds. H. Ehrig, M. Nagl, A. Rosenfeld and G. Rozenberg (Berlin: Springer-Verlag), vol. 291, 534–548.
36. Robinson, D. F. (1986), "A Symbolic Notation for the Growth of Inflorescences," *New Phytologist* **103**, 587–596.
37. Rozenberg, G., and A. Salomaa (1980), *The Mathematical Theory of L-Systems* (New York: Academic Press).
38. Salomaa, A. (1973), *Formal Languages* (New York: Academic Press).
39. Salomaa, A., and M. Soittola (1978), *Automata-Theoretical Aspects of Formal Power Series* (Berlin: Springer-Verlag).
40. Schmidt, U. (1983), "Syntaktische Inferenz von DTOL-Systemen," *Diplomarbeit, T. H. Darmstadt*.
41. Siero, P. L. J., G. Rozenberg, and A. Lindenmayer (1982), "Cell Division Patterns: Syntactical Description and Implementation," *Computer Graphics and Image Processing* **18**, 329–346.
42. Smith, A. R. (1978), "About the Cover: Reconfigurable Machines," *Computer* **11**, Nr. 7, 3–4.
43. Smith, A. R. (1984), "Plants, Fractals, and Formal Languages," *Computer Graphics* **18**, Nr. 3, 1–10.
44. Szilard, A. L., and R. E. Quinton (1979), "An Interpretation for DOL-Systems by Computer Graphics," *The Science Terrapin (University of Western Ontario)* **4**, 8–13.
45. Thompson, d'Arcy (1952), *On Growth and Form* (Cambridge: University Press).
46. Troll, W. (1964), *Die Infloreszenzen* (Stuttgart: Gustav Fischer Verlag), vol. 1.
47. Tutte, W. T. (1982), *Graph Theory* (Reading: Addison-Wesley).
48. Yokomori, T. (1980), "Stochastic Characterizations of EOL Languages," *Information and Control* **45**, 26–33.

Peter Oppenheimer
Computer Graphics Lab, New York Institute of Technology, Old Westbury, NY 11568

The Artificial Menagerie

WHAT IS ARTIFICIAL? WHAT IS LIFE?

Any precise definition of these terms is likely to be too narrow, and be vulnerable to a "What about such-and-such?" counterexample. The Euclideans considered *point, line* and *plane* as undefined terms in their new realm of study. In the same spirit, we could consider *Artificial Life* as an undefined. Rather than characterize these terms with axioms (e.g., "I don't now what life is. But if it's alive it had better do XYZ"), this paper will present a few isolated *examples* which are representative of a genre of artificial life-forms. These life-forms occupy one branch in the evolution of what will no doubt become a much larger taxonomic kingdom.

When one looks outside, one has little difficulty separating what is conventionally considered natural from artificial. The difference is not so much a theoretical one. In fact, a simple line of reasoning could lead one to conclude that in theory, all artificial forms are natural. Rather this subjective difference is one of practice. The principles of growth, production, and design of nature are different or more general than man's. Man-made objects seem more linear, geometric, or serial. If man could incorporate the principles of design by nature in his work, then the line between artificial and natural would become more blurred. One is likely to find a

Artificial Life, SFI Studies in the Sciences of Complexity,
Ed. C. Langton, Addison-Wesley Publishing Company, 1988

continuum between natural and artificial, whereas now the gap seems wide, due to the underdevelopment of natural methodologies in human constructions.

In attempting to create *artificial life-forms* on, for example, a computer, the more principles one borrows from nature, the more *natural* or *lifelike* the result, subjectively speaking. If, on the other hand, one borrows certain principles, but ignores or even violates others, the resulting life-forms will be hybrids, lying somewhere between our current notions of natural and artificial. These hybrids are perhaps the more interesting types of simulations. They are more unfamiliar, novel, exciting although often disturbing versions, than the existing extremes.

The genesis of these sometimes monstrous creations, is more than just an amusing exercise. By programming life to new specifications, we gain an understanding of its underlying principles. By breaking the rules (in a non-hazardous way), we learn their importance. By isolating a single aspect or parameter, we can perform a more controlled experiment *in silica*, than could ever be performed *in vivo*. Our ability to deduce cause and effect in the external world is tenuous. But in the case of a mathematical model, cause and effect can be reduced to the *if A then B* logic of the programmed representation. And perhaps the most subtle ramification of these hybrid creations points to evolution itself. Mutation is a vital mechanism in evolution. These *artificial mutants* are linked somehow to the birth of new species, that could have been, could become, or are in themselves living entities of organization.

In science, the process that is analogous to mutation, is *error*. If one takes an open-minded but controlled approach to examining results that at first look like mistakes, then new unexpected viewpoints can emerge.

Images that have an organic quality with a surreal twist, heighten our awareness. By seeing nature as it is not we learn to better appreciate it as it is.

100% NATURAL

To create images with an organic quality, surreal or not, we must begin with organic principles. These principles establish a natural foundation, a basis for a recognizable lifelike quality. Departures can then be made. Direct natural representation is a vital starting point.

Two such principles used in the images presented here are:

1. The notion of the gene

2. Self-similarity

1. NUMERICAL GENES

Our knowledge of the workings of genes in natural organisms is surely vast and expanding. The artificial model of these mechanisms can exist at many different levels. One concept that has already become something of a paradigm in the field of artificial life is that of a *numerical gene*. The meaning of this term has taken on many interpretations.

At the more abstract end of the scale, a numerical gene is a function p. Its argument g (a number, set of numbers or a bit string) is called the *genotype*, and represents the genetic material of the organism (or trait). The value $p(g)$ is called the *phenotype*, and represents the resulting organism. A *fitness function* f can be applied to $p(g)$ to produce a relative numerical evaluation of the viability of that organism (or trait) in a given environment.

The more theoretical, abstract study of *genetic algorithms* considers purely numerical relationships of the genotype, phenotype, and fitness function. Algebraic operations such as bit string concatenation, or bit mutation are applied to the genotypes. Constraints are applied to the functions. Physiology is *not* taken into account. The phenotype does not have any physical, or geometric properties, nor does the genotype behave like molecular DNA. They are only numbers. However, the mathematical structure imposed on the genotype, phenotype, and fitness function, when appropriately designed, can be biologically meaningful. By paralleling natural genetic logic (or not as the case may be), theoretical experiments can be performed. These experiments include game playing simulation, statistical analysis, or mathematical deduction.

By treating the phenotype as an organism with geometric and physiological properties, rather than just a number, the study of numerical genes becomes a notch less abstract. In this scenario, the genotype-phenotype relationship codes for such properties as size, shape, and color, ability to flee predators, ability to produce vital substances, etc. Fitness analysis is now applied to these more concrete physical properties.

Fitness, of course, is in the eye of the beholder. The population which evolves will greatly depend upon the choice of fitness paradigm. A fitness function that favors physically robust organisms will produce a different speciation than one based on aesthetic visual appearance.

At the totally opposite end of the spectrum from the purely mathematical genetic algorithms, one generates a "concrete, living, breathing, in the flesh organism" phenotype, based on a numerical genotype. This technology, commonly known as *genetic engineering*, is growing rapidly. Its development is both the dream and fear of anyone involved with artificial life.

The approach taken in this study assigns geometric or other visual attributes (such as color) to the numerical genes. The genotype for each organism is a collection of roughly 15 real number parameters. Each parameter relates to a geometric trait of the growth process of the phenotype. In the case of the branching tree organisms, examples of these traits include:

FIGURE 1   Self-similar set (Von Koch construction).

- The angle between the main stem of a tree and its primary branch(es).
- The size ratio of the main stem and its primary branch(es),
- The amount of helical twist in the branches,
- The curvature of the stem,
- The color of the bark, etc.

In general, two numbers correspond to each trait: a mean and a standard deviation. That is each trait is controlled by a random variable with uniform distribution. Physical properties such as mechanical tissue strength, or biochemical composition are not taken into account. The resulting phenotype is a computer-generated color video image of a tree or other organism. The fitness function is purely subjective. If the picture looks *right*, the organism survives, the genotype and phenotype being saved in the database. If not, the dials are turned, the dice of mutation are rerolled, resulting in a new genotype and phenotype.

In the earliest phases of this study, the principle motivation was to gain some understanding of the geometry of actual trees. The development of the genotype was an attempt at reducing the geometry of "real" trees to a collection of numbers. The fitness function favored images that "looked like the real thing." As the study evolved, and understanding of the significance of the parameters increased, the motivations shifted. The emphasis of individual parameters and the expression of their meanings as universal principles, or even their emotional qualities, became of increasing importance. To express such ideas required a fitness function that did not favor *realistic images*, but on the contrary one that contrasted natural selection. Through deliberate distortion of the parameters, interpolating animation of isolated parameters, or other *unnatural* processes, one created species that survived this artificial fitness function. Later in this paper we will look at some consequences of the genotype-phenotype paradigm in artificial life models.

FIGURE 2   Self-similar set (Von Koch construction).

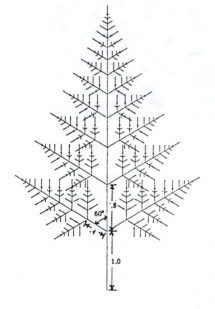

FIGURE 3   Tree skeleton with default parameters. Stem/stem ratio = .8; branch/stem ratio = .4; and branching angle = 60°.

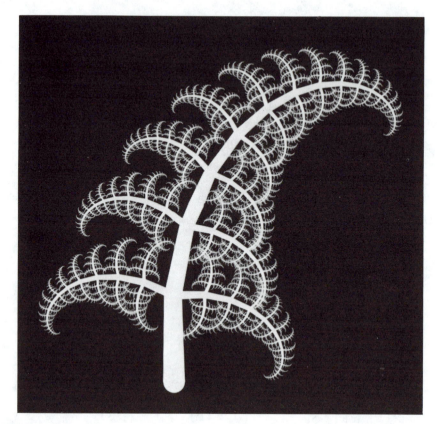

FIGURE 4   Fractal fern.

2. SELF-SIMILARITY

Another principle that has been borrowed from nature to lend a lifelike quality to artificial models is the principle of *self-similarity*. An object is *self-similar* if it is made of several pieces, each of which is a scaled down copy of the entire object. Figure 1, for example, is made of two halves, each of which is a scaled down copy of the entire shape. Each of these halves is likewise made up of two quarters which are each scaled versions of the halves and hence of the whole, *ad infinitum*. The component pieces need not be equivalent halves as illustrated in Figure 2.

A slight generalization of self-similarity is the notion of self-similarity with *residue*. An object that is made of several pieces, at least one of which is similar to the whole, is said to be self-similar with residue. The residue referring to the part that is not similar to the whole. Figure 3 illustrates such an object. It is made up of 3 pieces that are similar to the whole, plus a trunk, which is not. The leaf set of this tree is self-similar without residue. A spiral is a degenerate case of a self-similar object with a residue. (Figure 5).

The notion of *fractal* and the notion of *self-similar* are often considered synonymous. They are, however, different classes of object. Depending on the evolution of the accepted usage of these terms, the distinction will either become important, or it will become an insignificant fine point. The difference, again depending on one's definition, is as follows: an object is a *fractal* if as one considers finer magnifications of the object, the component pieces have non-trivial detail at arbitrarily small scales. A sphere is an example of a non-fractal. As one zooms in on the surface of a sphere, the component patches become flat, and without detail. The surface of a rough stone or the coastline of an island, however, has detail at arbitrarily small scales and hence is a fractal. If the small-scale detail is similar to the large-scale detail, the object is said to be self-similar. But of course this is not always the case. The geometry of the detail may vary as the scale diminishes. Pine needles don't look like the trunk of the tree. A rock does not look just like the mountain (not

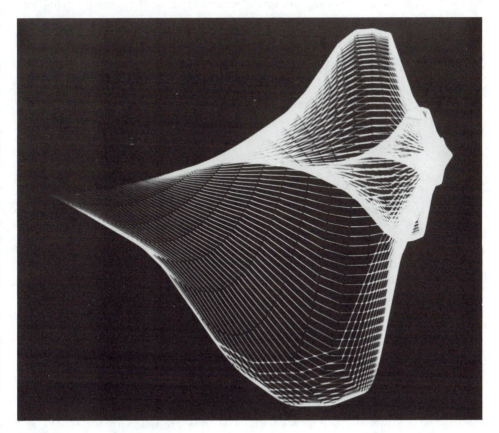

FIGURE 5 Conch shell skeleton.

even statistically). And yet rocks have apparent detail at arbitrary scales. (This of course depends on one's paradigm of the sub-sub-subatomic nature of matter).

If the geometry of the detail only varies very *slowly* as a function of scale, then we are nonetheless aware of a strong self-similar *component* in the object's construction. This self-similar quality is, in some definable way, a local function of scale change rather than a global property of the object. When one says a natural object is self-similar one is often referring to this locally defined self-similarity. Global self-similarity in general only occurs in abstract artificial mathematical constructions.

To understand fractals is to be aware of the relationships of large to small-scale detail.

ARTIFICIAL SELF-SIMILARITY

To create artificial images of self-similar objects, one can use the recursive definition of the object as transformed copies of itself.

In pseudo-code:

```
tree :=
{
        Draw Branch Segment
        if (too small)
                Draw leaf
        else
        {
                # Continue to Branch
                {
                        Transform Stem
                        "tree"
                }
                repeat n times
                {
                        Transform Branch
                        "tree"
                }
        }
}
```

Paraphrased, a tree node is a branch with one or more tree nodes attached, transformed by a 3 × 3 linear transformation. Once the branches become small enough, the branching stops and a *leaf* is drawn. The trees are differentiated by the geometry of the transformations relating the node to the branches and the topology of the number of branches coming out of each node. These branching attributes are controllable through the numerical genotype. Some of the individual genes are geometric, that is they determine the stem-stem and stem-branch transformations. Others are topological, that is they determine the number of branches emanating from each branch point. By executing this recursion, one can generate a self-similar image with arbitrarily small detail. The set of transformations uniquely determines the self-similar set.

On a serial computer, the time expenditure increases exponentially as a function of branching depth. On a natural tree, the recursive branching growth takes place

FIGURE 6 Xerox fractal tree.

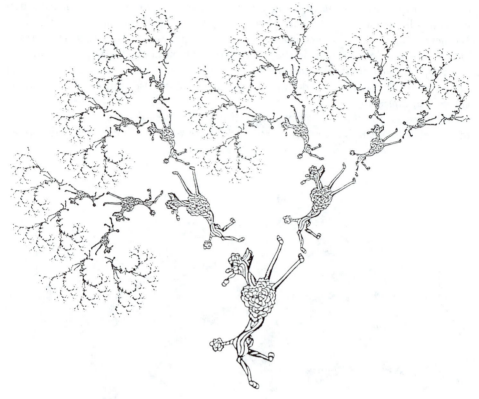

FIGURE 7   Dogwood.

in parallel. The time expenditure is a linear function of branching depth. A machine that can transform arbitrarily many branch points in parallel, can also attain this linear time cost. A xerographic machine is such a device. Each pass of a photo-reducing xerographic machine transforms the entire tree in one shot. Figures 6 and 7 illustrate xerographically generated self-similar trees with residue. To generate such a tree 10 levels deep with 2 branches per branch point takes 2×10 copies, not 2^{10}. At a nickel a copy, these pictures cost a buck each.

But perhaps more significant than the time it takes to grow the tree is the time it takes to design the tree parameters. The evolution of a genotype takes much longer than the execution of a single corresponding phenotypic organism. If the construction of a complex phenotype requires a complex genotype, then the evolution of that organism will be slow. If, however, a complex phenotypic structure can be generated from a simple set of genotypic rules, then evolution can be accelerated. The ratio between phenotypic complexity and genotypic simplicity in computer science is called database amplification.

Recursive self-similarity is a vital mechanism for achieving database amplification, both in natural and artificial organisms. Presumably, complexity has evolved to bestow benefits on an organism. But complexity must not be a burden. An organism must be simple to build, simple to describe. Like their natural counterparts, computer-generated artificial life-forms must accommodate and create complexity within demanding constraints—not only memory limitation but of data design cost as well. In generating artificial images of natural life-forms, the programmer faces a problem much like that which confronts nature: how to build something visually complex yet semantically simple. Indeed, we might expect the two would go about solving the problems in much the same way.

Since the origins of computer science, recursion has been commonly used to resolve the tension between the simple and the complex. *Tree-like* data structures are used in searching, sorting and a multitude of other computations. Applying a simple operation locally, efficiently solves the more complex global problem. The use of recursion to create artificial models of trees completes the cycle, applying a principle borrowed from nature to model nature. As a consequence of self-similarity, a complex image can be created from a small initial database, that is a small genotype can generate a complex phenotype. If a self-similarity methodology were not invoked, then distinct genetic data would be required for all nodes in the model. If, on the other hand, the model was self-similar, then genetic data could be shared by all the nodes in the tree. Since small-scale detail looks like large-scale detail, one simple set of parameterized rules relating the transformation from a parent stem to its branching offspring, could determine the global structure of the model. This economy of data, accelerates the evolving design, and implementation of the artificial life-form and suggests a possible explanation as to why self-similarity abounds in the natural world: evolution has resolved the tension between complexity and simplicity in the same way computer science has—with recursive fractal algorithms.

A simple organism such as a fern has only a picogram of genetic material with which to construct its entire structure. It economizes on DNA through some variant of recursive self-similarity, repeating the same simple rules at all scales, large and small. The apparent complexity is embodied in simple rules. The struggle to simplify its genetic requirements, determines the geometric self-similar structure of the plant. Form follows genetic economics.

The early stages of the computer model presented here contained only 3 changeable numerical parameters. The resulting images were of very simple fern-like plants. New species were generated by controlling the parameter values, rerolling the dice of mutation, and then selecting the forms that would be allowed to proliferate. As the model became more complex with the inclusion of more parameters, the program created images of more genetically complex trees such as cherry trees, higher on the evolutionary scale. This process seems to parallel the tendency in natural evolution for more complex species to have, in general, more genetic material, than simpler predecessors.

Hierarchical structures abound in human societies. They range from consciously man-made corporate hierarchies to spontaneously emerging community, neighborhood, family, individual relationships. From urban plans, to mental thought processes, and beyond—down to the innermost fabric of our flesh. Spanning a continuum, these societal groupings provide a bridge between natural and artificial. Recursively organized, they provide efficient conduits to resources, information, and other human needs.

ARTIFICIAL INTUITION

The above analogy between natural and artificial genetic economics must be viewed with the right perspective. Creating an artificial model of plant growth allows one to manipulate, analyze, and deduce properties of the artificial model only. One cannot infer outright that the process or principles are identical in the natural organism. One at times can derive some understanding of how a similar but different natural mechanism might behave. For example, if one notices a certain phenotypic pattern when turning the dials of the artificial genotype, one gets a feel for the geometric patterns in the natural organism although the internal mechanism remains unknown. One gains an understanding of the type: "I don't know the precise mechanism of this natural organism, but since it behaves similarly to this artificial model, it might share a similar structure."

This understanding is highly subjective. Often it cannot be translated into words or formulas but must be felt. One way to feel these sensations is to observe the artificial imagery and interact with it. By turning the dials and observing an instantaneous visual feedback, logically linked to the changed input, one *senses* the geometry. In this way, an interactive animation system that provides this kind of cause and effect isolation of input parameters—even if the images are not realistic—can be more valuable to our understanding than a highly realistic still image. A key step in this process is reducing the model to independent input parameters and providing modes of interaction with meaningful feedback. Realistic representation is orthogonal to this effort.

The paradox of this reductionist approach is that whereas the artificial model is controlled by independent, separable, orthogonal parameters, nature appears to be a flow of intertwined interdependent ingredients, pushing and pulling beyond any perceivable *if A then B* causality. Whereas science hopes to provide objective knowledge, these descriptive artificial models often provide subjective understanding bordering on intuition. Perhaps this places this study outside the realm of science proper, but perhaps the intuitive aesthetic meaning it provides is justification enough.

INTERACTING WITH THE TREE GENOME

Let's examine the types of intuitive understanding this artificial tree model provides, while explaining the techniques of the mechanism of interaction. Again, two of the key steps in the process are

1. Interactive Gene Design and Real-Time Image Feedback
2. Isolating the Effects of Individual Numerical Genes

1. INTERACTIVE GENE DESIGN AND REAL-TIME IMAGE FEEDBACK

Complex tree images can take 2 hours or more to render on a VAX 780. Editing tree parameters at this rate is not very effective. Near real-time feedback is needed to allow one to freely explore the parameter space, and design the desired tree. Since vertex transformation cost is high for a complex fractal tree, hardware optimized for linear transformations was used for the real-time editor. The Evans and Sutherland MultiPictureSystem generates vector drawings of complex three-dimensional display lists in near real time. The display lists on the MPS look a lot like our tree nodes: primitive elements, transformed by linear transformations, and linked by pointers to other nodes. For a strictly self-similar tree, the transformation is constant, therefore the entire display list can share a single matrix. To modify the tree, one only has to change this one matrix, rather than an entire display list. This makes updating the tree display list very fast. For non-strictly self-similar trees, the transformations are not the same. However, a look-up table of less than a dozen transformations, is adequate to provide the necessary randomness.[20]

Figure 8 illustrates the logic of the display list.

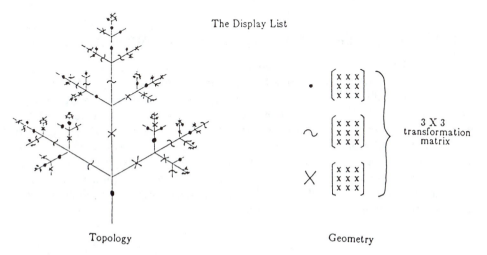

The Display List

3 X 3
transformation
matrix

Topology

Geometry

FIGURE 8 The display list.

The left side of the display list contains the topological description of the tree. Each node contains pointers to offspring nodes plus a pointer to the transformation matrix which relates that node to its parent node. This part of the display list is purely topological: it contains no geometric data. The geometric information is contained in the small list of transformation matrices on the right. To edit a tree, one can create an arbitrarily large topology list once and then rapidly manipulate only the small geometry list. Alvy Ray Smith in his research on *graftals*, recognized the separation of the topological and geometric aspects of trees. He calls these components the graph, and the interpretation respectively. His work deals primarily with specification of the tree topology, ignoring interpretation for the most part.[20] The model presented here emphasizes the geometric interpretation. The thesis of this paper is that the key to realistically modeling the diversity of trees lies in controlling the geometric interpretation. Many different topologies were used in this project. But by varying the geometric interpretation of a *single* topology, one could still generate a wide variety of trees each with its own distinct taxonomic identity.

The real-time generation of fractal trees has been packaged as an interactive editing system. This multi-window system allows one to edit the tree parameters, (both geometric and topological) via graphically displayed sliders. A vector image of the tree responds in real time. To see all the parameters change at once one performs keyframe interpolation of the parameters. Each tree parameterization is written to a keyframe file. A cubic spline program, interpolates these parameters to create the inbetween frames. In the resulting animation, the tree metamorphoses from key to key. A simple *tree growth* animation is achieved by interpolating the trunk width parameter, and the recursion size cutoff parameter. Modifying additional parameters makes the growth more complex and natural looking. For example, many plants uncurl as they grow. A metamorphosis animation is achieved by interpolating parameters from different tree species.

2. ISOLATING THE EFFECTS OF INDIVIDUAL NUMERICAL GENES

By fixing all but one of the artificial gene parameters, one isolates the effect of that gene, in a way that one cannot do in nature. One can exaggerate the effect of that gene or disable it entirely. For example, by zeroing the size ratio of the side branches to the main stem, one effectively strips all the branches. Left then with only the stem, one can better observe the effect of the stem-shape genes. By varying the transformation between stem segments, one derives the class of *spirals and helixes* and their random perturbations. These shapes appear in all forms of growth, organic and inorganic—from the inner ear, sea shells, and plant sprouts, to spiral galaxies. *Spirals and helixes* are in some sense degenerate self-similar sets. They are the atomic units that make up the fractal trees.

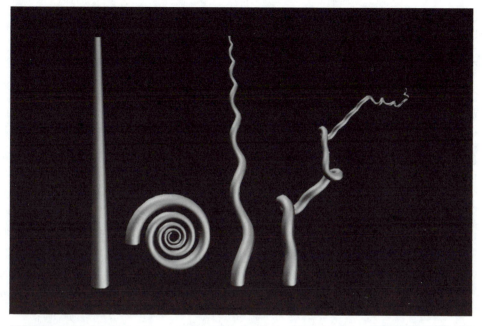

FIGURE 9   Helixes.

Figure 9 shows 4 typical *stem* shapes.

a.   *Cylinder*: the transformation is a translation and a scale.

b.   *Spiral*: one performs a rotation perpendicular to the stem axis, in addition to a scale and translation.

c.   *Helix*: one performs an additional rotation along the stem axis.

d.   *Squiggle*: by randomly changing the transformation from segment to segment, case c becomes case d.

Branches are simply stem shapes attached to the main stem and each other.

Another genotypic attribute that merits isolation is randomness. Since each genetic parameter is a random variable, one can control the degree of randomness by varying the standard deviations. If all the standard deviations are zeroed, one gets a very regular-looking tree such as a fern (see Figure 4). This tree is strictly self-similar; that is, the small nodes of the tree are identical to the top level, largest node of the tree. By enabling the randomness, one gets an irregular gnarled tree such as a juniper. The greater the standard deviation, the more random, irregular, and gnarled the tree (see Figures 10 and 11). The resulting tree is *statistically* self-similar; not *strictly* self-similar.

There are several reasons for the stochastic approach. First, adding randomness to the model generates a more natural-looking image. Large trees have an intrinsic irregularity (caused in part by turbulent environmental effects). Random perturbations in the model reflect this irregularity. Second, random perturbations reflect the diversity in nature. A single set of tree model parameters can generate a whole forest of trees, each slightly different. Database amplification is increased.

One could write a whole chapter about the relation between natural and artificial randomness. If randomness reflects our incomplete ability to measure and predict an otherwise deterministic world, does that make randomness a purely perceptual artificial quantity, or does natural randomness exist? Is randomness in nature a measure of the balance between the organized structure of an object struggling against the chaotic influence of its environment? If so, what determines the level of equilibrium? In the artificial model, a numerical gene determines the level of randomness. But what about in the natural case? Does the randomness parameter reflect the object or the environment? That is to ask, does the organism have some genetic mechanism that specifies how strictly its structural blueprint is to be transcribed in the phenotype? Such control could, after all, be an adaptive mechanism for evolution. Under certain conditions of natural selection, one might favor organisms that transcribe fairly precisely, that is randomness disabled, while under other scenarios, genetically controlled loose transcription could promote increased

FIGURE 10   Views 1

FIGURE 11   Views II

mutation rates and hence accelerate evolution through faster adaptation to rapidly changing environments.

WARNING: THE ABILITY TO SIMULATE NATURAL PHENOMENA IS EXTREMELY ADDICTIVE.

Over the past few years, members of the computer graphics community have focused their efforts towards visual representation of natural phenomena. These phenomena obey physical laws, and these laws can be expressed mathematically. A computer is, among other things, an engine for executing formulas, and expressing these results visually. By encoding an object's geometric and optical formulas in the computer, one can obtain a visual representation that is often considered *photographic*. As more of the laws of nature are translated into computer programs, the range of representable phenomena and the *photographic quality* of the visual expression increase. There are enough laws in the books, many of which have been *known* for centuries, to keep computer researchers busy for a long time.

In observing the improved image quality, one receives a gratifying sensation. To be the cause of the recapitulation of the evolution of species gives one a feeling of power, and, as with all forms of power, it is easy to become obsessed with its cultivation. Spending countless hours painstakingly refining the models, eliminating artifacts, and tuning parameters to achieve a more exact replica, has become a prevalent behavior among computer graphics researchers.

How shall we justify this obsession with realism? What purpose is served by mirroring nature as exactly as possible? These are, of course, questions about human nature. It is critical that we at least ask them from time to time, lest we lose sight of our underlying motivations.

Striving for perfection is an important human endeavor. We observe an apparent perfection in nature. By emulating nature, we praise it. It is a noble, humble activity, deeply rooted in our essential spirit.

Mankind has always sought a complete understanding of the external world. The generation of computer graphics replicas is a testimony to our increased understanding of nature. At a more human level, one needs no more justification for attempting to simulate nature than: *because it's there*. Nature is the standard, that drives the challenge.

One can avoid the pitfalls of addiction, and obsession, by maintaining the proper *awareness*. In particular, one must remain aware that an exact mathematical, analytic, simulation (whatever that means) is *not* the whole story. The representation of a phenomenon includes both the object and its perception. The obsession with realism, is the attempt to eliminate the involvement of the observer or the artifacts of the medium from the image, reducing the object to pure formula.

It is not enough.

By going beyond the laws of nature, by presenting perception as well as the object, one can create an image that says more than the object alone. One can create the possibility of meaningful communication, of image content. By representing an object, not as it is ordinarily seen, but by emphasizing, exaggerating selected aspects of a perception, one can express more of the object's underlying essence. Going beyond the laws of nature adds new meaning and spirit to the image. What in fact, does an exact mirror image say? We already have the *real thing*, and it is more *perfect* anyway. What has been added? By simulating nature exactly, we say nothing new about nature other than affirming our understanding of it.

The agreement of the image to nature, as judged by the eye, demonstrates that the creator of the image has captured some aspect of the object's essence. But which aspects? We cannot see them from the replica, if it is indistinguishable from the object. The image creator alone knows. He can tell us verbally how he reduced nature to formulas, numbers, methods. But by expressing nature in such terms, it becomes dry, stripped of meaning and spirit. The language of imagery is more universal than the language of words and symbols. Objectivity expresses less than subjectivity.

By steering away from the exact replica, the image(maker) can express meaning through contrast to nature. By showing less one expresses more through difference. By going beyond the laws of nature, we create a new reality, and that helps us appreciate nature more.

Consider, for example, the image of the raspberry bush (see color plate 25). One would not go so far as to call this an astonishingly realistic image. A few liberties have been taken no doubt. However, it does look more or less like a raspberry plant, with nothing terribly abnormal about it. The colors were chosen to agree roughly with the natural counterpart although they are a bit on the saturated *electronic* side (as are most computer pictures these days) Stylisticly, a Japanese wood-cut composition was sought, but in the end turned out resembling a French candy tin.

The laws of nature which generate a raspberry plant are represented, in some form, on the computer. The genetic code which determines the branching pattern of the stems and leaves, and the symmetry of the berry drupelets, have a numerical counterpart. On the computer, as in nature mutations are possible. The laws that produced normal raspberries, can be recombined at will, with other laws, in artificial ways. The result is raspberry stems and leaves, with eyeball fruits (see Figures 4, 10, 11 and color plate 26). For now the possibility of actually engineering such an organism can only be conjectured on the computer.

These two images say very different things.

The more edible version:

- Emphasis of three elements.
- Pretty, but not interesting,
- Form but little content. Says little about the object itself.

The more toxic version:

- Juxtaposition of unexpected elements.
- Not so pretty, but interesting.
- Realism is not avoided. The individual elements are realistic. The juxtaposition is surreal. In fact, the eyeballs were generated by mapping a scanned photograph of an actual eye onto the surface of a computer-generated sphere.

This surrealism is more than just a comment about perception. It expresses something about the object itself. The eyeballs give the plant a new kind of spirit, or life force. The plant has a personality. These feelings can be extended to the traditional normal raspberry as well. This increases our appreciation of nature. We take nature for granted. We take raspberries for granted. The eyeball picture makes us thankful for the fragile process of heredity.

One says something about a raspberry, by showing what it is not.

PROOF BY PICTURE?

Computer graphics has been considered a tool to develop and test new theories, with the eye as the judge. If the image *A looks like* the modeled object X, then the formulas, assumptions, and parameters contributing to the image, seem *more correct* than those used in image B, which does not look like X. The picture proves the theory.

Or does it?

The image may look *right* but for the wrong reasons. Images of mountains generated by recursive subdivision and random perturbation of polygonal surfaces look reasonably convincing. However, few would argue that mountains actually form this way. Such images say more about our perception of mountains, rather than expressing any theoretical *truth* about the mountains themselves.

If, however, the image looks *wrong*, then one can conclude that the underlying formulas, assumptions and parameters contributing to the image are *wrong*. Such a result is likely to be swept under the carpet, filed as bad data to be refined, tuned, corrected. However, in *proof by pictures*, these negative results can be more meaningful than positive ones.

Pictures cannot prove, but they can disprove.

By seeing nature as it might have been under different conditions, we question assumptions that we had previously taken for granted. The juxtaposition of the *right* and *wrong* images, can clarify what made the picture right, leading us to new conjectures and understanding. Animation, showing a continuum between right and wrong, illustrates the cause and effect meaning of the parameters even better. Showing nature as it is not, helps us appreciate the way it is.

BEYOND THE LAWS OF MOTION

One natural phenomenon that has attracted considerable attention in computer graphics is human movement. A typical computer model organizes the human into a hierarchy of joints. *The hip joint's connected to thaa knee joint... The knee joint's connected to thaa ankle joint...* Each joint controlled by three numerical parameters: x rotation, y rotation, and z rotation. By independently varying these parameters, the model can be made to walk.

Anyone who has ever tried manipulating such a model can tell you that it is nearly impossible to control effectively. It takes several hours of mere dial turning to get even a few seconds of awkward walking. It is fairly safe to conclude, that this paradigm for motion is *not* what's going on inside the human body. Humans have a much more complicated mechanism for coordinating the joints than simply independently varying the rotations of each joint. Since computer-generated three-dimensional objects *pass right through* each other, unlike *real* solid objects, it is very difficult to keep the model's foot from going right through the floor. Physics itself breaks down. The computer model does not sufficiently describe the natural phenomenon.

One reaction to this result is to improve the computer model. By encoding kinematic and dynamic relationships between the joint rotations, and object interactions, a more intelligent program could help orchestrate the movement better than a user turning dials. Research in this area is well underway, and already has and will continue to shed light on the topic of human motion.

At the same time, however, we should not dismiss the negative conclusion as being a non-result, by sweeping it under the table. It is a very meaningful result. It implies that actual human motion is very complicated and elegant—something we tend to take for granted. Rather than agonizingly suppress the artifacts and inadequacies of the computer model, let's exaggerate them, amplifying the awkwardness of the model. By seeing how we don't walk, we can better appreciate and understand how we do walk, and be led to the next stages in the model.

The evolution is accelerated, through not ignoring *failure*. So it is not a failure at all.

Let the foot go right through the floor.

The laws of physics are not *a priori*, universal, or obvious. Computer physics has its own set of laws that allow its characters to do things that humans cannot.

FIGURE 12   Moe Naleesa.

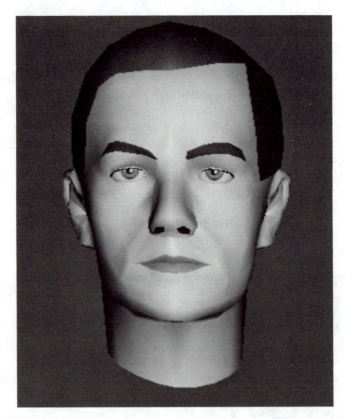

FIGURE 13   Ralf (image by Rebecca Allen, Steve DiPaola, and Robert McDermott).

THE ARTIFICIAL MENAGERIE

Each of us has a vision of the natural world in our minds, based on our perception. We strive to express this vision, to externalize it. To map these mental images onto an external canvas we must translate them into the language of some medium. The resulting artificial forms are distorted by the artifacts of our perception, the output medium, and the choice of representation in the language of that medium. The Sapir-Whorf hypothesis states that a culture's world view is correlated to the language of that culture. Similarly, the content of an artificial life-form is correlated to its internal linguistic representation in the language of its medium.

Figures 12 and 13, show two views of man. These artificial life-forms are based on the same natural species, and yet their forms and meanings differ. This is directly correlated to a difference in internal representation in the language of the computer. Figure 12 is based on the *Mr. Potato Head* approach of decomposing the form into individual isolated components. The internal decomposition uses words like *nose, eye, pupil, lips,* etc, arranged hierarchically. The primitive quality of this image is correlated to the *English-like* internal representation, pointing out limitations of our spoken language. Figure 13, based on a more mathematical representation, seems more objective.

Yet both images represent a world view. Artificial life-forms need not be objective. In fact, they are always subjective to some degree. Artificial life-forms need not adhere to a single reality, a single myth. Rather they can express a multiplicity of truths, none of which dominate any of the others—all of them coexisting on an artificial landscape.

If we keep an open mind and a heightened awareness, then the menagerie of artificial life-forms that we create will be rich, meaningful and beautiful.

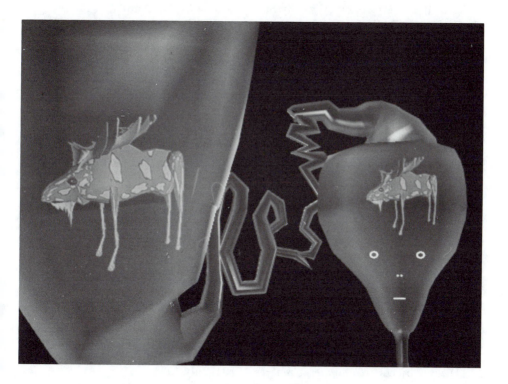

REFERENCES

1. Allen, R., and P. Oppenheimer (1985), *The Palladium* (Old Westbury, NY: New York Institute of Technology), video.
2. Aono, M., and T. L. Kunii(1982), "Botanical Tree Image Generation," *IEEE Computer Graphics and Applications* **4(5)**, May 1982
3. Bentley, W. A., and W. J. Humphreys (1962), *Snow Crystals* (New York: Dover Publications Inc.); originally McGraw Hill, 1931.
4. Bloomental, J. (1985), "Modeling the Mighty Maple," *Computer Graphics* **19(3)**, July 1985.
5. Cole, V. C. (1965), *The Artistic Anatomy of Trees* (New York: Dover Publications Inc., New York); originally Seeley Service & Co, London, 1915.
6. Demko, S., L. Hodges, and B. Naylor (1985), "Construction of Fractal Objects with Iterated Function Systems," *Computer Graphics* **19(3)**, July 1985.
7. Gardiner, G. (1984), "Simulation of Natural Scenes Using Textured Quadric Surfaces," *Computer Graphics* **18(3)**, July 1984.
8. Kawaguchi, Y. (1982), "A Morphological Study of the Form of Nature," *Computer Graphics* **16(3)**, July 1982.
9. Klee, P. (1948), *On Modern Art* (London: Faber & Faber Ltd.).
10. Mandelbrot, B. (1977), *Fractals: Form, Chance and Dimension* (San Francisco: W.H. Freeman and Co.).
11. Mandelbrot, B. (1982), *The Fractal Geometry of Nature* (San Francisco: W.H. Freeman and Co.).
12. Marshall, R., R. Wilson, and W. Carlson (1980), "Procedural Models for Generating Three-Dimensional Terrain," *Computer Graphics* **14(3)**, July 1980.
13. Oppenheimer, P.(1979), *Constructing an Atlas of Self-Similar Sets, undergraduate thesis, Princeton University.*
14. Oppenheimer, P. (1985), "The Genesis Algorithm," *The Sciences* **25(5)**.
15. Oppenheimer, P. (1986), "Real Time Design and Animation of Fractal Plants and Trees," *Computer Graphics* **20(4)**, August 1986.
16. Oppenheimer, P. (1986), *Son of Doctor Skitzenheimer* (Old Westbury, NY: New York Institute of Technology), video.
17. Reeves, W. (1983), "Particle Systems—A Technique for Modeling a Class of Fuzzy Objects," *Computer Graphics* **17(3)**, July 1983.
18. Reynolds, C. (1981), "Arch Fractal," *Computer Graphics* **15(3)**, August 1981, front cover.
19. Serafini, L. (1983), *Codex Seraphinianus* (New York: Abbeville).
20. Smith, A. R. (1984), "Plants, Fractals, and Formal Languages," *Computer Graphics* **18(3)**, July 1984.
21. Stevens, P. S. (1974), *Patterns in Nature* (Boston: Little, Brown, and Co.).

Charles E. Taylor,† David R. Jefferson,‡ Scott R. Turner,‡ & Seth R. Goldman‡
†Department of Biology and ‡Department of Computer Science, University of California, Los Angeles, CA 90024

RAM: Artificial Life for the Exploration of Complex Biological Systems

In this paper we describe RAM, an Artificial Life system developed at UCLA for modeling population behavior and evolution. RAM is a powerful new kind of tool for biological simulation, capable of modeling population behavior and evolution to a finer level of detail than any other tool we know of. It is also a vehicle for expressing, in program form rather than traditional mathematical form, theories of population behavior, ecological interaction, animal communication, adaptation, and natural selection. We believe it offers a new vocabulary and new metaphors for describing biological phenomena, and new ways of testing theories that are otherwise mathematically intractable and experimentally unthinkable.

RAM is based on the observation that the life of an organism is in many ways similar to the execution of a program, and that the global (emergent) behavior of a population of interacting organisms is best emulated by the behavior of a corresponding population of co-executing programs. Under RAM each individual organism is represented by a separate Lisp program with its own code and memory. Environmental processes (weather, seasons, nutrient replenishment, etc.) are each represented by separate programs as well. Just as organisms are born, live, learn, interact with each other and

Artificial Life, SFI Studies in the Sciences of Complexity,
Ed. C. Langton, Addison-Wesley Publishing Company, 1988

with the environment, reproduce with modification, and die, so too can animal programs initiate, execute, learn, communicate, interact with environment programs, replicate with modification, and terminate. RAM directly simulates a population, sometimes over many generations, by concurrently executing all of these programs (and their progeny).

In addition to describing the RAM system itself, we will describe three biological applications illustrating RAM's value and range: (1) lek formation among sage grouse, (2) mosquito control, with some of the simulated "animals" representing mosquito populations and others representing little expert systems attempting to control the mosquitos, and (3) evolution in a simple ecosystem containing predators, herbivores, and vegetation.

---

1. INTRODUCTION

Explicating the nature of mind and intelligence is one of the greatest scientific challenges, but until recently there were fundamental barriers to progress because of our limited ability to perform controlled experiments with the prime examples of intelligence: human brains. The advent of Artificial Intelligence has provided a new way to address difficult questions without *in vivo* experimentation, through computational modeling. AI and cognitive science owe much of their intellectual success to the fact that they impose a discipline on themselves, requiring that any new theory of intelligence, cognition, language comprehension, or perception, be expressed ultimately not in prose, but in explicit, unambiguous computational form. Such theories, in the form of programs, can then be the objects of controlled, repeatable experiments, and their strengths and weaknesses exposed by any investigator.

Understanding the behavior of living systems, particularly at the population level (ecology and evolution), is just as complex a challenge as understanding intelligence, and has suffered from similar barriers to progress. Because of the scale of space and time over which they behave we have only limited ability to experiment with the primary objects of study, i.e., evolving populations and ecosystems in their natural state. We believe, however, that the new field born at this conference and aptly named Artificial Life, may make it possible to address complex ecological and evolutionary questions in a new way through computational modeling. Artificial Life should follow the example of Artificial Intelligence, grounding itself in a similar discipline: theories of evolutionary dynamics and population behavior should be expressed and tested in the form of programs.

We are interested in viewing life at the *population level*, and in studying ecological and evolutionary systems that are too complex to study analytically, and are either too large or last too many generations to study experimentally. Accordingly, we have developed a simulation shell called *RAM* in which populations of organisms are simulated by populations of interacting programs on a one-animal/one-program basis. Each organism and each environmental process is represented by a separate program, with separate code and data. Each one, if so programmed, can:

- accumulate knowledge about the environment and about other organisms (i.e., sense, perceive, learn);
- modify the environment and other organisms (i.e., interact, communicate, cooperate);
- calculate (i.e., evaluate, decide, reason);
- exhibit time-dependent behavior (i.e., age, sense time);
- change location (i.e., move);
- create other organisms with mutation or recombination (i.e., reproduce); or
- terminate, or cause others to terminate (i.e., die, kill).

Animal programs, environment programs, and the RAM simulation shell itself, are all written in T (a dialect of Lisp) and thus have the full computational power of Lisp at their disposal. Any of these actions can be conditioned by arbitrarily complex genetic, ontological, temporal, environmental, or random criteria. Simulation of a population's behavior is simply the time-stepped concurrent execution of all these environmental and animal programs (and their progeny) in a shared two-dimensional environment.

We see RAM as embodying directly the view that living systems (cells, organisms, populations) are best understood as processes. It is a step toward addressing the concerns of those evolutionists who are dissatisfied with the formal limitations of the current mathematical tools of population genetics (differential equations, vector fields), and with conceptual limitations of the nonobservable abstractions (e.g., fitness functions, selection coefficients) that one is forced to invent in order to use those tools.

2. OVERVIEW OF RAM

RAM is a simulation shell: it provides time-stepped coordination, a two-dimensional grid space, the notion of animal-as-process, graphics and interaction tools, and a consistent set of biological metaphors. To construct a biological model within RAM the user must design and program three systems: the *environment*, the *species* that populate it, and the *observers* that gather and display statistics.

The basic environment is a two-dimensional grid. To design the behavior of the environment the user must declare the attributes (temperature, nutrient level, etc.) associated with each cell in the two-dimensional space, and then write environment procedures to express their dynamics. Observer programs are syntactically indistinguishable from environment programs. The difference is that an observer program only *reads* the states of animals and the environment, and displays results. It never modifies an animal or the environment.

Designing an animal species requires deciding *everything* about its abilities and limitations, its genetics, and its behavior. Animal representations must explicitly address the following issues:

- How does information from the environment affect the animal?
- How does the animal change the environment?
- How does it move around the environment (if at all)?
- How does it interact with other animals of the same or different species?
- What does it need to stay alive, and how does it obtain it?
- How does it interact with other animals?
- What information is genetic?
- How does different genetic information affect behavior?
- When and how does it reproduce?
- What are the characteristics of a newborn organism?
- How does its behavior change as it ages?
- How is genetic information modified upon reproduction?
- Under what circumstances does an animal die?

The answers to all of these questions must be embodied in the animal-behavior code written in Lisp, because RAM has almost no biological knowledge built into it. The user must introduce all such knowledge relevant to his problem in the form of animal and environment programs. The disadvantage of this is that almost nothing is assumed *a priori* about animals, or even about the laws of nature, and hence every realistic model must include code for even obvious things, e.g., the requirement that organisms must eat to live. The advantage, however, is that the modeler has unprecedented freedom to represent rich and complex systems naturally. It is straightforward (though not necessarily simple) to represent such complicated genetic issues as frequency- or density-dependent genetic effects; linkage and epistatic effects; many genes with arbitrarily complex, nonlinear interactions; small population effects; inbreeding effects; a wide variety of sexual or recombination systems (haplodiploidy, self-fertilization, hermaphroditism, environmentally-dependent sex determination, mitochondrial inheritance, mixed sexual and asexual systems, etc.); genetic drift; and cultural or other non-Mendelian inheritance (even Lamarckian). In addition, ecological and demographic issues are equally representable, including multispecies interaction; geographic structure; seasonal and other temporal effects; trophic, competitive, symbiotic, and other ecological relationships; age structure effects; and animal learning, communication, and cooperation. Any or all of these features can be present *simultaneously* in a RAM model if desired, and their interactions explored.

2.1 THE ENVIRONMENT

The *environment* consists of *cells* arranged in a two-dimensional grid. Each cell contains *cell variables*, *cell procedures*, and *animals*. Cell variables are used to represent biologically relevant conditions in that region (nutrient level, altitude, water depth, etc.) Cell procedures are Lisp functions associated with cells and used to model processes that change the environment independently of the action of animals. An example cell variable might be grass-amount, whose value represents the amount of "grass" available in the cell to feed grazing animals. In this example we would choose not to model "grass" as a full-blown organism with its own genetic endowment and behavior, but to treat it simply as an environmental attribute whose value is reduced by animal procedures and increased by cell procedures. A cell procedure grow-grass, for example, might be responsible for adding grass by modifying the grass-amount variable at regular intervals. In addition to cell variables and procedures, a cell may contain any number of animals of any number of species, each of which is free to move around and modify the environment or one another, and behave as circumstances dictate.

In addition to cell variables and cell procedures, we have found it convenient to introduce *environment variables* and *environment procedures* that are associated with the environment as a whole (for an example, see section 3.3 and Figure 3). A typical environment variable would be daynight, representing whether it is day or night in the region of the environment, and it might be controlled by an environmental procedure called daily that reverses the value of daynight every 12 time units. The meaning (scale) of a time step is arbitrary and determined by the user.

2.2 ANIMALS AND SPECIES

An *animal* resides in some cell of the environment and consists of a collection of *animal variables* and *animal behaviors*. (We use the term "animal" to refer to any entity represented this way, although it might actually represent a tree, or even a truck.) Animal variables are used to indicate an animal's internal state (genetic, phenotypic, and mental), and are divided into four kinds. *Static* variables are read-only, are set when an animal is created, and thereafter never changed. They are used for representing such attributes as birthdate or genotype that remain unchanged throughout the life of the animal. *Private* variables are modifiable, but are internal to the animal; no other animal can look at or modify these directly. They are used for representing such attributes as the animal's learned knowledge. *Public* variables can be read, but not modified by other animals (e.g., size, sex). *Mutable* variables can be read or modified by any animal or environment procedure (e.g., state-of-health). These four variable categories are crude software engineering aids to the design and debugging of animal programs; their biological significance is purely heuristic.

The structure of our code and our terminology used for describing it has been inspired by the object-oriented programming paradigm. One can easily identify *species* as *classes*, *animals* as *instances* of classes, etc. However, we do not adhere to any particular discipline, and we violate object-oriented programming philosophy in at least one major respect by using more of a shared-variable semantics than a message-based semantics.

Animal behaviors are represented by Lisp functions, and always include at least an initialization behavior (executed when the animal is born), a normal behavior (which the simulator executes every time step), and a termination behavior (executed when the animal dies). However, there can be additional behaviors, e.g., a fox wishing to attack a rabbit might invoke that rabbit's defense behavior, which otherwise would not be executed.

The normal behavior of an animal is the code that decides how the animal behaves during each time step. It will generally include code that does the following:

a. Examine the nearby environment, and possibly some of the animals found there.
b. On the basis of the nearby environment, the time, its own age, and its state (including genes and past history), decide, perhaps probabilistically, what actions to take next.
c. Take action, including any or all of the following: update statistics, move, update its own memory, modify the environment or other animals, reproduce, or die.

Hence, an animal's normal behavior can be viewed as a rule to "assess the current situation, and then take action." This rule is executed every time step, so the "life" of an animal is basically an iteration of this rule, repeated every time step from birth until death.

It is usually inconvenient to write separate behaviors for every individual animal in a simulation. We therefore make the convention that each individual is a member of a *species*, and all the individuals in a species share the same behavior code. However, it is important to realize that even though two animals of the same species share the same behavior functions, it does not follow that they behave identically: the behavior code may be, and usually will be, environment-dependent, gene-dependent, experience-dependent, time-dependent, age-dependent, and partly random as well. Those aspects of an animal's behavior that are inherited and not considered to be subject to variation in the course of the study are appropriate for coding directly into the species behavior.

2.3 RAM EXECUTION

A RAM simulation is time-driven, and proceeds in discrete *steps*. At each step the simulator executes *once* all of the environment and cell procedures and the normal behavior of each animal. The execution of a RAM simulation proceeds as follows:

```
for each time step do
    {     for each global environment pre-procedure
              execute it;
        for each cell in the environment
              execute cell's local environment pre-procedures;
        for each animal
              execute animal's normal behavior;
        for each cell in the environment
              execute cell's local environment post-procedures;
        for each global environment post-procedure
              execute it
    }
```

Global environment procedures are divided for convenience into two kinds, the pre-procedures and the post-procedures. In each time step, the pre-procedures are executed before the animal behaviors, and the post-procedures are executed after. There is no deep theory to this; it is just usually convenient to have some global environmental procedures (e.g., **grow-grass**) to execute before the animals, and others (e.g., **take-statistics**) to execute afterwards.

Animal behaviors are executed serially, but in an unspecified order. Similar comments apply to the global and local environment procedures. Clearly, the detailed behavior of the simulation will depend on this order, and it must be the responsibility of the user to consider this issue and design his model in a way that assures that this seriality will not introduce any systematic statistical bias into the model's behavior.

Right now there are no firm plans to build a parallel version of RAM. Ideally the animal and environment programs can all execute concurrently on a multiprocessor, but this might require the authors of animal programs to understand more about parallelism and synchronization than we believe is reasonable to expect at the present time. In any case, parallel RAM would probably not be built for a network or for a hypercube. It would more likely be built for a shared-memory (or shared virtual memory) machine because a system whose structure is so dynamic (replicating objects, moving objects, global objects, shared variables) is not easily decomposable into coarse-grained processes that communicate by message. Not only the animal simulation, but the Lisp environment as well would have to be restructured for parallel execution. For these reasons, we believe it would require a complete redesign of the RAM to make it suitable for parallel execution. However, we do plan to build a system somewhat similar to RAM in philosophy, but radically different in design, to run on a Connection Machine.

2.4 USING RAM

RAM runs only on Apollo workstations today, although currently we are porting it to the Macintosh II. The basic simulator (excluding graphics) can be used on any system that supports T (or soon, Common Lisp).

After programming the environment and the animals, all of the environmental and animal specifications are assembled into a single execution package under the control of the RAM system, which itself runs under Lisp. The last step is to set up the initial conditions of the environment, specifying its dimensions, the initial values of each cell and environmental variables, and the initial distribution of animals of various kinds within the environment. This is specified in a configuration file read at the beginning of a RAM execution, although it can also be modified by hand after initialization.

The user interacts with RAM primarily through a Macintosh-like interface based on menus, windows, mouse interaction, and graphics. One window displays a schematic view of the environment showing the spatial distribution of animals and of cell attributes. Each cell contains small pictures of the animals present and additional information, determined by the user, concerning cell variables. Other windows can be used to monitor the progress of the simulation, either graphically (bar charts, cluster, or line graphs) or textually (messages from animals or from environment procedures). RAM provides the ability to step the simulation, run it free, or interrupt it at any time. At any pause in the simulation the user can examine it, edit environment or animal variables or procedures, add or delete animals or even species, and then continue the modified simulation. Examples of RAM displays are shown in Figures 1, 2 and 4.

The name "RAM" is not an acronym, nor does it refer to sheep. It is a foreshortening of the coinage "programinals." The original version of RAM was written in 1982 as a Pascal program that executed animal programs in a calculator-like postfix programming language with output to a 24 × 80 character screen. It was subsequently rewritten in T, so that system commands and animal programs would all share a common programming language, and a common graphics- and mouse-oriented interactive interface.[1] Lessons learned from these earlier versions have influenced our current (third) implementation, which has improved windowing, better graphics, better modularity and portability, and better performance.[21] In these respects, and in others, RAM is still evolving (no pun intended). Although it is difficult to cite reliable performance figures, as a ballpark estimate we can testify that RAM, on an Apollo 3000 workstation, can simulate a population of 500 moderately complex animals (such as the foxes and rabbits described later) for between 100 and 1000 time steps per hour, depending on the amount of I/O the model performs (e.g., screen maintenance), and on the amount of instrumentation that is added to the simulation.

3. APPLICATIONS

In this section, we discuss three biological models we have built under RAM. At this early stage in our work, no application we have studied exercises the full range of RAM's capabilities. Therefore, we have chosen to describe three simple examples to illustrate its use. The first example, lek formation among sage grouse, is concerned primarily with trying to understand the rules underlying the peculiar spatial distribution of sage grouse at mating time. The second, mosquito control, is basically an ecological/population dynamics model concerned with the rise of mosquito populations as a function of weather, breeding site availability, and application of insecticide and other mosquito control measures. The third, foxes and rabbits, is a toy model designed to illustrate RAM's ability to combine ecological effects (predator-prey) with natural selection effects (among alleles of three genes in the rabbits) and model the resulting interactions.

3.1 LEK FORMATION AMONG SAGE GROUSE

Leks are areas where animals assemble solely to find mates and copulate. They are formed by a variety of bird and mammal species, and even some invertebrates. In sage grouse, dozens of males will assemble at one place during the early morning and strut about, displaying and thumping loudly. Females travel from their nests, sometimes several kilometers, to wander about the lek area and, after observing the males for some time, to mate with one of them. Typically one male in a lek will perform the vast majority, though not all, of the copulations at a given lek. Leks are biologically interesting because they serve as a model for the general evolutionary phenomenon of sexual selection.[4]

Little is known about what considerations the animals use for choosing a mating site. Current thinking has been that female preference is the most important factor. When choosing a lek to visit, the females probably weigh several factors: the number of males in the location, the distance of the site from her nest, and the length of the queue for the most desired males.[4] The relative importance of these factors is completely unknown, and it has even been speculated that the females are largely irrelevant for the formation of the leks, i.e., lek formation is a function solely of male dominance relations.[2] In collaboration with Robert GibsonGibsonRobert of UCLA, we have been using RAM to simulate lek formation in an effort to learn how the hypothesized factors might be weighted to reproduce observed grouse behavior. Such simulations cannot prove that one set of factors actually underlies the process, but they can test whether or not a specified explanation is able to describe what is observed, and they can demonstrate that certain hypothesized explanations are inadequate.

Gibson's study area is located in a caldera of about 100 square miles near Mammoth Lakes, in the Sierra Nevada. There are 200–300 males located there that assemble into 8–10 leks of varying size, from 5 to nearly 200 males per lek. For a detailed description, see Gibson and Bradbury.[8] We examined a variety of hypothesized female and male behaviors to see which ones could reproduce the lek size and distance distributions observed there.

For each simulation, 100 male and 100 female animals are created and distributed randomly throughout the environment. Each female's nest site is the same as her initial location. In each step, all animals survey all locations and then move to the most favorable of these, as judged by their variables and programmed behavior. Females weigh the several relevant features of each location (number of males, distance from their nest, and length of the queue) before moving. The males are thought to all choose the most favorable location available to them, based on the number of females per male in each location. Hence, both sexes apparently condition their behavior on that of the other. One variant we have studied presumes that the males have an intrinsic quality, something like degree of "macho-ness," which in turn determines their dominance in relation to other males. They weigh their own relative rank in the calculation, the higher ranking males expecting to get proportionately more matings. The assessment of each animal depends on where the others are located and on several other factors, so several iterations are typically required before convergence to stable leks. Equilibrium often depends on the starting conditions, and although convergence is not guaranteed; in practice we typically observe it within ten steps.

Figure 1 shows a screen from one such simulation. The number of males and females in each location are shown below their respective pictures. Note the formation of a major lek in the highlighted square. With 200 animals, each surveying 100 locations per step, each counting all the other animals there, perhaps estimating their position in the dominance hierarchy at the time and then moving to the most favorable cell, it takes RAM approximately 5 minutes per iteration on an Apollo DN 3000.

This study has shown, first, that the hypothesis attributing lek formation to male dominance alone without consideration of female behavior[2] is unable to account for observed results. Some consideration of female choice is necessary. Second, some consideration of length of queue, distance from nest, and number of males in the display area seem required as well; no one or even two of these variables seems adequate to reproduce the observed behavior. And third, the range of considerations of female behavior which give rise to the observed lek distribution is broadened when male dominance is also included,[4] and should be investigated experimentally.

A system as rich in capability as RAM is not absolutely necessary for a study like this one (e.g., see Bradbury et al.[5]). The advantage of RAM, however, is the straightforward nature of the representation and the ease of modifying or substituting alternative rules for decision-making by the animals.

3.2 MOSQUITO CONTROL

A second example is a simulation of mosquito populations in Orange County, California. The eventual goal is to find better ways for controlling mosquitos with insecticides while preventing (or managing) the evolution of insecticide resistance. Here we report only on the results of the first phase, without any model of insecticide resistance.

A variety of techniques have been used for past simulations of mosquitos. The most elaborate was described in Greever and Georghiou.[9] It was a large difference equation system which was found to model some features of the populations fairly well, but fell far short in others, principally because of problems with representing the geographic distribution of mosquito populations and the variety of breeding sources. While the young live exclusively in the water, the adults do not, occupying entirely different ecological niches.

Figure 2 shows a map of Orange County, California with the locations labelled according to the Mosquito Abatement District (MAD) control zones. The locations are designated by their grid coordinates (e.g., 4,5) and by their zone (e.g., W15). Each zone is approximately 4 miles square, and has been inventoried by MAD personnel for 100 types of aquatic breeding sources where immatures may develop. At present, only the four most important are represented in our RAM model: swimming pools, storm drains, flood control channels, and gutters. Precipitation and

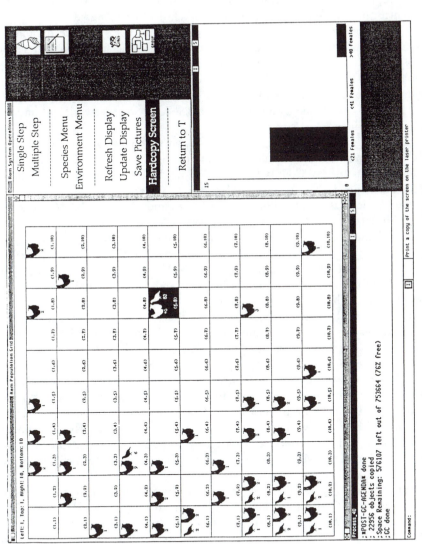

FIGURE 1   Screen display of simulated lek formation by sage grouse. The numbers of each sex are shown below the respective figures in that location. It may take several rounds for convergence to a stable arrangement. Note the beginning of a large lek in the highlighted location, (5,8).

temperature affect the availability and suitability of each of these differently. A global environment pre-procedure reads weather data for each day from files representing the seven U.S. Weather Bureau stations in the region.

There are currently several types of animal. It would be computationally unthinkable to represent each individual mosquito by a separate RAM "animal." Instead, the entire population of mosquitos in each class of breeding site in each cell is represented by one animal, as though that population were a single organism. Thus there are up to five animals per cell, representing the four populations of immatures (pool, drain, channel, and gutter) and one animal representing the population of adults. There are up to 250 such "animals" (populations) in our current implementation. Each is a separate, simple compartment model of its population,[20] where survival and development time depend on the local weather variables and on insecticide applications.

One "animal" in the model is special. The Supervisor, shown in the lower left of the figure, is an expert system animal that contains the rules for applying mosquito control measures. In Orange County the chemical and biological methods most used are juvenile hormones (Altocid), oil (Golden Bear), bacteria that are eaten by the larvae and are toxic to them (*Bacillus thuringiensis*), and mosquito fish (*Gambusia*) that eat the immatures. In the future, the supervisor "animal" will control a fleet of truck "animals" that will move around the environment combatting the mosquitos.

Local environment post-procedures tally the adults in 22 areas corresponding to the locations of real traps monitored by the MAD. They compare the predicted abundance of adults to what is actually observed, and then a global environment post-procedure sums the data from the traps for statistical analysis.

This model is still in the early stages of validation. For the years 1983 and 1984, it reproduces the populations quite well from early spring until late summer-autumn, when it then predicts a significantly higher population than is actually observed. We are confident this fit will be improved, based on the experience of others.[9,10] When we are able to reproduce the observed information reasonably accurately, or at least make better guesses than the actual supervisor does, then we will look at alternative rules for insecticide use (e.g., how early in the year insecticide treatments should begin, how long they should continue, how frequently they should be repeated, etc.), and incorporate a genetic model of resistance to insecticides, to evaluate the likelihood of resistance evolving under the chosen pattern of insecticide application.

The features that originally motivated this application of RAM were the ease of incorporating geographically structured populations with a variety of breeding sites and the naturalness of constructing an expert system "animal." Since that time, we have been struck by the ease with which complex interactions can be incorporated without radically rewriting any but a few procedures (also see Caulson et al.[6]). The naturalness of this representation results in such a large saving in programming time and ease of interpretation that RAM has substantial advantages.

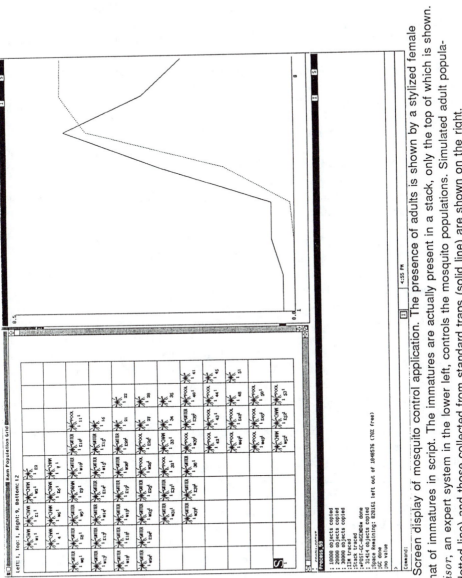

FIGURE 2   Screen display of mosquito control application. The presence of adults is shown by a stylized female adult, and that of immatures in script. The immatures are actually present in a stack, only the top of which is shown. The *supervisor*, an expert system in the lower left, controls the mosquito populations. Simulated adult population sizes (dotted line) and those collected from standard traps (solid line) are shown on the right.

3.3 FOXES AND RABBITS

As a third application we have built a simple ecosystem, exploring predator-prey relations where the prey have several genetically determined defensive strategies available to them. We wished to reproduce the essential features of predator-prey systems and to observe the evolution of behavioral strategies in action.

In our simple ecosystem, illustrated in Figure 3, there are foxes, rabbits, and grass. The grass is present in the model to act as rabbit food. It is simply an environmental attribute, and grows at a uniform random rate as programmed by an environmental procedure. The rabbits and foxes, however, are fully represented as RAM species with each individual having its own program and data.

Foxes and rabbits have several features in common. Both are endowed with a **weight** attribute (variable), which is intended as a general indicator of the animal's health. An animal whose weight goes below a lower threshold dies; one whose weight goes above a higher threshold reproduces, dividing its weight between itself and its offspring according to a fixed formula for the foxes and a genetically influenced formula for the rabbits. In both cases reproduction is asexual. Foxes are all genetically identical, but rabbit reproduction is accompanied by a low rate of genetic mutation. Neither animal ever dies except by starvation or predation.

Both foxes and rabbits can move one cell in any direction per time step. Both can "see," i.e., examine the environment for food and enemies, for a distance of two squares in any direction.

Rabbits gain weight (health) only by eating grass; foxes gain weight only by eating rabbits. Both animals are endowed with a characteristic "metabolism" rate, that causes them to lose a certain percentage of their weight each time step, regardless of what else happens in their lives. Hence each animal must eat regularly just in order to keep from starving to death. Only animals that are able to eat more than enough to offset their metabolism will ever reproduce, and in general the more eating they do, the faster they will reproduce.

In the absence of predation, the rabbit population will rise and stabilize around the carrying capacity of the environment which is entirely determined by the rate of grass replenishment. But with predation, the number of rabbits and foxes tends to oscillate, at least for a while. It is important to realize, however, that this is not the usual simple textbook predator-prey ecosystem. It is already complicated by a number of factors. First, there are three trophic levels (including the grass). The grass grows arithmetically (as we programmed it), not exponentially with time. Second, the rate of rabbit reproduction is not proportional to the number of rabbits, but to the fraction with **weight** above reproductive threshold. A large population of scrawny rabbits will have a low growth rate, while a small number of fat ones will have a high rate. Third, there are random small-population effects in this model that strongly affect the applicability of classical predator-prey theory. Finally, as we shall see, it is still further complicated by the fact that rabbits have genes affecting both their metabolic rate and their defense against foxes.

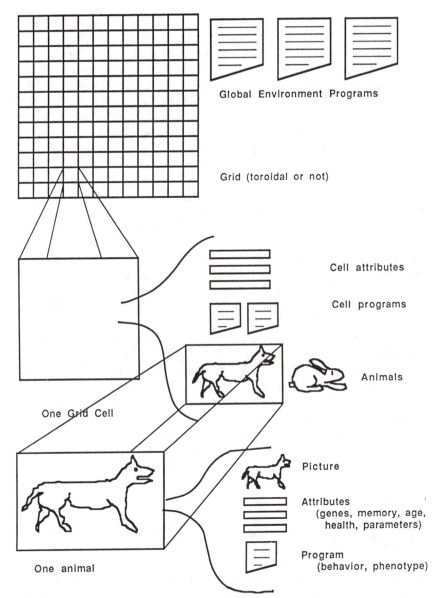

Global Environment Programs

Grid (toroidal or not)

Cell attributes

Cell programs

Animals

One Grid Cell

Picture

Attributes
(genes, memory, age,
health, parameters)

Program
(behavior, phenotype)

One animal

FIGURE 3 Structure of a simple ecosystem in RAM. The environment may contain global programs, capable of altering the attributes of all the locations and the animals with them. Each location may have both variables and programs, capable of altering its own variables and those of the animals it may contain. Each animal have have associated with it a picture, attributes with various sorts of protections, and a behavior encoded in its program.

Rabbits each have three genes: the speed gene, the birth-weight gene, and the multiple-birth gene. The speed gene has two alleles, *fast* and *slow*. Rabbits with the *fast* allele are better able to avoid being eaten by foxes, but have a higher metabolism and must eat correspondingly more food in order to avoid starvation. The multiple-birth gene also has two alleles, *single-birth* and *twin-birth*. Rabbits with the *single-birth* allele give birth to a single offspring at a time, while rabbits with the *twin-birth* allele produce two offspring (each weighing one-half of what the equivalent single offspring would weigh). The birth-weight gene is a number that represents the fraction of the parent's weight that is transferred to the offspring at birth. This gene can vary from zero (which would cause offspring to die immediately) to one hundred (which would kill the parent by using all the parent's available weight for the offspring). The original rabbit population consists entirely of *slow, single-birth* rabbits with the birth-weight allele of 0.50. Mutations from one allele to another of all three genes occur at a fixed rate coded into the rabbit behavior program.

The dynamics of this system are illustrated in Figure 4. The line graph displays the rabbit population, the fox population, and the frequency of the *fast* allele in the rabbit population, all as a function of time. The bar chart records the three gene frequencies for the current population. This simple ecosystem behaves similarly to laboratory ones, in that the rabbit population oscillates with the fox population, though slightly out of phase, and then the animals go extinct after a few generations.[11] As expected, the genetic composition of the rabbits varies with the predation pressure from the foxes. When the fox population is large and evolutionary pressure from the foxes outweighs the pressure from the limited supply of grass, then rabbits are dominated by the fast morph (i.e., rabbits having the *fast* allele). The opposite occurs when the fox population has been at a low. We believe that this application can prove an excellent teaching tool and will provide the opportunity for studying the evolution of complex behaviors in ways that are not otherwise possible.

4. DISCUSSION

In the preceding section, we illustrated how RAM provides a natural means of expressing the rules thought to govern biological populations. As such, it provides an excellent vehicle for teaching and for research simulations. However, its most important use is likely to be in providing the vocabulary to be precise and clear about systems that are too complex to be easily described by traditional mathematical means. By framing them in terms of RAM or similar Artificial Life models, it may be possible to greatly enlarge the richness, precision, and testability of evolutionary theories. That, at least, is our goal.

The traditional tools evolutionists have used are mathematical, principally the concept of a vector field in n-dimensional space, which

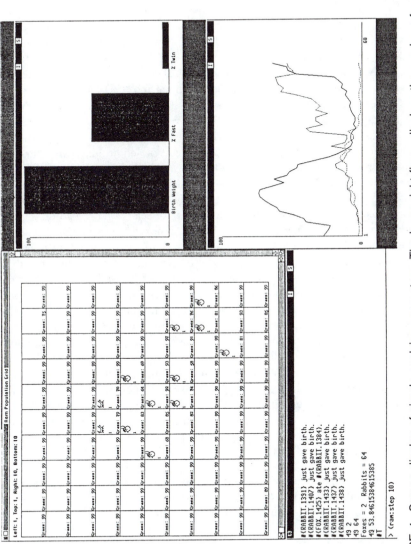
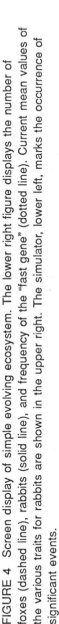

FIGURE 4 Screen display of simple evolving ecosystem. The lower right figure displays the number of foxes (dashed line), rabbits (solid line), and frequency of the "fast gene" (dotted line). Current mean values of the various traits for rabbits are shown in the upper right. The simulator, lower left, marks the occurrence of significant events.

"...comes to ecology and population genetics in large part from physics, by analogizing the changes through time for a population or community with the changes in the position of a particle in space, or, on a more generalized level, with the position of a population of particles in a phase space." [Lewontin,[13] p. 13]

There can be no doubt that this approach has served evolutionary theory very well in some regards. Successes include the demonstration that Darwinian evolution is compatible with Mendelian principles, accomplished in the 1920's and 1930's by Wright, Fisher, and Haldane (see, e.g., Mayr and Provine[19]). More recently, this approach has demonstrated that much, perhaps most, DNA change can be accounted for very effectively through non-selective, random genetic drift described by partial differential equations similar to those for Brownian motion.[12]

But, in other regards, evolution has been served less well by this type of theory, most notably when genotype-genotype and genotype-environment interactions are involved. For example, the theory of selection at one locus was completely described as early as the 1920's, but when there are even two loci, the classification of their possible equilibria has proven mathematically intractable. Other troublesome theoretical problems include cases where the "units of selection" are so difficult to identify and describe that they have been largely ignored,[7] and in the discussion of the inheritance of behavioral traits, where genotype-environment interactions contribute substantially to the expression of phenotypes.[14] Because of these difficulties some of the most fundamental issues in evolution, including cultural evolution, coadaptation, and speciation from genetic revolutions, have been treated only inadequately, or simply dismissed as "more rhetorical than scientific."[12] At this conference, we have begun to see the emergence of an approach that may correct some of these shortcomings to current evolutionary theory, based on the analogy between life processes and computer processes.

Evolutionists have long been aware of the difficulties encountered when formulating hypotheses in the traditional physics-inspired manner. Alternatives have been explored, for example, in Birch and Cobb,[3] Lewontin,[14,15] and Mayr.[18] Even twenty-five years ago in the early 1960's, Mayr in particular drew analogies to the unfolding of a computer process encoded by its program and the development of an organism. For example:

"The young in some species appear to be born with a genetic program containing an almost complete set of ready-made, predictable responses to the stimuli of the environment. We say of such an organism that its behavior *program* is *closed*. The other extreme is provided by organisms that have a great capacity to benefit from experience, to learn how to react to the environment, to continue adding "information" to their behavior program, which consequently is an *open program*." [Mayr,[17] p. 23].

"Every generation of organisms consists of innumerable slight variations in the DNA program of information that is characteristic for the particular species. Each individual phenotype is the product of one such program, and adaption of the phenotype to its surroundings is based on the information derived from this genetic program. The carriers of successful information will be the progenitors of the next generation. The individual is enabled to act purposefully because it is endowed with the proper information. Like an electronic computer, it is "programmed," permitting it to cope with the vicissitudes of development and of life." [Mayr,[16] p. 43]

In this paper, we have attempted to develop the life-as-process metaphor into a serious tool for studying population and evolutionary behavior. It appears to us that adaptation and evolution are treated most naturally by models that are constructed from interacting parts that exert their influence indirectly throughout the process of unfolding, not directly in the end structure. No notion of fitness has much value in this system, except after the fact or in abstract discussions of extreme cases.

The real value of RAM is not simply that it produces a better environment in which to conduct computer simulations, but that it, and the study of Artificial Life in general, provides a vocabulary for precise description of evolutionary processes that traditional mathematics cannot. What we have seen at this conference may be the start of a discipline that will lead to a more comprehensive and exciting theory of evolution.

ACKNOWLEDGMENTS

We are indebted to many people who have contributed to the ideas expressed here and to the evolution of RAM. In particular we wish to thank Brian Beckman, Charles Birch, Michael Dyer, Daniel Hillis, Richard Lewontin, and Ernst Mayr for their suggestions and encouragement. We also acknowledge the contribution of John Ampe, who wrote the predecessor version of RAM and John Fry, who developed the mosquito application. The W. M. Keck Foundation funded the UCLA Artificial Intelligence Laboratory, where much of the work was done, and the California University-wide Mosquito Research Program provided funding for some of the program development.

REFERENCES

1. Ampe, J. (1986), *Ram: A Simulation Tool for Biologists, Masters Thesis, Department of Computer Science, University of California, Los Angeles.*
2. Beehler, B. M., and M. S. Foster (1988), "Hotshots, Hotspots, and Female Preference in the Organization of Lek Mating Systems," *Am. Nat.*, in press.
3. Birch, L. C., and J. B. Cobb, Jr. (1981), *The Liberation of Life* (Cambridge: Cambridge University Press).
4. Bradbury, J. W., and R. M. Gibson (1983), "Leks and Mate Choice," *Mate Choice*, Ed. P. P. G. Bateson (Cambridge: Cambridge University Press), 109–138.
5. Bradbury, J., W. R. Gibson, and I. M. Tsai (1986), "Hotspots and the Dispersion of Leks," *Anim. Behav.* **34**, 1694–1704.
6. Caulson, R. N., L. J. Folse, and D. K. Loh (1987) "Artificial Intelligence and Natural Resource Management," *Science* **237**, 262–267.
7. Dawkins, R. (1982), *The Extended Phenotype: The Gene as the Unit of Selection* (Oxford: Oxford University Press).
8. Gibson, R. M., and J. W. Bradbury (1986), "Male and Female Mating Strategies on Sage Grouse Leks," *Ecological Aspects of Social Evolution*, Eds. D. I. Rubenstein and R. W. Wrangham (Princeton: Princeton University Press).
9. Greever, J., and G. P. Georghiou (1979), "Computer Simulation of Control Strategies for *Culex tarsalis* (Diptera: Culicidae)." *J. Med. Ent.* **16**, 180–188.
10. Haile, D. G. (1986) "A Discrete Life-Cycle Model for Computer Simulation of the Dynamics of a *falciparum* Malaria Transmission"; unpublished but presented at the 18th Annual Conference of the Society of Vector Ecologists, Riverside, California, November 19–21, 1986.
11. Hutchinson, G. E. (1978), *An Introduction to Population Ecology* (New Haven: Yale University Press).
12. Kimura, M. (1983), *The Neutral Theory of Evolution* (Cambridge: Cambridge University Press).
13. Lewontin, R. C. (1969), "The Meaning of Stability," *Diversity and Stability in Ecological Systems, Brookhaven Symposium in Biology* **22**, 13–24.
14. Lewontin, R. C. (1979), "Sociobiology as an Adaptationist Program," *Behavioral Science* **24**, 5–14.
15. Lewontin, R. C. (1983), "The Organism as the Subject and Object of Evolution, *Scientia* **118**, 63–82; reprinted in R. Levins and R. C. Lewontin (1985), *The Dialectical Biologist* (Cambridge: Harvard University Press).
16. Mayr, E. (1962), "Accident or Design: The Paradox of Evolution," *The Evolution of Living Organisms. Proceedings of the Darwin Centenary Symposium of the Royal Society of Victoria. Melbourne University Press. Melbourne*, 1–14; reprinted in E. Mayr (1976), *Evolution and the Diversity of Life: Selected Essays* (Cambridge: Harvard University Press).

17. Mayr, E. (1964) "The Evolution of Living Systems," *Proc. Nat. Acad. Sci. (USA)* **51**, 934–941; reprinted in E. Mayr (1976), *Evolution and the Diversity of Life: Selected Essays* (Cambridge: Harvard University Press).

18. Mayr, E. (1976), *Evolution and the Diversity of Life: Selected Essays* (Cambridge: Harvard University Press).

19. Mayr, E., and W. B. Provine (1980), *The Evolutionary Synthesis: Perspectives on the Unification of Biology* (Cambridge: Harvard University Press).

20. Taylor, C. T., and G. P. Georghiou (1982), "The Influence of Pesticide Persistence in the Evolution of Resistance," *Environm. Entomol.* **11**, 746–750.

21. Turner, S. R., and S. Goldman (1988), *Ram Design Notes, Technical Notes, UCLA Artificial Intelligence Laboratory, UCLA-AI-1988-2.*

P. Hogeweg
Bioinformatica, Padualaan 8, 3584 CH Utrecht, the Netherlands

MIRROR beyond MIRROR, Puddles of LIFE

PRELUDE

We call our approach to Artificial Life MIRROR modeling to indicate that Artificial Life should:

- Be "As large as life and twice as natural,"
- "Reflect our reflections" on life rather than life itself,
- Form almost independent, partly overlayed, sometimes interacting "puddles" which are represented at multiple levels of detail,
- Repeat simple structures,
- "Shape, unshape and reshape," and
- Be interesting to observe.

Artificial Life, SFI Studies in the Sciences of Complexity,
Ed. C. Langton, Addison-Wesley Publishing Company, 1988   **297**

1. BIOLOGICAL MODELING, BIOINFORMATICS AND ARTIFICIAL LIFE

Artificial Life could mean many different things. In this paper, we delineate a class of models, which in our opinion could profitably be termed "Artificial Life" models. "Artificial" because they are not designed to be in close "one-to-one" correspondence with previously observed life forms; "life" because they expand our observable universe with entities "which live their life" in which we can observe patterns normally preeminently associated with life. We sketch the position of such models among other biological and bioinformatic models.

1.1 OUTPUT-ORIENTED VS. STRUCTURE-ORIENTED MODELS

In conventional biological models, one usually takes as starting point some more or less well-described phenomenon, and tries to find a representation which reproduces this phenomenon. We call such an approach "output-oriented" modeling. An alternative approach, which we call "structure-oriented modeling," is to start with an *a priori*-defined model structure and study the types of behavior it generates. In the first approach, constraints may be (and usually are) added to the structure of the model; in the latter case, ideas usually exist about the type of behavior it at least should generate. Nevertheless, the two approaches are conceptually sufficiently different to warrant the distinction.

For the purpose of bioinformatic research (i.e., for studying informatic processes in biotic systems), we have emphasized the need for structure-oriented modeling[15,16] because:

1. Bioinformatic processes are inherently "local," and bioinformatic models should be in terms of local information processing.

2. Our insight in local/global transformation is entirely insufficient to "predict" the outcome of local rules; output-oriented modeling, therefore, is impractical, and if tried, will usually lead to much too complicated models.

3. Local micro-interactions can generate a set of qualitatively different macrophenomena. These phenomena would seem to be unrelated if studied at the macro level only. Thus, in an output-oriented approach, we would probably construct models for each of the phenomena separately, and fail to recognize their interrelationships.

4. It is not true that we "know" macro-level phenomena much better then micro interactions.

Thus, in bioinformatic modeling, we define the structure of a universe, requiring that the information processing of this universe is a "reasonable" representation of the informatic processes in the system under consideration, and "observe" the ensuing macro behavior of the system; this behavior may or may not coincide with behavior previously observed in biotic systems. Observation of the model system becomes the crucial, nontrivial part of modeling.[11,15]

"Artificial Life" models can belong to either class of models. In the output-oriented approach, one has to set up some (partial) definition of "life" and generate systems which do satisfy this definition of life, possibly in a way structurally, not only materially, quite different from the way biotic life does. In the structure-oriented approach, one can set up information processing universes and study the behavior. Both approaches were used in the pioneering work on cellular automata: Von Neumann's[28] self-reproducing cellular automaton is clearly an output-oriented model (and, consequently, turned out to be far more complex than necessary,[3,24] whereas Ulam's Cellular Auxology models[36] are clearly structure-oriented models (Ulam coined the term "imaginary physics"). Both approaches still flourish in cellular automaton and related contexts. Output-oriented models, for example, are cellular automaton models which are designed to "self-reproduce,"[24] network models designed for recognizing certain types of patterns (e.g., symmetry groups[32]) and L-systems designed to resemble certain plant species.[21,26] Structure-oriented models are the studies on the the patterns generated by classes of cellular automata,[6,25,30,37] networks[22,23] or L-systems.[7,9,33]

In this paper we will focus on structure-oriented Artificial Life models: the reasons mentioned above for bioinformatic models apply here as well. Moreover, when generating artificial life it seems more interesting to generate and observe new (lifelike but yet unknown) behavior patterns than to generate "more of the same," i.e., a priori definable behavior patterns from new ingredients.

1.2 MODELS, PARADIGM SYSTEMS AND ARTIFICIAL LIFE

The shift we are presently experiencing from analytically treated models toward models studied by simulation is much more profound than is usually acknowledged, and has profound methodological (and ontological, if we are to make Artificial Life) consequences. An important side effect of this shift from models allowing for analytical solutions to models which are studied by simulation is that instead of studying a class of model systems (i.e., the different parameter values and different initial conditions), one studies just one, fully specified system at a time. This is, of course, why analytical solutions are supposed to be more valuable than simulation results. However, there are a number of reasons why this latter conclusion is not necessarily correct:

1. The class of models studied in analytical solutions is shaped primarily by solvability, and not by problem-oriented considerations, and its relevance therefore can be limited.

2. Such a model selection is not at all an unbiased sample: this is forcefully illustrated by the late recognition of the chaotic behavior of very many systems which are simple generalizations of earlier studied models. Moreover, generalizations are demanded by the object studied. For example, before 1970 many

discretizations of the classical logistic-growth model were proposed for species with discrete generations, all of which avoided chaotic behavior. Nevertheless, such density-dependent growth models were criticized because of the absence of severe fluctuations which were observed in such populations.

3. Some behavior patterns only occur in models which do not allow for general "shortcuts;" their fate can only be resolved by letting a fully specified system "live its life" (e.g., behavior patterns associated with universal computation).

4. Parameter values are just as important in representing some other system as are structural properties and they are often better known.[2]

The usefulness of studying fully specified systems is evident when the model is supposed to represent one particular other system(e.g., a particular lake). Such "one-to-one" models are particularly important in semi-engineering context (ecosystem management, medicine, building airplanes), but are usually not the goal of scientific modeling. Fully specified model systems then function as "paradigm systems,"[11,15] i.e., they are specific examples of an almost always loosely circumscribed class of systems. Such paradigm systems should be located at "interesting" points within this class of systems. A paradigm system can derive its interest in several ways:

1. It represents closely some other ("real world") example, but is made up of different ingredients.

2. It represents some sort of "average" of the class of systems, i.e., is located close to the middle of the class of systems envisaged.

3. It is a very simple (the most simple known so far) example of the class, i.e., is located near the margin of the class of systems envisaged.

4. It lies clearly outside the class of systems but is related to it in an interesting manner; for example, it is very similar to a paradigm system which does lie within the class.

5. It exhibits unique features not known from any other representative of the class.

In fact, most scientific models also take the form of paradigm systems if parameter-independent solutions are sought: their structure and thereby the implicit parameters are paradigm-like. Simplicity (3 above) is very often the prime source of interest, for which close similarity is readily sacrificed.

Artificial Life models are clearly paradigm systems: nothing can "live" without being fully specified, and "life" is surely a loosely specified class of phenomena, if abstracted from its physicochemical basis. Artificial Life models can be "interesting" because of all the reasons mentioned above (maybe except 2, whose usefulness is doubtful in any case when none of the other points are satisfied), although points 4 and 5, (i.e., those exhibiting hithereto unknown behavior) are of special interest in this context, because these are often not recognized as interesting in "normal science." Nevertheless, studying examples outside the known class of living systems is very valuable for understanding life.

We conclude that "Artificial Life" models differ only gradually from other bioinformatic paradigm systems, and are indispensible in bioinformatic research. At the beginning of modern science Bruno, having heard Galileos theories, concluded, "Then there are many worlds like our own." Now at the beginning of flexible media to make paradigm systems, which once again revolutionizes science, we should consider that we may be able to make many (artificial) worlds, some quite unlike our own ("real") world which may yet support a form of "life," as worthy of observation as the biotic ("real") variety of life is.

2. MIRROR MODELING

2.1 BASIC STRUCTURAL CHARACTERISTICS OF MIRROR WORLDS

In our opinion there are, apart from the structural requirements set by any particular problem, a number of structural requirements which should be satisfied in any system in order to qualify as (structurally) lifelike (artificial or otherwise), although systems may be behaviorally lifelike without satisfying them. MIRROR modeling methodology is developed to make it easy to satisfy these requirements. They are discussed in this section.

2.1.1 LOCAL DESCRIPTION
The importance of local description is widely recognized, and it is the main motivation for cellular automata-like formalisms. Nevertheless, these formalisms do not realize a fully local description: implicit global structure may be responsible for the observed behavior. Such implicit global factors are, in particular, a global timing regime and a fixed network, which are discussed in turn.

1. Global Timing Regime (Synchronicity). A number of authors have shown that synchronicity generates global structure and that, therefore, cellular automata cannot directly model certain processes, e.g., morphological developments[9] or Ising systems,[37] because unrealistic long-range correlations are introduced by the synchronicity. Asynchronous versions may generate interesting macro structures, but for other local rules than their synchronous counterparts.[20,27,31] However, synchronicity changes what is "simple" and therefore profoundly influences our mapping of possible worlds. A local timing regime implies asynchronicity, but does not introduce this in a global way: timing depends on local circumstances, e.g., internal kinetics,[34,35] or triggering by events.[8-10] In MIRROR modeling, a local timing regime is enforced: there is no global monitor and entities either schedule their own next activity on a timing list (implementing internal kinetics), are explicitly activated by other entities, or are implicitly triggered by events through DEMONs, which are posted by themselves

or other entities. The self-structuring properties of MIRROR worlds make use of the local timing regime. In particular, when a (semi)invariant structure is recognized, which has a predictable behavior until something special happens, the initially explicitly happening but predictable events may be moved to the "internal dynamics" of a new entity. This new entity replaces (temporarily) the set of entities involved in the invariant structure. Henceforward, only the macroevents, i.e., the "something special" which changes invariant structures, take place explicitly and are timed on the basis of the experience gathered about the invariant structure when it was still explicitly simulated. We call these macro events "interesting events,"[8,10,14] interesting because they are not predictable.

2. Global Interaction Structure (Fixed Network). In many modeling formalisms which incorporate local dynamics, the interaction structure is either local but fixed and uniform (e.g., cellular automata), or global (every entity can interact with every other entity) and only quantitatively modifiable (e.g., (neural) network models). This restricts the choice of entities severely.[15] In the case of cellular automata, the fixed local structure enforces that the information processing units represent patches of space instead of biological information processing entities.[17] We think the latter should be the unit of description in bioinformatic models and also in (some) artificial life models; we call such models "individual oriented" (see section 2.1.2). In the case of network models, it implies the preexistence of all cells and interconnections, which is only partially true for neural systems and is not only unrealistic, but leads to begging the question in evolving systems. Thus, models of evolving systems should include the evolution of interaction structure. This is also true for models which simply involve a variable number of information processing units. We call such models "variable structure models" (compare references [4,5]; see section 2.1.3.).

2.1.2 INDIVIDUAL-ORIENTED MODELS Although it may be an ultimate goal to understand "life" as a property (state) of space/time (as is attempted in cellular automata models; see 2.1.1), it may be more directly interesting to understand how complex biological information-processing structures emerge in assemblies of simpler biological information processing units. Moreover, we think that the process of individualizing (and deindividualizing) is fundamental in lifelike processes, and therefore should be fundamental in (artificial) life models. Stated more simply, if "life" is generated as a property of space/time (in a manner described by Langton[25]), it still seems to need us to recognize it as alive; if complex individuals are generated from more simple ones, they themselves may recognize themselves as living individuals in space/time.

Thus, according to the first point, the basic entities in MIRROR models are "individuals," possessing an (extendable) behavioral repertoire. These individuals live in spaces. Their behavioral repertoire includes sensing their local environment, changing the environment locally, and changing their own position in the space (i.e., changing what constitutes their local environment). The space in which the individuals live can take the form of (non-sychronized) cellular automata. The state of the cells of the cellular automata can include, apart from regular state variables of the "patches" (cells), the individuals which inhabit it. Thus, cellular automata are incorporated as a MIRROR world without individuals (as yet).

MIRROR worlds can contain separate structure-recognizing entities (DWARFs). DWARFs recognize invariant structures, e.g., recognize that a certain set of individuals interact exclusively among each other. As a still primitive form of individualizing (2 and 3 above), DWARFs can replace such invariant structure with a new individual, whose behavioral repertoire corresponds to the behavior of the invariant structure it replaces (incorporates). However, certain events ("interesting events," see 1.1.1) can disrupt the invariance, and if such events occur, the new individual is again replaced by the set of interacting individuals whose behavior it incorporates, i.e., deindividualizing occurs.[10,16]

2.1.3 VARIABLE STRUCTURE MODELS

In order to be able to generate structure in a non-trivial way, (artificial) life models should be variable structured. Engaging in an interaction is part of the behavioral repertoire of an individual. However, in order to interact, it should "know" (sense, have a pointer to) the other individual. In MIRROR worlds, such knowledge is obtained by

1. Spatial Embedding. The individuals live in spaces. Through this spatial embedding they can select potential interaction partners (those that are "nearby"). Who is nearby depends, of course, on the behavior of many entities and on the topology of the space. Spatial embedding is the most important structure-generating device in MIRROR models. MIRROR worlds typically contain several spaces in which individuals live, and the above applies to each of them. Primarily there is "SPACE SPACE," i.e., the space in which the "organisms" (or cells or molecules) move about. Obviously nearness is very important for interaction among them. There are a variety of other represent national spaces, most importantly SKIN SPACES: with each individual a space may be associated which contains its representation of its world. It is inhabited by individuals which interact in ways similar to that of space space (see section 3.2). The topology of the latter spaces, of course, can be non-euclidean.

2. Acquaintance. The pointer to an individual (once obtained by spatial proximity) can be stored in its memory.

3. Ancestry Based. Such stored acquaintances may be passed on to the "offspring." Note that this is the way variable structures are realized in L-systems.

4. Pattern Based. Interactions are based on pattern similarity. This can be seen as a special case of spatial embedding, without movement through space. It is used by Farmer et al.[5] and in many message passing algorithms.

An important side effect of using variable structure models is that at all times the interaction structure can be minimized. Thus, relevant interactions can be more easily observed. Observation of the relevant interactions, both within the model and by us, is what life and modeling is about.

2.2 ENTITIES OF MIRROR WORLDS

MIRROR worlds are not only characterized by the types of interaction and control structure as discussed in the previous section, but also by the (proto)types of entities which form the world. These prototype entities are an important heuristic for shaping models.

MIRROR worlds consist minimally of a SPACE, subdivided in PATCH(es), in which DWELLERs live. DWELLERs posses a set of sensors with which they obtain information of their immediate surroundings, and can perform actions on the basis of this information. The sensors include those which react on signals created by the environment on an action of a DWELLER, e.g., an environment may generate "cracks" when a certain type of DWELLER moves, which warns other DWELLERs of its approach. These actions include changing position in space, changing local features of the space or of DWELLERS they meet, and creating new DWELLERs of the same type as itself (reproduction) or of other space inhabiting entities (like PATHs or ODOUR)s.

A MIRROR world often consists of several SPACES inhabited by DWELLERs. These SPACEs are largely independent, but are interrelated via DWELLERs and their behavior: with each DWELLER, a SPACE may be associated (its SKIN-SPACE) in which DWELLERs dwell. The behavior of the DWELLER can be dependent on the configuration of DWELLERs in its SKINSPACE (which it "observes") and thereby influences the configuration of DWELLERs in the SKINSPACE of other DWELLERs. An extremely simple example of the use of such multiple spaces is given in section 3 and Figure 1. We think, however, that more complex implementations of this structure may go a long way to creating the multiple, almost independent "puddles" which make biotic life fascinating.

Apart from this multiple level "real" world, MIRROR universes incorporate a "shadow world." This shadow world is also defined in terms of locally interacting individuals, although "local" may be defined in a way quite different from the "real" world. Entities of the shadow world include DEMONs, DWARFs, OBSERVERs and WIZARDs.[8,11,15,16] DEMONs activate entities on certain clues: they are extensively used by all types of entities of the MIRROR world (including the shadow world entities) to "notice" relevant events; also the above-mentioned "cracks" are generated by a specialized DEMON (CRACKER).

The other shadow world entities are more specialized: DWARFs detect invariant relations (see Hogeweg and Hesper[16,18] for an explanation of how they go about doing it). OBSERVERs find (e.g., by nonsupervised learning methods) interesting patterns in MIRROR worlds, and WIZARDs may change worlds, e.g., to maximize or minimize a certain type of invariant relation. Although presently DWARFs, OBSERVERs and WIZARDs are mainly used to generate multiple representations of the MIRROR world to the user, they are intended to be used by the entities of the "real" world (and the shadow world itself) for the same purpose.

2.3 NOTES ON THE IMPLEMENTATION OF MIRROR

The current implementation of our ideas on MIRROR modeling is called MIRSYS, and runs on the XEROX-1186 INTERLISP-D workstation.

2.3.1 INDIVIDUALS The individuals are (parallel) invocations of INTERLISP functions. Using the Spaghetti stack facilities (and downloading whenever possible), the entities exist simultaneously in their respective stack environments. An individual is represented in the system by an atom whose value is the stack pointer. Typically a DWELLER has access directly to its own stack environment and further up to the stack environment of the space in which it lives. Thus it inherits the space-defining properties which determine its sensing and acting on its environment (e.g., the same DWELLER can live in a two- or three-dimensional environment). Moreover, an entity can examine (and initiate functions in) the stack environment of its interaction partners.

2.3.2 ENTITY-DEFINING FUNCTIONS The function defining a type of entity can take any form: for example, it can be a regular function or it can be "rule based." It can be modified by adding rules, or by using the advice facilities of INTERLISP. The behavior of the individuals, of course, is modified all the time by the modification of the information used by the function. Moreover, individuals typically "extend" themselves by generating other entities, in particular DEMONs which are attached to certain variables or procedures of other individuals and "revive" the individual when this variable is accessed or changed or the procedure is called (by certain entities and/or with certain parameters).

2.3.3 REVIVAL OF INDIVIDUALS Control is passed from one individual to the next by explicit REVIVALS, DEMON-based REVIVALS, and time-based REVIVALS. Revivals pass control to the stack pointer representing the revived individual and flag the cause of revival. In explicit revivals, this flag refers to the individual doing the revival and possibly any "message" it cares to send, in DEMON-based revivals the event on which it was activated (DEMONs themselves are transparent for the other entities, although they can add "messsages" to further specify the event and can delay the activation following the event) and in time-based REVIVALs just time. Time-based revivals take place whenever nothing else is happening: the time then proceeds to the time at which some event is scheduled.

2.3.4 IMPLEMENTATION OF DEMONS The DEMONs are placed in property lists of individuals under the name of the variable or procedure which they "haunt." The basic MIRSYS procedures check the property lists of the individuals forming their current stack environment for property names corresponding to their parameters and revive the DEMONs stored there. For example, the MIRSYS function SETENV sets a variable in some explicitly referenced individual (stack environment) and, if the value of the variable is indeed changed, revives the DEMONS in the property <variable name> of the referenced individual as well as the individuals (e.g., its SPACE) in whose stack environment it occurs. Thus, DEMONs can haunt events at several levels of generality (e.g., eating of a specific individual, eating of any individual in a SPACE, etc.). Standard INTERLISP functions can also be "haunted" without being accessed via the MIRSYS local-function call functions: if so, these functions are automatically advised to check for DEMONs in

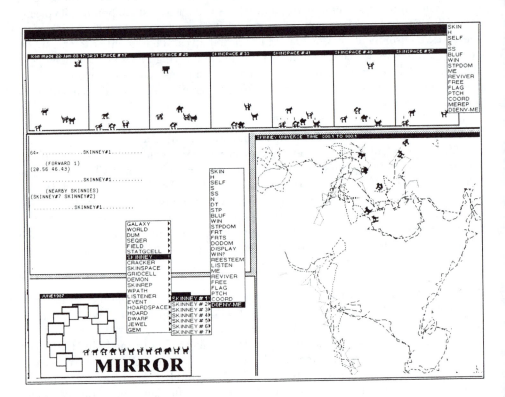

FIGURE 1 Interactive facilities of MIRSYS as functioning in the SKINNY world. MIRROR window (lower left) gives access to all entities of the MIRROR world through a hierarchical popup menu: all types of entities, all individuals of the clicked type, all variables of the individual clicked. Interactive window (middle left): By clicking DOENV-ME user control is passed to the stack environment of the clicked individual and can "do" things,e.g., move forward (returns the new coordinates and generates as side-effects the moving of the icon representing the individual, etc.) and inspect the world from the viewpoint of the clicked individual, e.g., inspect the world for nearby SKINNIES (returns the list). ICON popup menu (upper right): the same interactive facilities are available when clicking an icon representing an individual; here a SKINREP is clicked, i.e., an individual representing a SKINNY in mental space.

the property <function name> of the individuals in its own stack environment. DEMONs typically check for some conditions before passing on the revival to the individual tied to it (minimally it checks for its existence, if it does not exist anymore the DEMON kills itself).

2.3.5 GENERATION OF OUTPUT

All the generation of "output" is done by specialized entities (RECORDERS, REPORTERS and OBSERVERs), which are activated by the DEMON mechanism: in this way the model entities are not contaminated with output generation and a very flexible output-generating structure is established: any collection of information-gathering entities can be let loose in the world. RECORDERs are the simplest of these entities, they gather statistics on certain types of events and display them in real time, periodically or on user request. REPORTERS and OBSERVERs are progressively more versatile: REPORTERs report on properties of the system not explicitly represented in the model formulation but which can be gathered fairly easily, whereas OBSERVERs use their own "judgment" on interesting phenomena and on when (and how) to tell about them. The graphics representation of the various spaces (see Figures 1 and 2) is also done in this way: whenever an individual moves in a space, a DEMON activates a "DIS-PLAY PATH" entity which moves an icon representing the individual in the display space.

2.3.6 INTERACTIVE FACILITIES

MIRSYS makes full use of the INTERLISP-D graphics and interactive facilities (see Figures 1 and 2). At any time, a MIRROR universe can be interrupted and examined at any level of detail, can be changed (by changing the values of variables of individuals, by adding new individuals to the universe, or by changing the definition of (types of) entities) and subsequently can continue its operation. For example, a standard MIRROR window gives, via a hierarchical pop-up menu, access to all entities in the MIRROR world (Figure 1): all types of entities (i.e., entities being defined by a specific function), all individuals of that type, and all local variables of those individuals. The values of each of these can be displayed and changed by mouse clicking (and typing the new value). Moreover, by clicking the DOENV box, user control is passed to the stack environment of the specified individual; the user then can view the world from the viewpoint of the entity, e.g., by typing (NEARBY SKINNIES), a list of nearby SKINNIES being considered by the "possessed" individual is displayed. The user can also make the individual do things, e.g., by typing: (FORWARD 1), the possessed individual will move forward a unit length. This may cause several DEMONs to be activated; for example, a DEMON tied to the DISPLAY-PATH entity mentioned above, so that the icon representing the possessed individual will move, and the CRACKER which activates possibly other individuals who begin to fight with the individual possessed by the user (etc.). The same interactive facilities are available by directly clicking into an icon representing an entity in a display space.

3. FROM BIOINFORMATIC MODELS TO ARTIFICIAL LIFE, AN EXAMPLE; SPATIAL AND SOCIAL STRUCTURE FROM LOCAL INTERACTIONS

Our bioinformatic research on the emergence of social structures due to pairwise interactions between initially identical individuals[12,13,15] illustrates the use of several paradigm systems to map a set of behavior patterns found in a large variety of biotic systems. Some of these paradigm systems proved to represent the observed behavior of certain animals closely whereas others were never intended to do so, but were chosen so as to be representative for certain basic socioinformatic processes. Such "artificial life" models, apart from being interesting in themselves, generated the knowledge needed to attempt to construct paradigm systems for creating specific behavior patterns. In other words, this research illustrates the concerted use of models and artificial life models in bioinformatic research.

3.1 A MODEL OF BUMBLEBEES: A SOCIALLY REGULATED "CLOCK"

A pattern analysis study on the interactions of live bumblebees has shown that the workers in bumblebee colonies can be subdivided into two groups: "common workers" and "elite workers."[19] Once a worker has entered the elite, she remains in it until the end of the season when the queen is killed or kicked off the nest. After that, she will lay unfertilized (drone) eggs. We set up a paradigm system to find the requirements for the formation of the two types of workers under the assumption that all workers are identical when hatching.[12,13] To this end, we set up a MIRROR world consisting of a nest space in which BUMBLEs dwell. The behavior of the BUMBLEs was derived from:

1. The known population dynamic properties of bumblebees (i.e., development time of eggs, larvae, pupae)

2. The TODO principle, i.e., the BUMBLE's do what there is to do, not what they "intend" to do. Thus, if an adult BUMBLE meets a larval BUMBLE, it feeds it; if it meets a pupae of the right age, it starts building an eggcell; etc.

3. All social (i.e., non-maintenance) interactions are of the DODOM type. DODOM interactions involve three stages: (1) displaying/observing mutual dominance; (2) win/lose, determined on the basis of the mutual dominance, local factors and chance; (3) updating of the relative dominance based on 1 and 2 in such a way that expected outcomes reinforce the relative dominance only slightly, whereas unexpected outcomes give rise to a relatively large change in the dominance. Thus a damped positive feedback ensues.

4. A criterion of viability: is there enough food made available. This criterion is used to adjust unknown parameters.

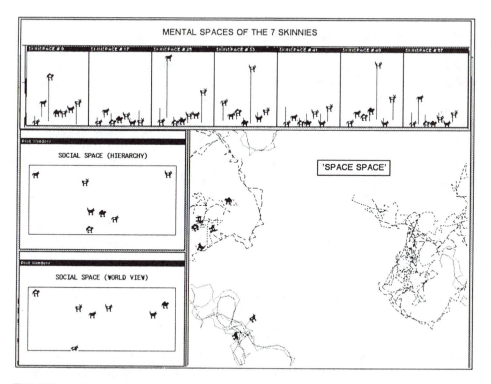

FIGURE 2 Display of spatial and social structure in a MIRROR world inhabited by SKINNIES. SPACE-SPACE(lower right): space in which the SKINNIES move about and interact. MENTAL SPACES (top row): relative dominances of the SKINNIES as perceived by each of them. SOCIAL SPACE (hierarchy) (middle left): Plot of the largest two principal components of the similarities of SKINNIES as measured by the estimate of the others. Corresponds to the hierarchy. SOCIAL SPACE (world view) (lower left): like the hierarchy, but based on their own estimate of the others: represents average spatial structure. The same icon is used in all spaces to represent the same individual.

It turns out that the resulting (simple) structure is sufficient to generate the stable class structure, provided that the nest space is subdivided into a CENTER (where the brood is and all interactions take place) and a PERIPHERY where inactive (common) workers dose part of the time. This, indeed, seems to be the case in live nests. Moreover, it turns out that the model generates a number of other phenomena observed in live bumblebee nests, most notably: (1) the queen was killed at the end of the season; (2) her departure resulted in chaos in the nest; and (3) the time at which the queen was killed was a stable, self-regulated property of the social structure which remained the same for a wide range of growth rates of the nest (a variable parameter in live nests). This time of departure of the queen is crucial for the "fitness" of the colony, as only afterwards generative offspring are reared. Interestingly, and counterintuitively, when this socially regulated clock breaks down because of stagnant growth of the nest, the small number of workers do kill the queen too early—not too late as a simple "force of numbers" argument would suggest. This is observed in live nests as well, and was used as an argument against a social influence of the timing of the "worker rebellion."[1]

Thus, the MIRROR world provided a fairly complete integrated view on the social structure of bumblebee colonies, although its initial goal had been much more limited. Moreover, it leads us to the hypothesis that a variety of self-regulating social structures which seem "adaptive" can be generated by the combination of TODO and DODOM. This implies that a change in the TODO (maintenance requirements, overpopulation, etc.) results automatically in a change in social structure. Thus, social structure is not an independent parameter which is optimized to fit certain circumstances (compare Oster and Wilson[29]), but an (emergent) property of its circumstances. Also intriguing was the important influence of spatial structure on social structure and vice versa.

3.2 AN ARTIFICIAL LIFE MODEL: THE CONCORDANCY BETWEEN SPATIAL AND SOCIAL STRUCTURE

The latter conclusions were further investigated in a MIRROR world which was not designed to represent any particular species but did incorporate the same type of interactions in a continuous spatial environment. The DWELLERs of this MIRROR world we called SKINNIES, to indicate that (1) they are the most meagre implementation of such a spatial TODO/DODOM structure and (2) the individuals know each other personally, i.e., have a representation of each other in their SKINSPACE (mental space).

Figure 2 summarizes the model as well as some model results. SKINNIES live in small groups in SPACE-SPACE (lower right window in Figure 2). If they meet an other SKINNY(or are disturbed by one in their neighborhood), they either initiate an overt DODOM interaction or fly. The choice depends on a DODOM interaction in their own SKINSPACE (which is represented in the upper left of Figure 2) in which their representation of self and interaction partner interact. When such an interaction is lost, they fly in SPACE-SPACE; otherwise, they DODOM in SPACE-SPACE, displaying their dominance estimate according to their SKINSPACE. Notwithstanding these interactions, they are inherently "social," i.e., when no other SKINNY is close enough, they will move towards the nearest (group of) SKINNIES.

This straightforward, rather simple-minded, interaction structure generates an interesting social structure: SKINNIES tend to form faithful pairs or small groups, which meet once in a while. On meeting a struggle ensues, after which it is most likely that the original pairings re-emerge intact. Moreover, it appears that for groups larger than two, the most dominant SKINNY tends to be in the middle and to interact with several submissive SKINNIES, whereas the latter ones only interact with the "boss." This interaction structure was not at all expected or even preconceived, and was only observed with the aid of DWARFs.

The lower left corner of Figure 2 gives two "social spaces" in which the group of SKINNIES live. The upper one is a representation of the hierarchy, the most dominant SKINNIES to the right. There is no absolute measure of hierarchy, but the ordering agrees well with all pairwise dominance relationships. The space is constructed from the estimate that the SKINNIES have of each other. More specifically, it optimally represents the similarity of the SKINNIES as viewed by other SKINNIES. Hierarchies are very common in groups of animals. The lower social space compares the "world view" of the SKINNIES, i.e., it optimally represents the similarity of the SKINNIES with respect to their estimate of the other SKINNIES. Interestingly, this does not correspond to the inverse analysis which represents the well-known social-concepts hierarchy; instead it gives the dominant SKINNIES a central position. In fact, it represents the average spatial structure of the group very well: the more dominant SKINNIES are located in SPACE-SPACE in the middle of larger groups and are together. Although this is not so easily detectable, it appears that such a spatial/social structure exists in many groups: the present research shows that this relationship between social and spatial structure is an intimate one (both are caused by the other and generated by the other). This is indeed reflected in our language: we are "close" to each other or "remote."

We conclude that, by studying an artificial world of SKINNIES, we can dig out relationships, which otherwise are obscured by all other things that are going on, but which nevertheless are an important force in shaping many worlds. We also conclude that one basic interaction structure, like the DODOM mechanism, can lead to a great variety of self-regulating macro structures depending on the "environment" in which it occurs. In particular, we conjecture that the DODOM/TODO mechanisms are (co)responsible for the variety of social structures observed in groups of animals. The more general implications of the occurrence of versatile, regulating, emerging structures in biotic systems are discussed in the following section.

4. SELF-STRUCTURING, ADAPTATION AND EVOLUTION

The above-described MIRROR worlds have non-trivial self-structuring properties: new, unexpected macrostructures emerge. These macrostructures are not only recognized by us, but are also recognized by the model. Recognition by the model can be explicit—OBSERVERs and DWARFs generate a representation of the macrostructure—or implicit—the emergence of the macrostructures is crucial for other properties of the model. For example, the elite group and the common worker group in the bumblebee world are recognized as such by the OBSERVERs, and their formation brings about the switch from worker offspring to generative offspring. Likewise the "friendship pairs" in the SKINNY world are recognized by DWARFs and determine the outcome of interactions and the spatial structure if pairs meet. The emerging structures represent temporary (semi)invariant relations: they remain embedded in the variable structure world and can (and do) evolve and dissolve. It is this dynamic aspect of the emerging structures which leads to "adaptive" behavior in the MIRROR worlds discussed. The behavior is adaptive in the sense that under a change of circumstances, the "shape" of the emerging structure changes so as to be "optimal" or at least "sufficient" in these circumstances to bring about some other property (or emerging structure) in the world.

This is clearly demonstrated in the bumblebee world: when a change of circumstances leads to a higher growth rate of the nest (in this case, the change is an external parameter, but a change of circumstances, of course, can be due also to a change in other emerging structures), the difference between elite workers and common workers becomes less pronounced with respect to the behavior of the individuals and with respect to membership. Conversely when the growth rate of the nest stagnates, a very pronounced small elite group is formed. By this mechanism, the switch between worker offspring and generative offspring are "optimally" timed. For a large range of "normal" growth rates, the timing is set to the end of the season: under the assumption of ergonomic constraints, in this way a maximum number of generative offspring is reared because the exponential growth phase is maintained as long as worker offspring are produced.[29] The earlier switch in the case of stagnant nest growth seems "optimal" as well: if there is no growth of worker force anymore, why wait for the end of the season?

Clearly, the fitness of a bumblebee colony depends crucially on the emerging structure. There is no concept in the model formulation akin to a switch from worker to generative offspring, and even less to a timing of such a switch. In other words, there is nothing like a gene representing this switch (or its timing) which by random mutation and selection is set to a sufficiently correct (or optimal) value. Instead, the combination of the DODOM interaction structure and the maintenance characteristics of a bumblebee nest creates the elite structure which, in combination with some other behavioral parameters (e.g., the fact that the queen tries to prevent workers from rearing new queens and laying (drone) eggs of their own), generates an adequate regulation of the generation of generative offspring. The point is not only that the switch is regulated by multiple "genes" and that each of these "genes" is involved in other (crucial) processes, but also that:

1. It is the regulation of the timing rather than the timing itself which is the interesting feature.

2. An emerging structure (which is likely to be self-regulating) can be easily exploited for a variety of "purposes" (fitness criteria) simultaneously.

3. Emergent structures determine what can be used as crucial fitness criteria (i.e., if no elite structure should arise from DODOM and maintenance, the ergonomic optimality would not be reached by a switch but along a quite different route, e.g., via physiological adaptation to, say, a multiannual colony structure).

In feedback mechanisms, cause and effect, of course, are distinguished with respect to time scale only (if at all). Nevertheless, the DODOM mechanism reverses what we would normally see as cause and effect. An individual does not primarily win because it is dominant but, rather, because it happens to win it becomes dominant (and therefore wins, etc.). Thus, a bumblebee becomes a member of the elite because she happens to interact with elite workers, therefore, happens to win once in a while from elite workers and, therefore, becomes elite and interacts with elite workers. Likewise SKINNIES happen to be near each other in space, therefore come to "know" each other well, and therefore will remain together in space.

A similar reversal is suggested by the above observations for evolutionary processes. A property is not selected because it has a high fitness value, but, because of the set of available properties, a semi-invariant structure emerges and these semi-invariant structures create new fitness dimensions, which can be optimized by adjustment of available properties. This slight change of viewpoint has a number of "nice" side effects; for example:

1. No *generatio spontanae*. Conventional evolutionary theory shuffles the interesting processes behind the curtain of unanalyzed (random) events: properties are created by mutation and only their selection is studied. The shift in viewpoint can open this curtain a little bit through studying conditions for the emergence of (semi)invariant relations and their exploitation.

2. "Arrow of complexity." Such a process may create an "arrow of complexity," instead of the "arrow of efficiency" created by conventional evolutionary concepts. This is because "self-sufficient" properties cannot keep up their self-sufficiency when (part of) their function is fulfilled by an emerging structure which is self-regulating and/or under the protection of some additional selective constraint. Indeed, the molecular record of evolution shows that conserved sites often have at least a dual function (e.g., the conserved sites in tRNA function in transcription as well as in translation).[18]

The great challenge for Artificial Life models is to obtain insight in the formation of self-regulating emergent structures and the landscape they create, rather than to try to create entities which evolve in externally supplied landscapes. If we will meet this challenge, our models may be caught in the arrow of complexity; if so, maybe the time will come when these models become more interesting to observe than life itself, and they might be observing us. If so, we will truly have: MIRROR beyond MIRROR, Puddles of LIFE.

ACKNOWLEDGMENTS

Mirror modeling was and is developed in close collaboration with Dr. B. Hesper.

REFERENCES

1. Blom, J. van der (1986), "Reproductive Dominance within Colonies of Bombus terrestris (L.)," *Behaviour* **97**, 37–49.
2. de Boer, R. J. and P. Hogeweg (1985), "Tumor Escape from Immune Elimination: Simplified Precursor Bound Cytotoxicity Models," *J. Theor Biol* **113**, 719–736.
3. Codd, E. F. (1968), *Cellular Automata* (New York: Academic Press).
4. Farmer, J.D., S. A. Kauffman, and N. Packard (1986), "Autocatalytic Replication of Polymers," *Physica* **22D**, 50–67.
5. Farmer, J. D., N. H. Packard, and A. S. Perelson (1986), "The Immune System, Adaptation & Learning," *Physica* **22D**, 187–204.
6. Hartman, H., and G. Y. Vichniac (1986), *Inhomogeneous Cellular Automata (INCA) in Disordered Systems and Biological Organisation*, Eds. Bienenstock et al. (Berlin: Springer Verlag), 53–57.
7. Hogeweg, P., and B. Hesper (1974), "A Model Study in Biomorphological Description," *Pattern Recognition* **6**, 165–179.
8. Hogeweg, P., and B. Hesper (1979), "Heterarchical Self-Structuring Simulation Systems: Concepts and Application in Biology," *Methodology in Systems Modelling and Simulation*, Eds. B. P. Zeigler et al. (Amsterdam: North Holland), 221–232.
9. Hogeweg, P. (1980), "Locally Synchronised Developmental Systems, Conceptual Advantages of the Discrete Event Formalism," *Int. J. General Systems* **6**, 57–73.
10. Hogeweg, P., and B. Hesper (1981a), "Two Predators and a Prey in a Patchy Environment: An Application of MICMAC Modeling," *J Theor Biol.* **93**, 411–432
11. Hogeweg, P., and B. Hesper (1981b), "On the Role of OBSERVERs in Large-Scale Systems," *UKSC Conference on Computer Simulation* (Harrogate: Westbury House), 420–425.
12. Hogeweg, P., and B. Hesper (1983), "The Ontogeny of the Interaction Structure in Bumble Bee Colonies: A MIRROR Model," *Behav. Ecol Sociobiol.* **12**, 271–283.
13. Hogeweg, P., and B. Hesper (1985a), "Socioinformatic Processes, a MIRROR Modelling Methodology," *J Theor Biol.* **113**, 311–330.
14. Hogeweg, P., and B. Hesper (1985b), "Interesting Events and Distributed Systems," *SCS Multiconference 1985* (La Jolla: Simulation Councils, Inc.), 81–87.
15. Hogeweg, P., and B. Hesper (1986), "Knowledge Seeking in Variable Structure Models," *Simulation in the Artificial Intelligence Era*, Eds. M. S. Elzas, T. I. Oren & B. P. Zeigler (Amsterdam: North Holland), 227–243

16. Hogeweg, P., and B. Hesper (1988a), An Adaptive, Self-Structuring Non-Goal-Oriented Modelling Methodology," *Modelling and Simulation Methodology: Knowledge System Paradigms*, Eds. M. S. Elzas, T. I. Oren and B. P. Zeigler (Amsterdam: North Holland), in press.

17. Hogeweg, P. (1988b), "Cellular Automata as a Paradigm for Ecological Modelling," *Applied Math. & Computation*, in press.

18. Hogeweg, P., and D. A. M. Konings (1985), "U1 snRNA: The Evolution of its Primary and Secondary Structure," *J. Mol. Evol.* **21**, 323–333.

19. van Honk, C., and P. Hogeweg. (1981), "The Ontogeny of the Social Structure in an Captive Bombus Terrestris Colony," *Behav Ecol & Sociobiol.* **9**, 111–119.

20. Ingerson, T. E., and R. L. Buvel (1984), Structure in Asynchronous Cellular Automata," *Physica* **10D**, 59–68.

21. Janssen, J. M., and A. Lindenmayer (1987), "Models for the Control of Branch Positions and Flowering Sequences of Capitulain Mycelismuralis (L.) Dumont (Compositae)," *New Phytol.* **105**, 191–220.

22. Kauffman, S. A. (1969), "Metabolic Stability and Epigenesis in Randomly Constructed Genetic Nets," *J. Theor Biol.* **22**, 437–467.

23. Kauffman, S. A. (1984), "Emergent Properties in Random Complex Automata," *Physica* **10D**, 145–156.

24. Langton, C. G. (1984), "Self-Reproduction in Cellular Automata," *Physica* **10D**, 135–144.

25. Langton, C. G. (1986), "Studying Artificial Life with Cellular Automata," *Physica* **22D**, 120–149.

26. Lindenmayer, A. (1968), "Mathematical Models for Cellular Interactions in Development I and II," *J Theor Biol.* **18**, 280–299 and 300–312.

27. Natamura, K. (1981), "Synchronous to Asynchronous Transformation of Polyautomata," *J. Comput System Sci.* **23**, 22–37

28. Von Neumann, J. (1966), *Theory of Self-Reproducing Automata*, Ed. A.W. Burks (Urbana: University of Illinois Press).

29. Oster, G. F., and E. O. Wilson (1978), *Caste & Ecology in the Social Insects* (Princeton: Princeton Univ Press).

30. Packard, N. H., and S. Wolfram (1985), "Two-Dimensional Cellular Automata," *J. Statistical Physics* **38**, 901–946.

31. Park, J. K., K. Steiglitz, and W. P. Thurnston (1986), "Soliton-Like Behaviour in Automata," *Physica-D* **19D**, 423–432.

32. Sejnowski, T. J., P. K. Kienker, and G. E. Hinton (1986), "Learning Symmetry Groups with Hidden Units: Beyond the Perceptron," *Physica* **22D**, 260–275.

33. Smith III, A. R. (1984), "Plants, Fractals and Formal Languages," *Computer Graphics* **18(3)**, 1–10.

34. Thomas, R. (1973), "Boolean Formalisation of Genetic Control Circuits," *J. Theor Biol.* **42**, 563–585.

35. Thomas, R. (1985), "Kinetic Logic as a Formal Description of Asynchronous Automata and Biological Models," *Dynamical Systems and Cellular Automata*, Eds. J. Demongeot, J. Cole, and M. Tchuente (New York: Aacademic Press), 269–282.

36. Ulam, S. (1970), *Essays on Cellular Automata*, Ed. A. W. Burks (Urbana: University of Illinois Press).

37. Vichniac, G. Y. (1984), "Simulating Physics with Cellular Automata," *Physica* **10D**, 96–116.

38. Wolfram, S. (1984a), "Cellular Automatas as Models for Complexity," *Nature* **311(4)**, 419–424.

Narendra S. Goel & Richard L. Thompson
Department of Systems Science, State University of New York, Binghamton, NY 13901

Movable Finite Automata (MFA): A New Tool for Computer Modeling of Living Systems

1. INTRODUCTION

Living systems are quite complex: they have many attributes with complex interdependencies and rules of interaction, which are not always obvious; they are usually nonlinear and have non-deterministic dynamics; and they have a strong behavioral component.

Understanding these systems requires experimental investigations which yield concrete results about specific cases. But these results alone are like pieces of an unassembled puzzle, and one usually can not generalize them. As soon as one wishes to complete the puzzle or generalize the results, one is engaged in model building. Models can help understand and better explain living systems and the phenomena associated with them. A complete model can also be used to control and predict the behavior of the living systems. (The detailed work on the functioning of biological clocks has suggested desirable pre-flight dietary and sleeping pattern that minimizes the effects of jet lag.) Modeling can also lead to the introduction of new concepts and new techniques.

Artificial Life, SFI Studies in the Sciences of Complexity,
Ed. C. Langton, Addison-Wesley Publishing Company, 1988

Computer simulation models involve the creation of an artificial system on a computer which displays a behavior close to that of the real system. They are simply a convenient way of expressing general principles in a fashion which is (1) precise, since the symbols in the computer language have precise defined meanings, and (2) complete, since, if essential steps or features are omitted from a computer model, it won't work. They require a scientist to think and communicate as clearly as possible. Computer models allow easy, rapid, inexpensive, and controlled experimentation in order to infer properties of the real system. They enable one to design alternate systems capable of performing better or different functions. For example, a model for the simulation of protein folding, were it available, could dramatically reduce the cost and time involved in experimentally determining the three-dimensional structure of a protein from its sequence of amino acids. It could be used to study the functions of mutations, drugs and inhibitors on protein function, to design a protein sequence with a particular shape (and, hence, presumably function), and to study the evolutions of proteins and life. In some cases, computer models also serve as theories in biology and in others they are the only means of thoroughly testing and examining a large and intricate theory.

Biological systems have a great richness of behavior, which needs to be modeled and understood. However, there is one aspect which is more amenable than others to successful computer modeling. This is the phenomenon of self-organization, or the emergence of behavior/structure out of the interacting components of the living system. This paper is devoted to computer models of self-organization.

The existing computer models can be divided into four categories as described in Section 2. They include: analog models, physically realistic models, reaction-diffusion-equations-based models, and cellular automaton models. These models have played an important role in understanding living systems but they have some limitations. In Section 2, we point out these limitations. We then describe a new class of models, *Movable Finite Automata (MFA)* models, which overcome some of these limitations.

In Section 3, we describe the use of MFA models for simulating the interaction of protein molecules in the self-assembly and operation of the T4 bacteriophage which infects *E. coli*. This description is brief and adapted from the detailed ones given elsewhere.[8,9,18,19] In this initial study we have made some simplifications in the architecture of the phage, namely, our model phage does not possess a head capsule and the neck which attaches the head to the tail. We describe the basic model, which includes a description of cell wall and phage molecules (subunits), their mutual interactions, and their rules of movement (simulation dynamics). These molecular subunits are the finite automata of our model. We describe a specific design for a phage, by which we mean a list of parameter values specifying the properties (dimensions, bond site locations, and interactive properties of these sites) of particular components making up the phage. We show that for this, molecules will spontaneously self-assemble to form a complete structure that is like a phage contractile tail which incorporates a test molecule (representing the phage DNA), attaches itself to the cell wall, penetrates the wall, and places the test molecule on the other side of this wall.

In Section 4, we review another use of the MFA model, namely, the simulation of the elongation cycle in the protein biosynthesis, in which successive amino acids are added to a growing polypeptide chain, in the order determined by the genetic information encoded in a messenger RNA molecule. The details of this application are given elsewhere.[9,10]

In Section 5, we make a few concluding remarks.

2. MODELS FOR SELF-ORGANIZATION IN LIVING SYSTEMS

Various existing approaches to modeling of natural living systems can be divided into the following four categories.

1. ANALOG MODELS. Here instead of modeling the specific living system, one models an "analog system," without necessarily incorporating into the model working mechanisms of the living system. Some recent examples of such models are the use of fractals, L-systems, and other computer graphics techniques to simulate plant growth,[12] the movements of birds in a flock,[15] and of fishes in a water tank.[2] As an example, the simulation of growth of plants using L-systems does not involve either the plant physiology or the morphogenetic phenomena, but the simulated plant could be quite realistic. In its present form, the computer simulation could only be classified as fancy and creative computer graphics. Although this type of model does not realistically use the biology of plant growth, they are still useful. For example, when one makes small changes (at the local level) in the instruction set for generating the plant, one could obtain a different morphology of the plant. This cause-effect study could be used to understand the genotype-phenotype relationship. One hopes that in future one would be able to obtain a one-to-one correspondence between the computer instruction set and the plant growth instruction set (the genetic map), making this type of modeling quite fruitful.

2. PHYSICALLY REALISTIC MODELS. If one is to make a reasonable assumption that living systems obey the laws of physics, a desirable model of a living system will be the one based on the analytical or numerical solution of the Schrödinger equation. Unfortunately, inherent mathematical difficulties make this approach impractical for the study of living systems at all but the simplest molecular level.

    A somewhat more tractable type of model has been constructed, for example, to study the folding of globular proteins,[11] by making semiclassical approximations to the potential function of a molecular system, and then using various numerical techniques to study the behavior of the system under the influence of this potential. However, this approach has not been very successful because of the difficulties in specifying the potential function and the enormous amount of computation required for the simulations.

3. REACTION-DIFFUSION-EQUATION-BASED MODELS. In these models the concentrations of various chemicals are represented by continuous variables which in general are functions of position in three-dimensional space. The changes in the concentrations as a function of time are modeled by first-order differential equations containing reaction terms specifying the rates of various chemical reactions, and diffusion terms representing the migration of molecules through random thermal motion. This type of model has been used to explain cellular rearrangement,[13] assuming that such rearrangement is determined by the spatial and temporal concentration of a certain chemical which is continuously produced and/or destroyed and also diffuses in space.

Although such models are often reasonably amenable to mathematical analysis and computer simulation, they cannot directly represent the three-dimensional structures generated by molecules. Since molecular structures are represented in such models simply by real (or integer) variables, each molecular structure which is formed must be explicitly represented by a variable. This makes such models difficult to set up in cases where very large numbers of intermediate structures are possible (as is true in the protein folding problem, for example). In such cases a model representing structural relationships between intermediates may be needed to generate the reaction rate coefficients for the approach. Thus, models based on reaction-diffusion equations are incomplete in general and may need to be supplemented by models that can explicitly deal with structure.

4. CELLULAR AUTOMATON MODELS. These models tend to greatly lack direct physical realism. In general, such a model consists of an array of simple automata situated at the sites in a one-, two-, or three-dimensional lattice of integers. The automata all obey the same rules, each one can be in one of a finite number of different states, and each one changes its state in accordance with information obtained from the automata which are its immediate neighbors in the lattice.

These models have the advantage that they can be readily analyzed and simulated on microcomputers. As a result, they are quite popular[1,20] and have been used in simulating many types of living (and non-living) systems. They also make it possible to rigorously raise and answer many questions that can at best be considered only vaguely in the absence of definite models. For example, using such models, von Neumann was able to address the question of whether or not self-reproducing machines capable of arbitrarily complex functions can be constructed. Models of this type also make it possible to discover many unexpected phenomena through numerical experiments, as has been seen with Conway's game of life.[1]

Realizing the advantages and limitations of types of models discussed above, we[8,9] have recently introduced a new class of models, the *Movable Finite Automata*

(MFA) models which are similar to cellular automata models but are endowed with rules of operation that mimic as closely as possible some of the key biophysical principles governing the interaction of biological macro-molecules, cells, and other natural subunits. These models are based on finite state automata that undergo discrete changes in a step-by-step fashion with the passage of time, and thus they share some of the features of cellular automaton models that make the latter easy to handle mathematically and computationally. The key feature allowing for greater biophysical realism in MFA models is that the automata are allowed to move about and interact with one another.

The nature of these models will become clearer in the next section where we describe their use for simulating phage assembly and operation.

3. T4 BACTERIOPHAGE ASSEMBLY AND OPERATION

Bacteriophages (or simply phages) are viruses which infect bacterial cells, take over their replicating mechanisms, and use them to generate more viruses.[3,14,21] T4 bacteriophage infects the bacterium *E-coli* strain B. It has played a major role in the progress of molecular biology over the past 50 or so years; it has become an impressive paradigm for how a combination of genetic, biochemical, and ultrastructural

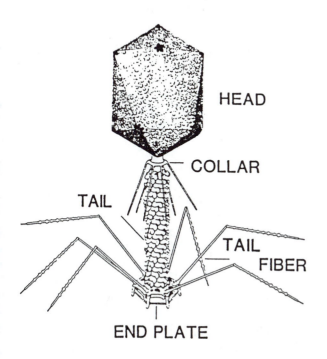

HEAD

COLLAR

TAIL

TAIL
FIBER

END PLATE

FIGURE 1   A diagram of the structure of a T4 bacteriophage.[6]

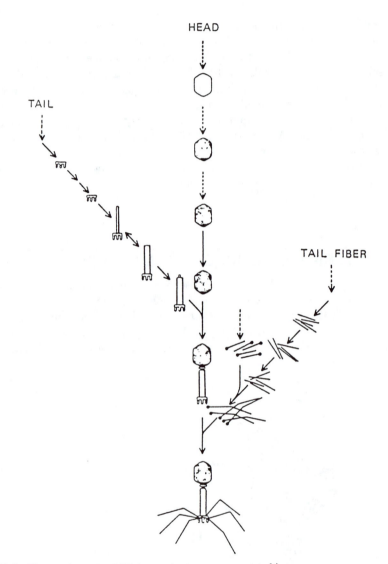

FIGURE 2   The pathways of T4 bacteriophage assembly.[14]

approaches can be employed to dissect and then reconstruct a complex biological system.[3,4,21] One of the fruits of studies on this phage has been an insight into the general problem of how linearly arranged information in genes controls the formation of three-dimensional biological ultrastructures and, ultimately, morphology on a larger scale.

The T4 phage is a complex assembly of protein components (see Figure 1). It consists of an elongated icosahedral head, formed out of protein, and filled with

DNA. It is attached by a neck to a tail consisting of a hollow core surrounded by a contractile sheath and based on a spiked end plate to which six fibers are attached. The spikes and fibers affix the virus to a bacterial cell wall. The sheath contracts, driving the core through the wall, and viral DNA enters the cell.

Two distinct approaches have been used to get insight into the T4 phage assembly process. In one approach, the assembly process was studied in the absence of one or more of the subunits. In the other approach, genes were mutated and the aberrant assemblies (e.g., "monster phages" with one head and two tails) were used to identify the structure defining genes. These approaches led to the pathway for T4 phage assembly as shown in Figure 2.

BASIC PRINCIPLES OF SELF-ASSEMBLY

The detailed experimentation with phage assembly suggests the following principles[7,9] involved in its assembly.

1. SUBASSEMBLY. As the name implies, this principle requires that the assembly of subunits into biologically functional units be done in stages. For example, if the unit consists of 1000 subunits, then instead of putting all 1000 subunits together at once, it is more desirable to first assemble, say, ten subunits into one big unit, ten of these big units into a bigger one, and ten of these bigger units into the final one. Figure 2 clearly shows subassembly at work for a T4 phage.

    It is fair to assume that this principle is used in phage assembly for the same reasons that it is used in the industrial world, namely, efficiency (subassembling can be carried out simultaneously), reliability (bad intermediate units can be discarded), and variety (one kind of battery in different models of cars).

2. MINIMIZATION OF FREE ENERGY. The driving force for organization may vary with each specific biological system, but most biological organizations involve optimization (minimization or maximization) of a certain basic property of the system. Thermodynamic free energy seems to be minimized in the phage self-assembly.

3. SEQUENTIAL ASSEMBLY AND CONFORMATIONAL SWITCHING. The minimization of free energy drives the process of phage selfassembly, but it does not determine the specific pathway followed by that process. The T4 phage assembly process requires that certain steps occur in one, rather than another, temporal order. For example, the sheath subunits must not associate with one other until the tube-baseplate is assembled and is ready to receive them; likewise, the head-tail connector must not attach to the tail until the sheath is assembled around the tube, and so on. Also, the phages contain a number of proteins that are capable of polymorphic assembly into different configurations of approximately equal thermodynamic stability, and thus the assembly process must select certain configurations in preference to others.

These controls appear to be built into the structural proteins themselves. The order of assembly of these proteins is highly specific, and it seems to be controlled through conformational changes or switching.[4] According to this idea, proteins C and B may not bond together until B has combined with another protein A. This is achieved by a conformational change in B which occurs when it combines with A, and which makes it able to combine with C. This will lead to a sequence ABC of formation and not any other sequence. By putting together such conformational change rules, complex assembly pathways can be specified.

MFA MODEL FOR T4 PHAGE

In the MFA model, each of the movable automata can have a number of bond sites x on its perimeter. All parameters in the model are integers, including the position coordinates of bond sites and the size of the automata. Each bond site x is associated with a bond site number $b(x)$, which defines how that site will interact with sites on other automata.

In one simple form of this model, $b(x)$ consists of three parts: $b(x) = L : VV : II$. Here L denotes the bond length which is allowed to take on two values, 0 and 1, to simulate the stretching and compression of a bond between two molecules. VV denotes an arbitrary scale of "strength" for the site. A high value of VV for a site indicates that a bond formed at such a site will be strong. II provides a label for the configuration of the site and is used to determine whether or not two sites have "complementary" configurations, thus allowing them to form a bond. This label for a bond site may change due to changes in the configuration of other sites on the molecule (caused by the formation or dissolution of bonds). This change in label

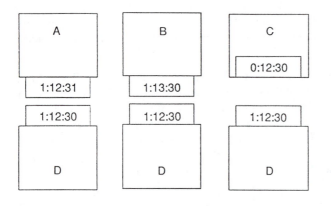

FIGURE 3   Illustration of subunit bonding rules. See text.

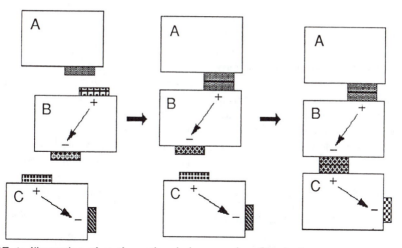

FIGURE 4   Illustration of conformational change rules. See text.

is introduced to emulate the conformational changes occurring in one part of the subunit as a result of changes made in another part (conformational switching).

The following assumptions are made about the interactions between two automata: (1) Two automata cannot overlap one another, and they generally repel each other in proportion to the area of direct contact. (2) Two bond sites on different automata can form a bond only when the two sites are directly opposite one another on facing surfaces. For two sites x and y, if $b(x) = b(y)$, a strong bond of strength $4VV$ will be formed. A bond can also be formed if $b(x)$ and $b(y)$ differ by 1 but it will be a weaker bond, of strength $2VV$. On the other hand, if they differ by 2 or more, no bond will be formed.

With these rules two sites can bond only if their bond lengths (L and site strengths (VV) are identical, and their configuration labels (II) do not differ by more than 1. The key justification for this assumption is that the affinity for bond formation between two subunits depends on how well their shapes match (the lock-and-key concept).

Figure 3 illustrates the operation of these rules. There, four subunits are represented by two-dimensional boxes, and bond sites are indicated by rectangles containing values of $L : VV : II$. These rectangles lie outside of the subunits in the case of bonds of length 1 ($L = 1$), and they lie inside in the case of bonds of length 0 ($L = 0$). According to the rules, A and D can form a bond of strength 24, whereas neither B and D nor C and D can form bonds.

Figure 4 illustrates the rules for conformational change. Here the bond site numbers $b(x)$ are represented by different hatching patterns. Initially A and B are able to form a half-strength bond, but B and C have no affinity for one another. Once the bond between A and B has formed, the upper bond site on B undergoes the change $b(x) \rightarrow b(x) + 1$, bringing the A-B bond to full strength. This change in the upper bond site causes the lower bond site to make a corresponding change of the form $b(y) \rightarrow b(y) - 1$. As a result, B and C are able to form a half-strength bond, and similar conformational changes occur in the bond sites of C.

A total configurational energy U is defined for the complex of all subunits. This consists of the sum of all repulsions minus the sum of the strengths of all bonds which have formed in accordance with the bonding rules. Subunits are allowed to move and interact with other subunits. However, changes in the total complex of subunits are allowed only if they do not result in an increase of U (free energy minimization). This fully defines the dynamics of the model.

The MFA models can be used to simulate on a microcomputer and display in color the self-assembly of a phage-like entity from its components. This artificial "phage" is depicted in Figure 5, and a few steps in its self-assembly are shown in color plates 33 to 40. Here, for simplicity, we have made the following assumptions.

1. All subunits are assumed to be boxes with rectangular sides and fixed shapes. Also, we specify a simulated bacterial cell wall, which is taken to be a array of rectangular boxes that tend to stick together to form a flexible sheet.

2. The phage neither possesses the head capsule nor the neck, which is needed to attach the head to the tail. The main function of the head is to store the DNA molecule, and to channel it into the tail tube after the tube has penetrated the bacterial cell wall. To simulate this function, a standardized test molecule is

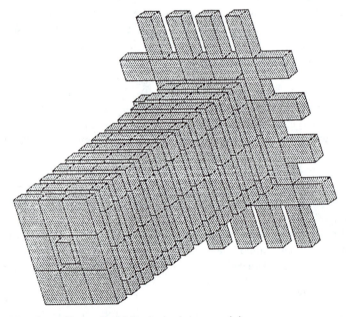

FIGURE 5   The three-dimensional bacteriophage model.

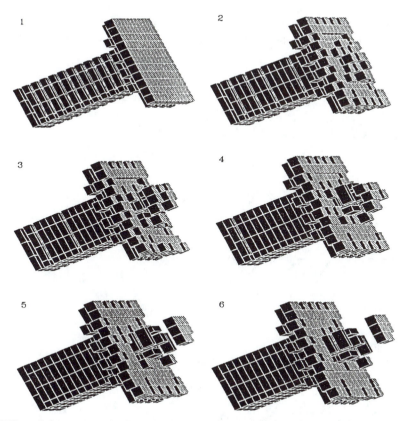

FIGURE 6   A few stages in the penetration of the simulated cell wall by the phage of Figure 5. A section of cell wall is represented by the grid of boxes.

introduced, which will be injected through the cell wall. We note that the self organization of the head is not at well understood, and seems to be much more complex than that of the relatively simple tail mechanism that we are considering.

3. The phage does not possess flexible tail fibers. Thus, the initial attachment of the fibers to the cell wall and their bending to allow base plate attachment to the cell wall are not simulated. Instead, the base plate is allowed to make direct contact with the wall.

As indicated in color plates 33 to 40, it is possible to characterize the bonding sites on each of the subunits so that the phage will self-assemble. In addition, the attachment of this phage to the cell wall triggers conformational changes which propagate through the body of the phage, forcing the central tube to penetrate the cell wall and release the "DNA" test molecule on the other side of the wall. A few stages

in the penetration process are shown in Figure 6. Thus, the basic rules used to specify self-assembly can also ensure that the assembled phage will carry out the operation

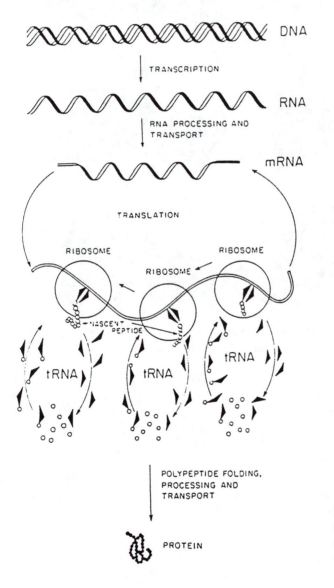

DNA

TRANSCRIPTION

RNA

RNA PROCESSING AND
TRANSPORT

mRNA

TRANSLATION

RIBOSOME   RIBOSOME

RIBOSOME

NASCENT
PEPTIDE

tRNA   tRNA   tRNA

POLYPEPTIDE FOLDING,
PROCESSING AND
TRANSPORT

PROTEIN

FIGURE 7 General
scheme of protein bio-
synthesis (DNA → RNA
→ Protein).[17]

it is supposed to perform. This allows us to simulate a complete phage life cycle, and it opens up the possibility of studying the fate of a population of phages over many generations.

4. PROTEIN BIOSYNTHESIS

The process of protein biosynthesis provides another example in which sophisticated editing mechanisms serve to assure the accurate production of cellular proteins.

The biosynthesis of proteins uses the genetic information of the DNA molecule to determine the sequence of amino acids in the protein synthesized. It involves two major processes[17]: transcription and translation, schematically depicted in Figure 7.

In the transcription process, the sequence of nucleotide bases in one strand of the DNA molecule is enzymatically transcribed into a messenger RNA (mRNA) with a sequence of nucleotide bases (A, U, C, or G) complementary to that of the DNA strand being transcribed.

Translation is the process by which the genetic message is decoded, with mRNA used as a template in determining the specific amino acid sequence during protein biosynthesis. It begins with the initiation phase, in which the mRNA chain becomes associated with a ribosome. For *E. coli* and other procaryotic cells the ribosome is made of two subunits: a large (50 S) one and a small (30 S) one, each containing RNA and a number of proteins.

The sequence of nucleotide bases in mRNA can be divided into ($4^3 = 64$) groups of three nucleotides, the codons. In accordance with the genetic code, each codon represents either a particular amino acid or termination of the polypeptide chain. The section of the mRNA chain encoding a particular protein must begin with an initiation codon (typically AUG).

The genetic code is physically expressed through transfer RNA's (or tRNA's) and aminoacyl-synthetases. The tRNA molecule serves as an adapter connecting a particular kind of amino acid to the mRNA molecule through a triplet of nucleotides, the anticodon, on this tRNA molecule. This triplet is recognized by the codon on the mRNA molecule corresponding to the amino acid. Since there are 64 codons, there must be at least 64 different types of tRNA's (and actually there are more). For each type of tRNA there exists a specific aminoacyl-synthetase which acts as an enzyme for joining the tRNA to an amino acid of the kind corresponding to its anticodon. Once an amino acid has been joined to a tRNA, the resulting complex is known as an aminoacyl-tRNA (or Aa-tRNA).

In the initiation phase, the mRNA is aligned within the ribosome in such a way as to place the initiation codon in the proper position for the beginning of protein synthesis. This involves the action of three proteins called initiation factors, and a special kind of tRNA called initiator tRNA. Once the initiation phase is completed, the initiator tRNA is positioned on the ribosome in a location known as the P (for peptidyl) site, and its anticodon is attached to the initiation codon of the mRNA. At this point the elongation cycle begins.

In this cycle, the ribosome consecutively reads the mRNA codons from the 5' end to the 3' end, and concomitant synthesis of the polypeptide chain takes place through sequential addition of amino acid residues (at the carboxyl terminus—the C-terminus—of the growing polypeptide chain). At an intermediate stage of an elongation cycle, the nth codon of the mRNA is connected to the tRNA in the P-site, and in turn, this n is connected to a polypeptide chain consisting of the first n amino acids of the nascent protein. (A polypeptide chain linked to a tRNA which has donated the last amino acid residue to the peptide is known as a peptidyl-tRNA complex or simply a Pept-tRNA complex.) The $(n + 1)$ codon of the mRNA is positioned in what is known as the A (for aminoacyl) site of the ribosome. An

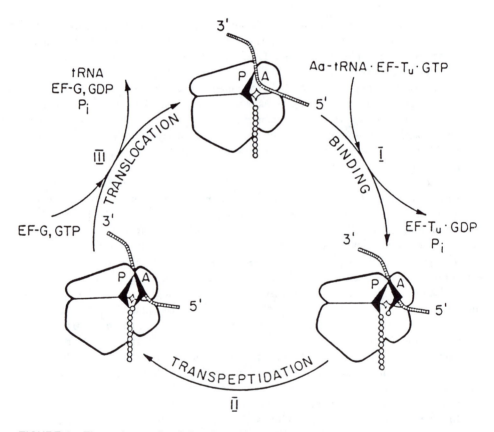

FIGURE 8   Elongation cycle of the ribosome. mRNA is shown in the form of a bent straw and polypeptide chain as hanging chain.[17]

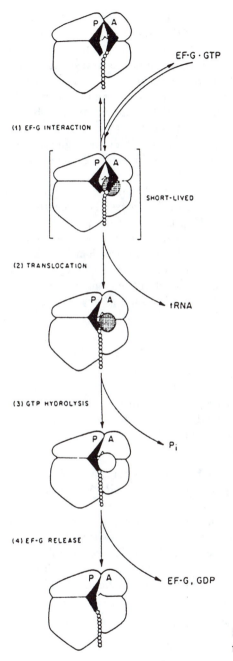

FIGURE 9 Sequence of events during translocation.[17]

Aa-tRNA is now bound to the *A*-site by an enzyme known in procaryotic cells as elongation factor T^u (or EF-T^u). If the anticodon of this *Aa*-tRNA properly matches the codon exposed in the *A*-site, the polypeptide chain is transferred from the *P*-site tRNA to the *Aa*-tRNA's amino acid. This reaction is known as transpeptidation, and it converts the *Aa*-tRNA to a Pept-tRNA with a chain of $n + 1$ amino acids.

The tRNA in the *P*-site is now removed from the ribosome, and the new Pept-tRNA is moved to the *P*-site. At the same time, the mRNA chain is advanced by one codon. This step is called translocation, and it is catalyzed by an enzyme known as elongation factor G (or EF-G). Once this step is completed, a new elongation cycle can begin. In Figure 8 is shown the elongation cycle and in Figure 9, the sequence of events during translocation.

Once the polypeptide chain is complete, it is released from the ribosome along with the mRNA in a process called termination. This process is initiated by a termination codon (*UAG*, *UAA*, or *UGA*), marking the end of the protein's amino acid sequence, and is mediated by a number of enzymes known as release factors.

The elongation cycle involves several complex mechanisms which are now known in reasonable detail and are described in a recent text on protein biosynthesis.[17] We have used these details as the basis for our MFA model of the elongation step. We should note that this material pertains to protein synthesis in procaryotic cells—the process of protein synthesis in eucaryotic cells is similar, but there are differences in details and nomenclature.

MFA MODEL OF ELONGATION CYCLE

The MFA model of the elongation cycle of protein biosynthesis is similar in many aspects to that for a T4 bacteriophage described in the preceding section. However, there are ways in which the model differs significantly.

The basic model consists of two parts:

1. Models for the key structures of the protein biosynthesis machinery: a ribosome, an mRNA chain, a collection of amino acid molecules, two tRNA molecules, and one molecule each of the elongation factors EF-G and EF-T^u. These models include assumptions about the shapes and sizes of subunits which constitute these structures and their mutual interactions. All the key structures of the protein biosynthesis machinery are represented by individual rectangular boxes or by groups of such boxes cemented together along their surfaces. All groups of boxes which are allowed to move are linked together by bonds. In Figure 10 are shown models for these key structures.

2. Rules for bonding and conformational changes in the shape of various molecular subunits, required for their coordinated functioning.

3. Special rules for transitions between two energy states separated by an activation energy barrier. These transitions require random thermal fluctuations.

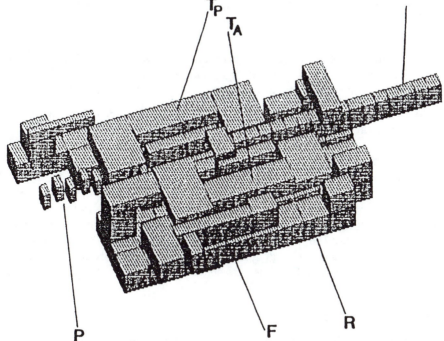

FIGURE 10   Some of the important elements of the protein biosynthesis model: (R) the ribosome, (T_A) and (T_P) the tRNA's in the A and P-sites, respectively, (P) the partially completed poly-peptide chain, (M) the mRNA chain, and (F) enzyme EF-T^u.

RULES FOR BONDING AND CONFORMATIONAL CHANGES

Two subunits cannot overlap one another and they may attract or repel each other when they come in direct contact. The potential associated with this attraction or repulsion is assumed to be proportional to the size of the shared surface, and the proportionality constant depends on the surfaces involved.

Subunits can bind with each other at their bonding sites. Any such site x on a subunit is assumed to be characterized by a state parameter $b(x)$ which is allowed to take integer values from -999 to 999, and which we will refer to as the ID number for the site. We define a bond between two sites x and y to have full strength if $b(x) + b(y) = 0$ and zero strength otherwise. The requirement that the bond ID numbers must add up to 0 makes it possible to easily represent the complementary base pairing between the codons of the mRNA and the anticodons of the tRNAs. If one defines the ID numbers for the RNA bases U, C, A, and G, as 5, 10, -5, and -10, respectively, one can see that the combinations U-A, A-U, C-G, and G-C result in bonds, while others do not.

The conformational changes in shapes of the subunits and the changes in bonding which may be coupled with the formation or breaking of bonds are simulated through the following two rules (although there are some exceptions, as one can see below).

1. The ID number of bond site x in a subunit may vary, depending on whether another site y on that subunit is positioned to form a bond with a site z on another subunit. The shape of the subunit may similarly vary.

2. The ID number of a bond site in a subunit may vary in accordance with deformations in the subunit's shape.

When a change involving rotation, translation, deformation, or bonding is to be made in a group of subunits, the total energy of the group is first computed. The change is then made, and all further changes required by rules (1) and (2) are also made. The total energy of the group is then re-evaluated, and the decision to retain or reject the changes is based on the change in total energy. [We note that according to rule (1), if the juxtaposition of y and z has allowed x to form a bond by changing $b(x)$, then a move changing that juxtaposition must also break that bond by resetting $b(x)$.]. The bond energies are defined using standard units of kcal/mol.

TRANSITION BETWEEN STATES

In protein biosynthesis, bonds are formed and dissolved between various structures. That is, the system makes transitions between one energy state s_i and another energy state s_j. The rate of transitions r_{ij} depends upon the free energies G_i and G_j of these states and the energy G_A of an unstable intermediate state A, with G_A greater than G_i. $\Delta G = G_A - G_i$ is known as the activation energy. For a transition to occur, this activation energy must be provided. In protein biosynthesis, random thermal fluctuations play an important role in providing this energy and, hence, in the transition between two states.

According to the Arrhenius equation,[5] for a simple unimolecular reaction, the rate of reaction is given by

$$r_{ij} = (kT/h) \ \exp(-\Delta G/RT) \tag{1}$$

where k is Boltzmann's constant, R is the gas constant, h is Planck's constant, and T is the absolute temperature. The term $kT/h \approx 6 \times 10^{12}/sec$ at 25°C. Since ΔG is expected to be positive, kT/h provides an upper limit on the rates of chemical reactions.

Since the free energies G_A for various transitions, in general, are not known, we have found it necessary to express them somewhat arbitrarily in terms of G_i and G_j. We assume that

$$\exp(G_A/RT) = \frac{\exp(G_i/RT) + \exp(G_j/RT)}{z} \tag{2}$$

where z is a factor which is ≤ 1. With this assumption, the rate of transition is given by

$$r_{ij} = \frac{p_{ij}}{(N-1)\Delta t},\qquad (3a)$$

where

$$\Delta t = \frac{h}{kT(N-1)},\qquad (3b)$$

$$p_{ij} = z\,\frac{\exp[-(G_j - G_i)/RT]}{1 + \exp[-(G_j - G_i)/RT]} < 1 \qquad (3c)$$

and N is the total number of states.

For purposes of computer simulation, it is convenient to choose Δt to be the discrete time step for any transition to occur, and to define a Markov chain[16] giving the probability that the system will move from state s_i to state s_j in one time step. This probability is given by

$$M_{ij} = r_{ij}\Delta t$$

and the probability that the system will stay in the state s_i in one time step is given by

$$M_{ii} = 1 - \sum_{j \neq i} M_{ij} \qquad (3e)$$

The energetics of the various processes involved in protein biosynthesis, which determine the transition from one state to another, are given elsewhere.[10] These energetics are used in carrying out the simulations described next.

SIMULATION DYNAMICS

With the MFA model discussed in the preceding section, one can simulate the elongation cycle and display it in color on a microcomputer. Figure 11 gives 12 frames, produced by the simulation based on the MFA models, showing incorporation of an amino acid in a partially made polypeptide chain. A brief explanatory description of these frames follows, with emphasis on the differences between successive frames.

Frame (1) shows mRNA, a tRNA in the P-site, a growing polypeptide chain with 5 amino acid residues, and the ribosome.

Frame (2) shows that an Aa-tRNA complex and an EFT complex have been added to the ribosome. The newly arrived tRNA, with its attached amino acid Aa, is in the A-site. This extra amino acid is later incorporated into the polypeptide chain through the elongation process. We assume that the codon-anticodon pairing is correct. Therefore, the Aa-tRNA complex binds to the ribosome. This binding is followed with the ejection of EFT as shown in frame (3).

In frame (4) the center subunit of the tRNA in the A-site has expanded, the tRNA has entered the locked phase, and everything is ready for the transpeptidation phase.

Frame (5) shows the transpeptidation reaction. Note that the polypeptide chain has shifted from the *P*-site tRNA to the *A*-site tRNA, and both tRNAs have undergone conformational changes. Note that subunit of the tRNA nearest to the polypeptide chain is thicker for the tRNA in the *A*-site than in frame (4) and thinner for the tRNA in the *P*-site. Thus the two tRNAs have switched their conformations.

Because of these conformational changes, the EFG complex (*G*) is now able to bind to the ribosome. This is shown in frame (6).

In frame (7), the *A*-site tRNA has contracted again due the presence of EFG complex. Because of this contraction, it is free to flip into the vertical position shown in frame (8).

In frame (9), the *P*-site tRNA has left the ribosome and the length of subunit *G* of the EFG complex undergoes a random variation. Here we can clearly see that

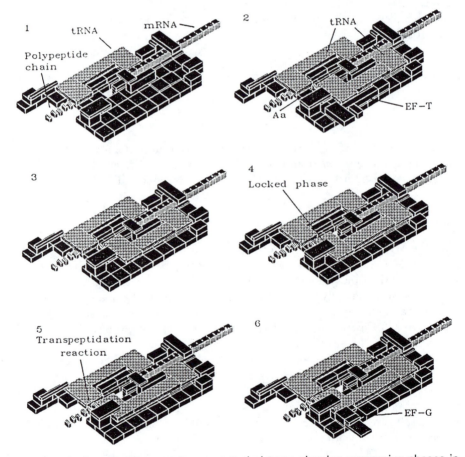

FIGURE 11A   A series of computer-generated pictures showing successive phases in the operation of the protein biosynthesis model.

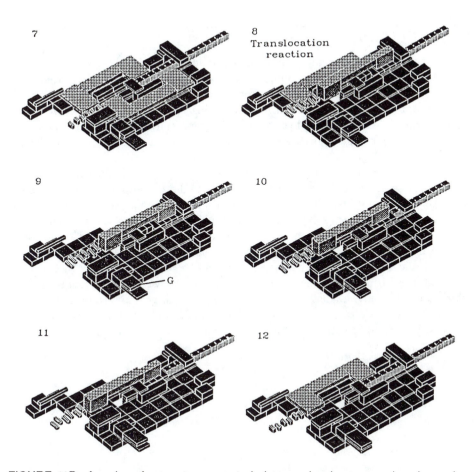

FIGURE 11B A series of computer-generated pictures showing successive phases in the operation of the protein biosynthesis model.

the amino acid brought by the Aa-tRNA·EFT complex has been incorporated into the polypeptide chain, which now has six amino acids.

The tRNA and mRNA are now free to move back and forth together, and they have moved one unit to the left in frame (10). In frame (11), this movement has continued, and the tRNA and mRNA both have moved two more units to the left.

In frame (12), the tRNA molecule has flipped from the vertical position into the P-site, and the EFG complex has left the ribosome. This frame is similar to frame (1) except that the polypeptide chain has six amino acids. This configuration is ready to accept another Aa-tRNA·EFT complex.

5. CONCLUDING REMARKS

In this paper we have presented a new class of models, the Movable Finite Automata (MFA) models for simulating living systems on a microcomputer. We used these models to simulate bacteriophage assembly and operation, and elongation cycle in the protein biosynthesis. Although models of this kind cannot represent molecular interactions on a detailed biophysical level, they can faithfully represent the logical steps in the conformational changes that govern the behavior of macromolecular complexes in biological systems.

The two models by no means represent the respective biological systems as realistically as one would like. With sufficiently realistic computer models, one can study the effects of design changes either by (1) simply running the model's program for different designs, or (2) by performing a general mathematical analysis of the model. Such investigations can be useful in studying the origin and evolution of biological systems, since this is a field where general theoretical principles play an essential role. They can also be useful in efforts to control infectious diseases and to breed new forms of organisms. The question is can we devise computer models which are sufficiently realistic to be applicable to such general studies? To answer this, it is necessary to pursue both model building and experimental research in parallel. We are optimistic that the answer will be in the affirmative.

REFERENCES

1. Berlekamp, E. R., and J. H. Conway (1982), *Winning Ways for Your Mathematical Plays* (London: Academic Press).
2. Broadwell, P. (1988), "Plasm - a Fish Sample," demonstrated at the conference on Artificial Life held at Los Alamos National Laboratory from September 21 to 25, 1987.
3. Casjens, S. (ed.) (1985), *Virus Structure and Assembly* (Boston: Jones and Bartlett).
4. Caspar, D. L. (1980), "Movement and Self-Control in Protein Assembly," *Biophys. J.* **32**, 103–135.
5. Fersht, A. (1977), *Enzyme Structure and Mechanism* (San Francisco: W. H. Freeman).
6. Fuller, W. (1969), "Physical Contributions to the Determination of Biological Structure and Function," *Biophysics*, Eds. W. Fuller, C. Rashbass, L. Bragg, and A.C.T. North (Menlo Park,CA: W.A. Benjamin).
7. Goel, N. S., and R. L. Thompson (1986), "Organization of Biological Systems: Some Principles and Models," *Int. Rev. Cytol.* **103**, 1–88.
8. Goel, N. S., and Thompson, R. L. (1987), "Microcomputer Modeling of Biological Systems," *The World & I* **2**, 162–173.
9. Goel, N. S., and R. L. Thompson (1988a), *Computer Simulation of Self-organization in Biological Systems* (London: Croom-Helms, and New York: MacMillan).
10. Goel, N. S., and R. L. Thompson (1988b), "Movable Finite Automata (MFA) Models for Biological Systems II: Protein Biosynthesis," *J. Theor. Biol.* (in press).
11. Karplus, M., and J. A. McCammon (1986), "The Dynamics of Proteins," *Sci. Amer.* **254(4)**, 42–51.
12. Lindenmayer, A., and P. Prusinkiewicz (1988), "Developmental Models of Multicellular Organisms: A Computer Graphics Perspective," these proceedings.
13. Meinhardt, H. (1982) *Models of Biological Pattern Formation* (New York: Academic Press).
14. Primrose, S. B., and N. J. Dimmock (1980), *Introduction to Modern Virology* (New York: John Wiley).
15. Reynolds, C. W. (1988), "Flocks, Herds and Schools - A Distributed Behavioral Model (Proceedings of SIGGRAPH '87)," *Computer Graphics* **21(4)**, 25–34.
16. Ross, S. M. (1985), *Introduction to Probability Models* (Orlando, FL: Academic Press), 3rd edition.
17. Spirin, A. (1986), *Ribosome Structure and Protein Biosynthesis* (Reading MA: Benjamin/Cummings).
18. Thompson, R. L., and Goel, N. S. (1985), "A Simulation of T4 Bacteriophage Assembly and Operation," *BioSys.* **18**, 23–45.

19. Thompson, R. L., and N. S. Goel (1988), "Movable Finite Automata (MFA) Models for Biological Systems I: Bacteriophage Assembly and Operation," *J. Theor. Biol.* **131**, 351–385.

20. Toffoli, T., and Margolus, N. (1987), *Cellular Automata Machines - a New Environment for Modeling* (Cambridge, MA: Massachusetts Institute of Technology Press).

21. Wood, W. B. (1980), "Bacteriophage T4 Morphogenesis as a Model for Assembly of Subcellular Structure," *Quart. Rev. Biol.* **55**, 353–367.

Marek W. Lugowski
Indiana University Computer Science Department,Lindley Hall 101, Bloomington, IN 47405;
marek@iuvax.cs.indiana.edu

Computational Metabolism:
Towards Biological Geometries for Computing

Artificial life is not just a field of study but also a new way for accomplishing old things. Computational Metabolism (ComMet) is a class of abstract machines that embody some principles of artificial life (e.g., emergence) in order to study or simulate certain intricate processes. The processes best matched to ComMet are those that may be too unwieldy to render using standard computers (e.g., metaphor) or those that yield good insights when cast as a liquid-like animation (e.g., the formation of inorganic membranes). After sketching ComMet, I discuss its capacity for adaptive behavior and consider its relationship to cellular automata. Next, I follow with the emergence of ComMet hierarchies that comprise several layers of composite structure. Last, I collect the motivations behind ComMet and close with remarks on the nature of artificial life and ComMet's ability to encode evolution. An appendix contains illustrated though modest examples of the already implemented instances of ComMet.

Artificial Life, SFI Studies in the Sciences of Complexity,
Ed. C. Langton, Addison-Wesley Publishing Company, 1988   **341**

ARTIFICIAL LIFE: A NATURAL FOR COMPUTING HARD THINGS?

Artificial life, to a computer scientist, is one possible future: a symbiosis of discrete mathematics with the "ecophysics" of life. This paper is a step in that direction. It describes a class of computers based on localized recognition, localized changes and consequent emergence—called Computational Metabolism, ComMet for short.

Traditional thought in computer science calls for figuring out an algorithm (usually sequential, or one that combines sequential strands) that reflects a process of interest.[5,6] One thus writes a top-down specification stipulating certain processing steps at appropriate times. This big picture comprises a collection of little ones, each one chained to its superior via marionette-like dependence. There are exceptions to this picture, but not many.

Sometimes the marionette framework clashes with the realities of the modeled world. It may be difficult to serialize a process. Or it may be intractable to view it as a hierarchy of subprocesses. Or it may be impossible to specify the process *a priori*. Some processes depend in complex ways on their history. Sometimes we don't even know enough about the dynamics involved: have we cut the world up along the dotted lines?[3,10,12] How does one compute insufferable processes? Traditional tools and methodologies don't always suffice.[14]

Enter artificial life, more abstractly, biology. Biology showcases—and most computer science presently ignores—evolution as a computing engine.[4] Not much rigid sequencing appears in biology. Little computational brittleness appears here. No master control program synchronizes everything. Instead, cooperation and competition drive changes in vast, heterogeneous data structures (called ecosystems). Embedded there, unpredictable yet clearly nonrandom processes literally run around (organisms). These processes communicate through many channels, often via form-fitting or via co-location within an ecosystem (predator/pray, host/parasite, enzyme/substrate). Processing (existence) is contingent on many independent variables. The interaction of processes is both simultaneously localized (the characteristic behavior of an individual) and globalized (the characteristic behavior of species). In biology, the perceived data structures often become processes and vice versa. Processes are usually not predictable as individuals, becoming more fathomable as aggregates, when they effect a higher organization. Yet, these individuals are already hierarchies in themselves. What do we have here? It is a self-similar yet scale-dependent physics of information (cells, organs, organisms, groups, societies). It is hardly algorithmic, in computer sense, yet is steeped in computation.[9,11]

In this paper, I am merging the vocabularies of biology and computer science as I demonstrate the benefits of combining the two fields. Certain familiar terms thus will acquire very specific meanings. For one, *phenotype* will refer to an individual process's behavior. *Genotype* will denote shared information within a population of like processes that describes their possible behaviors. Also, the notion of *bonding* will refer to the effects taking place within a process: state-change, as contrasted to any physical linkage between two agents. The expressions *"recognizing a neighbor"* and *"reacting to a neighbor"* will be carefully distinguished, as they happen to lie at the crux of the system described herein. *Recognizing* is when input is acquired. *Reacting* is the consequence of recognizing, and involves output, such as bonding. Only some recognizing leads to reacting because recognition events compete with one another for realization. Finally, the notion of *physics* of computation embraces both the rules and the configuration of a computer. Our blurring of the distinction between the two, as our blurring of the distinction between data structures and processes, or for that matter, between the interpreter and the interpreted, is the essence of Computational Metabolism.

A MOVING PICTURE IS WORTH A THOUSAND EQUATIONS

A moving picture has a lasting impact on the viewer. We are good at understanding pictures and when the pictures move, we're splendid.

Computational Metabolism, or ComMet for short, is a class of abstract machines that are reminiscent of fluid in motion. ComMet embodies principles of artificial life such as emergence, neighbor recognition, and chance in order to study or simulate certain intricate processes. The processes best matched to ComMet are those that may be too unwieldy to render using standard computers, e.g., metaphor, or those that yield good insights when cast as a liquid-like animation, e.g. the formation of inorganic membranes. Another type of computer, the cellular automaton, is similar to ComMet in these regards. The two are compared later.

Computational Metabolism does computation so that we may view it in progress and find it immediately revealing, even the intermediate steps—especially the intermediate steps. Computational Metabolism starts up as a pattern of mobile elements. This pattern is an ontological approximation to an initial state in a physical, biological, chemical or mathematical process. The mobile elements encode all the explicit information in ComMet. All the implicit information resides in elemental interactions and their spatial distribution. There is no master control program. There is no sequencing. If sequences arise, they do so through emergence, however carefully sought, anticipated and arranged for. The mobile elements affect only their neighbors (phenotype, as defined above), but they contain species-wide information which allows them to change internal configuration and thus behavior in carefully prescribed or carefully left unsaid ways (genotype, again, as defined above).

The resulting interactions add up to cooperation and competition. Skillfully, we may set up elemental interactions so that the emergent flux of patterns takes on a "life" of its own, yielding the organ, organism, group, etc. hierarchy. In the implemented ComMet, the mobile elements do not change in unordained ways (natural selection), and the ordained ways amount to simple substitutions chosen from a small list of alternative states.

COMPUTATIONAL METABOLISM: A LIQUID COMPUTER

Computational Metabolism, or ComMet, is a tessellation whose elements are roving processors, called *tiles*. It is a liquid whose molecules—the tiles—*recognize* and *react* to their *neighbors*. Each tile is stamped with a unique identification: a list of coordinates for its starting position. (This information is used only in debugging, or analyzing the system, although it admits interesting computational uses, such as gradient descent in displacement space or tile metamorphosis as a function of path.)

The concept of neighborhood needs defining. Plainly said, two neighbors are any pair of tiles sharing an edge (or a fragment of an edge, for out-of-alignment tilings). Geometrically said, a tessellation may yield corner-adjacency and edge-adjacency. Neighborhood is edge-adjacent only. For example, a three-dimensional tile has edges that are exactly all of its two-dimensional boundaries. This definition generalizes to n-dimensional systems, and with a bit of interpreting, to fractional-dimensioned systems. Incidentally, many tessellations yield edge-adjacency alone, because of symmetry considerations, e.g. spheroid symmetries, phase shift, and aperiodical tilings.

Now that neighbors are discernible, we give a bottom-up description of what a tile is and just exactly how it *recognizes* and *reacts* to neighbors. We in turn will consider how properties of collections of tiles emerge from tiles' individual states and from their aggregate interactions, and how mobile tile populations may give rise to massively parallel, biological geometries for computing.

All the while, let us remember that the system being described, already implemented in full, is exceedingly simple modulo the class of conceivable ComMets. This system is merely a first worked-out instance. It hints at still richer possibilities; it defines an envelope of ideas meant to be pushed.

Let us refer to Figure 1 for the length of this description: the exploded view of an individual tile shows the entire substructure of a tile.

First and foremost there is the *color*, which is what a tile examines in its neighbors as it attempts to recognize them. Color is the group identity, or the species, for tiles. Each tile comes in one of the user-defined colors. All tiles sharing a color also share the bulk of their initial state, the reminder being the starting coordinates mentioned above. (I have considered tiles that belong to more than one color, change colors, develop new colors, learn to recognize new colors, and tiles which denote colors idiosyncratically as opposed to poll neighbors with a predicate. One can clearly do a lot with color!)

Next of interest in Figure 1 is the set of black and white rectangles, one of each on every edge. These are the *templates*, or lists of colors currently recognizable on a given edge. If a neighbor's color is listed in the black template on the edge shared with the neighbor, the neighbor is automatically recognized. To be recognized is no more than matching an entry in a neighbor's template on the shared edge. The same goes for white templates, except that *any* black recognition taking place

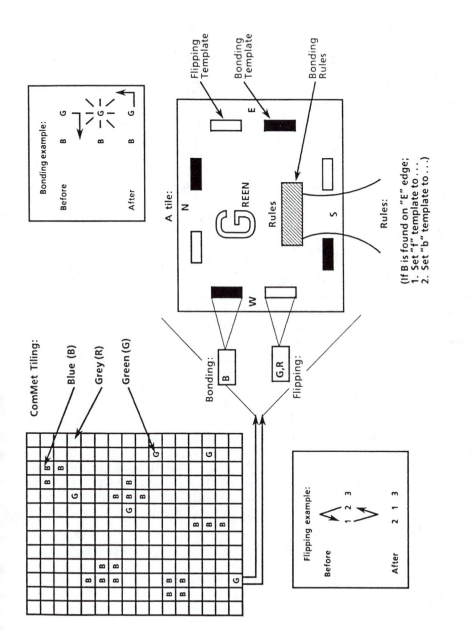

FIGURE 1 Computational Metabolism, ComMet, at a glance.

within a tile immediately shuts down *all* white recognition within that tile, for a fixed duration.

Given this set-up, it follows that a tile may simultaneously *recognize* its entire neighborhood. However, a tile may not *react*—at once—to its entire neighborhood. After all, the parallel-universes cosmology has not yet been incorporated into Com-Met! However, Stanislaw Ulam's Monte Carlo methods have. In such ponderous situations, ComMet merely tosses dice, conceivably loaded ones, although we've kept them fair so far.

Let us see why ComMet features two kinds of recognizing. The black recognition, the preempting kind, is known as *bond recognition*. As defined earlier, *bonding* is the replacement of a present tile state with another. Bonding *is not* tiles sticking together or vibrating in tandem. In fact, bonding is more like transitions in finite-state automata. The term itself is meant to suggest lasting changes to the content of a tile's templates, and to stress the tiles' role as bearers of changing messages. The modification of message content *is* bonding. Strikingly different behaviors come about from bonding: mobilizing and immobilizing a tile, biasing its direction of propagation and, beyond the implemented envelope, associative memory, learning and natural selection.

Tiles bear information simply because they rove within the tiling. When they interact with other tiles, their presence *is* that information. This roving is an epiphenomenon of reciprocated white recognition. White recognition is called *flip recognition*. The mechanics of flip recognition parallel those of black recognition, or bond recognition: the neighbor's color when present within the shared-edge template of the recognizing tile triggers automatic recognition. The competing flip-recognition events are arbitrated by Monte Carlo, with one winner. If the tile's intended flip partner reciprocates, i.e., when there is mutual flip recognition, the two tiles *react* to one another, automatically, by swapping places. This volume-conserving transaction is the elementary event in tile propagation. Taking place simultaneously all over the tiling and over a period of time, it gives rise to the liquid-in-motion dynamic that is computation in ComMet, while expressing the underlying mathematics as one succinct and elegant formalism: localized perturbations on a tessellation, usually encoding a "trap-door" function.

Although we have amply described flip reaction, we have more to say about *bond reaction*. Bond reaction is expressed by the *rules*. The rules are the same for tiles that share color. The rules express the *reaction* of a tile that just bond-recognized another. In Figure 1, the rules are symbolically depicted as a box. In actuality, they are a list of individual rules, each one listed first by edge and then by color. For example, here is a rule encoding a bond reaction to Blue:

```
;;; A sample rule for an edge that bond-recognizes Blue, b.
(b nil nil nil nil (g) (g) (b g) (g)))
```

This rule encodes the following information:

```
If bond-recognized a Blue then react as follows:
      set the North, East, South and West flip-templates to empty (nil),
      set the N, E and W bond-templates to bond-recognize Green (g),
      set the S bond-template to bond-recognize either Blue or Green.
```

Note that in the implemented ComMet being described, there are four edges, each one labeled for a direction. The tiles are aligned and in phase, like graph-paper squares. Also note that a rule, every rule, is a nine-item list, starting with the color it reacts to. For every color entry in an edge's bond template there is a corresponding rule listed for that edge.

Having seen the content of rules and the limited amount of processing their substitution involves, we now consider the entire state of a single tile. A tile is represented as a list with sublists, of which the rules lists, one per edge, are the last four:

```
;;; A Gray (.) tile initialized at position x = 4, y = 5.
((4 5) . (b g) (b g) (b g) (b g) nil nil nil nil DONE
      nil
      nil
      nil
      nil)
```

The above is a fifteen-item list. The starting coordinates are first. They are followed by the color identification. Then come the 4 flip templates, listed in the usual N, E, S, and W order. Same is true for the bond templates, next. Their "nil" values stand for empty lists. Next, a "DONE" marker follows. It toggles to nil as the simulator marks tiles to keep track of the already processed ones. Last, the four sets or rules, one per edge, bring up the rest of a tile's state.

Having seen a complete tile, we now jump one more layer of abstraction and contemplate an exemplar of the complete state of ComMet *before* tile initialization, 4 colors in all, in this instance. The complete state of ComMet at that time is just a list of color specifications. Once that is given, tile initialization is just copying one of them, prefixing it with the coordinates of a chosen tile location and placing the result, the tile itself, at that location.

```
;;; Color specifications for the ''How to invent gravity'' example.

;;; See the appendix for the discussion and illustrations.

;;; Color Gray (.), which features no bond-recognition.
((. (b g) (b g) (b g) (b g) nil nil nil nil DONE
    nil
    nil
    nil
    nil)
```

```
;;; Color Green (g), which flips with Gray and Green on all edges.
(g (. g) (. g) (. g) (. g) (b) (b) (b) (b) DONE
    ((b (. g) (. g) (. g) (. g) nil nil (b) nil))
    ((b (. g) (. g) (. g) (. g) nil nil nil (b)))
    ((b (. g) (. g) (. g) (. g) (b) nil nil nil))
    ((b (. g) (. g) (. g) (. g) nil (b) nil nil)))

;;; Color Blue (b), which bonds with both Green and Blue on all edges.
(b (.) (.) (.) (.) (b g) (b g) (b g) (b g) DONE
    ((g (.) nil nil nil nil nil (b g) nil)
     (b nil nil nil nil (g) (g) (b g) (g)))
    ((g nil (.) nil nil nil nil (b g))
     (b nil nil nil nil (g) (g) (g) (b g)))
    ((g nil nil (.) nil (b g) nil nil nil)
     (b nil nil nil nil (b g) (g) (g) (g)))
    ((g nil nil nil (.) nil (b g) nil nil)
     (b nil nil nil nil (g) (b g) (g) (g))))

;;; Color Empty ( ), which is used to line the outside of the tiling.
(" "))
```

At this point, the reader is urged to read ahead through two of the several examples of ComMet included in the appendix. "Abacus sort" is a deterministic, pure flip-recognition ComMet. "How to build a wall" introduces simple bonding within a stochastic world. Each example has much less state than the "How to invent gravity" example cited above, yet both are good approximations to their real-life physical analogues.

Since to flip is to migrate via local transactions, one might assume that the tiles are free to do so asynchronously and that the flipping time varies. This is what the mathematical model allows, but the vicissitudes of conjuring a working implementation on a fairly powerful yet hopelessly serial machine dictate otherwise. The presently implemented ComMet is a lock-stepped system; all flips are one cycle long.[18] This gives rise to a phenomenon of generations, as witnessed in John Conway's Game of Life.[1]

The above are simple mechanisms for encoding and changing state. There are more efficient methods that could replace these mechanisms. There are also good reasons for retaining the ones used: their physical realism, their extensibility to adaptive behavior and the generality of list-oriented pattern matching.

---

[1]Speaking of implementations, the ComMet simulator turns out these generations from 21 kilobytes of naive Common Lisp (including comments and utilities). It runs compiled, though without using any of the Common Lisp-available optimizations, on a TI Explorer Lisp machine with 8 megabytes of memory. Within a 5000-tile ComMet, a generation is born every 45 seconds: 30 to compute and 15 to print out (with graphics, the simulation slows down by a third). As for memory requirements, one 5000-tile ComMet weighed in at 6.5 megabits of state. Clearly, locality of computation has its price. Nonetheless, faster, more efficient, better optimized, and less rigid incarnations of ComMet are forthcoming, contingent on the improving understanding of the system's dynamics and pending the availability of better simulators and faster simulation engines.

This completes the architectural description of Computational Metabolism. All interactions within the system are a juxtaposition of the mechanisms described above. ComMet is a purely bottom-up system, although top-down inference (more aptly: interference) may be used in conjunction with it. For example, a graft or a wholesale replacement of a region of tiles could be used to perturb the system non-locally. Still, even such drastic tinkering does nothing to its elemental interactions.

The system's dynamics give rise to a choice function. This function first records all chosen bond-recognition events. It then carries them out. Then, it first records and later carries out all reciprocal flip desires that were not preempted by bonding. The unrequited flips are presently ignored. (Instead they could be counted and accumulated to influence the choice function; such memory effects ought to have interesting adaptive ramifications.) The choice function, emergent from the combined interaction of many tiles—including the proximate non-neighbors everywhere—maps to perfect determinism given an appropriate initial configuration of ComMet. Thus, flexibility incarnate, ComMet may be configured not just as a stochastic (Monte Carlo) engine but also as a deterministic (this includes chaotic) machine. In fact, it may switch modes in midstream or combine them—even convolve them.

When running as a stochastic device, ComMet's local indeterminacy multiplies itself manyfold throughout the tessellation as tile clusters and time dependencies arise and dissolve. This truly is a metabolism of unsupervised activity. But it should not imply unwieldiness, however. As Stanislaw Ulam was fond of showing, using compounded randomness is a natural way to make something come out predictably.

Although capable of encoding reversible computation, especially when enumerating a chaotic system, generally ComMet instances are not set up to run in reverse.[1] Most initial states do not specify a unique path. However, later we will see how appropriate selections of tile colors, templates, bonding rules and initial spatial configurations can exploit, control or remove ambient indeterminacy, yielding deterministic computations on the level of composite structure—and if desired—on the level of individual tile paths.

We use stochastic processes as substitutes for decision in ComMet's control structure. Explicit boundary conditions (color, templates, starting formation, etc.) will comprise our implicit algorithm specification. Probabilistic choice as control structure and boundary conditions as algorithm specification help produce a computation reminiscent of fluids.

One archival note: unlike the original specification of ComMet,[15] on which this paper builds and expands, the flip-templates and the bond-templates discriminate directly between colors. Originally, templates acted as simple enzymes, recognizing only one *pattern-in-time*. This pattern may have been shared by several colors, thereby blurring color identification. This consequence complicated and slowed down the computer simulation efforts, thus hampering all ComMet research. The pattern-in-time specification may yet be reincorporated, but not before more pressing issues are addressed. It is as compelling as ever that association on patterns-in-time is the recognition of distinct entities in real-life eco-computing with physical shapes: consider substrate substitution (blocking) and recognition through form-fitting or replicative scanning.

TILES COULD LEARN OVER TIME

The present implementation of ComMet does not yet feature learning. However, nothing stops us from including adaptive behavior in the model. The simplest way (and there are many, many others!) is to permit random mutation to occur within the rules internalized in every tile. The range of mutation can be easily controlled via look-up tables in order to permit syntactically correct outcomes only. Alternatively, ungrammatical rule bodies could become inert for the duration of their ungrammaticality.

Specifically, we could regard each rule as a list of bits. We could impose a robust interpretation such that all possible bit values map to well-formed rules. The sublists of particular interest to us include color, edge and template identifications. In fact, the entire rule is expressible as a concatenation of color identifications, listed with positional significance.

For example, a rule of the form (01 0 101...) may be interpreted as "On edge 1 (East), template 0 (flip), given the presence of either Red or Blue tile but not White (color combination 5), the following action is taken: ..." Random bit changes could allow individual tiles to change behavior. We could control the probability and locality of such mutation by encoding this information as another rule. This would allow the design of richly adaptive behaviors contingent on tiles' histories and neighborhoods.

We could attempt pattern recognition by first training mutable tiles on reference inputs by configuring them in certain initial groupings. Later we would group them in novel ways, retaining accumulated expertise intact. These initial groupings could correspond to images in need of interpretation. Resultant configurations could correspond to the desired categorizations. Similar inputs (whatever measure of similarity is used) would map to similar resultant configurations. There are many similar possibilities, of which this scheme is only one.

Suffice to say that tasks ranging from visual pattern recognition[10] to mixing metaphors[13,19] may be well-suited for implementation via ComMet because transforming contextual information is the bulk of all its activity, and because ComMet scales well since all processing and infrastructure act strictly locally. Furthermore, the two-dimensional graph-paper instance of ComMet is but a point in its configuration space. Who is to say what a 65,536-dimensional ComMet with an asymmetrical hyper-hexagonal neighborhood might be a cinch at encoding? After all, N. J. A. Sloane has found interestingly densest sphere packings at exactly 65,536 dimensions, all in the search of practical applications. One may well inquire about the general computing properties of tessellations of identical dimensionality.

COMPUTATIONAL METABOLISM AS A CELLULAR AUTOMATON

Another mechanism that acts locally is the cellular automaton.[17,21] Is ComMet a member of the cellular automata family? Yes, provided that (1) we equip cellular automata with migrating cells, (2) let them dynamically change cellular rules, (3) let many such rules be active at once, possibly overlapping, (4) let cells have persisting internal state, and (5) let new cells grow or existing ones mutate.

However, the current focus on cellular automata does not appear to strive for the above characteristics. Furthermore, ComMet does not postulate cellular automata's lock-step execution, so long as mutual exclusion facilities are provided for successful tile flips. For instance, the very concept of a migrating tile in ComMet, although reminiscent of propagating wavefronts within a cellular automaton, is fundamentally different. Its purpose is to encapsulate a quantity of state, such as adaptive memory, and physically communicate it over distance without process control, global interactions, routing or wire.

Furthermore, the migrating tiles in ComMet have, in aggregate, properties similar to those of migrating molecules in a fluid or of transmutable atoms in a crystal. These quasi-physical properties make ComMet tiles a compelling vehicle for devising computations expressible with actual fluids and crystals. ComMet's semantics mimics those physical systems that provide the tessellating mechanism in another aggregate-level equivalent way.

This gives hope to using ComMet as a laboratory for experimenting with truly nonsymbolic computing, a workbench for the spatially extended concept of information. ComMet-debugged processes could then be transferred to a crafted "informational fluid," rich with designer phase transitions and chemical adaptive mechanisms.[8] Alternatively, a biochemical soup or a matrix of tuned clay latices[2] could be the fast engine for computations pioneered with Computational Metabolism. These futuristic applications are somewhat divergent from those laid out for cellular automata (e.g., nanoelectronics, or quantum-effect solid-state supercomputers.), but some overlap exists.

An additional reason for distinguishing ComMet from present cellular automata is the emerging research on formal systems that exhibit conservation of particles (generalized permutations).[21] Formal systems of this sort closely model many observed physical systems (latices, liquids, membranes), thus offering hope for the emergence of computational principles directly applicable to such systems.

Nevertheless, ComMet is only one of many possible attempts towards creating computing substrates that use the spatially extended concept of information. Biology offers many such inspirations. The essence of undertaking attempts such as ComMet is to explore what it means to compute. Certainly, limiting computing to the generalized von Neumann model and to his cellular automata (or any small set of models) will not accomplish this exploration.

Closer to the present, today's cellular automata are a matrix of uniform cells obeying fixed rules. The rules exist beyond the matrix and are first selected and then applied to it "from above," stipulating each cell's state transition. On the other hand, in Computational Metabolism, rules are internalized by the individual cells. That means that cells are ultimately in a position to evolve their rules on the basis of experience. In any case, ComMet's system behavior results from the interactions of individual edge-neighboring cells, tiles. Still, the interaction "neighborhood" for ComMet has been set arbitrarily at nearest neighbor. All sorts of fun field effects such as attraction, repulsion and magnetic-like induction may be had if we were to generalize this stipulation. Although ComMet is conceptualized with purely local interaction in mind, departures from this localist paragon seem tempting and may probably be rewarding. Just the same, a purely local connectivity can give rise to thoroughly distributed aggregates when the cells are in flux.

In summary, Computational Metabolism appears not to fit the description of cellular automata, but maybe a more intriguing conclusion is that cellular automata are not yet computational metabolisms. The area of their overlap is likely to expand all the same.

PATTERNS COMPOSED OF PATTERNS—HIERARCHIES AND ABSTRACTIONS

For every tile in ComMet, there is set of relationships defined by its spatial proximity to, and template-match potential with, all the other tiles. This is a very rich collection of building blocks from which to build. Even though there is nothing in ComMet that expressly displaces tiles in spatially coherent bunches—the principle of locality discourages it—the user is free to notice structure on arbitrary levels of abstraction. A single tile and its potential dynamics define level 1. But what happens when we have several tiles forming an interacting whole?

For instance, as a token level-2 structure, let us consider a blob of tiles, all Blue and one Green. These tiles are in a Blue-Green flux, but never moving off the initial blob's footprint. To a turquoise eye—which sees all Blue and Green as one—there is no flux motion. Let us now set up an accelerating mechanism which "turns on" any randomly walking Orange tile that happens to drop in the neighborhood of our turquoise blob, provided that it squares off with our solitary Green. A turned-on Orange accelerates in a preferred direction.

We just sketched a 2-level organization: (1) a contained collection of loose tiles, and (2) a device composed of tiles that acts as an accelerator. Now let's consider an arrangement of accelerators whereby the hastened Orange tiles travel in a ring. We now have a three-level hierarchy: tiles, a tile-accelerator, and a tile-cyclotron. Let us then proceed to describe the activity within our cyclotron: the flux of accelerated particles and the availability of raw material. We may now invent performance measurements for the ring structure and could even talk about the ring's behavior at equilibrium, at capacity, etc.

Instead, let us step up to level 4: we choose to view the ring as a memory device. We do so by ascribing meaning to numbers of particles found in the ring, their flux density, average interparticle gap. We then choose to associate with certain measurements certain conventional memory contents. For instance, 25 exhausted Orange tiles maintaining a roughly equidistant spacing represents something—and when bunched up, something else.

We can now incorporate our level-4 abstraction, a memory device, as part of computation on level 5: a predicate. After all, a memory device lends itself to drawing distinctions amongst memory states, and associating actions with these distinctions. Distinct states arise as a consequence of manipulating the system in a reproducible way. When the reproducible coupling is strong, level 5 implements a look-up table. Otherwise, it implements a choice bag (probabilistic selector).

Once endowed with a look-up table or a choice bag, we can observe it over time. This innocent extension signals another jump in abstraction. It is so because level 5 alone—a look-up table or a choice bag—is not itself an inductive inference of the state of its environment. It is merely a subcomponent, a tool for such operation, just as a cyclotron is not a memory device but a subcomponent of one. Level 6 computing uses level 5's contribution to implement a sensor/counter. The sensor/counter measures, as permitted by the ontology of levels 1 through 6, a postulated global property of the tiling, as revealed by the ambient flux of Orange tiles. This measurement in turn may test a predictive theory about Orange tiles and their behavior, modulo the ability of the sensor to tell us about it (which itself embodies another theory, one in experimental design). In 6 levels we're up to physics and the whole bit, including philosophy of science, yet our world is a small array full of displacing tiles.

From a small blob of Blue and Green tiles to theoretical physics—what should we make of the above chain of abstractions? It demonstrates the heart of Computational Metabolism: the idea of constructing multi-layered dynamical descriptions, without marionette control, in exquisite 100% local representation, complete with plurifunctionality,[11] stochasticism and chaos—each entirely user-tunable, where memory, processing, control and display are one.

The reality of accomplishing the above in a computationally painless way still eludes us, although years of research have shown it to arise spontaneously not just in biology but also in mathematics.[18] However, succeeding in this endeavor is of importance to many fields of study and engineering, particularly artificial life.

WHY BOTHER WITH COMPUTATIONAL METABOLISM

Computational Metabolism was invented with dynamical systems in mind. One design criterion is the empirical study of a class of dynamical systems, not the modeling of its behavior via differential equations or statistical means. Another concept figuring prominently in the ComMet design is the strict adherence to locality of computation for the sake of unlimited scalability and ease of formal description. Most important, however, is my firm conviction that fluid-like local computation expressed by the very boundary conditions defining it offers an intriguing laboratory capable of sustaining intricate behavior such as metaphorizing or setting up artificial biochemistries.

More basically, ComMet is a part of our quest for the answer to the question: What is computation? ComMet tries to endow computation with the adjective "molar", loosely speaking. More precisely, ComMet is a programmable digital liquid, and as such, a test bed for computing with fluids. Ultimately, ComMet extends the concept of data structure and of the synonymity of function with structure, at high enough ontological magnification—plurifunctionality.[11] ComMet postulates computational processes complex enough that they alone are their best descriptions.

Computational Metabolism is a new kind of computer, drawing from cellular automata, neural networks, Monte Carlo methods, finite state machines and statistical mechanics. Computational Metabolism is a computation where the running program is the change in the computer that executes it: formally, there is no interpreter vs. the interpreted. There is no process vs. the process control. Of course, in order to simulate ComMet on a von Neumann machine, these dichotomies are resurrected. But in principle, each tile is a processor and it is the displacement of processors that defines all computation here; their input/output and state-changing behavior is merely a means to carry out this displacement. The input to ComMet is a particular initial arrangement of tiles, and so is the output. The computation is a transition of ComMet between the input and the output configurations. Theoretically, even the inter-processor transaction of flipping does not require an interpreter to shunt around the processors; it may be carried out via negotiation. Incidentally, it has been noticed that the two-dimensional ComMet lends itself to fast implementations in semiconductor hardware as smart memory.

The description of this computation is a recipe for emergence, a specification of an initial state that will undergo possibly stochastically constrained transformations. It is the programmer's job to control the amount of randomness or chaos permitted in the computation and thus vary the "algorithmicity" quotient of the specification. It may be perfectly desirable to make the algorithmicity miniscule so that the computation may uncover the basins of attraction that may in turn draw the interest of the programmer who may then perturb or redesign the computation.

ComMet has very explicit physics and, in fact, one programs ComMet by modifying its physics (physics, per our earlier warning, is rules *and* configuration, for we choose to blur the distinction between the two). Its start-up geometry intimately reflects on the emergent computation. Its computation gives rise to hierarchies of events. It is perfectly possible to have a static equilibrium on a high level achieved through a considerable flux on a lower one. This makes it possible to nudge the high-level equilibrium into a different high-level state either by diffusion or by "infection" (substitution of lower-level components). This geometrical sensitivity to nudging may prove instrumental in making the fluid data structures that Douglas R. Hofstadter claims as necessary for constructing human-like perception.[9]

Incidentally, my original motivation for inventing Computational Metabolism was to set up a computational ecosystem for constructing a viable model of metaphor.[13,16,19] My metaphor for metaphors, unlike most computational, psychological, linguistic and semiotic metaphor models, demands meta-stable dynamical systems that shift amongst contextually cued attractor states, with most shifting taking place via diffusion, infection and convolution (combining separate ComMet tessellations). Simply, nothing but a "smart fluid" computation seems capable of capturing metaphor. As corollary, other uses of ComMet, either for direct emulations of fluid computations or for spatially extended statistical and chaotic computations appear ComMet bound.

But the biggest benefit yet of ComMet may well be motivating the synthesis of actual "smart liquids." These multi-state, multi-phase substances would undergo physical interactions that act out in mass ComMet's cybernetics. Viewed in this light, ComMet is a design specification for material scientists.

CLOSING REMARKS

Let me close by explicitly answering the question: Is ComMet good for modeling artificial life? Certainly the hierarchical correspondences between ComMet's features and those of ecosystems suggest some potential. But first, let me defend an exploded view of what artificial life is.

Artificial life is the study of quasi-biological systems and the construction and classification of the same. As a systems science, artificial life ought to investigate computations not yet exhibited in nature—or not yet noticed. To be in artificial life's domain, something merely needs to involve a process that may be modeled or constructed using the organism/ecosystem or metabolism metaphors. I would even include here the yet-to-be investigated proof techniques in logic that could be based on quasi-ecosystemic open-system formalisms. Similarly, I would include the study of grammar-driven adaptive processes, specifically language acquisition and metamorphosis. At the same time, I welcome the clearly biological simulations of flock behavior or wasp populations. We should think of extending those descriptions to attempt process-cooperation-driven computer operating systems and self-organizing manufacturing methods. In short, artificial life may involve entirely inanimate, metallic, even purely abstract objects, so long as the processes involved remind us of life. Seen in this light, we should anticipate artificial life in future methods of graph theory and computer science—and be ready to attempt them. To sum up the exploded view, artificial life is anything with "free wheel," because its self-organization and its environment shape it contextually and continually. The "control" metaphor need not apply. Watch the idiom "to be in control" one day become a quaint anachronism, a victim of artificial life, as was phlogiston of chemistry.

A field of artificial life so broadly construed is tailor-made for ComMet. A purely formal system, ComMet takes full advantage of its pictorial, spatial and temporal assets to model processes as transformations on a tessellation. We are free to regard such computations as corresponding to all sorts of models. We may view ComMet as a computing engine in its own right or as a self-propelled data structure in flux. We may assign significance to a specific tile's position, or to the presence or absence of a tile formation, or its displacement. We may count tile contacts or measure the flux of tiles over particular geometries. We may even substitute new tiles for old ones, thereby computing via infections.

However, even the narrowly construed view of artificial life—strictly mimicking the *in vivo* biology—makes ComMet of interest. Evolutionary behavior, though so far not incorporated on the level of single tiles for want of tile reproduction, can be engineered for assemblies of tiles. Also, autocatalytic reactions within a group can be modeled with tile colors whose rules are arranged so as to close the catalytic loop over a chain of "reactions" (say, tile-bonding encounters). We can use biological principles to design clever processing for ComMet. We can also study biological principles per se, using ComMet as a test bed. In such embeddings biological theories can become falsifiable using techniques and measures of physics, engineering and computer science. Competing biological theories could thus be tested.

Last, ComMet is but an instance of a much larger and richer family of massively parallel computers.[3,7,17,21] Fusing these formalisms with the empirical richness and holistic descriptions of biology is a tantalizing undertaking. The nascent fields of computational biology and biological computer science could use more workers.

APPENDIX: SOME MODEST (NOT YET ALIVE) EXAMPLES

The following are basic examples of ComMet behavior. These were chosen for simplicity of encoding, description, short runtime and fun content. They're clearly preliminary exercises.

HOW TO BUILD A WALL (FIGURES 2, 3, 4)

Start with three colors, Gray, Ivory and Umber. Arrange ComMet into a rectangular tessellation. Distribute the colors in equal numbers like so: put all the Umber on one side. Put all the Ivory on the other. Fill the middle with Gray.

The Gray bond to no one, flip with all, in any direction. The Ivory bond to the Umber on all four edges, and flip with each color, also on all edges. The Umber mirror the Ivory (they bond to the Ivory). When an Ivory or an Umber bonds it becomes immobilized. This implies that the encountered tile (or tiles) also become immobilized themselves.

Hoped for result: When we run this instance of ComMet we expect to see a wall arise roughly in the middle, separating an Umber-Gray mix on one side from an Ivory-Gray mix on the other. In chemical terms, we wish to create an impermeable membrane between two substances in a common solvent.

Observed result: as hoped for (Figures 2, 3, 4). If the Gray are not allowed to flip amongst themselves, the wall arises much sooner.

HOW TO INVENT GRAVITY (FIGURES 5, 6, 7)

Start with three colors, Gray, Blue and Green. Arrange ComMet into a rectangle. Distribute the colors like so in number: 80% Gray, 12% Blue, 8% Green. Distribute the colors isotropically in space (Figure 5).

The Gray flip with the Blue and the Green in all directions. The Gray bond to no one. The Blue bond to other Blue, and to the Green. The Blue flip with Gray in all directions. When a Blue bonds to another Blue, it becomes flipless (stationary). However, when a Blue bonds to a Green, it ceases to flip on all edges except the bonding edge (it wants to go in the direction of contact). A bonding Blue ceases to bond on all edges except the opposite one to the contact edge. The Green bond only to the Blue, but they flip with Gray and Green in all directions. When a Green bonds, it shuts down flipping everywhere save on the opposite edge of its bond (it wants to runaway from the contact edge). Green's bonding templates do not change.

Hoped for result: When we start this instance of ComMet with the isotropic distribution of the three colors, the Blue-Blue bonds should produce clumps of Blue, displacing Gray. The Blue-Green bonds should produce Blue-Green-Blue sandwiches because Blue-Green bonds are meant to produce chasing Blue-Green

FIGURE 2   Computational Metabolism. How to build a wall. Clock = 0

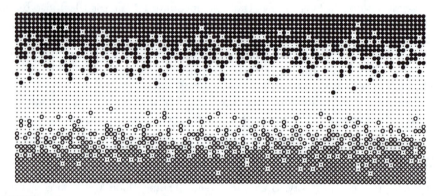

FIGURE 3   Computational Metabolism. How to build a wall. Clock = 16

FIGURE 4   Computational Metabolism. How to build a wall. Clock = 220

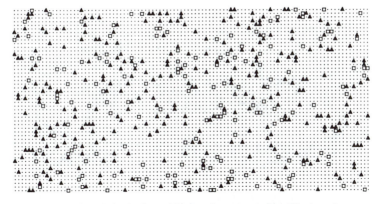

FIGURE 5   Computational Metabolism. How to invent gravity. Clock = 0

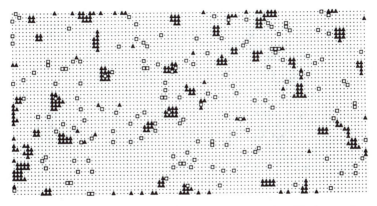

FIGURE 6   Computational Metabolism. How to invent gravity. Clock = 673

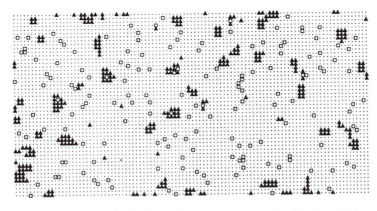

FIGURE 7   Computational Metabolism. How to invent gravity. Clock = 787

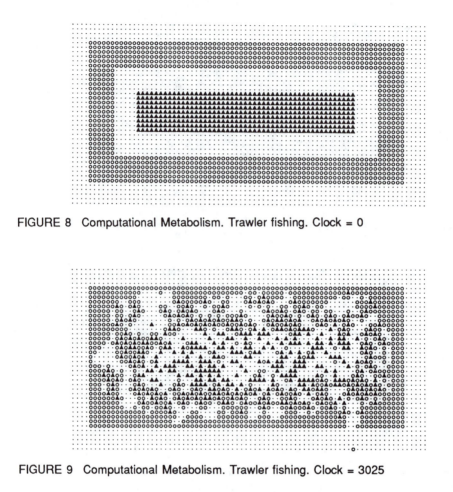

FIGURE 8   Computational Metabolism. Trawler fishing. Clock = 0

FIGURE 9   Computational Metabolism. Trawler fishing. Clock = 3025

FIGURE 10   Computational Metabolism. Better trawler fishing. Clock = 0

pairs likely to run into another Blue. In the limit, we should see large clumps of Blue, with some Gray and Green trapped inside, surrounded by a sea of Gray, with free Green deflecting from encountered Blue. Some short-lasting Green twosomes should form epiphenomenally whenever oppositely minded Green meet.

Observed result: indeed, clumps of Blue develop, albeit not particularly large clumps. A surprising number of Green remains free. Some predicted Green twosomes were observed (Figures 6 & 7). Some Green bumped into congregating Blue and rebiased them. While this changes Blue clump layouts slightly, it does not have much bearing on the stability of those clumps. Note the Blue wedged at the perimeter of the tessellation. They were biased by the Green and ran out of room to run. This is a very pretty example of a simple ComMet reducing the entropy of the initial disordered tessellation (Figures 5, 6 & 7). It took only three passes to get the rules and the color ratios right.

TRAWLER FISHING (FIGURES 8 THROUGH 13)

Here we use Red, White and Blue. The Blue form a block surrounded by a thick ring of Red (Figure 8). The Red bond-recognize Blue on all edges. The Red are immobile at first, but once a Red recognizes a Blue whenever the Blue diffusion reaches it, the bonding Red attempts to follow its contact Blue by shutting off the flip template on the opposite edge. Same applies to the Reds' bond templates. The Red and the Gray do not flip amongst themselves. Blue do.

Hoped for result: actually, this is an open question. Will this configuration produce a "trawling" effect in which Red chase the Blue without allowing many to escape outside the initial Red perimeter? How effective a "net effect" will it be? Will many Gray be trapped with the Blue tiles?

Observed result: the "net" precipitated some of the Blue out of the sea of Gray, yielding curiously layered laminar clumps of "Red-Blue-Red-Blue," with an occasional "Red-Blue-Blue-Red" defect (Figure 9). This heterostructure forms an atol around a lagoon of free Blue, into which some Red have managed to seep in. This outcome is most unexpected, and not particularly net-like. Other Blue, the successful escapees, show up pinned by the Red to the edge of the tiling. This is an artifact of the presence of ComMet boundaries—one highlighting ComMet's geometric programmability for computation as animation. Most Blue do stay within the original Red perimeter, albeit that depends on the number of holes in the net, which varies with each run. Altogether, the resulting dynamic is disappointingly far removed from the expectation. But the disappointment of seeing most Red fail to pursue the Blue can be remedied by adding a new color, Green. Coat the initial block of blue with a Green perimeter (Figure 10). The Green are set up to recognize the Blue and bolt in the opposite direction (Figure 11). Then, the outward-bound Green come in contact with the Red, which now bond to them and react by streaming in the direction of contact (Figure 12) which most often is the diffusing school of Blue. Naturally some Red bond-recognize a Green tile

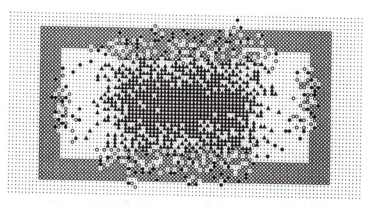

FIGURE 11   Computational Metabolism. Better trawler fishing. Clock = 29

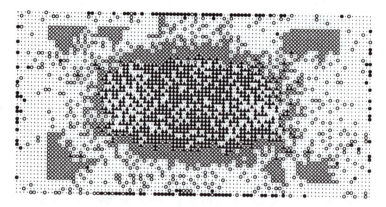

FIGURE 12   Computational Metabolism. Better trawler fishing. Clock = 130

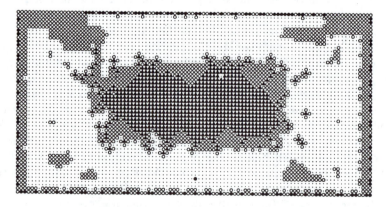

FIGURE 13   Computational Metabolism. Better trawler fishing. Clock = 8001

"obliquely". These Red tiles move away to the side and are effectively prevented from contributing to the net. They end up littering the perimeter of the tiling. I think of the Green as the Blue echo that Red pick up and act upon. The long-term outcome (taken from a re-run, actually): an ever more perfectly compressed mass of Blue, and a perfect seal (Figures 13). Note the sole wandering Green in the lower middle. It alone amongst the Green retains the initial Green behavior: same as Gray. This came about because as a corner tile, it never got a chance to face a Blue. The thick dusting of rebiased Red that lines the edge of the tiling is mostly attributable to this Green's random walk.

THE N-DIMENSIONAL DUTCH FLAG SOLUTION (FIGURES 14, 15, 16)

This is after Dijkstra's celebrated sort on a one-dimensional array of pebbles.[5] Pebbles come in one of three colors. My pebbles, or tiles, do as well. Assume a rectangular tiling with Red, White and Blue. No bonding templates whatsoever. All tiles flip against any tile. The color distinction reflects flipping biases in the following way. White flip isometrically. Red do not flip towards the bottom and Blue do not flip towards the top.

Start the machine with an equal number of tiles for each color, preferably in a rectangular tiling whose width is a divisor of a color's number of tiles (for the optimal flag effect). The distribution does not matter, though the most convincing starting formations are the upside down version of what you want to obtain or a thoroughly random mix (Figure 14).

The interesting fact about solving the Dutch flag problem by means of ComMet is that the generalized problem (to three or more dimensions) is trivial: each extra dimension results in a new pair of edges per dimension, both set to flip with any tile. In fact, the restricted problem (to one dimension) results in subtraction of all those edges. So the complexity of specifying the sorting algorithm is constant, merely necessitating the addition/reduction of constant-valued subscripts.

ComMet may not be the fastest way to sort a roomful of red, white and blue cubes, but when the cubes sort themselves, it sure is hard to beat just sitting back and watching. Programming them to do so wasn't exactly labor, either. One could say that we traded algorithm cleverness for massively distributed processing where data structures arise naturally as patterns of processor connectivity. This sounds like a neural network, and it is, except for the lack of distinction between nodes and connections. Instead, it is a swimming sea of nodes, each liable to effect a change in another's state merely by wandering by.

Observed result: getting more Oranje by the minute (Figures 15, 16). Note: in ComMet, the n-dimensional Dutch Flag problem reduces to the generalized Dobosz Tort lemma. The construction is left to the reader.

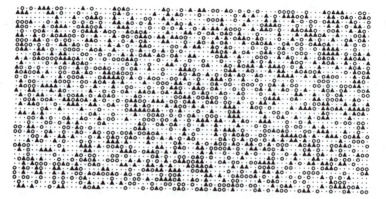

FIGURE 14   Computational Metabolism. Dutch flag. Clock = 0

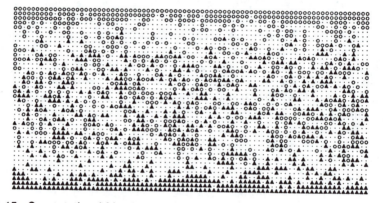

FIGURE 15   Computational Metabolism. Dutch flag. Clock = 124

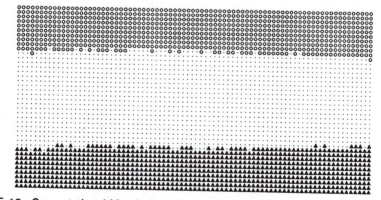

FIGURE 16   Computational Metabolism. Dutch flag. Clock = 805

ABACUS SORT (FIGURES 17, 18, 19)

This computation was suggested by Tony McCaffrey at Indiana University. Picture an abacus with red beads. Line up the beads as columns representing magnitudes, the lowest bead always on the lowest wire, and no gaps within a column. For example, a five would be represented by beads on wires one, two, three, four and five. Represent a zero with an absence of column. Now, how do you sort your list of numbers?

You slam the beads against the frame. Watch them line up sorted (the zero, of course, at the tail end). Interestingly, most of the time numbers do not reconstitute themselves with their original component beads. But no information was lost, thanks to the wires of the abacus and the uninterrupted column representation, both combining to perform an information-preserving, component-scrambling, meaningless intermediate-state, parallel sort.

To do this in ComMet, we need Red tiles to play the role of beads and White tiles to function as air, or the absence of bead. Allow no bonding. Shut down all the flip templates, too, except for permitting the Red tiles to flip on one horizontal edge with the White. For White, set up a mirror-image arrangement. White flip with Red. Red flip with White.

Arrange ComMet in a rectangle and encode numbers as large as the height of your tiling (Figure 17). Now start the machine and watch the tiles migrate (Figure 18) into a sorted list (Figure 19). Note that you're able to use the partially sorted list while the sorting is still in progress for the larger values. Although as parallel sorts go, this is a rather slow one; one can use it as a look-up table, because it has potential to encode much more information than just a list of numbers.

This is an example of a highly algorithmic ComMet instance. In fact, every intermediate state of the machine is replicated on successive reruns—perfect determinism, which is nice to see, once in a while, from a stochastic device.

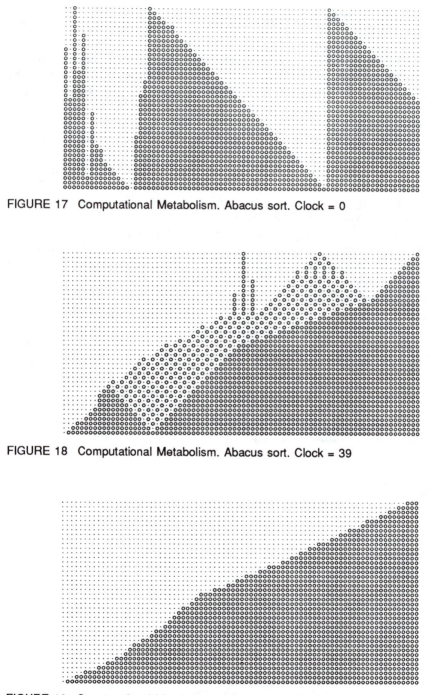

FIGURE 17 Computational Metabolism. Abacus sort. Clock = 0

FIGURE 18 Computational Metabolism. Abacus sort. Clock = 39

FIGURE 19 Computational Metabolism. Abacus sort. Clock = 75

ACKNOWLEDGMENTS

This work was conceived as my dissertation topic in the fall of 1985.[16] I would like to thank my advisor at the time, John Barnden, and the other faculty members at Indiana University who have expressed interest and encouragement, including Dirk Van Gucht, Jim Burns and Steve Johnson. Tony McCaffrey, Anand Deshpande, Marlies Gerber and Elma Sabo have contributed in various ways to the ongoing effort. Douglas Hofstadter's teaching and friendship and my old fascination with John Conway's Life made it possible that I'd end up thinking of something such as ComMet. I owe Geoff Hinton and Dave Touretzky at Carnegie-Mellon University as well as Terry Sejnowski at Johns Hopkins gratitude for permitting me to attend the 1986 CMU Summer School: Connectionist Models on the merit of this idea. There, kind encouragement from Rik Belew, Richard Durbin, Timothy Harrison, Bartlett Mel, Mike Malloch, Katie O'Hara, Jordan Pollack and Paul Smolensky, among others, has kept me thinking about Computational Metabolism. I would like to thank all my co-participants of the 1st Artificial Life Workshop, Los Alamos, for inspiration and for the confirmation of ComMet's aspirations, as well as for all the new ideas. All of us owe Chris Langton a great debt for having the good sense and perseverance to bring us all together, and then to put up with us and our contributions.

In Dallas, Jini Fuller, Mike Gately, Bill Frensley, Gary Frazier, Kendra Hoppie Penny Romell and Derek Smith helped me prepare this paper. Derek Smith not only helped prepare, but actually took ComMet to heart, offering crucial insights and tirelessly interceding on behalf of the abused reader. If this paper is at all readable, it's mostly Derek's fault.

REFERENCES

1. Bennett, C. H., and R. Landauer (1985), "The Fundamental Physical Limits of Computation," *Scientific American*, July 1985.
2. Cairns-Smith, A. G. (1985), "The First Organisms," *Scientific American*, June 1985.
3. Churchland, P. S. (1986), *Neurophilosophy: Towards a Unified Science of the Mind/Brain* (Cambridge, MA: MIT Press/Bradford Books).
4. Dawkins, R. (1986), *The Blind Watchmaker* (New York: W. W. Norton).
5. Dijkstra, E. (1976), *A Discipline of Programming*, (Englewood Cliffs, NJ: Prentice-Hall).
6. Findler, N. (1979), *Associative Networks: Representation and Use of Knowledge by Computers* (New York: Academic Press).
7. Hinton, G., and J. A. Anderson, eds. (1981), *Parallel Models of Associative Memory* (Hillsdale, NJ: Lawrence Earlbaum and Associates)

8. Hofstadter, D. R. (1982) "Who Shoves Whom Around Inside the Careenium?," *Indiana University Computer Science Technical Report No. 130*; also in Hofstadter, D. R. (1985), *Metamagical Themas* (New York: Basic Books).

9. Hofstadter, D. R. (1985), *Metamagical Themas: Questing for the Essence of Mind and Pattern* (New York: Basic Books).

10. Hofstadter, D. R. (1979), *Gödel, Esher, Bach: An Eternal Golden Braid* (New York: Basic Books).

11. Holenstein, E. (1985), "Natural and Artificial Intelligence", in *Descriptions*, Eds. D. Ihde and H. Silverman (Albany, NY: SUNY Press).

12. Innis, R., ed. (1985), *Semiotics: An Introductory Anthology* (Bloomington, IN: Indiana University Press)

13. Lakoff, G. (1987), *Women, Fire and Dangerous Things: What Categories Reveal about the Mind* (Chicago: University of Chicago Press).

14. Lem, S. (1974) *Summa Technologiae* (in Polish) (Kraków, Poland: Wydawnictwo Literackie).

15. Lugowski, M. (1986) "Computational Metabolism," *Indiana University Computer Science Technical Report No. 200.*

16. Lugowski, M. (1986) "Metaphor As Computational Metabolism," *Indiana University Computer Science, Dissertation Proposal, March 1986.*

17. Preston, K., and M. Duff (1985), *Modern Cellular Automata: Theory and Applications* (New York: Plenum Press).

18. Poundstone, W. (1985) *The Recursive Universe* (New York: William Morrow).

19. Ricoeur, P. (1975) *The Rule of Metaphor: Multi-Disciplinary Studies of the Creation of Meaning*, trans. R. Czerny (Toronto: University of Toronto Press)

20. Sloane, N. J. A. (1984) "The Packing of Spheres" *Scientific American,* January 1984.

21. Toffoli, T., and N. Margolus (1987), *Cellular Automata Machines* (Cambridge, MA: MIT Press/Bradford Books).

Harold C. Morris
University of British Columbia, 23147 - 20th Avenue South, Des Moines, WA 98198

Typogenetics:
A Logic for Artificial Life

This paper develops and explicates a formal system, Typogenetics, first proposed by Hofstadter,[1] intended for the investigation of the propagation of artificial, symbolic entities that are realized at multiple levels of logical description. In Typogenetics a string ("strand") codes for a sequence of operations that work to transform that very strand (qua operand) into successor strands. The structure of such strands and the procedure by which descendants are generated is thoroughly explained and illustrated. Various interesting types of strands are then instanced and classified according to their propagative propensities. Among these are self-replicators and self-perpetuators, and a strand with infinitely many descendants different from one another and their progenitor.

Meta-logical proofs *about* the system deal with, among other things, the implications of a Russellian-like paradox.

To reach a diverse audience the exposition has been kept relatively informal, and presupposes almost nothing in the way of specialized knowledge of biology or formal logic. The thrust of the article has been in keeping with the expressed view that the next generation of Artificial Life workers will build, not on the artifacts we have constructed with our tools of thought,

Artificial Life, SFI Studies in the Sciences of Complexity,
Ed. C. Langton, Addison-Wesley Publishing Company, 1988

but on those generalizable insights of *approach* we have extracted and made accessible through our researches.

INTRODUCTION

A new science of the artificial is emerging, dedicated to the invention and study of *Artificial Life* ("AL"), i.e., entities that simulate, emulate, or instantiate structures and processes characteristic of life forms. One of the main characteristics of life forms is of course their capacity for propagation. In this paper a unique logic for the study of some varieties of propagation is presented.

Typogenetics was first introduced by Hofstadter,[1] in connection with his discussion of the "tangled hierarchy" of DNA's replicative processes. A DNA strand contains, among other things, instructions prescribing the production of enzymes capable of operating on (destroying, repairing, copying, converting, purging, manufacturing, etc.) any of the contents of the cell—including the DNA strand. In fact, a part of the information in the DNA base sequence prescribes the synthesis of the enzymes (principally, polymerase) that make a copy of the DNA strand itself, in preparation for mitotic self-replication.

Hofstadter saw this logic of the living state, whereby DNA instigates the formation of operators that will, in the natural course, treat that very DNA source as an operand, as analogous to a formal system with axioms that generate theorems that can come back and operate on the axioms. Typogenetics was the result of his efforts to create a formal system fulfilling the analogy. With it, one was supposed to be able to invent artificial propagative entities reminiscent of DNA strands, though obedient to their own "natural laws" and possessing their own special potentialities. It was an inspired notion.

Typogenetics, however, as presented by Hofstadter, was "ill-formed"—necessary rules and definitions were lacking—and its author hastened on to other topics in his book without deriving a single interesting result in the system.

In this article a developed version of Typogenetics is offered, with exhibition of some of the results that can be obtained with it. In the final sections the relevance of Typogenetics to the AL researcher will be discussed. The exposition will be relatively informal (a fuller and more formal presentation is in Morris[4]).

TYPOGENETICS

A Typogenetics string, or *strand*, has a double aspect: it is a coded message prescribing operations, and it is the very operand or data those operations will work on (see Figure 1). A typical strand codes for operations that will result in its transformation into one or more *daughter* strands different from their progenitor—that being the special sense of "propagation" involved here.

In introducing his system Hofstadter challenged readers to create a strand coding for a sequence of operations that would yield two or more same generation descendents identical to their progenitor. An example of that kind of strand, among other interesting varieties, will be presented below after the system has been more fully explained.

STRANDS AND THEIR CONSTITUTION

A *strand* is composed of *units*. Each unit is of a *base* type, symbolized by one of the four letters A, C, G, or T. The base type of an individual unit is always known—there being no variables in this system—so there is no need for a special symbol for the unit qua individual atom, its base denotation and position in the string sufficing to distinguish it from other units. A strand is any permutation of

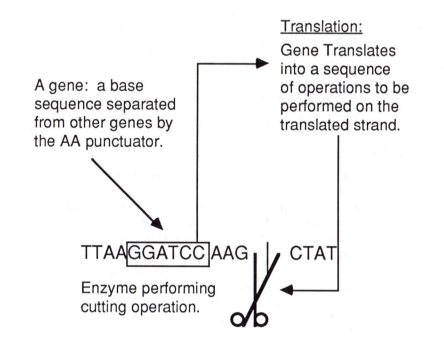

FIGURE 1 A Typogenetics strand translates into a sequence of operations; then it serves as the operand upon which those operations are performed.

the letters A, C, G or T, with the restriction that, to be well-formed, it must have at least one duplet (defined below) other than AA.

The bases themselves have fixed higher order predicates: A and G are *"purines"*; C and T are *"pyrimidines."* The bases also have a fixed relationship of *complementarity*: A is complementary to T, C to G.

HORIZONTAL AND VERTICAL UNIT RELATIONS

A strand may temporarily become a "double-decker" during its growth and transformation. In such a temporary double strand, a unit may be in a vertical relationship with another unit (i.e., a bottom row unit of a double decker may have a unit above it). Units in a vertical relationship must be complementary to one another. For example, an A base unit can be in a vertical relationship only with a T base unit.

Pairing from the left horizontally, units constitute *duplets*. For example, ACTAT is a strand consisting of the duplets AC, TA, and the left over bit "T"—part of the strand but not in a duplet relationship.

GENES TRANSLATE INTO ENZYMES

Every duplet save AA codes for an "amino acid operation," i.e., an operation to shorten or add to the strand, or relocate the current worksite on the strand. The Typogenetics user *translates* a duplet into a corresponding operation by looking up the appropriate heading in the Translation Table (Table 1). For example, AC translates as the cut command. There are 16 possible duplets, 15 of which translate as operations, one of which, AA, translates as a punctuator.

The left-right order of the duplets in the strand corresponds to the order in which operations are executed; the operation corresponding to the leftmost duplet is the *first* to be executed.

The first "gene" of a strand begins with the leftmost non-AA duplet and extends until an AA duplet or the strand's end is reached. If there is more than one gene, then each next gene begins with the next non-AA duplet and continues to the next AA duplet or strand's end. AA, then, which does *not* code for an operation and is *not* part of a gene, serves as the punctuator, or boundary, between the genes of a multi-gene strand. For example, AACTGTAACC consists of two genes: CTGT and CC. TAAT, though, is a single gene strand; duplets are paired from the left, so TAAT's adjacent AA units are not read as a duplet. A single duplet suffices as a gene; a leftover bit, as the T at the end of GGAAT, is not a gene.

Since each duplet codes for an amino acid embodying a specific operation, there corresponds to a gene a series of operations. This series of operations constitutes an "enzyme" program. An enzyme can be visualized as a robot arm operating on the

TABLE 1 Duplet/Amino Acid Operation Translation Table

Duplet	Command	Operation Described
AC s	cut	cut the strand to the right of the present unit; through both levels is double strand.
AG s	del	Delete this unit, then move one unit right.
AT r	swi	Switch enzyme to unit (if any) in vertical relationship with present unit.
CA s	mvr	Enzyme moves one unit right.
CC s	mvl	Enzyme moves one unit left.
CG r	cop	Turn on copy mode. Until turned off or detached, enzyme produces complementary units vertical to all units it touches or inserts.
CT l	off	Turn off copy mode.
GA s	ina	Insert A to right of this unit.
GC r	inc	Insert C to right of this unit.
GG r	ing	Insert G to right of this unit.
GT l	int	Insert T to right of this unit.
TA r	rpy	Find nearest pyrimidine (C or T) to right.
TC l	rpu	Find nearest purine (A or G) to right.
TG l	lpy	Find nearest pyrimidine (C or T) to left.
TT l	lpu	Find nearest purine (A or G) to left.

strand, carrying out the commands its corresponding gene codes for. An enzyme's present position (point of attachment to the strand) is noted typographically by reducing that unit's base to lower case, e.g., CTTaT. Order of enzyme activation follows the left-right order of genes; the enzyme corresponding to the leftmost gene is activated first.

ENZYME ATTACHMENT

Procedurally, once a strand is posed, e.g., GCCACT, it is analyzed into its genes—here there is only one, consisting of the duplets GC, CA, and CT—and the duplets translate into amino acid operations that may transform the strand. The first duplet of GCCACT codes for the command to insert a C base unit. But it now should be apparent that the point of attachment of the enzyme is a key consideration, because it will insert the unit at precisely that point (that is, to the right of the unit to which it is bound). Each enzyme has a "tertiary structure" which points to a *binding preference*, for choosing for that enzyme an *initial* binding site, a place to begin work.

Binding preference is determined by the relative orientation (in the tertiary structure) of the *first* and *last* amino acids of the enzyme (which could be the same, in a one duplet gene). This determination is really not as difficult to do as to explain—though a mite cumbersome in the case of a longer gene. Although the understanding to be gained by going through the procedure on one's own is invaluable, the convenience of automation for those longer genes or multiple-gene strands is overwhelming. Code for a BASIC program for personal computers which accepts any input strand (to length 255 characters), isolates the strand's genes, and for each determines and displays the corresponding enzyme's binding preference, is printed in Appendix I.

On the right of a duplet's entry in Table 1 is displayed the *folding inclination* for that duplet's amino acid, indicated with a letter ("l" for left-turning, "r" for right, or "s" for straight). Once all folding inclinations are known the tertiary structure can be graphically represented two-dimensionally, as in Figure 2.

Notice that the first arrow in this graphing of the tertiary structure is always pointed right by convention. This does not mean that the first amino acid's folding inclination is ignored, however. The first amino acid's inclination determines which subtable of Table 2 is used. In any case, once started to the right the tertiary structure may turn left or right, continue straight for a ways, then turn again, etc. A typical enzyme of many amino acids will fold back and over itself in a complex fashion. Ultimately the last arrow points in one of the four directions. To determine

TABLE 2 Tertiary Structure/Binding Preference Table

1st Amino Acid Inclination	Last Amino Acid	Binding Preference
1. Straight	$\Rightarrow$	A
	$\Uparrow$	C
	$\Downarrow$	G
	$\Leftarrow$	T
2. Left	$\Downarrow$	A
	$\Rightarrow$	C
	$\Leftarrow$	G
	$\Uparrow$	T
3. Right	$\Uparrow$	A
	$\Leftarrow$	C
	$\Rightarrow$	G
	$\Downarrow$	T

$$(1) \qquad \text{cop} \Rightarrow \text{cut} \Rightarrow \text{mvr} \Rightarrow$$

$$\Uparrow$$

$$(2) \qquad \text{cop} \Rightarrow \text{swi} \qquad \text{mvl}$$

$$\Downarrow \qquad \Downarrow$$

$$\text{lpu} \Rightarrow \text{lpu}$$

FIGURE 2   Enzyme tertiary structures of strand CGACCCAACGATTTTTCAT's two genes, CGACCC and CGATTTTTCAT.

binding preference, one then goes to Table 3, finds the right subtable on the basis of the first amino acid's folding inclination, then locates in that subtable an arrow whose direction corresponds to that of the final arrow of that enzyme's tertiary structure. That tabular entry will have associated with it a base—which is the preferred base for the enzyme.

ELECTION OF BINDING SITE

The binding preference is stated as a base. Where more than one unit in the strand is of that base type, an election must be made as to where first to attach the enzyme; in this important respect, Typogenetics is *non-deterministic*.

For example, in the strand ACAAACAA both enzymes have a binding preference for the base A—and there are no less than six different units of that base type. An enzyme must attach to one—and only one—of those units to effect its cut command. Since different choices of binding site will very frequently lead to different results, much of the "gamesmanship" of Typogenetics is in making a fruitful selection (in the present case, the right choices will result in ACAAACAA cutting itself into identical twin daughters). A preferred binding site on a strand can be indicated by reducing its unit to lower case, e.g., GAGa. Such elections are comparable to the conditional assumptions one makes in natural deduction logics. Once the enzyme is attached, it remains there unless and until relocation is explicitly dictated by a command.

OPERATIONS

There are 15 operations, summarily stated in Table 1.

1.   *Relocation* operations are of three sorts:

    a.   *Searches* send the enzyme down the strand right or left to find a unit of the specified base type (pyrimidine or purine). There is no limit to the distance it can travel in this search.

    b.   *Moves* have the enzyme move horizontally exactly one unit.

    c.   *Switching* is the only mode of vertical relocation; a switch attaches the enzyme to the other strand of a double strand; hence two consecutive switches would return the enzyme to its starting place.

2.   *Effective* operations diminish or augment the strand:

    a.   The *insertion* commands have the enzyme insert exactly one new unit of a specified base into the strand immediately to the right of the enzyme's point of attachment. The enzyme "makes room" even in a double strand. In the following example a T will be inserted:

AGTcCGT   ACTcTCGT

    b.   *Deletion* commands are the reciprocal of insertion, except that the one unit deleted need not be of any particular base type, and is the very unit the enzyme is currently attached to. After deleting the unit the strand (at that level) rejoins, the enzyme attaching itself to the unit to the right.

    c.   *Cutting* separates the strand into two parts. The part "cut off" is that lying immediately right of the enzyme's point of attachment; it is from that moment a daughter of the original strand and not subject to any further operations of this or of any subsequent enzyme's operations. The part of the strand the enzyme has remained attached to (which might conceivably be only one unit long, if the enzyme performed its cut while attached to the leftmost unit) is now the operand for any and all further operations.

    d.   *Copying.* When copy mode is turned on, the immediate result is the vertical insertion of a *complement* of the base of the unit the enzyme is presently attached to. Copy mode remains on until the enzyme detaches, exhausts its run of operations, or the copy off command is given (whichever comes first). When it is on, the enzyme doubles the strand (in that peculiar fashion of manufacturing *complements*) whenever and wherever it moves, with this exception: if commanded to conduct a search that will take it clear off the strand the enzyme succeeds merely in detaching itself (see next section) if it cannot find the pyrimidine or purine that is the object of its search (no copying is done).

QUALIFICATIONS ON OPERATIONS
DETACHMENT

In some circumstances operations will result in detachment of the enzyme; subsequent commands supposed to be carried out by that enzyme thus cannot be executed and the next enzyme is activated. This happens when the enzyme is commanded to move or search off the end of the strand; when it is to switch to a nonexistent second level strand; and when it moves off the end of a strand pursuant to the second part of a delete command. Arranging for timely detachment of an enzyme is yet another important aspect of Typogenetics design strategy.

BENIGN IMPOSSIBILITY

Benign impossibility occurs when the enzyme is commanded to cut, copy, or turn off copy mode when it cannot. (If the enzyme is already on the far right-hand end base it can't cut the strand to the right of the present unit; if the copy mode is already on, it can't be turned on again; and when off, can't be turned off again). In these circumstances, the enzyme simply ignores the command and proceeds to execute the next operation in the amino acid sequence.

OPERATIONS ON DOUBLE STRANDS

Units of an upper level of a double strand are recorded upside-down (see illustrative derivation under "procedure illustrated" section). When the enzyme switches to that upper level, its sense of right and left will be in reverse to right and left on the lower level. Formally, for all units w, x, y, and z, if x and y are vertically related, and w is left of x and z is vertical to w, then z is right of y. An enzyme switching from x to y and then moving right will attach to z.

GAPS IN STRANDS

If strand segments remain separated by a gap after all enzymes translated out of the original strand are exhausted, then the segments are permanently separated. Double strands, either through deletion or intermittent copying, may have a gap at one but not both levels; such a gap affords no bases onto which an enzyme may be bound, hence presents an obstacle to switching and moving (unless and until plugged). However, such a gap can be traversed by a searching enzyme (the enzyme requires only something to "slide along" when searching, and the intact level of the double strand provides this). Just as an enzyme can traverse an indefinite number of units in a search across an intact strand, it can span an indefinitely wide gap so long as the complementary strand upon which it is sliding is continuously intact.

An enzyme traversing a gap while in copy mode copies the bases of the intact level of the double strand it passes over (doubles the strand as it moves horizontally, filling the gap it passes through with the complements of the bases above or below the gap).

DAUGHTERS

One or more strands (or, at bare minimum, a single bit) will be left over after all operations have been carried out by all enzymes. If a double strand has resulted, it must separate at this point so that the upper and lower levels are distinct strands. Strands left after all transformations have occurred, including segments cut off by an enzyme, are the *daughters* of the original strand.

Each daughter can be translated and transformed to produce daughters of its own (granddaughters to the original strand). Typogenetics can be carried on literally for generations!

TABLE 3 Steps in Strand Formation

Stage	Strand Transform	Operation
1	CGACCCAACgATTTTTCAT	1st enzyme attached
2	CGACCCAACgATTTTTCAT	Copy on
3	CGACCCAACg ATTTTTCAT	Cut
4	CGACCCAAcG	Move left
5	CGaCCCAACG	2nd enzyme attached
6	CGaCCCAACG	Copy on
7	CGACCCAACG	Switch
8	CGACCCAACG	Find first purine left
9	CGACCCAACG	Find first purine left

PROCEDURE ILLUSTRATED

Derivation of the descendants of a Typogenetics strand is given below in a format resembling a derivation in a standard logical system. This strand was designed to serve as a didactic vehicle and is of no special interest beyond that capacity.

Notice that the point of enzyme attachment on the current transform is always indicated by reducing that unit's base to lower case (see Table 3).

The strand:	CGACCCAACGATTTTTCAT
Translation:	1st Gene Duplets: CG AC CC
Amino Acid Operations:	cop cut mvl
Folding Inclinations:	r s s (See Figure 2)
Binding Preference:	G
2nd Gene Duplets:	CG AT TT TT CA T
Amino Acid Operations:	cop swi lpu lpu mvr
Folding Inclinations:	r r l l s (See Figure 2)
Binding Preference:	A

At step 9 of Table 3, the enzyme moved off the strand; no further commands can be carried out. There being no further enzymes to be activated, strand transformation is completed. The double strand separates into the two strands: CGTTGGGT and CGACCCAACG.

The fragment of double-strand ATTTTTCAT cut off from the rest of the strand at stage 3 is also a daughter.

There are thus three daughters of the strand, given the binding site choices made above; as they are well-formed strands, for each of them too descendants could be derived.

TYPES OF STRANDS PRODUCIBLE

The following represents a classification of strands producible in Typogenetics, with examples usually chosen for their simplicity. Since, for most of these strands, a particular binding site of the several available must be elected for the strand to demonstrate its peculiar capabilities, the lower-case letter convention has been employed to designate which is the appropriate unit to be elected.

Two types of *sterile* strands are incapable of any reproductive activity:

1. *Duds* never get off the ground because they do not offer their enzyme(s) any binding site to begin work. An example: GGTT (Binding preference: A).

2. Enzymes of *trivial* strands shuttle about moving, searching, or switching, without ever settling down to do any real work (copy, insert, delete, or cut), e.g., TACCAT (Binding preference: T; it is trivial on either binding site election).

Monstrous strands beget descendants nonidentical to themselves, hence are not self-replicating, e.g., our strand illustrating Typogenetic derivation above, with its three daughters. These are the most commonly encountered strands.

A particularly remarkable subtype here can be called *infinitely fertile* because it has literally an infinity of descendants, each different from the parent and from each other. The simplest example is GAGAa. GAGA adds by insertion a pair of A bases to its right side. That descendant, GAGAAa adds another pair of A's to become GAGAAAAa. So for each descendant, *ad infinitum* (cascading), so long as the A chosen as the binding site in each case is the rightmost of the strand. The AA duplet does not affect binding preference, so every descendant of GAGA will have the same binding preference as the original progenitor. That binding preference is for A, the type of base that keeps being inserted onto the right side of each descendant down the strand line. These added AA duplets, of course, do not code for operations. So the two operations, the A inserters, passed on from generation to generation, are solely responsible for making each descendant a duplet longer; no descendant of GAGA has any other kind of operation (so long as the rightmost A is selected at every generation as the enzyme's initial binding site).

A *self-perpetuator's* enzyme activity alters its operand at some stage(s) in the game, only to undo whatever was done, e.g., units are deleted only to be replaced. Example: TCAGCGATGTCTAGAGCT.

Recurrors—rare creatures—come in two types:

1. A self-recurror is like a self-perpetuator, but returns to itself inter-generationally rather than intra-generationally.

2. An incidental recurror is produced at regular intervals as the by-product of a self-recurror's reproductive activity.

Example: The strand cGATGTGGCG is a self-recurror; it yields as daughters the strands GGT and AcCGATGTGGCG. The latter yields in turn (another) GGT, AACC, and (the self-recurror) CGATGTGGCG. Thus CGATGTGGCG recurs along its own line of descent, every other generation. Also recurring along that line at every generation is the strand GGT—purely as an incidental by-product (GGT does *not* recur in its *own* line of descent, and thus cannot be considered a *self*-recurror; its only daughter is the dud GGTT).

SELF-REPLICATORS

There are in principle two ways that a self-replicator can be made in Typogenetics: one type would extend itself horizontally along one level (perhaps through many generations) and then cut itself into two pieces which are either already replicas of their parent or will beget same. No such strand is known to have been created—the reader is challenged to create one.

The other type of self-replicator, an example of which has been found and which will be displayed, uses the copy function to create a double strand that will separate into two daughters that are either already replicas of their parent or will grow into same.

A simple *self-copying* strand illustrates an important consideration to be borne in creating this latter type of strand. CGTTCCCCCCa succeeds in copying it-self in its entirety. However, because copying in Typogenetics involves making a "negative," or *complement* of what is copied, the resultant double strand separates into these two strands: CGTTCCCCCCA and TGGGGGGCACG. These strands are in the Typogenetical sense copies (complements) of one another, but clearly TGGGGGGAACG is nonidentical to its parent.

There is reason then to make the original strand *self-complementary*; its copy is both complementary to *and* a replica of itself. Four of the 16 possible duplets of Typogenetics are self-complementary: AT, CG, TA, and AT. Fortunately, it is not necessary to restrict one's choice of duplets to just these special ones in assembling a self-replicator (the four commands those duplets code for may not provide an adequate repertory). Rather, one induces from such examples the generalization that any strand which is left/right complementary is self-complementary. Once this is seen, it becomes possible to conceptualize one's target strand as consisting of two parts, right and left. To one of these parts is assigned the duty of copying the *entire* strand. As for the other part, it need only be (1) the precise complement of the copying half, and (2) harmless—i.e., does not code for any commands disruptive of the overall objective.

A strand S–R meeting these requirements may now be given. This strand, with 64 units and 4 genes, generates perfect replicas through copying, replicas which of course possess their parent's capability of generating further perfect replicas. The self-replicating strand:

CGCGCGCGTAATATAACGATCGCGCGTATTAATTAATACGCGCGATCGTTATATTACGCGCGCG

1st enzyme binding preference: C

2nd enzyme binding preference: G

3rd enzyme binding preference: C

4th enzyme binding preference: A

The first enzyme, attached to the leftmost unit of S–R, simply copies the first three bases of the strand.

The second enzyme copies all the rest of S–R. It begins by binding to the rightmost unit. Copy mode is turned on and that unit (G's) complement (C) appears overhead. The next command, switch, moves the enzyme up to the top level of this forming double strand. Three following copy on commands are superfluous (the corresponding duplets were required to give the gene the correct binding preference).

Then we come to the final command of the enzyme's sequence, which is: "Find the nearest pyrimidine right." The enzyme's position, and the structure of the double strand at this point, is this:

←(Gap enzyme passes over in its search)

ᗡᗡᗡ
CGCGCGCGTAATATAACGATCGCGCGTATTAATTAATACGCGCGATCGTTATATTACGCGCGCG ᴐ

The enzyme, which is in copy mode still, doubles the strand en route to finding a pyrimidine on the far side of the gap. The two remaining genes' corresponding enzymes prove on activation to be "trivial," i.e., shuttle about harmlessly searching, switching, and superfluously turning copy mode on.

After all enzymes have performed their programs, the double strand separates into two identical twins, each of course capable in turn of bearing identical twins.

TYPOGENETICS AND THE REGRESS PARADOX OF SELF-REPLICATION

The Typogenetical self-replicator shows how a notorious regress can be avoided.

It seems that an entity with reproductive capacity must possess information descriptive of what it is to generate, as a factory's automated chemical-synthesizing system is programmed with the recipe for the compound it is to synthesize. But now an entity $S0$ which is going to replicate *itself* would seem to need information descriptive of itself; we might suppose it is equipped with a blueprint of itself. Yet its offspring $S1$, being identical to $S0$, also has the capacity to self-replicate, so $S1$ must too possess a self-descriptive blueprint. $S0$'s blueprint then *describes* a blueprint-equipped self-replicator, which means the self-replicator's blueprint describes, among other things, a blueprint. Yet it doesn't stop there, because $S1$'s blueprint must contain a description of *its* blueprint-equipped self-replicating offspring $S2$. Following this transitive line of reasoning out it appears the blueprint of $S0$ embeds an infinity of self-descriptions.

Is reproduction paradoxical, i.e., does it defy logical analysis? Can only "vitally" endowed natural organisms self-replicate? No, and no: a "trick" of self-reference turns out to be the key to solving the problem of the regress. The self-replicating Typogenetics strand is, first, a sequence of operations (a program or recipe) directing the construction of its copy from a model; it is, second, uninterpreted data forming a passive operand, the model to be copied. There is no description of the strand in the strand because the commands the strand codes for are addressed not at the construction of a design contained in a blueprint, but rather at the making of a copy of the strand itself. This solution may sound like a simple, even obvious one, and its principle has been well grasped by biologists and logically analyzed by Von Neumann; perhaps the main value of Typogenetics in this connection then is to demonstrate that it is really no mean trick to arrange it so that the uninterpreted-information-as-model side of things *squares precisely* with the interpreted-information-as-commands-to-copy-the-model side. A similar lesson can be gained from study of the comparable self-replicating entity Langton[3] designed in a cellular automaton environment.

It may seem remarkable to learn that an undirected nature arrived at the same kind of arrangement. For in natural self-replication DNA has the double character noted in the Typogenetics strand. The different uses of information, interpreted and uninterpreted, correspond respectively to the biological processes of *translation* and *transcription*. But, of course, in actuality nature here operates as it always must, in accord with the principle of least action; it has no design objectives, and hence needs no design solutions—in sharp contrast to we who would create artificial life.

META-TYPOGENETICS

Many important things cannot be demonstrated *within* the system of Typogenetics. For example, if a negative assertion is made that such and such a well-defined strand does not exist, its truth cannot be proved *in* the system short of listing every possible strand and showing that each is *not* the strand claimed not to exist. But results obtained for or about Typogenetics, using formal logic, simple algebra, and the exhibition of general formulas or algorithms with probative value, can be as important and interesting as those obtained in the system. Some representative "meta-Typogenetic" results and methods are given below.

REDUCTIO STRATEGY FOR DISPROVING A HYPOTHETICAL STRAND

A *reductio ad absurdum* strategy can take one from an initially plausible conception of a strand with certain desired structural and functional characteristics to the conclusion that the conception is self-contradictory.

The initial conception of the strand must be very detailed. The usual contradiction to be obtained contrasts the structure of the hypothetical strand as operand with its structure as implied by the operators its propagative function requires.

To illustrate a meta-Typogenetic proof using this strategy, consider whether it is possible to devise a "self-erasing" strand. A self-erasing strand would be a strand leaving no daughters or bits. In other words, after the operations that the strand coded for have been carried out, absolutely nothing is left of the strand. The following argument demonstrates that total self-erasure is impossible.

There are only two operations having the effect of diminishing a strand. The cut operation though always leaves a bit or daughter—that part which was cut off, which is beyond the reach of the enzyme to further reduce. So the hypothetical self-erasing strand must not involve cutting. That leaves the delete operation. The self-eraser must delete itself into oblivion.

AG is the duplet coding for deletion. Since the hypothetical self-erasing strand cannot diminish itself by cutting, there must be at least as many deletions carried out as there are units in the strand, and that requires as many AG duplets as there are units in the strand. But this is clearly problematical. A duplet is two units. We have said that the strand of n base length must have n AG duplets, i.e., must be at least $2n$ units long. Of course a strand cannot both be n and $2n$ units long.

In fact, the single duplet *strand* aG is the closest thing to a self-eraser possible, because it can at least leave only a single bit, G, behind. It is a non-trivial strand with no daughters.

NO MOTHERLESS CHILDREN IN TYPOGENETICS

As recognized in our classification of types, there are sterile Typogenetic strands, i.e., strands that have no propagative capacity. Such strands are the dead ends of their strandlines. But what about the converse? Are there "motherless" strands?

The answer is no. Algorithms can be made (a flowchart for one is given in Appendix II) which will take any input strand and create for it a mother strand. One such algorithm scans the target daughter strand; as each next unit's base is determined a duplet coding for the insertion of that base is added to the growing mother strand. The mother strand then possesses the insertion duplets sufficing to insert the sequence of bases constituting the daughter. No doubt a similar algorithm could do the trick using copying. When looped back on itself (previous output strand becomes current input strand), such "mother-maker" algorithms will produce an arbitrarily long line of ancestry for the original strand.

The existence of any such algorithm proves there are no motherless children in Typogenetics—no "Garden of Eden" strands, to borrow from the lexicon of cellular automata. In this respect Typogenetics is like nature—all living entities have progenitors. The existence of similar algorithms shows that, unlike natural entities, any Typogenetics strand has *more than one mother*. For example, the self-copying strand already introduced, CGTGCCCCCCa, has as its daughter TGGGGGGCACG. Now, if this daughter is run through the mother-maker algorithm, the mother outputted for it would not be CGTGCCCCCCA. And different mother-maker algorithms (e.g., the one using copying) could make different mothers for it. Any strand therefore has indefinitely many different mothers!

Equally strangely, every strand is a sibling to any other strand. This is established by an algorithm that will accept as input any finite set of strands and creates for the entire membership of that set *one mother common to all*. The simplest such algorithm merely concatenates the input strands (each strand embedded as a gene) and includes cut commands sufficient to dismember this parent into the inputted strands, its "daughters."

A PARADOXICAL STRAND

Notwithstanding the aforementioned algorithms, there are limits on how many little monsters one mother strand can bear, as we now see. First, let us lump together as "self-reppers," the types of strands previously distinguished as self-perpetuators, self-replicators, and self-recurrors, since all these non-trivial strands share the attribute of having a descendant identical to themselves. Now, consider the set of all strands that are non-self-reppers (that being the large majority of all strands). Could there be made or found for this set a single mother common to all the set's members, such that on the intended election of binding sites for its many enzymes, this strand (call it p) would have as its progeny all and only non-self-repping strands?

It seems not. The conception of such a strand is paradoxical, as can be seen by asking of it whether or not it is a self-repper. Suppose it is a self-repper, which implies that it has a descendant identical to itself, $p1$. But $p1$, being identical to p, also would be a self-repper. This descendant violates the proviso that p has *only* non-self-reppers for descendants.

It may then be assumed that p is a non-self-repper. According to the definition of p, p has *all* the non-self-reppers as its progeny. So p, here hypothesized to be a non-self-repper, must be among its own progeny. Yet that makes it a self-repper!

One is inclined to blame "infinity" here—for the number of non-self-repping strands is certainly infinite. To be sure, our algorithm above, that makes a mother common to all members of any set of input strands, could not be expected to accept an infinite set, for its mode of operation involves imbedding in the parent one gene for every input strand, which would lead to an infinitely long strand in this case— not a typographical entity. But the non-applicability of that algorithm does not dispose of the matter.

We saw earlier that there are "infinitely fertile" strands, like GAGA, capable of having infinitely many *different* descendants (that, incidentally, is proof of the assertion that there is an infinitude of non-self-reppers). So we must also conclude that either no such infinitely fertile strand *can* give rise to all the non-self-reppers, or that if it does, then it must also bear at least one self-repper—for *no* routes to paradox can be left open in the world of logic.

URSTRAND

Now consider the obverse side of the coin. There is nothing patently paradoxical about a strand—call it the *urstrand*, after Goethe's *urpflanze*, the primal plant from which all other plants stem—which manages to have all and only the self-replicators for its progeny. It may or may not exist; and it may or may not have to have infinite descendants, depending on whether there are infinitely many self-replicators. But even if there are infinitely many, the existence of infinitely fertile strands allow us to preserve our hopes for the urstrand.

Actually, there may at first appear to the reader to be a problem. A strand cannot have infinitely many *daughters*, since that would require infinitely many duplets coding for the infinitely many operations to build all those strands, and infinitely many duplets cannot be accommodated in a typographical entity. While if it has a *finite* number of self-replicating daughters, and each self-replicating daughter produces only its own kind, how do we get the *infinitely* many self-replicators? But in fact there is nothing in the definition of a self-replicator which says it cannot have descendants *different* from itself, so long as it has some that are identical. The urstrand then would have to have daughters, some of which are themselves self-replicators but having, *in addition* to descendants identical to the daughters, other descendants non-identical to the daughters but themselves self-replicators, so that all the infinitude of self-replicators can be unpacked out of the finitude of daughters (even if that takes infinitely many generations).

Does the urstrand exist? It is a pregnant question, for if it does, *its form captures the entire domain of self-replication in Typogenetics.*

Indulging some highly speculative transfer to the world of natural biology, it seems very likely a DNA strand could be created that is able to direct the manufacture of DNA strands non-identical to itself and to each other (through design, not random deviation as in mutation). Somewhat less certainly, there might some day be developed means of generating, for any set of DNA strands D, another DNA strand P capable of manufacturing every member of D. Following out this parallel to our meta-Typogenetics results, could there be a single DNA urstrand able to synthesize all possible DNA-based self-replicating forms? Such an entity for Goethe would be the *Urphanomen* of organic morphological science, and its discovery would surely constitute the ultimate achievement of the wetware side of the AL enterprise.

ARTIFICIAL LIFE IN A TYPOGENETICAL WORLD

Typogenetics perhaps offers a few analogies, hypotheses and directions for molecular biologists. But it does not shed in itself any direct light on nature. Our platonistic strands have value, if at all, chiefly for a science of the artificial.

Can Typogenetics, or some modification of it, be used to design artificial organisms? It would not be easy to find a suitable chemical or mechanical interpretation by which these purely symbolic artifacts could become physical beings. The computer, though, affords a receptive domain for physically realizing symbolic structures. One could implement Typogenetics in an electronic automaton environment. Of course, these strands do nothing but propagate. To make such an automaton more interesting, one would want to give the strands' world "laws of nature" governing them and, preferably, enabling them to "behave" in some senses. Fortunately, as one can see in the case of the self-replicating strand presented here, there is often abundant structural information to "spare" (i.e., information other than that coding for the replicative copying) which one could tap into for this purpose. With the right interpretation, strands might output signals or even become programmable entities reminiscent of Laing's[2] "molecular machines."

Strands—at present asexual beings—might mate and recombine genes. They might exploit environmental resources (including other strands) to give them the raw materials out of which units for inserting or copying daughters into existence would be drawn. With limited space or resources acting as selective pressures, evolution could occur. In furtherance of that the non-deterministic aspect of binding site selection could be rendered probabilistic, permitting stochastic "mutation" of a kind. In time, strandlines beginning with simple progenitors "planted" in the environment at time 0, might develop some very sophisticated descendants embedded in a complex ecology.

ARTIFICIAL INTELLIGENCE FOR TYPOGENETICS

Enhancements of Typogenetics promise more elaborate, lifelike artifacts. But the system *as is* certainly has great depth and complexity—and complexity is perhaps the foremost challenge facing designers of Artificial Life.

Imagine for a moment what it is we are hoping to make in the way of Artificial Life. An entity—software or hardware— as adept as the dragonfly at aerial interception would represent a tremendous technological advance, with obvious military and civilian utility. It would also be more complex than any existing artifact, including present-day computers. To produce such an entity we are probably going to have to lean heavily on our complexity-managing tool, the computer. One could put it that we will need AI to help us design AL!

A good project on which to cut our teeth in developing such Artificial Intelligence might involve designing a program capable of automatically constructing interesting Typogenetics strands (e.g., self-replicators), or the urstrand itself. Let a warning be appended here though: brute force searches, though easily devised, are not practical in this vast recursive space. As is true for DNA, for a Typogenetics strand of n units length there are 4^n permutations of the four bases possible, and nearly all of those are well-formed *strands*. That means there are, give or take many trillions, 3.403×10^{38} distinct strands equal in length to the (64 unit) self-replicator presented in this paper. Presumably no large percentage of that huge space of relatively short strands is taken up by other self-replicators; those that do exist must be sought or fabricated *intelligently*.

THE VALUE OF TYPOGENETICS

The transfer of the lessons learned from undertaking the projects described above would not be insignificant to the endeavors of future biotechnologists and *in silico* AL workers alike. That point is worth expanding on.

It is a guiding article of faith of this author that the main value of Typogenetics, Laing's molecular machines, or the like, lies not in *what issues from them*, but simply in the fact that their utilization contributes to a certain intellectual atmosphere that is, over the long term, conducive to emergence of refined methods and concepts pointing to useful applications. Let us look to (recent) history for substantiation of this position.

George Boole's *The Laws of Thought*, featuring his now famous and much used "Boolean logic," appeared (1854) before a technology adequate for construction of logic machines was in place. But, when that (electronic) technology did become available, computers and their programming languages sprang up almost overnight, because pioneers like Wiener, Turing, Simon, and Burks were steeped in the "pure" researches that had been made with the Boolean and post-Boolean symbolic logics. To men quite accustomed to formal systems, and the rigorous modes of thought that produce and manipulate them, the construction and use of computer technology at hardware and software levels came as second nature.

We are, as regards AL, in that "basic research" period, inventing and trying out various tools of thought. And I would contend—hard as this is for the result-oriented pragmatist to understand—that the most general insights we can extract from our present efforts are obtainable from focusing on *how* we establish our problem space, set and work toward our goals, measure our progress and learn from our failures, rather than on what is produced. The *artifacts* we pioneers have created with our tools of thought have less enduring value than the crudest calculators of the 18th and 19th centuries. The next generation will not build on them; they *will* build on the insights disclosed in the *approaches* we have taken. No one is apt to try to "do" anything with, e.g., the strands displayed in this paper, or Langton's self-replicator. But those who follow, through surveying the results that have been obtained, and/or by working out their own, may come via Typogenetics to an appreciation for the design challenges posed and opportunities provided by the fusing of operators and operand in propagating entities. Or they may appropriate a concept such as that of the "urstrand," and, entirely beyond the bounds of Typogenetics, make it bear fruit.

They, in sum, will advance and eventually embody the concepts and modes absorbed from our present-day, seemingly esoteric exercises, in applications that will constitute Artificial Life worthy of the name. In the meantime, we can aid the cause by taking pains to make explicit and accessible the lessons we think we have won from our efforts and envisionings. That is not always easy or even possible for us to do, of course. But if we do not make our best effort in that regard, the troubling thought that anyone working in the technical fields today experiences, to the effect that much that is relevant to one's own endeavors is probably already "in the literature," albeit latently and in some state of encryption, will seem overwhelming to those who succeed us. The bottom line is that, if we cannot bring to light insights that are deeply buried in our own results, they may effectively be lost; for others who come later will be reticent to dig for them on the strength of faint suspicions that "something worthwhile *might* be hidden there."

APPENDIX I

GENE ISOLATING AND BINDING PREFERENCES DETERMINING PROGRAM

Written in IBM BASIC by Harold C. Morris, 1988. A fully annotated version of this program is available from the author.

```
10      PRINT "ENTER ANY TYPOGENETICS STRAND. ITS GENES WILL
        BE IDENTIFIED AND, FOR EACH CORRESPONDING ENZYME, A
        BINDING PREFERENCE DETERMINED.";: PRINT
15      DIM G$ (64)
20      INPUT S$ ;: PRINT
30      LS=LEN(S$ )
40      PRINT "THIS STRAND IS ";: PRINT LS;: PRINT " UNITS
        LONG.";: PRINT
50      FOR P1=1 TO LS
60      C1$ =MID$ (S$ ,P1,1)
70      IF C1$ <>"A" AND C1$ <>"C" AND C1$ <> "G" AND C1$ <>"T"
        THEN PRINT "THIS IS NOT A STRAND; A TYPOGENETICS
        STRAND MAY CONSIST ONLY OF THE ALLOWED LETTERS
        A,C,G, OR T": END
80      NEXT P1
90      LT=INT(LS/2)*2
100     T$ =LEFT$ (S$ ,LT)
110     FOR P2=1 TO LT STEP 2
120     C$ =LEFT$ (T$ ,2)
125     LT=LEN(T$ )
130     IF C$ ="AA" THEN T$ =MID$ (T$ ,3,(LT-2)) ELSE GOTO 150
140     NEXT P2
150     FOR P=1 TO LT STEP 2
160     IF MID$ (T$ ,P,2)="AA" THEN GOTO 240
170     NEXT P
180     IF LT>1 THEN G$ =T$
190     FOR K=1 TO LT
200     K$ =MID$ (T$ ,K,1)
210     IF K$ <>"A" THEN GOTO 290
220     NEXT K
230     END
240     AA=AA+1
250     G$ =LEFT$ (T$ ,P-1)
260     W1=LT
270     W2=LEN(G$ )
280     W3=W1-W2-2
290     REM INPUT GENE
```

```
300    G=G+1: G$ (G)=G$
310    FOR I=1 TO LEN(G$ ) STEP 2
320    D$ =MID$ (G$ ,I,2)
330    IF LEN(D$ )=2 THEN GOSUB 530
340    IF LEN(D$ )=1 THEN GOTO 380
350    IF I=1 THEN GOTO 480
360    T=T+X
370    IF T=4 OR T=-4 THEN T=0
380    NEXT I
390    IF TB=1 THEN GOSUB 710
400    IF TB=2 THEN GOSUB 770
410    IF TB=3 THEN GOSUB 830
420    IF LEN(G$ (G))=0 THEN IF G>1 THEN G=G-1: GOTO 450
430    PRINT "THE BINDING PREFERENCE FOR GENE # ";: PRINT
       G;: PRINT G$ (G);:PRINT:PRINT " IS ";: PRINT BP$
440    IF AA=0 THEN END
450    IF AA>0 THEN AA=0: X=0: T=0
460    T$ =RIGHT$ (T$ ,W3)
470    GOTO 150
480    IF X=0 THEN TB=1
490    IF X=-1 THEN TB=2
500    IF X=1 THEN TB=3
510    X=0
520    GOTO 380
530    REM TABLE ASSIGNING EACH DUPLET ITS "TERTIARY VALUE".
       RIGHT TURN: 1; LEFT: -1; STRAIGHT: 0
540    IF D$ ="AA" THEN X=0
550    IF D$ ="AC" THEN X=0
560    IF D$ ="AG" THEN X=0
570    IF D$ ="AT" THEN X=1
580    IF D$ ="CA" THEN X=0
590    IF D$ ="CC" THEN X=0
600    IF D$ ="CG" THEN X=1
610    IF D$ ="CT" THEN X=-1
620    IF D$ ="GA" THEN X=0
630    IF D$ ="GC" THEN X=1
640    IF D$ ="GG" THEN X=1
650    IF D$ ="GT" THEN X=-1
660    IF D$ ="TA" THEN X=1
670    IF D$ ="TC" THEN X=-1
680    IF D$ ="TG" THEN X=-1
690    IF D$ ="TT" THEN X=-1
700    RETURN
710    REM TABLE 1 FOR BINDING PREFERENCE
```

```
720     IF T=0 THEN BP$ ="A"
730     IF T=1 OR T=-3 THEN BP$ ="G"
740     IF T=3 OR T=-1 THEN BP$ ="C"
750     IF T=2 OR T=-2 THEN BP$ ="T"
760     RETURN
770     REM TABLE 2 FOR BINDING PREFERENCE
780     IF T=0 THEN BP$ ="C"
790     IF T=1 OR T=-3 THEN BP$ ="A"
800     IF T=3 OR T=-1 THEN BP$ ="T"
810     IF T=2 OR T=-2 THEN BP$ ="G"
820     RETURN
830     REM TABLE 3 FOR BINDING PREFERENCE
840     IF T=0 THEN BP$ ="G"
850     IF T=1 OR T=-3 THEN BP$ ="T"
860     IF T=3 OR T=-1 THEN BP$ ="A"
870     IF T=2 OR T=-2 THEN BP$ ="C"
880     RETURN
```

APPENDIX II
PROCEDURE FOR CONSTRUCTING A PARENT STRAND

INPUT ANY STRAND. The input strand will be the daughter of the strand to be constructed. Call it d for "daughter."

Sample input strand: TTGAATAGC

SELECTING INSERTION COMMANDS. The parent strand p we are constructing will have, for every unit of the d strand, a duplet coding for insertion of an identical unit. Order of insertion will follow order of the units in d.

FOR the ith (next right) unit of d:

1. If ith base is

    a. A then select duplet GA

    b. C then select duplet GC

    c. G then select duplet GG

    d. T then select duplet GT

2. Concatenate the duplet selected above to the right end of the current strand (when $i = 1$ the current strand is simply the first selected duplet).

NEXT I

OUTPUT current strand, o.

For our sample strand TTGAATAGC, there are nine units; hence there must be nine insertion duplets. Those would be GCGGGAGTGAGAGGGTGT—and that is the duplet sequence o outputted from the loop. Notice the insertion duplets left-right order is reverse of the left-right order of the units to be inserted.

ADD TIP. The main body of the to-be-constructed parent strand consists of those duplets commanding the insertion of all bases needed for making the daughter. Now we add a small "tip" that will serve to (1) direct the enzyme to the proper binding site, and (2) cut off the finished daughter strand when the time is ripe.

1. Determine o's binding preference (the program in Appendix I may be used).

2. Construct tip. To o will be added the duplet AC (coding for the cut command) which will be the rightmost duplet of p. AC is "straight turning," so it does not affect the binding preference of o's tertiary structure. If o preferred C, then AC is all there is to the tip. If o's tertiary structure preferred a base other than C, then CT duplets, which are left turning, will have to be sandwiched in between the o segment and the final AC duplet, until the resulting tertiary structure does prefer C. (That would not require more than three CT duplets). CT was chosen because the operation it codes for, turning off the copy function, is "benign"—i.e., noninterfering.

OUTPUT IS THE PARENT STRAND P. Now, the p strand is ready. From left to right it consists of (1) the insertion commands; (2) the CT duplets (if any) needed to give it the necessary binding preference for C; and (3) as its rightmost and last executed operation, the AC cut command.

When p carries out its propagative program, the elected binding site will be the rightmost C unit, to the right of which all insertions will be made. Then, what has been inserted will be cut off. The portion of strand cut off—a daughter of p—should be identical to d.

Continuing our illustrative example, the binding preference for our o is A. One CT duplet sandwiched in between the final AC duplet of our tip and the o sequence will give us the correct binding preference for C. Thus p is

$$\text{GCGGGAGTGAGAGGGTGT+CT+AC.}$$

When the final C is chosen as initial binding site the single enzyme of this strand inserts a 9-unit sequence, which is then cut off. That 9 unit sequence is TTGAATAGC—our original input strand. We have created a strand which has TTGAATAGC for its daughter.

REFERENCES

1. Hofstadter, D. (1979), *Godel, Escher, Bach* (New York: Basic Books).
2. Laing, R. (1975), "Artificial Molecule Machines: A Rapproachment Between Kinematic and Tesselation Automata," *Proceedings of the International Symposium on Uniformly Structured Automata and Logic, Tokyo, August, 1975.*
3. Langton, C. (1984), "Self-Reproduction in Cellular Automata," *Physica* **10D**, 135–144.
4. Morris, H. (1988), *Typogenetics: A Logic of Artificial Propagating Entities, Ph.D. Dissertation, University of British Columbia, Vancouver, B.C., unpublished.*

Mitchel Resnick
Media Laboratory, Massachusetts Institute of Technology, 20 Ames Street, Cambridge, MA
02139

LEGO, Logo, and Life

ABSTRACT

LEGO/Logo is a type of Artificial Life Toolkit, designed primarily for children. Using LEGO/Logo, children can build artificial "animals" out of LEGO building pieces—including gears, wheels, motors, and sensors. Then, after connecting their creatures to a computer, they can write computer programs to control how their animals behave. Through these activities, children can explore some of the central themes of Artificial Life research. In this paper, I describe the LEGO/Logo system and discuss our experiences using LEGO/Logo with elementary-school children.

INTRODUCTION

Scientific ideas have their greatest force when they begin to seep out of the scientific community and into the general culture. This has happened with computer-science ideas during the past decade. As computers have proliferated throughout society, so have computational ideas and computational metaphors. Mental mistakes that were once seen as "Freudian slips" are now viewed as "information processing errors."[7]

Indeed, computational metaphors are influencing the way people think about things in nearly all disciplines, in nearly all parts of society.

Ideas from the emerging field of Artificial Life could have a similar influence in the next decade. Artificial Life is not merely a field of research, but a set of powerful intellectual ideas. The ideas of emergence, bottom-up design, and systems-oriented thinking have relevance and applicability far beyond the Artificial Life research community.

As ideas from Artificial Life infiltrate the culture, it is important to develop ways to share these ideas with children. Giving children access to ideas from the forefront of scientific research is, in general, a powerful pedagogic approach. The excitement of a new field can be contagious. When children sense that they are sharing in a new and dynamic enterprise, they are more likely to invest themselves in the process of learning. Thus, children can both benefit from and contribute to the dissemination of new scientific ideas.

In certain fields, such as high-energy physics and much of modern mathematics, it is difficult to make cutting-edge ideas accessible and relevant to children. But that is certainly not the case with Artificial Life. Children all share a strong interest in living things and animal behavior. Moreover, the general approach to Artificial Life research is one that meshes nicely with children's natural ways of learning. Artificial Life researchers generally learn by building things. They start with simple, easily understood rules or units, and they study how complexity emerges from interactions among these units.

This approach, which moves from the simple to the complex, from the concrete to the abstract, is well suited to the way children learn. Papert[5] argues that children learn best when they are building and inventing things that they believe in and care about. In working on Artificial Life projects, children do precisely that: they create things that they care about.

In this paper, I describe a computer-based system, called LEGO/Logo,[3] that serves as a type of Artificial Life Toolkit for kids. Using the system, children can build "animals" out of LEGO building pieces—including gears, wheels, motors, and sensors. Then, after connecting their creatures to a computer, they can write computer programs to control how their animals behave. For example, a child might program an animal to move toward a light, but to change direction if it bumps into any obstacles.

With the LEGO/Logo toolkit, children can explore how (artificial) animals behave in different environments, and how they interact with other (artificial) animals. In the course of such projects, children explore some of the central themes of Artificial Life research.

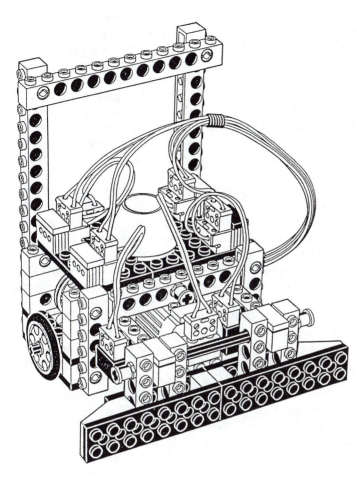

FIGURE 1 A
LEGO/Logo ani-
mal with touch
sensors on front.

LEGO/LOGO

LEGO/Logo has historical roots reaching back (at least) to the late 1940's, when British cybernetician Grey Walter[8] began experimenting with mechanical "turtles." Walter's turtles, so named because of the characteristic hemispherical "shells" protecting their electronics, were simple robotic devices. Walter equipped his turtles with a variety of sensors and programmed them to react to their environments. In one experiment, for example, Walter programmed a turtle to seek out a power source when it sensed that its power was running low.

In the late 1960's, Seymour Papert, then co-director of the Artificial Intelligence Laboratory at Massachusetts Institute of Technology, began putting robotic turtles into the hands of children. To control the turtles, children wrote programs in Logo, a language that Papert developed based on ideas from Lisp, the language of choice in the artificial-intelligence community. Logo includes commands like forward, back, left, and right to control the movements of the turtle. Papert[4] argued that activities with the turtle could bring children in contact with some of the central ideas of artificial intelligence. In programming the turtle, children naturally reflected on their own cognitive processes. The turtle thus served as an important "object to think with"—an object that enabled children to think concretely about thinking itself.

With the advent of video display terminals, the Logo community shifted its focus to "screen turtles." Screen turtles are much faster and more accurate than mechanical floor turtles, and thus allow children to create more complex graphics. Logo is currently used in more than one-third of all elementary schools in the United States, typically with an emphasis on turtle graphics.

LEGO/Logo brings Logo and the turtle back to three dimensions, but with several important differences. First of all, LEGO/Logo involves *two* types of building, building LEGO structures and building Logo programs. Second, children are not restricted to turtles; they can build and program many different types of "animals" and machines. Students in our LEGO/Logo classes have built and programmed everything from roller coasters to robots, from conveyor belts to walking machines.

The LEGO/Logo system is based on an enhanced set of building materials, including new types of LEGO blocks for building machines and animals, and new types of Logo "blocks" for building programs. On the LEGO side, there is an assortment of gears, pulleys, wheels, motors, lights, and sensors. The system includes specialized pieces such as worm gears, universal joints, and differentials. We have used a variety of sensors, including touch sensors, light sensors, and temperature sensors.

As its programming language, LEGO/Logo uses an extended version of Logo. The language includes the traditional Logo primitives and control structures (such as if and repeat), plus a set of new primitives added specially for the LEGO environment. For example, the primitive on turns on a LEGO motor. The primitive setpower sets the speed of a motor (or the brightness of a LEGO light). The primitive talkto designates which motor or light the computer should send commands to, while listento designates which sensor the computer should get information from.

Logo is a procedural programming language. Children create new animal behaviors by writing Logo procedures describing the behaviors. For example, the following procedure is written for a LEGO animal with a digital touch sensor. The procedure continually checks the touch sensor. Whenever the sensor is pushed, the animal reverses direction.

```
to bounce
listento  :touch-sensor
if sensor?  [reverse-direction]
bounce
end
```

ARTIFICIAL LIFE AND CHILDREN

For the past three years, we have been using LEGO/Logo with children (ages 10 and 11) at the Hennigan Elementary School in Boston. As the children worked on their projects, we acted as consultants and catalysts: we suggested new types of experiments, and we made connections to broader scientific ideas. In this section, I use examples from the Hennigan School to highlight the types of Artificial Life projects that children can work on with LEGO/Logo—and what they can learn in the process. In particular, I focus on how LEGO/Logo activities can help children develop new ways of thinking about systems and new ways of thinking about themselves.

THINKING ABOUT SYSTEMS

Using LEGO/Logo, children have created machines much like the early Braitenberg vehicles.[1] Some children start by building a LEGO turtle with two motors (one controlling each wheel) and two light sensors (pointing forward). To recreate Braitenberg's Vehicle 2, motors and sensors must be connected such that the more a sensor is excited, the faster the corresponding motor turns. In Braitenberg's models, this relationship is built into the hardware. In LEGO/Logo, the child must program the relationships in the software. For example, the following procedure links each sensor to the motor on the same side of the vehicle (as in Braitenberg's Vehicle 2a):

```
to fear
talkto    :left-motor
listento  :left-sensor
setpower  sensor-value
talkto    :right-motor
listento  :right-sensor
setpower  sensor-value
fear
end
```

The **fear** procedure sets the speed of the left motor according to the value of the left sensor, then sets the speed of the right motor according to the value of the right sensor. Then it runs the **fear** procedure again.

The procedure is named **fear** since, as Braitenberg points out, the resulting vehicle behaves as if it were afraid of light. The vehicle will spend most of its time in dark places. If it starts approaching a light source, the wheel closer to the light source will turn more quickly and the vehicle will turn away from the light.

Other Braitenberg vehicles can be created with slight modifications to the **fear** procedure. To create Braitenberg's *aggression* vehicle (Vehicle 2b), we need to "cross the wires" between the sensors and the motors. That can be done by interchanging the two **listento** lines (or the two **talkto** lines). The resulting vehicle seems to seek out lights and attack them violently. As Braitenberg puts it: the vehicle "resolutely turns toward (lights) and hits them with high velocity, as if it wanted to destroy them."

We can also create inhibitory connections, as in Braitenberg's Vehicle 3. One way to do that is to modify the **setpower** lines in the **fear** procedure, so that motor speeds vary inversely with sensor readings. The resulting vehicle, like Braitenberg's Vehicle 3a, seems to love the light source, "staying close by in quiet admiration from the time it spots the source to all future time."

When children begin writing procedures like these, they almost never give them names like **fear** or **aggression** or **love** (or other psychologically evocative names). Children, like Artificial Life researchers, generally work from the bottom up. They make "connections" between sensors and motors, then watch what happens. In the beginning, their LEGO animals almost always act in unexpected ways. Through these activities, children begin to get a sense of *emergent* properties.

After playing with their creatures for a while, children often begin to view the creatures on different *levels*. While building their creatures and programs, children typically think about the creatures on a mechanistic level. But as the creatures begin moving in the world, children begin to think in psychological terms, sometimes attributing intentionality to the creatures. Often, children will jump back and forth between the levels. Consider, for example, the comments of Sara, a fifth-grader, who was considering whether her machine would sound a signal when its touch sensor was pushed:

> "It depends on whether the machine wants to tell... if we want the machine to tell us... if we tell the machine to tell us."

Within a span of ten seconds, Sara (not her real name) described the situation in three different ways. First she viewed the machine on a psychological level, focusing on what the machine "wants." Then she shifted intentionality to the programmer, and viewed the programmer on a psychological level. Finally, she shifted to a mechanistic explanation, in which the programmer explicitly told the machine what to do.

Initially, Sara wasn't sure which levels of description were most appropriate for which situations; she needed to try out different perspectives and different levels to see which felt right. Sara's playful shifting is a sign of a good learning environment: good learning environments allow children to "play" with new ideas, to test them out, to see how they feel. By playing around with new ideas, children gradually appropriate the ideas, turning them into tools that they can use with comfort and assurance.

Children can run many types of experiments with their LEGO/Logo animals. Besides changing an animal's program, children can change the environment in which the animal lives. For example, children often try adding more lights to the environment, or making the lights flash. Then they begin to wonder: will the animal behave differently in the new environment? Which light will the animal seek out? Why? In many cases, the animal, placed in a new environment, reacts in unexpected ways, revealing new aspects of the animal's behavior—and driving home the importance of thinking about animals in the context of their environments.

Children can also change the hardware of the animal itself. Some students have tried turning the light sensors around so that the sensors face backward rather than forward. This reversal turns a light-seeking animal into a light-avoiding animal—without any changes in the software. This experience often leads children to reflect on the differences between hardware changes and software changes.

One group of students made a different sort of hardware change: they attached a light to the side of a light-seeking animal. The animal began running around in circles—in effect, chasing its own tail. Next the children programmed the light to flash. Even with a regular pattern of flashing, the animal moved about erratically, with no apparent pattern. The children were shocked that such a simple change could lead to such chaotic and unpredictable behavior.

As the children play with their LEGO/Logo animals, we challenge them with questions. For example, we often ask: what if a LEGO animal had only one "eye" (that is, one light sensor) instead of two? Could it still seek out light? At first, most children think the task is easy. Send the animal forward. As soon as the light level begins to fall, turn towards the light. Then they realize the bug: the animal doesn't "know" the direction of the light, so it doesn't know which way to turn. Children start searching for alternative approaches. One popular solution: make the animal veer in one direction as it moves forward. Then, if the light level starts to fall, veer off in the other direction until the light level starts to fall again. And so on. Through this activity, children learn that different sensor configurations require different algorithms to achieve the same behavior.

Not all LEGO/Logo activities are based on seeking or avoiding light. Some children have attached touch sensors to their animals and "taught" them to escape from mazes. Others have programmed their animals to follow lines of black tape on a white tabletop.

In all of these activities, children explore and play with fundamental ideas about systems:

- they build systems composed of many simple elements (both in hardware and software), and they observe new behaviors that emerge from the interactions of those elements;

- they begin to view systems on different levels, from the mechanistic to the psychological, and they learn that certain levels are more useful and appropriate than others in certain situations;
- they become familiar with common feedback algorithms that seem to crop up over and over, in situations that seem, on the surface, to have little in common.

In short, LEGO/Logo activities help children appropriate a new systems-oriented paradigm. By building and programming artificial animals, and by discussing and thinking about the results, children begin to view the world in a new way.

THINKING ABOUT THEMSELVES

This systems-oriented paradigm could influence the way children (not to mention adults) think about systems of all kinds: physical systems, political systems, economic systems. But perhaps most important is how it influences the way children think about *themselves* as systems.

While building and programming artificial animals, children naturally ask questions about themselves. In some cases, Artificial Life activities encourage children to reexamine things that once seemed obvious. As children try to make their creatures move, they reexamine how they themselves walk. And they reflect on the advantages and disadvantages of walking versus rolling, of legs versus wheels.

As children add light sensors to their creatures, they wonder: what would it be like to have only one eye? Or three? As children work with sensors, they also gain a deeper understanding for how their own senses work. At first, many children think that LEGO sensors will work "by themselves." They put a touch sensor on the front of a creature, for example, and they expect the creature to automatically change direction when it bumps into something. They don't realize that they must *program* that behavior into the creature.

This mistake is not all that surprising. Johnson and Welling[2] showed that fifth-grade children generally view the brain as the organ of mental life. Thus, children see the brain as essential for thinking, dreaming, and remembering, but they deem it irrelevant to walking, sneezing, or sensing. After building and programming LEGO/Logo animals, children generally have a much different view. They recognize the important role that the brain plays in processing sensory inputs and orchestrating actions by the body.

Through our work with LEGO/Logo and children, we hope to gain a sharper sense of how children think about their own minds. As children build and program artificial animals, how will their images of the mind change? So far, we do not know; we are just starting to study this question. But we expect that Artificial Life activities like LEGO/Logo will influence children's ideas in a fundamental way. No one expects children to come up with answers to ages-old questions about the mind, but it is important for them to think about these questions, to toy with new ideas and try them out. At the least, LEGO/Logo will give children a better set of tools for thinking about an important set of questions.

FUTURE DIRECTIONS

Our work with LEGO/Logo has just begun. There are many directions to go in the future. Among them:

NEW SENSORS AND ACTUATORS. The current LEGO/Logo system is rather limited in both its input and output devices. In the future, we plan to add ultrasonic rangefinders and a more sophisticated vision system. On the output side, we would like to add more sophisticated sound and speech systems for communications, and new materials to more accurately model muscle systems.

NEW PROGRAMMING PARADIGMS. The current LEGO/Logo software is based on a sequential programming language. Thus, it is often awkward to program concurrent activities, such as interactions among several animals—or interactions among multiple "agents" inside a single animal. To address this problem, we have developed a concurrent programming extension to Logo, which we call MultiLogo.[6] MultiLogo is based on a metaphor of interacting agents. Each agent is a separate computational process, similar to a traditional sequential Logo process. Agents communicate by sending messages to one another. With this extended language, children can model concurrent activities in a natural and intuitive way.

EMBEDDED COMPUTERS. In the current LEGO/Logo system, animals must always be connected to a personal computer. Thus the animals drag wires behind them as they move. One way around this problem is to develop a wireless communication system. We are taking a different approach: we are miniaturizing the computer so that each LEGO animal can carry its own computer. Just as children build sensors and motors into their animals, they should also be able to build computers into the animals. We call the miniaturized computer the Programmable Brick. We have already built a prototype the size of a deck of cards. To program the current prototype, you must connect it to a personal computer. Eventually, you will simply need to plug a keyboard and monitor into the Brick.

With these enhancements, the LEGO/Logo system could become a useful tool not only for children but for Artificial Life researchers as well. Indeed, one of our long-term goals is to create a general-purpose Artificial Life Toolkit that could be used by children and scientists alike. Already, researchers could use LEGO/Logo to test out various animal-interaction models. LEGO/Logo is different than most computer-based simulation tools in that it automatically includes much of the "messiness" of the real world—a feature that could be either an advantage or a disadvantage, depending on the experimental context.

Eventually (and more speculatively), LEGO/Logo could evolve into a type of universal constructor. One can imagine a LEGO-based automated factory, in which LEGO machines construct and program new LEGO animals—and new LEGO factory machines. Or, perhaps, the factory machines and the animals should not be viewed as separate categories; perhaps the factory machines are the animals. Obviously, such projects would require complex programs and new types of building materials, including specialized pieces outside of the LEGO repertoire. Such projects will probably not be feasible for many years—if ever. But in the meantime, LEGO/Logo can serve as a valuable catalyst for new ideas, sparking us to think about the tools we would like in the ultimate Artificial Life Toolkit.

ACKNOWLEDGMENT

Many of the ideas in this paper are based on collaboration with my colleagues in the LEGO/Logo research group: Seymour Papert, Stephen Ocko, Fred Martin, Edith Ackermann, Brian Silverman and Philip Ivanier. In addition, Mike Eisenberg and Franklyn Turbak provided helpful comments on a draft of the paper.

Our research on LEGO/Logo is supported by grants from LEGO A/S and the National Science Foundation. LEGO Systems Inc. markets a version of LEGO/Logo under the product name *LEGO TC logo*. Logo Computer Systems Inc. implemented the software for this commercial version.

REFERENCES

1. Braitenberg, V. (1984), *Vehicles* (Cambridge: Massachusetts Institute of Technology Press).
2. Johnson, C. N., and H. M. Wellman (1982), "Children's Developing Conceptions of the Mind and the Brain," *Child Development* **53**, 222–234.
3. Ocko, S., S. Papert, and M. Resnick (1988), "LEGO, Logo, and Science," *Technology and Learning* **2:1**.
4. Papert, S. (1980), *Mindstorms: Children, Computers, and Powerful Ideas* (New York: Basic Books).
5. Papert, S. (1986), "Constructionism: A New Opportunity for Elementary Science Education," *a proposal to the National Science Foundation*.
6. Resnick, M. (1988), *MultiLogo: A Study of Children and Concurrent Programming, Master's thesis, Massachusetts Institute of Technology*.
7. Turkle, S. (1984), *The Second Self* (New York: Simon and Schuster).
8. Walter, W. G. (1950), "An Imitation of Life," *Scientific American* **May 1950**, 42–45.

Bill Coderre
Media Lab, Massachusetts Institute of Technology, 20 Ames Street, Cambridge, MA 02139

Modeling Behavior in Petworld

This paper describes a system called *Petworld* for modeling aspects of animal behavior. Behavior is modeled as a rigid hierarchy of simple agents that make recommendations to their superiors. Since recommendations are in the form of rankings of alternatives, agents can resolve conflicting behaviors either by straight choice, compromise, or displacement. Substantial improvements have been made since the Artificial Life conference in September, including a simple perception model and a method for concentration on long-duration tasks. Plans are discussed for the successor to Petworld, in which learning will be incorporated, among other new features.

INTRODUCTION

The Vivarium Project is a collaborative effort between Apple, MIT, and the Open School in Los Angeles. The aim of Vivarium is to teach grade- and high-school students animal behavior as an approach to learning about thinking. As part of Vivarium advanced research, Petworld is not specifically geared for school use, but

Artificial Life, SFI Studies in the Sciences of Complexity,
Ed. C. Langton, Addison-Wesley Publishing Company, 1988   **407**

bears in mind the issues of complexity, intuitiveness, and user interface. I claim that Petworld has enough sophistication to produce interesting behavior, without being so complex as to make creation of that behavior difficult.

Petworld is a system for modeling non-species-specific behavior in an intuitive environment. I divide the system into four parts.

- how the pets *perceive* the world;
- pet *locomotion*;
- the *physics* of the world; and
- pet *intellect*.

Although all the parts are important, I have concentrated on intellect, especially the structuring of knowledge to produce behavior. Therefore, I tried to keep the first three aspects simple while still providing a reasonable environment.

The first section of this paper contains a description of the system and its commands, and descriptions of a sample brain. The second section contains an analysis of Petworld's intellect system in light of other systems, including more traditional knowledge-based systems, dataflow systems, and Minsky's Society of Mind model.

SECTION ONE: THE PETWORLD SYSTEM
ABOUT THE PETWORLD

Petworld is a world of pets, rocks, and trees which inhabit a two-dimensional, limited cartesian plane. Time passes in discrete quanta, and simultaneous action by all the inhabitants is simulated. Each action executed in the world takes one tick, and brain calculations are instantaneous. This is similar to a discrete-time feedback system. Another possible model, one with longer-duration actions and computer-style interrupts, was considered. Although both offer about the same power, the feedback-style approach was chosen as more intuitive.

Pets have a body orientation. The pets have a limited field of view, typically 90 degrees. Also, they can move for a limited distance in the direction of their body orientation, typically about one unit. Pets cannot push or throw things, but can carry one rock at a time. (Pets often use rocks to build nests.) Trees are food sources, and pets are browsers. Every time a pet eats, the tree loses some meal points and vanishes when the count reaches zero. Trees also grow (accrue meal points) as time passes. New trees appear spontaneously around the play field. Famines are prevented by creating a new plant when the last one dies.

Several pets are usually living in the world at once. In general, pets are mutually antagonistic, and can attack each other. Pets cannot reproduce, but can die from starvation or wounds.

ABOUT PETS

Each pet has a very limited set of internal states which indicate the condition of the animal. These are the location of its nest, the variables HUNGER, FEAR, and IN-JURY, which have as values numbers between 0 and 100, and the flag PAYLOAD. A hunger of 0 indicates being sated, and a hunger of 100 indicates death by starvation. Eating reduces the count, and the passage of time increases the count. Fear is determined by the distance of other pets. A low fear indicates that others are far away, and a high fear tells proximity. Injury is caused when one pet attacks another. Each attack increases the pet's injury count, and with time the count goes back down. If an animal's injury count exceeds 100, the animal dies from its wounds. If the payload flag is true, the pet is carrying a rock.

Since time passes in quanta, pets perform a "SEE—THINK— DO" loop similar to a "READ—EVAL—PRINT" loop in Lisp. First, every pet is given a chance to perceive the world, then a chance to decide what to do, then all actions are performed. Since there is still an order in which pets act (which might be important when, say, there is very little food left), this order is shuffled occasionally.

The pets' actions are MOVE-TOWARDS, TURN, LIFT, DROP, EAT, and ATTACK.

ABOUT PET BRAINS

A pet brain is a hierarchy of modules that I call *experts*. Each expert has inputs from its subordinates and the world, and provides as output a *ranking*—a list of possible actions with numeric weights attached—telling how good those actions are in the opinion of the expert. The action with the greatest weight of the ranking that the topmost expert recommends is the one that the pet executes.

Decision information (in the form of rankings) can be thought to flow up from the bottom-most experts, getting processed along the way. Typical expert processing strategies involve assigning weights to rankings, filtering rankings to remove unwanted elements, merging two rankings while possibly emphasizing one more than another, and reducing the elements of a ranking to a single object. Thus, the ethological constructs of competitive, compromise, and displacement behaviors can be directly implemented.

A SAMPLE BRAIN

As an example of the Petworld system, I offer a sample pet which is capable of several basic behavior patterns: interacting with other pets, foraging for food, building a nest and staying in it once it is built, and exploring areas not visible.

The strategy of the pet is decided by the topmost expert, *brain5*. Finding food and interacting with other pets are most important, followed by building a nest, exploring the world, and homing into the nest. In the case of conflicting recom-

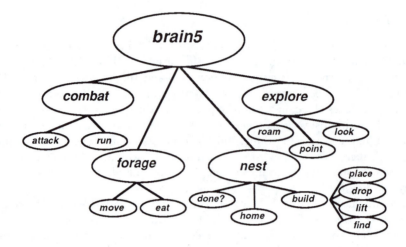

FIGURE 1   The sample brain.

mendations, several decision strategies are pursued. If there is a conflict between foraging and combat, a compromise is made, dependent on how serious the danger of starvation or attack is. Thresholds are also used to ignore experts. Assuming no life-or-death situations are imminent, nest building takes priority over exploring.

The basic *combat* strategy is to avoid other pets. Another pet is attacked only when the other is close enough and damage is low.

Foraging consists of either eating what is readily edible, or moving toward the closest available food.

Nesting is actually the most complex of the behaviors, incorporating several important functions. A functional expert called *done?* reports true, false, or can't tell dependent on the completeness of the nest. *Home* tries to get the pet to stay in the nest. And *build* models a strategy sometimes suggested for bird nest building: "If you are in your nest and have a rock, put the rock in the nest; otherwise get a rock and bring it back."

Last is the *exploring* expert. It is called by default, when nothing else works. Since pets have limited sensory range, they will often be in a spot where no useful objects are visible. In that case, they should first turn around, and if nothing becomes visible, go somewhere else. This mechanism makes good use of Petworld's feedback style of control, and of its history mechanism. As soon as the exploring expert brings something useful into view, another expert will be able to take control. And since the exploring expert is able to remember what it was doing previously, it can act as if it were performing long-duration actions.

Below are the actual rules from the brain, translated into English.

BRAIN5

1. If both HUNGER and FEAR are high, effect a tradeoff between COMBAT and FORAGE.

2. If FEAR is high, COMBAT.

3. If HUNGER is high, FORAGE.

4. If FORAGE is recommending that there is a food element immediately available, then FORAGE.

5. If NEST has some non-trivial action to perform, then NEST.

6. Otherwise, EXPLORE.

COMBAT

1. If you have an available attack, and your damage is low, then recommend a tradeoff of attacking and running away.

2. Otherwise, recommend running away from any visible pets.

ATTACK Construct a ranking of things you can attack standing right where you are.

RUN Construct a ranking which scores movements on how well they bring you away from other pets. *(This includes the current positions of the other pets, and does not take into account their potential movement.)*

FORAGE

1. If you are standing next to food, then recommend EAT.

2. Otherwise, recommend moving toward visible trees (MOVE). If none are visible, no recommendation is made.

MOVE Construct a ranking which scores movements on how well they bring you toward trees.

EAT Construct a ranking of things you can eat standing right where you are.

NEST

1. If DONE? returns true, then HOME.

2. Otherwise, BUILD.

DONE? *(The nest is done when an internal, hardwired template representation of the nest, when matched against the state of the world, appears complete.)*

HOME Return a ranking that gets the pet into the center of the nest.

BUILD *(In any of the following cases, if no forward progress can be made toward some destination, a displacement behavior of wandering randomly is undertaken.)*

1. If you have a rock, and the nest is completed, drop the rock. *(This in effect will drop it outside the nest.)*

2. If you have a rock, and you are standing in the right spot, then drop the rock. (DROP)

3. If you have a rock and and are not in the right spot, then move toward the right spot. (PLACE)

4. If you don't have a rock, and are standing next to one that is not part of your nest, then pick it up. (LIFT)

5. If you don't have a rock, and are not standing next to one that is not part of your nest, then move toward the closest rock that is not part of your nest. (FIND)

FIND Return a ranking scoring movements on how effectively they bring you towards the closest rock that is not already part of your nest. If no rock is visible, no recommendation is made.

PLACE Return a ranking of movements on how effectively they bring you towards the top-ranked spot in the nest to place a rock. *(The nest is defined as a ranking of squares in which to place rocks.)*

LIFT Return a ranking of possible lifting commands. Empty if no possible lifts.

DROP Return a ranking of possible dropping commands. Empty if no possible drops.

HOME Return a ranking scoring movements on how effectively they bring you towards "home" (the position on the board in which you started the simulation). The pet always knows where home is.

EXPLORE Exploration is allowed when no other strategy fires, probably due to a lack of visible objects. I have chosen a very simple exploration strategy: first turn around to look in all directions, then move in a random direction a few units and try again.

1. Recommend LOOK enough times in a row so that one complete revolution is made.

2. Recommend POINT once, to attain a random direction to travel in.

3. Otherwise, recommend ROAM several times in a row.

LOOK Turn clockwise by one field of view.

POINT Turn some random amount.

ROAM

1. If you can't move forward, turn right 90 degrees.

2. Otherwise, move forward.

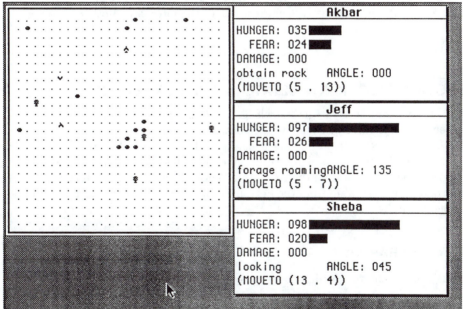

FIGURE 2 Petworld screen dump. Each pet has a window showing its current state, which part of its brain hierarchy is being utilized, and its next move. Pets, rocks, and trees are indicated in the world window with a distinguishing character set.

ANECDOTES

The screen dump in Figure 2 shows a sample pet simulation run. Three pets are shown in the example, all with identical brains and all with display windows. One may run a simulation with virtually unlimited numbers of pets, brains, rocks and trees, and the world can be made bigger. The practical limit seems to be about 10 pets and a world of about 50 by 50 elements. (The current graphic windows do not scroll, and the time taken for recalculation at that point would be on order of five seconds per round.)

I have run simulations for extended periods of time, finding that pets can be made that live for quite some time (thousands of rounds). Occasionally, pets exhibit unexpected or emergent behaviors.

When two pets build nests close to each other, for example, the nearest rocks to add to the nest are part of the other pet's nest, leading to many conflicts over rocks. These conflicts are usually stalemates, and get resolved when one pet gets hungry and heads off to forage.

Also, because the pet cannot see its nest when it is far away from it, the current nesting strategy causes pets to pick up extra rocks and stash them near the nest, a very advantageous strategy which arose out of the interplay of other behaviors.

SECTION TWO: KNOWLEDGE IN THE PETWORLD

Pets' brains contain strategic knowledge. In this sense, pets are knowledge-based systems. Unlike most other knowledge-based systems, the knowledge is organized in a rigid hierarchy of interacting agents called "experts." And unlike a decision tree, where the tree is used to score different moves, or a branching tree, where program control flows from top to bottom, Petworld utilizes a dataflow tree, where information is passed up from the bottom of the tree, being processed at each node in the hierarchy.

WHY A HIERARCHY?

Using a hierarchy to encode knowledge offers useful features:

Managing complexity: In a knowledge-based system, it is easy to end up with thousands of knowledge elements (ideas or rules) that must be chosen from. Some sort of structuring is needed—not only for efficiency in running the system, but also for clarity maintaining it. If we were to group ideas, they would share lots of information, and it makes sense to put them in an enclosure both to reduce the size of the knowledge file, and to indicate to the knowledge engineer the structuring of ideas. (In expert systems, these enclosures carry common preconditions for batches of rules.)

CONTROLLING EXECUTION: If we have grouped ideas together, it's easy to imagine some groups controlling other groups. For example, one group of rules could resolve conflicts between other groups. (This is the root of meta-rules as described by Randall Davis.[8]) It could resolve conflicts not only by having a built-in preference for one of the subordinates, but also by effecting a compromise between agents. If the animal is both scared and hungry, it might choose to run away in a direction that is favorable for finding food. Another possible resolution is called "displacement behavior": if two experts conflict, both are ignored and a third is employed. (Dogs, when unable to decide between running from an enemy and fighting, sometimes abruptly sit down and scratch an itch.)

INCREASING LEVERAGE: By taking the grouping strategy one step further and arranging experts into something resembling a hierarchy, a phenomenon similar to mechanical leverage arises: a little knowledge at the top of the hierarchy can have the same effect as attaching preconditions to many rules lower in the hierarchy.

WHAT KIND OF HIERARCHY?

The organization of the hierarchy of experts controls the flow of decisions. I examine several approaches to hierarchical structuring below.

1. *Decision trees*, from operations research, are similar to how many computer games work. A move generator feeds the hierarchy. Each expert passes to its superior a numeric rating for each move based upon world state, pet state, and the opinions of its subordinate experts. The action that gets the best numeric rating executes.

   This approach has two drawbacks. It evaluates a lot to find out a little—an entire tree might be calculated for dozens of possible moves. It also requires that compromises be finagled as weightings of the scoring functions, rather than being written out as rules. So, some of the decision knowledge is hidden in the rating scheme, a circumstance I have chosen to avoid.

2. *Branching trees* are similar to the meta-rule and grouping approaches extended to their logical extreme: control branches down the hierarchy by running rules in each expert along the path until an expert at the bottom of the tree is chosen. One of its rules then fires and action is finally taken.

   This approach, although theoretically as powerful as any other, in practice still has the reconciliation problem. To effect a compromise between two experts, a "compromise between a and b" expert will have to be constructed, and more rules will have to be added to the superior expert to decide when to choose it.

3. *Society of Mind*, a new theory by Marvin Minsky, seems the most powerful hierarchical structuring yet. Agents take control, and can invoke other agents

to assist them. Other agents also attempt to take control, and agent conflicts are an interesting part of behavior.

This approach has some difficult implementation problems, however. Conflict resolution between agents is currently an open question in SOM. For this reason, for simplicity, and because my research started as SOM was being finished, I opted for the simpler strategy below.

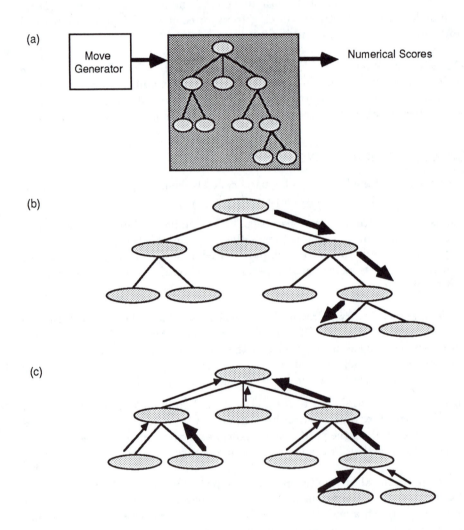

FIGURE 3   (a) Decision trees, (b) branching trees, and (c) dataflow trees.

4. *Dataflow trees*, the method I chose, has low-level experts "burble up" rankings from which higher levels choose actions. Each expert will be concerned with compromising between actions presented to it, either by chosing one action, or by creating a new action. Knowledge about making decisions is stored explicitly in the experts. The experts output only a relative ranking, not an absolute one.

LOCAL STATE IN PET BRAINS

One of my other design constraints was to build pets with as little local state as possible. Although originally undertaken as an exercise in seeing how much behavior could be obtained with how little a brain, I have since decided the decision was a good one since it echos the constraints of most of the animals I saw my system as capable of simulating.

One of the problems I have encountered is dealing with the concept of concentration: the ability to stay focused on one problem-solution path for some length of time. For example, because pets have limited vision, they must concentrate to find food they cannot see. Because experts do not have local state, and because the entire behavior network is reevaluated each time, food finding is impossible without some extra mechanism. To this end, I have incorporated a notion of recent history to allow concentration. Each expert can remember a history of its recommendations. In the case of the exploring expert above, the mechanism allows it to turn around precisely once, and then move several times in a row in a straight line. This history mechanism allows almost all of the power of Minsky's Society of Mind, with a simpler mechanism.

SECTION THREE: FUTURE WORK

Future work in Petworld will take place as my Master's thesis, including some of the topics below.

Since Petworld is part of Vivarium, issues of user interface, novice programming, and system complexity are crucial for any research to be used in a school setting. Several opportunities for user interface enhancements suggest themselves:

- A graphical dataflow programming language;
- A brain display showing the shape of the hierarchy and depicting the flow of information with animation and allowing for value display windows;
- Parameter controls that allow concepts such as "aggressiveness" to be added to brains; and
- A history recorder and forward-backward debugger, graphically similar to a video cassette recorder control panel, with a history line and markers for crucial moments.

The most recent addition to Petworld, the "recent history" mechanism, is currently incomplete and kludgy. I am designing cleaner functionality for a future Petworld.

LEARNING

But the most crucial missing feature is learning. Just about every living creature learns to some extent or another. Bacteria, for example, can learn to associate light level with temperature. Higher animals can remember locations and paths. Still higher level animals can make complex plans for action, and improve and radically change strategies.

As part of my Master's research, I have been developing a way to make pets learn somewhere on the third level mentioned above. Briefly, if there are many agents potentially employable for a given task, one could create managers to choose and remember agents that are more capable. Thus, managers can be expected to optimize behavior over time. If one then postulates an agent generator similar to those of Douglas Lenat,[12,13,14] pets could radically alter their behavior patterns. Research of this strategy is just beginning. (As a side note, one could then have pet brains reproduce sexually, since a lumping together of the two agent masses could be resolved by a simple mechanism. This does not happen in nature, but might be interesting to study.)

CONCLUSION

Petworld is a viable system for modeling some aspects of animal behavior. Pets can be created with simple, easily understandable brains that produce interesting, sometimes unexpected behavior. Although pets cannot reproduce or learn, they can undertake complex long-term actions because of the recent history mechanism. Work is underway to add learning to Petworld.

ACKNOWLEDGMENTS

Petworld 2.0 runs on Apple Macintosh computers with at least one megabyte of memory. It is preferable to use a machine with more memory and a faster 68020 processor, such as a Macintosh II. Petworld is written in Allegro Common Lisp. I wish to thank the people of Coral Software for their support. Inquiries about the code should be directed to the author.

Thanks go to Mario Bourgoin, Jim Davis, Marvin Minsky, Mike Travers, and the late Tom Trobaugh, all from the MIT Media Lab, for their crucial insights. Thanks also to the bizarre gang: BoB, Chris, Gypsy, Trrrisha, and even Terri.

REFERENCES

1. Agre, Phil (1985), "Routines," *MIT AI Laboratory Memo 828, May 1985.*

2. Alkon, Daniel L. (1983), "Learning in a Marine Snail," *Scientific American* **6/83**, 70–84.

3. Batali, John (1983), "Computation Introspection," *MIT AI Lab Memo No. 701, February 1983.*

4. Braitenberg, Valentino (1984), *Vehicles: Experiments in Synthetic Psychology* (Cambridge: Massachusetts Institute of Technology Press).

5. Camhi, Jeffrey M. (1982), "The Escape System of the Cockroach," *Scientific American* **12/82**, 158–172.

6. Chapman, David, and Philip E. Agre (1987), "Abstract Reasoning as Emergent from Concrete Activity," *Reasoning about Actions and Plans: Proceedings of the 1986 Workshop* (Los Altos, CA: Morgan Kaufman).

7. Davis, James R. (1983), "Pesce," *a computer program modeling fish behavior;* a paper describing his system is in preparation.

8. Davis, Randall (1980), "MetaRules: Reasoning about Control," *MIT AI Lab Memo number 576, March 1980.*

9. Groelich, Horst (1986), *Vehicles* (Cambridge: MIT Press); a software package for Apple Macintosh.

10. Hofstadter, Douglas R. (1984), "The Copycat Project: An Experiment in Nondeterminism and Creative Analogies," *MIT AI Lab Memo No. 755, January 1984.*

11. Kehler, Thomas P., and Gregory D. Clemenson (1983), "KEE, The Knowledge Engineering Environment for Industry," *Intelligenetics, Inc., 1983; paper available from Intelligenetics, 124 University Ave, Palo Alto, CA.*

12. Lenat, Douglas B. (1975), "Beings: Knowledge as Interacting Experts," *Proceedings of the Fourth IJCAI, Tbilisi, U.S.S.R.*, 126–133.

13. Lenat, Douglas B., and Gregory Harris (1977), "Designing a Rule System that Searches for Scientific Discoveries," *CMU Department of Computer Science.*

14. Lenat, Douglas B., and John Seeley Brown (1984), "Why AM and EURISKO Appear to Work," *Artificial Intelligence* **23**, 269–294.

15. Lorenz, Konrad (1952), *King Solomon's Ring* (New York: Thomas Y. Crowell Company).

16. MacLaren, Lee S. (1978), *A Production System Architecture Based on Biological Examples, Ph.D. thesis, U. Washington Seattle, 1978* (available as University Microfilms Order No. 79–17604).

17. Maruichi, Uchiki, and Tokoro (1987), "Behavioral Simulation Based on Knowledge Objects," *Proceedings, ECOOP '87, Springer-Verlag Lecture Notes in Computer Science No. 276* (New York: springer-Verlag).

18. Minsky, Marvin (1986), *The Society of Mind* (New York: Simon and Schuster).

19. Scientific American, eds. (1979), *Brain: A Scientific American Book* (New York: W. H. Freeman Company).

20. Stefik, Mark, D. G. Bobrow, S. Mittal and L. Conway (1983), "Knowledge Programming in Loops: Report on an Experimental Course," *AI Magazine* **Fall 1983**, 3–13.

21. Travers, Michael (1988), *Animal Construction Kits*, these proceedings.

Michael Travers
Media Laboratory, Massachusetts Institute of Technology, Cambridge, MA 02139

Animal Construction Kits

1. ABSTRACT

This paper describes a series of systems, implemented and projected, for the simulation of animal behavior. This work is part of the Vivarium Project, which has the goals of developing computational models for ethology, investigating situated action approaches to Artificial Intelligence, and providing an educational environment in which grade-school children can experiment with behavioral mechanisms and create autonomous animated characters.

2. INTRODUCTION

An *animal construction kit* is an interactive computer system that allows novice programmers to assemble active artificial animals out of prefabricated components. These components include sensors, muscles or other effectors, and computational elements. They also include structural components such as bones and joints. To be useful, an animal construction kit must also provide a world for the animal to live in, and must support the co-existence of multiple animals of different species.

Artificial Life, SFI Studies in the Sciences of Complexity,
Ed. C. Langton, Addison-Wesley Publishing Company, 1988   **421**

Building artificial animals serves several research goals from different fields. The goals of ethology are furthered by constructing computationally feasible models of behavioral control systems. Programming technology and interactive system design are furthered by the use of animals as a metaphor for computational processes. Artificial Intelligence and cognitive science are furthered by recasting their goals and questions in terms of intelligent situated action. Computer-generated animation is furthered by creating objects that compute their own motions according to behavioral rules.

Simulating animals in a world can be done at an arbitrarily detailed level. The requirements of interactivity necessitate that the level be carefully chosen so that the necessary computation can be done in real-time. The level of detail achievable within the range of real-time simulation will increase as more powerful computers become available. In particular, special-purpose graphics processors will greatly improve the quality of real-time animated images.

We hope to bring the capability of programming artificial animals to novice users, including grade-school children. To that end, user interface techniques are being developed that make the abstract processes of behavior more concrete and accessible. These include graphic interactive programming, force-feedback for tactile interaction, and real-time three-dimensional graphics for heightened realism.

Two systems are described in the following sections. The first system, Brain-Works, uses a neural model of computation with an interactive graphic interface, and a physical structure based on the Logo turtle. The second system, Agar, uses agent-based computation, more realistic physics, and provides a more flexible modularization of worlds, creatures, and components. Both systems are written in Common Lisp with object-oriented extensions.

3. BRAIN WORKS: NEURAL GRAPHIC PROGRAMMING

BrainWorks is a computer system that allows a user to construct a nervous system for a simple animal, using an interactive graphic interface. The user starts out with a brainless animal body that resembles a Logo turtle, equipped with several sensors as well as motors for movement. The sensors include eyes and touch bumpers, which respond respectively to food and to obstacles in the animal's world. The user also has a supply of neurons of various types, and tools for bringing them into the turtle's body and connecting them to the sensors and motors (see Figure 1). When the turtle is placed into its world, it responds according to its wiring. A turtle can be wired to display a variety of elementary behaviors such as seeking or avoidance, but usually its goal is to catch its food while avoiding being trapped by obstacles (see Figure 2). BrainWorks was largely inspired by Valentino Braitenberg's *Vehicles: Experiments in Synthetic Psychology*.[3]

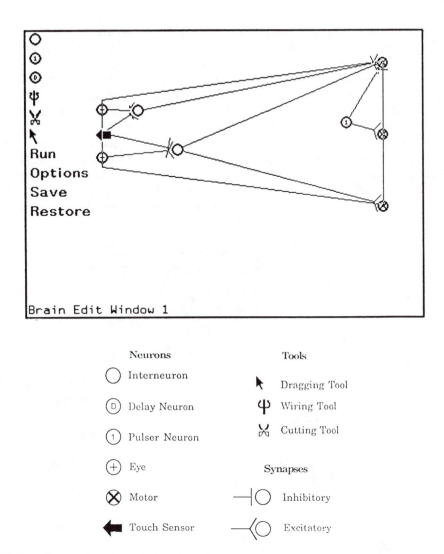

FIGURE 1   The turtle-wiring window from BrainWorks

3.1 NEURONS

BrainWorks uses modified McCulloch-Pitts neurons[20] as its computational units. They implement a mostly Boolean logic, in that neurons can be in one of only two states (on or off), connections similarly transmit a boolean value, and there are no registers. Connections can be inhibitory or excitatory. A neuron's activation is computed by summing up the number of activated connections, weighted appropriately (inhibitory connections have a negative weight). If the result is over a fixed threshold (usually zero), the neuron is considered activated.

The set of neuron types includes interneurons (which implement the threshold logic described above), the sensorimotor neurons that serve to interface the nervous system with the world, and two special types: pulser and delay neurons. A pulser acts as a constant. It is always activated, and can be used to provide default behaviors or biasing of other neurons. The delay neuron works like an ordinary interneuron, except that its output is based on the input values from the *previous* simulation cycle. Chaining delay neurons allows building delay lines of arbitrary length. This feature enables the turtle to control the timing of its reactions, but removes the nervous system from the class of strictly stateless Boolean machines.

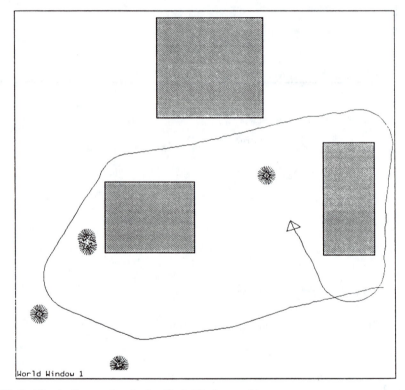

FIGURE 2   The turtle moving in the BrainWorks world

3.2 THE TURTLE

The basic design of BrainWorks creatures is based on the Logo turtle[25] and its autonomous antecedent.[38] They implement a form of local computational geometry. The locality is critical: all the turtle's operations are relative to its present position and orientation, rather than to a global coordinate system. This is appropriate for simulating animals, all of whose information and action must necessarily be local.

The Logo turtle takes commands from a programming language, such as FOR-WARD or RIGHT. In BrainWorks, *motor neurons* take the place of commands. There are three motor neurons, for moving forward, right, and left. Right and left movements change orientation only, while forward causes a translation along the current body axis. All the motors can be activated simultaneously, but left and right will cancel each other out. The magnitudes (angular or linear) of the movements are controlled by user-defined parameters (turn-factor and forward-factor), but are always constant for a given turtle. Movements take place in discrete time steps, but in a continuous coordinate system.

BrainWorks turtles have three sensors: two eyes and a bumper. Eyes detect food within a conical area of sensitivity that is controlled by several parameters: eye-angular-aperture, eye-linear-range, and eye-binocular-angle (the latter controls the angular offset of each eye from the turtle's body axis). The bumper detects an obstacle or wall in the turtle's path.

A turtle's basic task is to catch food while avoiding being blocked by a wall or obstacle. This can be accomplished by a variety of networks and parameter combinations. There are tradeoffs in the values of the parameters such as turn-factor: a small value will result in a large turning radius, making it hard to home in on nearby food, but a large value may result in food being missed due to oversteering. This interacts strongly with the eye-angular-aperture parameter, indicating the importance of matching sensors to motors, and appropriate coupling of both to the environment.

3.3 THE WORLD

The world of BrainWorks is very simple, but still complex enough to generate interesting behavioral problems. Essentially it is a 2-D Cartesian plane, with objects taking on a continuous range of positions (in contrast to worlds that use a grid or other discrete geometry). The objects in it are turtles, food units, and obstacles. Turtles are dimensionless, but have an orientation. Food units, on the other hand, have a non-zero radius so that the turtles will be able to intersect with them (and so eat them). Obstacles are rectangles into which the turtle cannot move. Any attempt to move into them results in no linear motion on that simulation cycle, and activates the turtle's touch bumper.

The properties of the world and the turtle were the result of compromises between the goals of keeping computation time low (to preserve the real-time feel of the system) and making a rich, realistic, and interesting world. For instance, using a Cartesian geometry rather than a grid allows a much more realistic simulation of orientation, a fundamental concept in animal movement. On the other hand, restraining the world to two dimensions allows faster movement and redisplay than is possible in a full-blown 3-D simulation.

3.4 EVOLVING TURTLES

BrainWorks was originally designed to allow building designed animals, but evolved to support evolved animals. An additional class of turtles was created that supports reproduction, mutation, and survival contingent on food-gathering performance. When many of these turtles coexist in a world, they compete for limited resources and evolve under the resulting selection pressure. Survival and reproduction are based on maintaining an energy reserve by finding and eating prey.

The evolving turtles maintain an energy reserve that is depleted by movement and the passage of time, and increased by eating prey. If the energy drops below zero, the turtle dies. Reproduction is asexual, and conserves system energy, in that the parent organism can pass on a certain amount of its energy to its offspring.

In these mutable turtles, the neural behavior model is replaced with a matrix that defines a function from the vector of sense inputs to the vector of motor outputs. A mutation consists of a small change to one of the elements of the matrix or to a small set of additional parameters. The fact that matrix elements and parameters are continuous rather than discrete is important—it enables the evolutionary process to proceed by incremental hill-climbing.

The two properties of differential survival and reproduction with mutation should be enough to produce adaptation. This proved to be the case. A random population of turtles showed improvement in their ability to catch prey over a long period of time. Unfortunately the world of BrainWorks is not rich enough to provide a great variety of possibilities for successful behavior. The turtle can get better at homing in on food units, and get faster at turning around after striking an obstacle, but there is not much room for improvements beyond these. Creating an artificial world that has enough combinatorial flexibility to permit innovative behavior is a still an unmet challenge.

The point mutations used in the evolution simulation generate incremental exploratory moves in creature-space. More powerful techniques such as the crossover rules used in genetic algorithms[16] could improve the ability to adapt to novel circumstances. Such a technique might be implemented in BrainWorks by a form of sexual reproduction that involves swapping matrix elements between two creatures.

3.4.1 EVOLVING TO EXPLOIT BUGS While the BrainWorks world as specified did not have enough diversity to support novel behaviors, the evolutionary process did succeed in generating novel and amusing ways of exploiting unintended bugs in the simulation system.

One strain of turtle took advantage of the fact that energy consumption was being computed as a multiple of the translation distance by moving *backwards*, gaining energy with each step. It did very well without being able to see where it was going, until the bug was fixed by using the absolute value of the translation vector in the energy usage computation.

Another strain exploited a more subtle bug. Understanding this bug requires some knowledge of the details of reproduction. A creature has several parameters controlling reproduction, including `reproduction-limit`, which specifies the minimum amount of stored energy a creature can have before it will even consider reproducing, and `reproduction-gift`, which specifies how much of this stored energy will be passed on to the offspring. Both of the parameters mentioned above are subject to possible mutation.

The bug was this: `reproduction-gift` was assumed to be lower than `reproduction-limit`, so that the turtle first checked its energy level against `reproduction-limit`, then, if reproduction was called for, passed on `reproduction-gift` energy units to the offspring, decrementing the parent's level accordingly. If, in fact, `reproduction-gift` was the *higher* of the two parameters, this could result in a negative energy balance for the parent, causing it to die immediately. Meanwhile, however, it had introduced some energy into the system from thin air. The new offspring now had more energy than the parent had had, and it immediately pulled the same trick and reproduced itself.

This, in fact, happened during an overnight run. Since the system creates a new data structure when a new creature is born, and doesn't garbage collect the dead ones (it leaves them around for analysis), these short-lived creatures quickly used up all of the virtual address space of the machine they were running on, causing an ecological collapse (i.e., a crash). So we have evolved a very unusual creature, one that reproduces and dies at the limit of the simulation's clock, and survives solely on the energy it manages to extract from a bug.

---

4. AGAR: PROGRAMMING WITH AGENTS

Agar is a newer system that builds on BrainWorks and attempts to go beyond its limitations. Agar's extensions include provisions for a variety of worlds and creatures, and a more flexible programming system based on agents.

4.1 SIMULATING A REAL ANIMAL

Agar is designed to simulate more realistic animals. One possible target is the three-spined stickleback, a small fish that was the subject of Tinbergen's classic ethological work[36] (see 4.3.1 and 5.1.1).

To implement a real creature, we must implement a realistic world for it to live in. The stickleback has to be able to sense the onset of mating season, the presence of other animals (including their social signals such as the red belly of the male), the presence of eggs in the nest, and the actions of its offspring. If it is to dig its nest, the physics of digging must be simulated to some degree of detail. Evaluation of

BrainWorks showed that simulating the world was by far the most computationally intensive part of the simulation. If the system is to run in real-time or close to it, efficient algorithms for object-object interaction and sensor simulation must be developed.

4.2 WORLDS

People and computers can demonstrate their intelligence by manipulating symbols. Animals, with their limited or nonexistent ability to manipulate symbols, can only display their intelligence by *action* in the world. The appropriateness of an animal's actions to its situation constitutes its intelligence. Winograd refers to this close relationship between an intelligent agent and its environment as *structural coupling*, and suggests that it is a fundamental root of human intelligence.

This topic is discussed further in Section 5.4. Here, the point is that for a simulated animal to have a variety of possibilities for action, and hence a need to develop intelligence beyond minimal reflexes, it must have a sufficiently rich world to act in. Since real-time operation is also a goal, full simulation of animal-motion dynamics is ruled out. Thus we must make trade-offs between realism and real-time operations. This can be done by designing worlds and creature-world interfaces that preserve essential characteristics and constraints of animals in the real world while bypassing computationally expensive levels of processing such as early vision.

The creatures interact with the world they inhabit, and with each other through the intermediation of the world. They need to gather information and perform actions. These capabilities are provided by the sensor and motor agents of creatures.

Some possible classes of worlds include Cartesian worlds with simplified physics (as in BrainWorks), more realistic worlds that simulate more details of locomotion such as balance and limb placement,[31] less realistic worlds such as grids,[28,37] or abstract game-theory worlds for simulating problems of behavioral strategy.[1]

There is also the possibility of bypassing the simulation problem and using the real world by means of robotics. This approach is strongly advocated in by Brooks[4] and brought to pedagogical reality in the Lego/Logo system.[24]

4.2.1 **EXAMPLE: A FISH WORLD** Agar's fish live in a finite two-dimensional world that looks like a front view of a fishtank. The fish keep track of their position and velocity. The fish also have a drag coefficient $d < 1$ that simulates the force of drag on a body moving in a fluid by the simple approximation $v_{t+1} = dv_t$.

Fish have a motor agent called `kick` which provides an impulse in the form of a velocity increase, and a sensor agent called `too-slow` which notices when the fish's forward velocity drops below a threshold. By connecting these two agents, a periodic motion is generated that suggests swimming. We have ignored all the details of fish swimming, which would involve expensive fluid-dynamic calculations to simulate accurately, in favor of generating an expressive motion by simple means.

4.3 AGENTS

An Agar creature's behavioral control system is made up of entities called *agents*, loosely based on those described in Minsky's Society of Mind theory.[21] In Agar's implementation of the theory, agents are computational objects that can:

- Execute concurrently,
- Maintain local state,
- Access the state of other agents through connections, and
- Take actions automatically when environmental conditions are met.

An agent is to be thought of as semi-autonomous. This means that an agent is not necessarily doing the bidding of some other agent or outside entity such as the user, but is responding to conditions in its environment, which can include other agents as well as the world. In this sense, an agent is similar to a production rule.

An agent's *condition* can be stated as a boolean function of sensor predicates and other agents' activations. An agent's *action* can be an arbitrary behavior expressible in Lisp. But most agents adhere to certain conventions that enable them to interact with their neighbors in well-defined ways. These conventions will be encapsulated in object classes. For instance, the agents that trigger a motor function will be instances of the class **motor-agent** or specializations thereof.

Actions that an agent may want to perform include:

- Activate or suppress other agents,
- Activate a motor function,
- Activate a script,
- Remember the current activation state of other agents (K-line creation, see 4.4.1), and
- Create a new agent or alter an existing one.

4.3.1 EXAMPLE: SOME STICKLEBACK AGENTS In the current preliminary implementation of Agar, an animal's behavior is specified by a set of **defagent** forms that create the agents and specify their relationships. A **defagent** form can include a number of different clauses.

- The **:if** clause specifies the conditions upon which the agent will be activated. It can be any boolean formula (in Lisp-like form) usually with other agents as its primitive terms.
- The **:do** clause declares some motor action for the agent to do. Here, it is simply a string that gets printed out, but in later versions of Agar it can be a command to a graphics system to alter position of the creature.
- The **:sensor** clause declares that this is a sensor agent. The actual sensing is performed by the **:if** clause, which specifies Lisp code to interface to the world.
- The **:self-inhibit-after** clause takes a numerical argument t that indicates that the agent is to be activated for at most t ticks, after which it will shut itself off.

Here are a few agents that make up the three-spined stickleback's reproductive behavior system:

```
(defagent see-conspecific :sensor)

(defagent see-female-with-gravid-belly :sensor)

(defagent male-reproduction
  :if (and male springtime))

(defagent fight-conspecifics
  :if (and male-reproduction see-conspecific (not court))
  :do "Fight conspecific")

(defagent court
  :self-inhibit-after 1000
  :if (and male-reproduction see-female-with-gravid-belly))

(defagent zigzag-dance
  :if court)

(defagent zig
  :if (and zigzag-dance (not zag))
  :do "Zig away from F"
  :self-inhibit-after 10)

(defagent zag
  :if (and zigzag-dance (not zig))
  :do "Zag towards F with mouth open"
  :self-inhibit-after 10)
```

From this sample of agents, we have gleaned a number of useful constructs to be included in the language. These include the ability of an agent to schedule its own activation according to a rule, the ability to interface to a motor subsystem, and the ability to impose limits on how long it may be active. We may find additional useful constructs, such as hysteresis (remaining active for a period of time after the condition has ceased to be satisfied) or latency (imposing a "resting" period between bouts of activity).

It should be noted that the hierarchical behavior structures of classical ethology (see 5.1.1) are implicitly encoded here by the interagent references in :if clauses, which provide a more general linking mechanism.

4.4 LEARNING

Learning, broadly defined, is the process of changing the behavioral mechanisms of an animal in order to promote its survival. Agar uses *K-lines*, an agent-based mechanism for learning described by Marvin Minsky.[21] We extend the notion of a K-line to include recognizers that learn (*R-trees*) and agents capable of recording temporal activation histories (*scripts*).

4.4.1 K-LINES AND R-TREES In the Society of Mind theory, a learning mechanism called a *K-line* figures prominently. A K-line has the ability to record and recall the activation state of a set of agents that are known to be useful for performing a specific task.

Here we describe a mechanism that is slightly different from K-lines. Whereas the original K-lines remember which other agents are useful for achieving a goal, this mechanism is used by an agent to learn when to activate itself. They can both be built out of the same associative mechanism. We'll call this mechanism an *R-tree*, for its function of recognition and for the tree-like shape that its connections form. An R-tree is in some sense the inverse of a K-line, operating on the input side of an agent rather than on the output side.

For example, imagine a system that lets an animal learn when something is dangerous. The animal already has a mechanism for knowing when it is being attacked, and for fleeing. The R-tree mechanism helps recognize an *impending* attack by learning to recognize the characteristics of an attacker. When attacked, the **attacked** agent activates the imprinting mechanism of the R-tree, causing the connections between the active sensory agents and the R-tree to become stronger. The R-tree will now be activated by similar sensory input in the future, helping the animal to anticipate and prevent future attacks (see Figure 3).

R-trees as described above are *ad hoc* learning mechanisms, attached to a specific agent and hence fulfilling a fixed, specific purpose. The existence of a one-shot special mechanism for learning seems to correspond to the ethological phenomenon of imprinting. A more general mechanism that incorporated the same idea might depend on agents that were capable of *creating* R-trees when a need for better recognition was detected.

On the neural level, an R-tree can be built out of Hebb synapses. These are connections that increase their weight when both sides of the connection are active simultaneously. An "imprint" signal activates the neuron, causing the synapses corresponding to appropriate sensory agents to increase their weights. In Agar's model, an R-tree simply has a list of agents that can potentially activate it, and polls them when given an imprint command.

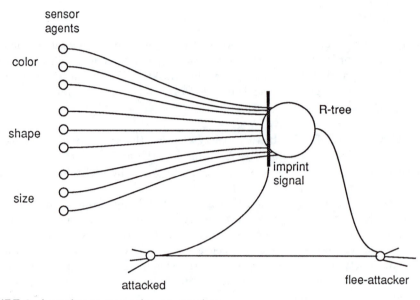

FIGURE 3 Learning to recognize an attacker

4.4.1 SCRIPT AGENTS

Temporal relations between events often contain important information. In the above example, it was important to record the state of the sensory agents as it was immediately *before* the attack. Other forms of learning may depend on noticing temporal relations between events and inferring a causal relationship. Here we sketch a theory of temporal learning based on simple mechanisms.

We postulate a type of agent that is capable of recording the activation history of other agents, called a **script-agent**. In the usage of Artificial Intelligence[34] a script encodes general knowledge of how scenarios unfold, in the form of descriptive symbolic relations. For example, a restaurant script would contain slots that specify that the actors involved are customers, waiters, cooks, etc; that there are subscripts such as ordering, eating, and paying, and that significant objects include food, utensils, and money.

Agar's scripts are less like propositions in a knowledge base and more like a recording of event happenings. A script is a recording of agent activations, which could correspond to sensory events, actions on the part of the creature, or the activation of internal units. A script agent can play back this record to recreate a pattern of events. Since it is rarely desirable to recreate a past event exactly, a script will only record agents at a fixed level of detail (a *level-band* in Minsky's terminology; see Minsky,[21] p. 86), representing, for instance, general patterns of action but not specific muscle firings.

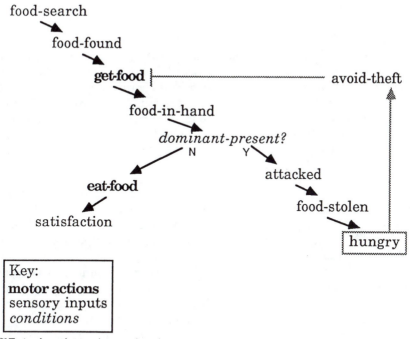

FIGURE 4 A script and associated agent.

To build a script mechanism, we first generalize the notion of a K-line to record the activation states of agents over time, rather than at a particular instant. Activating the script will activate agents in the original recorded sequence. We also assume that some generalization mechanism exists that can form branching structures to represent different possible paths that events may take.

To make these scripts useful, we will have to postulate some additional machinery. We will need agents that can look at the branching paths of a script and influence action based on the causal chains they imply. We want to activate script paths without actually causing actions to be taken. This means that we will also need gating mechanisms that will allow actions in a script to be activated without actually taking action. This may be another function of level-bands, in that the lower (motor) levels can be shut off in order to perform planning-like computations.

As an example, we will take an anecdotal account of an instance of deceptiondeception in a chimpanzee colony (from de Waal,[12] p. 73)

"Dandy has to offset his lack of strength by guile. I witnessed an amazing instance of this...We had hidden some grapefruit in the chimpanzees' enclosure. The fruit had been half buried in sand, with small yellow patches left uncovered. The chimpanzees knew what we were doing, because they had seen us go outside carrying a box full of fruit and they had seen us return with an empty box. The moment they saw the box was empty they began hooting excitedly. As soon as they were allowed outside they began searching madly but without success. A number of apes passed the place where the grapefruit were hidden without noticing anything – at least, that is what we thought. Dandy too had passed over the hiding place without stopping or slowing down at all and without showing any undue interest. That afternoon, however, when all the apes were lying dozing in the sun, Dandy stood up and made a bee-line for the spot. Without hesitation he dug up the grapefruit and devoured them at his leisure. If Dandy had not kept the location of the hiding place a secret, he would probably have lost the grapefruit to the others."

It seems that Dandy has the capability to manipulate a great deal of knowledge about the actions and intentions of other apes, displaying intentionality of a high order, that is, the ability to have beliefs about the beliefs of others, and to reason about them.[13] But equivalent mechanistic explanations are possible, and needn't be unduly complicated.

Suppose that Dandy has a script agent that has recorded previous experiences of other apes stealing his food (see Figure 4). We also assume that a learning mechanism has noticed a possible unpleasant outcome of the script (*hungry*) and has created a new agent (*avoid-theft*) that suppresses the action that leads to this consequence. Now suppose Dandy finds some food, and there are other apes present. The script is activated with the connections to motor activities turned off by the gating mechanism. When the **hungry** node is activated, the *avoid-theft* agent causes Dandy to suppress the action-taking portion of the script. When the apes are gone, the conditions of the agent are no longer met, and he is free to gather the food.

Many details are left out of this explanation. Some notable remaining questions are how the script recording mechanism knows which agents are important; how it knows how to generalize events into branching structures, how the agents know which actions to suppress, and how the connection between agent activity and motor activity is managed.

This scheme has no symbolic representations as such and no general reasoning mechanism. Instead, it depends only on a simple associative mechanism for recording temporal relationships between events as they occur, and some simple control functions. Brooks has argued[5] that representation can be dispensed with altogether for intelligent situated creatures. We take a similar but less radical position that representation should be dethroned from its central position, emphasizing instead the mechanisms that underlie representation and are responsible for its generation and use.

5. BACKGROUND AND RELATED WORK

5.1 MODEL-BUILDING IN ETHOLOGY

An ethologist's task is to observe animal behavior and form theories about it. Questions about the causality of behavior can be classed as questions about *proximate* cause ("how" questions) or *ultimate* cause ("why" questions). Proximate cause deals with the analysis of the workings of behavioral mechanisms. Ultimate cause deals with the evolutionary advantages that such mechanisms contribute. This section describes various efforts at building models to answer both types of questions.

5.1.1 CLASSICAL ETHOLOGICAL MODELS The early ethologists were interested in putting theories of animal behavior on a unified basis. Tinbergen developed a theory of hierarchically structured centers,[36] which was extended by Lorenz.[19] An example of such a hierarchy (for part of the stickleback reproductive cycle) is illustrated in figure 5.

In this model, "drive energy" moves downwards from the top unit to lower units under the control of *innate releasing mechanisms* (IRMs). These are sensory

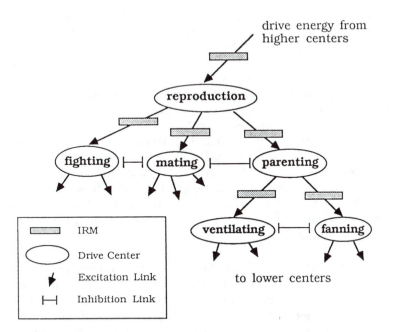

FIGURE 5   A partial behavioral hierarchy

mechanisms which are triggered by stereotypical fixed stimuli (such as the red belly of the male stickleback, which serves to trigger an IRM in the female that initiates mating behavior). The hierarchy of drive centers descends from general to specific units, terminating in units that are directly responsible for specific muscle contractions. Lateral inhibition links ensure that only one unit on any given level is active at a time.

This hierarchical scheme was later generalized to be a heterarchy (in which a low-level unit may be activated by multiple high-level units) and to include feedback mechanisms.[2] This style of modeling provided an organizing paradigm for ethology until the late 1950's, when the concept of drive energy fell into disrepute. It was later revived by Dawkins[11] who pointed out that the idea of hierarchical behavior structure could be salvaged by changing the interpretation of interlevel connections from energy transfer relations to control relations. This is the basis for the model used by Agar.

5.1.2 COMPUTATIONAL ETHOLOGY

Early cyberneticians built simple, reflex-based artificial animals to demonstrate the possibility of mechanical behavior and learning. Grey Walter built turtles that responded to light and could do conditioned learning.[38] This style of theorizing has been continued by Braitenberg,[3] whose work inspired BrainWorks.

More recent systems emphasize ecological interaction by supporting more than one species of artificial life. RAM[37] and EVOLVEIII[28] fall into this category. The latter is particularly interesting because it simulates several levels of biological organization (genetic, behavioral, and ecological) and the relationships between them.

Stewart Wilson has studied learning in artificial animals.[39] He utilized Holland's classifier systems, which learn by mutating rules in a manner derived from biological genetics. This system had some success in learning simple regularities in a grid-type world. Wilson suggests that in practice, classifiers must be arranged in hierarchical modules that learn to control each other. This is reminiscent of Society of Mind theories and suggests a method for learning within individual agents.

Petworld is another animal simulation system being developed under the auspices of the Vivarium.[9] Its features include a system of ranking alternative behaviors through a hierarchical structure of modules, and performance of the relatively complex task of nest-building.

5.1.3 COMPUTATIONAL EVOLUTION

Richard Dawkins has recently demonstrated a software package called *The Blind Watchmaker* (after his book of the same name[10]) that demonstrates the evolution of form. This system does an excellent job of creating a diversity of forms from a few simple parameters. However, the resulting form (called a "biomorph") does not behave, and its only criterion for survival is its appeal to the eye of the user.

Niklas[23] also simulates evolution parametrically, in the realm of plants. The parameters control the form of the plants, which then compete with each other for available sunlight.

Hans Moravec[22] has done some interesting theorizing about the nature of existing silicon life to biological life. Essentially, he claims that computer hardware and software together constitute a new form of life, which survives and reproduces by being useful to humans. Humans also mutate software, usually by better heuristics than natural selection's generate-and-test methodology.

5.2 PROGRAMMING METHODOLOGY

The goals of the Vivarium Project include not only simulating animals, but putting the ability to do so within the reach of grade-school children. To accomplish this, we must make the abstract operations of programming more tangible, thus engaging children's existing knowledge of how physical objects behave and interact. Several existing techniques for doing so have been incorporated into Vivarium systems. We also hope that we can contribute a new approach to the problem of visualizing computational processes by using animals as metaphors for them.

5.2.1 GRAPHIC PROGRAMMING
Graphic programming[32,33,35] refers to techniques of specifying programs by manipulating diagrams rather than text. The advantages of graphic programming are that tangible graphic relations replace the more abstract grammatical relations of traditional languages, and objects and options can be displayed in menus and palettes to guide the user in making choices. Graphic programming tries to use commonsense knowledge about connectivity as a basis for a language, alleviating the need to learn a textual grammar.

The disadvantages of graphic programming are that diagrams can often be as unreadable and complex as textual programs, and that not all relationships are easily translated to graphics. BrainWorks used graphic programming technology successfully because the possible complexity of programs, and hence diagrams, was inherently limited.

5.2.2 PROGRAMMING BY EXAMPLE
Programming by example refers to a class of techniques that generate programs from the user's manipulations of tangible objects. In such a system, the user performs a series of concrete operations on some computational objects, then generalizes these operations into routines, usually by replacing constants with variables.[14,18]

This technique is applicable to programming animals. Imagine an animal in a situation that it doesn't know how to deal with. We show the animal what to do via a graphic gesture, such as moving the mouse. Then we attach the resulting script to an agent that is triggered by the current conditions. This is very similar to the technique of K-line learning outlined in 4.4.1, but under control of an outside teacher.

Script-agents as described in 4.4.2 may be considered an attempt to do *learning by example*. That is, we want animals to program themselves by taking examples from life and altering their behavioral procedures based on them.

5.2.3 THE ANIMAL AS A PROGRAMMING METAPHOR What is a programming meta-phor, and why should animals be considered as such? Any language or system for manipulating information must provide some basic kinds of objects to be ma-nipulated and operations for manipulating them. Traditional computer languages (FORTRAN, C, PASCAL, etc.) provide abstract formal objects such as numbers, strings of characters, and arrays. Novices and children are not experienced in deal-ing with such objects. Good user interfaces allow the novice to draw upon existing knowledge, by presenting the computational objects via metaphors with tangible objects in the real world.[15,25] The behavioral rules of the system are presented metaphorically, by an implied analogy to familiar objects.

Examples of user interface metaphors include iconic operating interfaces (Mac-intosh, Xerox Star) that presents files as tangible graphic objects that can be phys-ically dragged from one file to another or thrown in a trash can; the HyperCard system which presents units of information as index cards; and the Logo computer language, which presents abstract geometric operations in terms of a turtle that can perform local movements.

Young children tend to think in animistic terms. They tend to attribute action to intentions of objects rather than mechanistic forces. Thus a rock rolling down a hill is alive to a child, while a stationary rock is not.[6] Computational processes are "moving" objects that have inherent goals and intentions (whether or not explicitly represented). Thus it should be possible to exploit a young child's existing animistic framework to give the child an understanding of computer processes.

Animals are complex, autonomous, and animate. They have particular re-sponses to stimuli, different modes or states, they can have goals, they can have emotions, they can be unpredictable. We hope to exploit the user's ideas about animals to allow her to build and understand computer processes that share some of these characteristics.

5.3 SELF-SCRIPTING ANIMATION

Traditional computer animation technology requires specifying explicit trajectories for each object in a scene. Current work focuses on making *self-scripting animation* animations, in which each object is responsible for determining its own motion by application of a set of rules to particular conditions in a simulated world.

Reynolds developed an actor-based animation system called ASAS.[26] It intro-duced the idea of generating animation by having active objects that are controlled by their own scripts. This idea was later extended to allow the objects to have scripts that could alter their behavior based on external conditions.[27] The result-ing system was used to produce animations of animal group motion such as bird flocking. It uses a true distributed behavioral model (each bird computes its own motion). It does not attempt to simulate sensorimotor constraints (birds have es-sentially direct knowledge of the locations of other birds and obstacles) and the behavioral model is very simple.

Kahn developed a system called ANI that creates animations from story descriptions.[17] It is more of a "top-down" system than others discussed here: it takes a description of a story and then computes how the characters should move to best enact that story, whereas we are more interested in specifying responses to particular situations and letting the stories unfold as emergent behavior. But ANI does use a distributed actor model to compute its behaviors, and has a great deal of relatively high-level knowledge. Some of its actors correspond to particular emotions or behaviors, i.e., `shy`, `angry`, `evil`. Others refer to particular scenes that occur in a movie: `kept-apart`, `justice`. In addition to affecting the actions of particular characters, agents can manipulate the film medium itself to aid the story telling process, such as introducing a close shot of two characters to suggest intimacy between them.

Zeltzer outlines a scheme for intelligent motor control.[41] It distinguishes between two domains of problem solving, cognitive and motor. Motor problem solving does not use symbolic representations. Instead it uses a lattice of behavioral skills (similar to Tinbergen's hierarchy of drive centers; see 5.1.1) that can be used to reduce differences from goal states by behavioral rules that result in immediate difference reduction.

5.4 THE SITUATED ACTION APPROACH TO ARTIFICIAL INTELLIGENCE

Traditionally, the treatment of action in Artificial Intelligence has centered around the notion of *planning*. The task of planning is to compute a sequence of steps for reaching a given goal. This data structure is called a plan, and is passed to another module (the executor) to be carried out.

This notion of planning has come under attack as unrepresentative of the way in which intelligent beings actually act in the world.[34] We do not compute a plan ahead of time and then carry it out; instead we are in an interactive relationship with the world, continuously exploring possibilities for actions. This is what gives us the flexibility to deal with a world that cannot be fully represented in memory.

Another criticism of planning as a means of generating action is that it assumes perfect information not only about the current state of the world, but about the changes induced in the world by the agent's operations. The problem of keeping track of the results of change is significant enough to be given its own name (the "frame problem"[30]). One suspects that this problem is a result of the logicist framework used for the analysis of planning rather than an inherent problem.

The situated action approach advocates a more interactive relationship with the world. Action is driven primarily by current conditions in the environment and only secondarily by internal state. This is the approach used by ethologists and by the systems described in this paper. Agre and Chapman describe an artificial agent that operates inside a cellular world using only combinatorial logic to connect a set of sensory and motor primitives.[8] They also suggest how abstract reasoning might exist as an emergent function of situated activity.[7] Brooks is working on theories that will allow robots to navigate and perform tasks without an internal representation of their space.[4]

Artificial animals are an ideal domain to test situated action ideas. There is less temptation to grant non-human animals either perfect global knowledge or unbounded rationality. If we are successful in modeling animal behavior, the question posed by situated AI is: What modifications or additions to these simple animal systems enable humans to perform their prodigious feats of everyday intelligence?

6. CONCLUSION

We have built systems that simulate some of the essential properties of animal behavior in simplified worlds: perception, locomotion, orientation, foraging, predation and evasion. We have seen that animals can accomplish their tasks using relatively simple distributed mechanisms.

Building artificial animals serves research goals from different fields. We can use them to help understand real animals (including humans), to provide accessibility and dramatization of computational processes, and to simplify the task of generating animated characters.

Future directions for work include increasing the realism of the simulated world, increasing the flexibility and complexity of the animal's behavioral control systems, and expanding the simulation to include more evolutionary and ecological context. We also hope to make friendly and transparent user interfaces to the agent languages to allow children to construct behavioral systems.

ACKNOWLEDGMENTS

This work was supported by Apple Computer Inc. Thanks to Phil Agre, Bill Coderre, Alan Kay, David Levitt, Ann Marion, Margaret Minsky, and Marvin Minsky for enlightened and enlightening comments.

REFERENCES

1. Axelrod, R. (1984), *The Evolution of Cooperation* (New York: Basic Books).
2. Baerends, G. P. (1976), "The Functional Organization of Behavior," *Animal Behavior* 24, 726–738.
3. Braitenburg, V. (1984), *Vehicles: Experiments in Synthetic Psychology* (Cambridge: Massachusetts Institute of Technology Press).
4. Brooks, R. A. (1986), "Achieving Artificial Intelligence through Building Robots," *AI Memo 899, Massachusetts Institute of Technology Artificial Intelligence Laboratory.*
5. Brooks, R. A. (1987), "Intelligence without Representation," unpublished.
6. Carey, Susan (1985), *Conceptual Change in Childhood* (Cambridge: Massachusetts Institue of Technology Press).
7. Chapman, D., and P. Agre (1986), "Abstract Reasoning as Emergent from Concrete Activity," *Proceedings of the 1986 Workshop on Reasoning About Actions & Plans* (Los Altos, CA: Morgan Kaufmann), 411–424.
8. Chapman, D., and P. Agre (1987), "Pengi: An Implementation of a Theory of Situated Action," *Proceedings of AAAI-87.*
9. Coderre, W. (1988), "Modelling Behavior in Petworld," these proceedings.
10. Dawkins, R. (1987), *The Blind Watchmaker* (New York: W. W. Norton).
11. Dawkins, R. (1976), "Hierarchical Organization: A Candidate Principle for Ethology," *Growing Points in Ethology*, Eds. P. P. G. Bateson and R. A. Hinde (Cambridge: Cambridge University Press).
12. de Waal, F. B. M. (1982), *Chimpanzee Politics: Power and Sex among Apes* (New York: Harper & Row).
13. Dennett, D. C. (1983), "Intentional Systems in Cognitive Ethology: The 'Panglossian Paradigm' Defended," *The Behavioral and Brain Sciences* 6, 343–390.
14. Finzer, W., and L. Gould (1984), "Programming by Rehearsal," *Byte* **June 1984,**
15. Goldberg, A., and A. Kay (1976), "Smalltalk-72 Instruction Manual," *Technical Report SSL-76-6, Xerox Palo Alto Research Center.*
16. Holland, John H. (1986), "Escaping Brittleness: The Possibilities of General-Purpose Learning Algorithms Applied to Parallel Rule-Based Systems," *Machine Learning II: An Artificial Intelligence Approach*, Eds. T. M. Mitchell, R. S. Michalski, & J. G. Carbonell (Los Altos, CA: Morgan Kaufmann), chapt. 20, 593–623.
17. Kahn, K. (1979), "Creation of Computer Animations from Story Descriptions," *AI Lab Technical Report 540, Massachusetts Institute of Technology.*
18. Lieberman, H. (1984), "Seeing What Your Programs are Doing," *International Journal of Man-Machine Studies* 21, 311–331.
19. Lorenz, K. Z. (1981), *The Foundations of Ethology* (New York: Simon & Schuster).

20. McCulloch, W. S., and Walter H. Pitts (1943), "A Logical Calculus of the Ideas Immanent in Nervous Activity," *Bulletin of Mathematical Biophysics* **5**, 115–133.

21. Minsky, M. (1987), *Society of Mind* (New York: Simon & Schuster).

22. Moravec, H. (1988), "Mind Children," forthcoming, and Moravec's article in these proceedings.

23. Niklas, K. J. (1986), "Computer-Simulated Plant Evolution," *Scientific American* **245(3)**, 78–86.

24. Ocko, S., S. Papert, and M. Resnick (1988), "Lego, Logo, and Science," *Technology and Learning* **2(1)**.

25. Papert, S. (1980), *Mindstorms: Children, Computers, and Powerful Ideas* (New York: Basic Books).

26. Reynolds, C. W. (1982), "Computer Animation with Scripts and Actors," *Proceedings of SIGGRAPH '82* ([publisher?]).

27. Reynolds, C. W. (1982), "Flocks, Herds, and Schools: A Distributed Behavioral Model," *Proceedings of SIGGRAPH '87.*

28. Rizki, M. M., and M. Conrad (1986), "Computing the Theory of Evolution," *Physica* **22D**, 83–99.

29. Schank, R. C. (1982), *Dynamic Memory: A Theory of Reminding and Learning in Computers and People* (Cambridge: Cambridge University Press).

30. Shoham, Y. (1986), "What is the Frame Problem?," *Proceedings of the 1986 Workshop on Reasoning about Actions & Plans* (Los Altos, CA: Morgan Kaufmann).

31. Sims, K. (1987), *Locomotion of Jointed Figures over Complex Terrain, Master's Thesis, Massachusetts Institute of Technology.*

32. Sloane, B., D. Levitt, M. Travers, B. So, and I. Cavero (1986), *Hookup: A Software Kit,* unpublished software.

33. Smith, D. C. (1977), *Pygmalion: A Computer Program to Model and Stimulate Creative Thought* (Basil und Stuttgart: Birkhauser).

34. Suchman, L. A. (1987), *Plans and Situated Actions: The Problem of Human-Machine Communication* (Cambridge: Cambridge University Press).

35. Sutherland, I. (1963), *Sketchpad: A Man-Machine Graphical Communications System, Ph.D. Thesis, Massachusetts Institute of Technology.*

36. Tinbergen, N. (1951), *The Study of Instinct* (Oxford: Oxford University Press).

37. Turner, S. R., and S. Goldman (1987), "Ram Design Notes," *UCLA Artificial Intelligence Laboratory.*

38. Walter, W. G. (1950), "An Imitation of Life," *Scientific American* **May 1950**, 42–45.

39. Wilson, S. W. (1987), "Classifier Systems and the Animat Problem," *Machine Learning* **2(3)**.

40. Winograd, T., and F. Flores (1987), *Understanding Computers and Cognition* (Reading, MA: Addison-Wesley).

41. Zeltzer, D. (1987), "Motor Problem Solving for Three Dimensional Computer Animation," *Proc. L'Imaginaire Numerique.*

Conrad Schneiker
Advanced Biotechnology Laboratory, Department of Anesthesiology, College of Medicine, University of Arizona, Tucson, AZ 85724

NanoTechnology With Feynman Machines: Scanning Tunneling Engineering and Artificial Life

Dedicated to the *original* NanoTechnologist, *Richard P. Feynman* (11 May 1918—15 Feb 1988) and to a major pioneer of *electronic* NanoComputing, *Forrest L. Carter* (29 April 1930—20 Dec 1987)

> In the year 2000, when they look back at this age, they will wonder why it was not until *the year 1960* that anybody began seriously to move in this direction.
>
> — *Richard P. Feynman*

> It is difficult to suppress one's enthusiasm for the development of a viable *molecular technological base* when one recognizes the possible scientific, industrial and economic spin-off opportunities.
>
> — *Forrest L. Carter*

Artificial Life, SFI Studies in the Sciences of Complexity,
Ed. C. Langton, Addison-Wesley Publishing Company, 1988   **443**

1. INTRODUCTION

Artificial life involves the replication of the processes associated with natural life, but possibly in new media, or on new size scales, or with new organizing principles, etc. One intriguing set of possibilities for creating artificial life is based on Richard Feynman's suggestions in 1959 for ultraminiaturization and extension of our industrial manufacturing capabilities all the way down to the molecular level. Other early pioneers developed related concepts and further generalized them. We will describe several important technological developments along these lines later. Such technology is now termed *nanotechnology* (1 *nanometer* = 1 nm = 10^{-9} m). These approaches to artificial life involve operating directly at the molecular level.

There are many ways that nanotechnology can eventually be applied to the development of artificial life. The following two possibilities are selected to illustrate (1) the possibility of a fine grained continuum of structures spanning natural and artificial life, and (2) the scale invariance principle of artificial life applied to hybrid size scales and hybrid technology life forms. (1) We can start with a completely natural life form and gradually transform it (bootstrap it) into a totally artificial life form by using molecule-by-molecule replacement. (2) We can develop a hybrid living system that incorporates some nanotechnology for computing functions, some natural biology for material synthesis functions, and some microtechnology for artificial replication. We will return to these topics after reviewing the history and state-of-the-art of nanotechnology.

The term "nanotechnology" encompasses much of the subject matter of older terms such as "ultraminiaturization" and "molecular engineering." A similar situation exists for "nanomachines" ("molecular machines"), "nanocomputers" ("molecular electronic devices," "quantum computers," "biochips,"), "nanoreplicators" ("molecular replicators"), and so on. Franks' excellent review paper[88] on nanotechnology notes that the term "nanotechnology" was defined in 1974 by Taniguchi[224]; Franks describes nanotechnology as,

> ... *the technology where dimensions and tolerances in the range 0.1–100 nm* (from the size of the atom to the wavelength of light) *play a critical role* ... *The field covered by nanotechnology is narrowed down to manipulation and machining within the defined dimensional range by technological means,* as opposed to those used by the craftsman ... Nanotechnology is an 'enabling' technology, in that it provides the basis for other technological developments, and it is also a 'horizontal' or 'cross-sectoral' technology, in that one technique may, with slight variations, be applicable in widely differing fields ... Nanotechnology is seen to be of particular importance, and of *immediate relevance*, in areas such as materials science, mechanical engineering, optics and electronics.
>
> *[in this and all other quotes, emphasis is added.]*

Taniguchi[225] notes that "'Nanotechnology' is the term used to classify the integrated manufacturing technologies and machine systems which provide ultra precision machining capability in the order of 1 nanometer...[It] might also be called 'extreme technology' because the theoretical limit of accuracy in machining of substances must be the size of an atom or molecule..." He examines the general trends in ultraprecision machining, which should continue on down into the *substance synthesizing* domain, sometime around the year 2020. This trend is similar to that for very advanced lithography and corresponding experimental electronic integrated circuit devices. Taniguchi[226] describes atomic bit machining as *atom-by-atom processing* using energy beams. He discusses several types of atomic bit processing including separation, consolidation, deformation, cutting, polishing, and surface treatment—each of which would be useful for making experimental nanotools.

By extension, "picotechnology" (1 *picometer* = 1 pm = $10^{-12}m$) deals with the manipulation and modulation of individual atomic bonds and orbitals plus the special equipment where atomic or molecular measurements are made to sub-nanometer precision. There are already special cases where mechanical positioning and measurements of spacing between individual atoms has been pushed to a resolution of 1–10 picometers! The instrument with this phenomenal capability was

FIGURE 1   An early STM constructed with the author's coworkers. Photograph courtesy of Mark Voelker, Optical Sciences Center, University of Arizona, Tucson, Arizona.

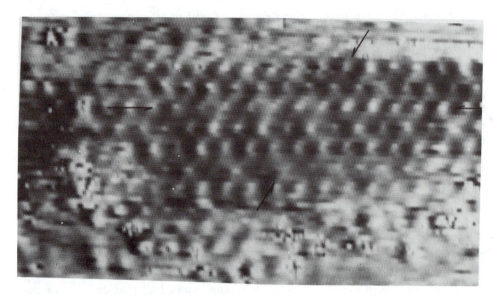

FIGURE 2   A STM image of sulfur adatoms on an Mo(111) substrate at atmospheric pressure. Photograph courtesy of Wigbert J. Siekhaus, Lawrence Livermore National Laboratory, Livermore, California.

invented early this decade; it is the scanning tunneling microscope—and it is also a fundamental tool for nanoengineering.

Scanning Tunneling Microscopes (STMs) can "image" atomic level (i.e., nanometer) structural, dynamic, and electronic properties of materials including metals, semiconductors, and biological molecules. In STMs, an electronic servo system driving a piezoceramic positioner controls highly localized, quantum mechanical, electron tunneling between ultrasharp electrode tips and chosen materials. Tip movement, which maintains constant tunneling current during surface scanning, translates to a surface map with atomic level resolution under optimal conditions. By holding constant different combinations of voltage, current, and tunneling distance, different types of information and material effects occur. Such versatility permits imaging and manipulation of molecules and molecular devices and endows STMs with nanoscale electromechanical interfacing and machining capabilities.

An early STM project initiated by the author and developed with the assistance of others at the University of Arizona (based on lots of useful information provided by Paul Hansma of the University of California at Santa Barbara) produced the STM shown in Figure 1. See Figures 2 and 3 for examples of STM images taken with more modern STMs.

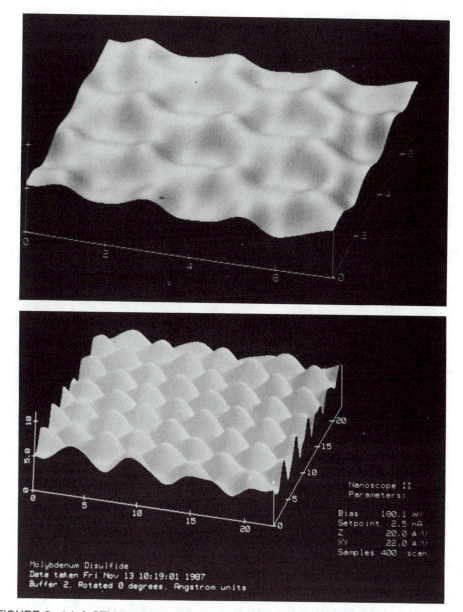

FIGURE 3   (a) A STM image of graphite. Nanoscope II Parameters: Bias—14.0 mV; Setpoint—1.1 nA; Z—21.0 Å; X,Y—21.0 Å; and Samples—400/scan. Captured Tues., Aug. 25, 15:15:38, 1987. (b) A STM image of molybdenum disulfide; parameters and date included in photo. Both images were taken with a Digital Instruments Nanoscope II STM. Photographs courtesy of Virgil Elings, Digital Instruments, Santa Barbara, California.

The possibility and great value of molecular level machines was first noted by Nobel physicist Richard Feynman, who nearly 30 years ago proposed tools to construct nanoscale mechanisms and devices; such tools and mechanisms are *Feynman Machines.* Emergence of STMs and their applied hybrids should greatly facilitate implementation of a general purpose nanoscale manufacturing capability (*nanotechnology*) with major implications for the development and testing of molecular electronic devices and molecular photonic devices.

The application of STM technology to implement Feynman's program was first described by Schneiker[121] in order to: (1) refute claims that Feynman's "top-down" approach to nanotechnology was not viable, (2) to provide an indication of the true scope of his ideas, and (3) to note previously unacknowledged contributions by other early nanotechnology pioneers. Since then, the relevance of Feynman's early work (and that of others) has become much more widely recognized. Hansma and Tersoff[108] note the possibilities of using STMs to realize Feynman's nanotechnology vision. In their Nobel Prize lecture, Binnig and Rohrer[18] note that the capabilities of STMs include the possibility "ultimately to handle atoms and to modify individual molecules, in short, to use the STM as a Feynman Machine." Franks' review of nanotechnology[88] notes the tremendous potential of STM-derived tools for *"scanning tunneling engineering."*

A "nanotechnology workstation," suitable for scanning tunneling engineering, is proposed in Schneiker and Hameroff.[202] It is illustrated in Figure 7, and will be described later. Nanotechnology workstations and related instruments should greatly enhance the fabrication, testing, and realization of molecular devices and structures needed for artificial life and artificial evolution.

As anticipated by Feynman in 1959, nanotechnology is now becoming one of the most exciting areas of research and development. A brief history of nanotechnology is now in order.

A BRIEF HISTORY OF NANOTECHNOLOGY TO 1980

Heinlein[111] nearly invented the concept of nanotechnology in 1940 by suggesting a process for manipulating microscopic structures (see appendix). However, he completely overlooked the full implications of his idea and it had no impact on technological development. However, by the early 1960's, *several* scientists had reinvented similar approaches to micromanipulation and miniaturization, but this time extending all the way down to the nanotechnology domain. Since then, nanotechnology has advanced substantially, although some related developments have sometimes been hidden by proprietary considerations and classification or hampered by lack of funding. Just as visions of future space faring technology include projections of human travel across the galaxy, so too have visions of future nanotechnology possibilities greatly surpassed any reasonably probable near-to-medium-term technical development; these more speculative possibilities are summarized in the appendix.

FEYNMAN: ORIGINATOR OF NANOTECHNOLOGY

In his remarkably prescient 1959 talk "There's Plenty of Room at the Bottom," Feynman[72] proposed using machine tools to make smaller machine tools, to be used in turn to make still smaller machine tools, and so on all the way down to the atomic level. Feynman prophetically concluded that this is "...a development which I think cannot be avoided." Such nanomachine tools, nanomachines and nanodevices are termed *Feynman Machines* (FMs). FMs can ultimately be used to develop a wide range of submicron instrumentation and manufacturing tools, i.e., *nanotechnology*. Feynman's suggested applications for these tools included producing vast quantities of ultrasmall computers and micro/nanorobots. A wide range of other nanotechnology applications have been anticipated.

Others, including Shoulders,[206] realized the technological value of operating in this submicron nanoworld in the early 1960's. In one of the earliest examples of an FM, Shoulders went beyond theory and in 1965 reported the actual operation of a micromanipulator able to position tiny items with 10 nm accuracy while under direct observation by field ion microscopy.[210]

Feynman's definitive source paper on nanotechnology is also reprinted (with Feynman's permission) in Schneiker.[200] It is worth quoting a few of the highlights from the original paper since it forms the basis of many later conceptual developments:

[Consider] the final questions as to whether, ultimately... *we can arrange the atoms the way we want, the very atoms, all the way down!*...[When] we have some control of the arrangement of things on a small scale we will get an enormously greater range of possible properties that substances can have, and of different things that we can do... We can use, not just circuits, but some system involving the quantized energy levels, or the interactions of quantized spins, etc....Another thing we will notice is that, if we go down far enough, all our devices can be mass produced so that they are absolutely perfect copies of one another...[If] your machine is only 100 atoms high, you only have to have it correct to one-half of one per cent to make sure the other machine is *exactly* the same size—namely, 100 atoms high!...Ultimately, we can do chemical synthesis...Put the atoms down where the chemist says, and so you make the substance.

[A] point that is most important is that *it would have an enormous number of technical applications*...

A biological system can be exceedingly small...Consider the possibility that we too can make a thing very small, which does what we want—that we can manufacture an object that maneuvers at that level!...

So I want to build a *billion* tiny factories, models of each other, which are manufacturing simultaneously...

These possibilities for making and manipulating nanostructures are keys to artificial life. Krkumhansl and Pao[140] presented an overview of microscience that was interspersed with germane references to Feynman's classic paper on nanotechnology. They noted [in 1979] that we have not [yet] quite achieved the possibility of fabricating structures at the molecular scale.

In later years, Feynman[75] mentioned giving other talks on his ideas in this area; however, he was unable to locate his notes for them. He also indicated that he had originated his ideas in the early 1950's and that others had independently suggested similar top-down approaches for nanomanipulation in the late 1950's and early 1960's. Freeman Dyson[59] mentioned that Tommy Gold also "was talking about nanotechnology as early as Feynman. But I have nothing in writing." Another one of the people that Feynman alluded to was probably our next remarkable pioneer, K. R. Shoulders.

SHOULDERS: NANOTECHNOLOGY EXPERIMENTALIST

Shoulders' early work (1960–1965) paralleled Feynman's in several important respects. However, Shoulders expended considerable effort in pioneering *experimental* work. Although he thought about nanoscale structures, he generally limited the scope of his writing to scales he thought he *personally* could accomplish in the lab. Shoulders[207] rejected the use of biological building blocks, even though biological "processes do work, and they can do so in a garbage can without supervision." He saw them as too limited environmentally and too difficult to control with available technology; instead, he sought to *directly* produce much simpler, more powerful, and much more rugged nanostructure arrays, at video frequencies rates, which in turn could ultimately aid in their own replication. These early efforts may be viewed as an early approach to a nonbiological proto-life system suitable for vacuum environments.

For *robotic artificial life forms*, superabundant computing power is one means to manage the complex organizational functions of the sensing, analysis, construction and manipulation tasks required for the replication and evolution of increasingly advanced systems. Both Shoulders and Feynman saw the possibilities and importance of building *immensely* more powerful computing machines and robots that would rival the human brain in numbers of parts and complexity. These would be important for advanced robotic replicators and for controlling nanomanipulators. Shoulders[206] states: "We want to build electronic data processing systems that have the complexity of human neural networks but are capable of operating with electronic speed." Feynman[72] also thought that the "possibilities of computers are very interesting...if they had millions of times as many elements..." Shoulders[207] elaborates:

We propose...a component based upon the quantum-mechanical tunneling of electrons into the vacuum...The transit time for electrons would be about 10^{-13} sec...

Ultimately, we would use a vacuum-tunnel effect cathode array for our electron source. The emission from discrete areas would be controlled by local grids...Thus we have components made by electronic micromachining responsible for the building of new systems by the same method. In the end, *self-reproduction would be a distinct possibility* without the use of a lens system, because all copies would be made on a one-to-one size basis.

As foreseen by Morrison,[171] this technology clearly had potential for an artificial life form. Shoulders[208] maps out an integrated circuit technology based on field-emission vacuum-tunnel-effect devices using this technology and adds:

Our over-all component size is constrained by the [electron beam] construction techniques [currently available at the time] to be within the limits of one-tenth of a μm and two μm... *The lower limit is set by the resolution of the machining process. A single metal and a single dielectric...seem to be the only materials needed... The geometry is extremely simple...Electrostatic relays, electromechanical filters, light generators and detectors, all of similar scale and simplicity are proposed for I/O channels and memory.*

Note that since the resolution of electron beam technology has improved by one to two orders of magnitude, many of Shoulders' proposals could now be scaled down to the 1–10 nm range. Indeed, Feynman[73,75] suggested using the end of a conical STM tip to make experimental nanoscale triode structures. Since then, Fink[76] has developed a technique to make monoatomic STM tips in which the last three atomic layers in the tip form an atomically perfect pyramidal structure. Thus, the key part for experimenting with absolutely minimum scale (and perhaps maximum performance) nanotechnological vacuum diodes and triodes now exists (Figure 8). Fink[79] has recently made field-emission measurements on the exceptionally narrow and intense electron beams from single atoms on monoatomic tips; those results and the analysis of Garcia et al.[94] indicate that it may be the key to the "micro-microscope" (a greatly improved scanning electron microscope which Feynman[73] wanted developed as part of his nanotechnology program.

Shoulder's proposed electrostatic relays could also incorporate wear-free, non-contacting, subnanometer, vacuum-tunneling gaps, incorporating just a few atoms in the limiting case.[203,204] These could also be used for electrostatic nano-actuators and mechanical power sources. Shoulders[209] later reports testing of prototype devices and structures.

Another innovative concept that might eventually bear on artificial life and nanotechnology is outlined in Shoulders,[209] including the discussion of plasma machines:

> This is the most *fantastic* tale. I will discuss how to make an intelligent organization without any [mechanical] parts at all ... I am primarily interested in man- or machine-made *plasmas* organized so as to constitute intelligent machines ... This ... leads potentiality to great structural richness, provided that we can organize the process to the critical level of complexity, which perhaps separates man from other animals, beyond which the machine can assist itself at organizing at a rate greater than the rate provided by more limited man ... So, the question is—how do we get started and take advantage of all of these potentialities? ... In the simplest case, the machine should be an iterative array of components that are isolated from each other, but have the ability to intercommunicate via sinuous electron paths. The components must be able to launch, adsorb, receive and steer electrons or groups of electrons. Such an organization would allow the creation of order within the machine at the dictate of other organized areas. This newly-made order could be propagated physically through the machine and then be destroyed if found wanting. The components could then be used again for higher levels of organization. This arena concept (using fundamental elements at fixed or movable loci but having flexible connectivity and a reversible change of state without leaving a residue) would greatly modify most present concepts of machine organization ... We will claim that field-emission devices could be used to fabricate a "wireless" machine using any interesting organization ... and that a complete machine could be built on this principle.

Shoulders describes some experiments in electrodynamic particle confinement and notes that "it will be a long time to the product phase of this kind of machine." He later adds,

> By learning the laws of organization through fixed structures we may eventually be able to organize plasma structures to the point that they provide enough inherent stability to become their own container. The number of interactions possible in a plasma system and the speed and energy density of these interactions make it highly desirable to seek a way of organizing them.

Finally the connection with nanotechnology:

> Nothing would be more desirable to me, as a researcher on intelligent structure organization than to dispense with the need for tediously carving microscopic shapes by taking advantage of the almost infinite elasticity and plasticity of plasma structures by organizing our machines in the large and

then squeezing the structures down to size by changing the confinement parameters.

Shoulders[210] indicated that he is still interested in self-organizing systems that could increase their complexity in order to evolve into artificially intelligent, self-conscious entities.

In discussing the brain's computing abilities, Feynman[72] also considered computing from a mechanical perspective: "But our mechanical computers are too big; the elements in this box are microscopic. I want to make some that are submicroscopic." Shoulders[208] considered a similar electro-mechanical alternative: "...we may reconsider a mechanical electrostatic relay..." Although (electro)mechanical nanocomputers are an interesting idea, both Shoulders and Feynman noted that when it came to speed,...the real promise of nanocomputing lay in utilizing quantum mechanical effects.

Feynman[72,76] suggested quantum effects might be used for computing and Carter[22] has proposed a chemical structure that would use quantum mechanical, electron tunneling effects for implementing a computer logic element. These and several other tantalizing possibilities are discussed later.

ETTINGER, WHITE, DARWIN, DONALDSON: MEDICAL NANOTECHNOLOGY

Ettinger[63,64] suggested using nanotechnology for life extension and artificial evolution, and "*nano*miniaturization" of robots was later suggested by Ettinger.[65] White,[239] Darwin,[38] and Donaldson[48,49] envisioned the use of genetic engineering and other possibilities for making nanorobots that could function as cell repair machines. These possibilities are discussed in the appendix.

VON HIPPEL, VON FOESTER, ZINGSHEIM, ELLIS, FULLER: MOLECULAR ENGINEERING

Forrest[85] notes, in connection with Von Hippel, that "... nanotechnology...will evolve from the integration of such diverse disciplines as genetic engineering, biophysics, robotics, artificial intelligence, computer-aided design and manufacture (CAD/CAM), biology, chemistry, physics, computer science, materials science, and many others. The concept of integrating these technologies is not new. Arthur von Hippel foresaw it at least as early as 1963, and others probably even before that."

Von Hippel[236] noted the fantastic possibilities that better materials technology holds for the world if only current science and engineering limitations could be conquered: "Suddenly all this is changing. 'Molecular science'...has made a more powerful approach possible: 'molecular engineering,' the building of materials and devices to order."

Von Foester[235] commented that von Hippel's anticipation of "engineering of molecules according to specification" may "also show us some novel manifestations of life." Von Foester derived some mathematical constraints for the molecules needed for such a "molecular bionics" modeled on chemical modifications to a particular macromolecular structure called a *macromolecular sequence computer.*

Zingsheim[252,253] was interested in molecular engineering using nanometer surface microstructures. He noted that "The aim of molecular engineering...is the design and construction of man-made complex supramolecular systems from building blocks of molecular dimension." He further noted that it concerns the "development of molecular components and assembly tools allowing manipulations at molecular dimensions."

In a paper on microteleoperators, Ellis[60] reports:

> Electrically operated micromanipulators add automatic high-speed movement to normal manual control...
>
> In addition to their use in biological investigations, piezoelectric micromanipulators may find important new uses in the development of semiconductor devices and microcircuits. For example, semiconductor junctions could be formed by microetching and electroplating with microelectrodes. Complex circuit paths could be formed by etching through a conducting layer deposited on an insulating substrate...With these techniques, complex circuits of unprecedentedly small size could be fabricated automatically.

Its interesting to note that these very same functions could be performed at or near the ultimate limits of miniaturization using STMs. These discussions of control methods, linkages, dynamics, etc., for teleoperated microtools still have relevance to Feynman Machines.

Buckminister ("Doing More With Less") Fuller proposed a nanoarchitecture having an interesting recursive or fractal structure (Fuller and Applewhite[89]). Its macro level structure is simply a tensegrity mast, which is a rigid structure constructed from an ingenious configuration of interconnected tension (cable) and compression (strut) members, with the unique feature that the struts are isolated from each other. In one version of the macro-tensegrity mast, each individual solid strut and cable may be replaced by a miniaturized version of the macro-tensegrity mast. And then each one of the miniature solid struts may itself be replaced by a still smaller subminiature tensegrity mast. And so on recursively, down to the atomic level. The end result is an extremely light, mostly empty, yet rigid structure. This same idea may be applied a large variety of other structures.

CHANGEUX AND KUHN: MOLECULAR MACHINES AND DEVICES

The feasibility of artificial molecular machines is implied by the view that biology is based on molecular machines. Changeux[28] notes that,

> The analogy between a living organism and a *machine* holds true to a re-markable extent *at all levels* at which it is investigated ... An organism can be compared to an *automatic factory* ...
>
> [The] cell is a mechanical microcosm: a mechanical machine in which the various structures are interdependent and controlled by feedback systems quite similar to to the systems devised by engineers who specialize in control theory ...
>
> Regulating the production lines are control circuits that themselves require very little energy ... The elementary machines of the cellular factory are the biological catalysts known as *enzymes* ... Built into [their] structure, as into a computer, is the capacity to recognize and integrate various signals.

Langmuir-Blodgett films are one molecule thick films that are made by spreading fatty acid molecules across an air-water interface. These films may have guest molecules inserted into them, and a simple dipping process may be used to stack up such film into multilayer structures. Kuhn[141] describes his pioneering work (starting in the 1960's) in applying Langmuir-Blodgett films to molecular scale devices:

> The construction of an artificial system acting as a complex machinery with cooperating components of molecular dimensions is a great challenge. First attempts in the early 1960's to approach this aim were governed by the idea that the Langmuir-Blodgett technique ... might be suitably modified and could then offer a way to arrange appropriate functional molecules of different species in an adequate fixed structure where the molecules could cooperate in a complex and purpose-oriented manner. Many different techniques to reach that goal have been developed and are summarized: methods to control monolayer formation and transfer and to study monolayer absorption, reflection and fluorescence spectra; special techniques to check the architecture of monolayer organizates by combining energy and electron transfer; synthesis of sterically interlocking and functionally cooperating molecules ...
>
> [It] should be mentioned that the search for the basic mechanisms in the origin of life can be strongly stimulated by studying possibilities of the formation of artificial machineries of molecular size and that quite unorthodox views are obtained from such studies. Conversely, the study of these mechanisms should indicate new ways of obtaining artificial devices of molecular size ...

The synthesis of molecules for use as components in *designed* assemblies to be used as tools of molecular size should be a challenging new field.

Kuhn[142] adds, "To our knowledge, the first demonstration of arranging molecules by means of an electron beam using a STEM (not STM) was realized in Kuhn's group by H. P. Zingsheim."[252] Nature provides us with a wealth of prototypical molecular machines; some basic mechanisms of some of these machines are examined by Mitchell[167] and McClaire.[162]

LAING: ARTIFICIAL NANOREPLICATORS

In a series of papers with direct relevance to artificial life, Laing[144-148] wrote about artificial replicators. In particular, his 1974 and 1975 papers are motivated by considerations of artificial molecular replicating machines, constrained to be "biologically reasonable" (including *future* biological possibilities and, hence by implication, artificial life). These papers have direct relevance to artificial nanoreplicators and microreplicators. While Von Neumann[237] formulated a general theory of replicating automata in abstract form, Laing has considered some important design possibilities for molecular realizations of such automata. One form of these nanoreplicators described by Laing[145] utilized molecular (data) tapes (based on the idea of universal Turing machines) to implement replicators (based on Von Neumann's generalization of self-replicating Universal Turing machines). He then showed three ways that such molecular machines might replicate themselves. His exploration of "artificial organisms" was "a vehicle for the exploration of broad biological possibilities."

Laing's[146] studies of kinematic replicators results in a new self-inspecting design that leads to the interesting conclusion that "contrary to von Neumann's surmise, a prior description is not essential to the nondegenerative machine self-reproductive process." Among other interesting features, this new replicator design has interesting self-repairing capabilities as well.

ROTHSTEIN, AVIRAM, RATTNER, CONRAD, CARTER, LIEBERMAN: MOLECULAR NANOCOMPUTING

Scientists extrapolating the development of electronic devices in the 1950's and 1960's noted that the next century should see the development of electronic devices the size of individual molecules and much faster than neurons. Rothstein[194] examined some fundamental limits on chemical information storage systems. In 1974, Aviram and Ratner[4] presented one of the earliest specific proposals for a molecular electronic device, in this case a molecular rectifier. Conrad[30,31] has studied information processing in molecular systems, molecular automata, and molecular computing in general. Forrest Carter[20,21] of the Naval Research Laboratory started searching for means to assemble entire computers from molecular devices. Lieberman[153] examined some of the possibilities for "molecular computers," including their phenomenal memory capacities, noting that while on a modern computer there are about "10^{10} simple memory elements," that on the other hand, "on an excellent machine of the distant future...there might be 10^{30} elements."

YOUNG, WARD, SCIRE, TEAGUE: PROTO-STM

The STM was very nearly invented 10 years earlier, but success was prevented by equipment vibrations and other problems. Young, Ward and Scire[249] reported metal-vacuum-metal tunneling experiments and described a machine they had developed[250] for these experiments at the National Bureau of Standards:

> A noncontacting instrument for measuring the microtopography of metallic surfaces has been developed to the point where the feasibility of constructing a prototype instrument has been demonstrated...In the MVM [metal-vacuum-metal] mode, the instrument is capable of performing a noncontacting measurement of the position of a surface to within about 3 Å. The instrument can be used in certain scientific experiments to study the density of single and multiple atoms steps on single crystal surfaces, absorption of gases, and processes involving electronic excitations at the surfaces.

And, as in most current STMs, piezo scanning systems were used in their system. They were so very close to constructing a full STM—just before their funding was cut in 1972 (Gadzuk[92])!

THE SECOND ERA OF NANOTECHNOLOGY: DIRECT NANOMANIPULATION TOOLS

The period from 1980 onwards is the second era of nanotechnology. It is characterized by the development of *tools for direct molecular and atomic scale "imaging" and manipulation*, including "scanning tunneling engineering" (Franks[88]) and "atomic bit machining."[226]

THE SCANNING TUNNELING MICROSCOPE

One extremely desirable nanotechnology tool has been missing: a nondestructive, relatively inexpensive and easy-to-use nanometer resolution imaging capability— hence the value of STMs. The STM was invented at the IBM Zürich Research Labs[11,12] in 1981. Binnig and Rohrer[12] describe how the STM originated from their attempts to study thin oxide layers on semiconductors in late 1978.

The STM (Figure 4(a) scans an ultrasharp conducting tip, using tungsten, for example, over a conducting or semiconducting surface. When the tip is within a few angstroms of the surface (Figure 4(b)), a small voltage applied between the two gives rise to a tunneling current of electrons, which depends exponentially on the tip-to-substrate separation (about an order of magnitude per angstrom). A servo system uses a feedback control that keeps the tip-to-substrate separation constant by modulating the voltage across a piezoelectric positioning system. As the tip is scanned across the surface, variations in this voltage, when plotted, correspond to surface topography. Depending on the substrate, typical tunneling currents and voltages are on the order of tenths of nano-amperes and hundreds of millivolts, respectively. Binnig and Smith[17] have recently developed a simpler, more compact piezoelectric scanner mechanism consisting of a single tube (Figure 4(c)) which should prove useful for multi-tip operations.

An added advantage is that STMs can function over wide ranges of temperature and pressure, and in liquid as well as gas environments. With STM-derived technology, the capabilities for building FMs can now be much more easily realized, for they reduce the originally proposed "building down" sequence to *just one step*, while providing an extremely useful feedback mode. Needless to say, Feynman was delighted when I first informed him about STMs and their capabilities.

The basic modes of operation of an STM, described in Hansma and Tersoff,[109] can be summarized as in Table 1. Here i, v, and h are the tunneling current, the voltage across the gap, and the gap size. Mode I is used to measure the topography of the surface of a metal or semiconductor and is the slowest mode since the electro-mechanical servo system must follow the shape of the surface during the scanning operation. The scanning speed here is determined by the response of the servo system. Modes II and III are faster since the tip maintains only a constant average height above the surface. The scanning speed in these modes is determined by the response of the preamplifier only. Mode IV measures the joint density of states which, for a small tip, is a measure of the local density of states of the substrate. For this mode, one dithers the tip-to-substrate bias voltage with a small ac signal and monitors i/v. One can actually do spatially resolved tunneling spectroscopy using this mode.

TABLE 1 STM operating modes[1]

Mode Number	I	II	III	IV
Quantity held constant	i, v	h, v	h, i	h, i, v
Quantity measured	h	i	v	i/v

[1] i, v, h are tunneling current, voltage and height respectively.

Binnig and Rohrer[15] note that the STM is,

...even energy selective (only electrons lying within the energy window Fermi energy ± applied voltage contribute to the tunneling) and spin selective (this in the case of a polarized tip). Thus, STM is not simply a surface structural method, but also, say, a surface chemical method with the same atomic resolution, which at present reaches 0.05 Å and 2 Å laterally... Electron energies and electric fields on the surface lie in the mV and 10^4 V/cm range, respectively, (typical values used at present, but, in principle, there are no difficulties going to considerably lower energies and fields and, if desired, also to higher ones)... Preliminary experiments on DNA on carbon look very promising...

There is, of course, an abundance of further applications of STM in science and technology... Just imagine, for instance, what could be done with a highly focussed beam (say with a diameter of 10 Å and upwards) of low-energy electrons (meV and upwards).

McCord and Pease,[163] who had done previous studies of fine electron beams generated by other techniques, followed up on this last suggestion. Fink[78] reports beams of 10 μ amps with his *mono-atomic* STM tips when operated in field emission mode. This is an enormous current density for one atom. Fink[79] has recently reported some other remarkable characteristics of his tips. If such tips could be mass produced, then the devices studied by Shoulders at last might be tested on the atomic scale, with potential performance exceeding that of most proposed molecular electronic devices.

STMs are emerging as remarkably versatile tools. They have been operated in air, water, ionic solution, oil, and high vacuum.[51,166] Their scanning speed may be pushed into the real-time imaging domain.[19] They may be used for high-resolution potentiometry and as atomic level force transducers.[172] Nanolithography

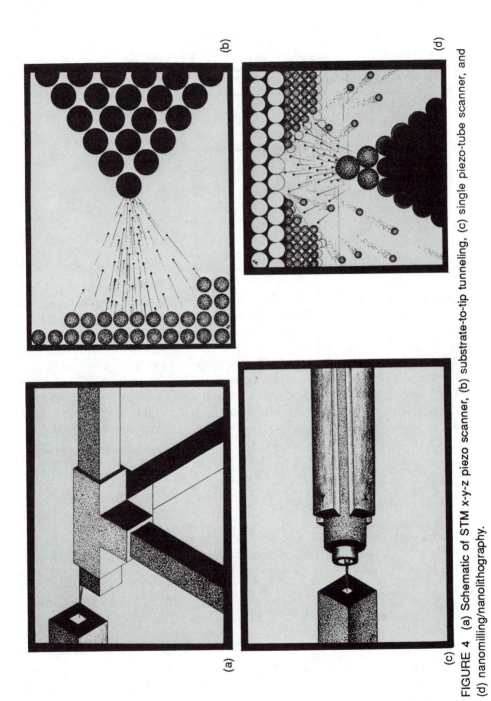

FIGURE 4 (a) Schematic of STM x-y-z piezo scanner, (b) substrate-to-tip tunneling, (c) single piezo-tube scanner, and (d) nanomilling/nanolithography.

with STMs has also been pursued.[190] STMs can be used to map out surface work functions and image charge-density waves. Reviews of these and other state-of-the-art applications are given in Binnig and Rohrer[15] and Golovchenko[99]; Morita et al.[169] have used an STM for electrochemical studies.

Techniques to make very sharp STM tips are described by Dietrich, Lanz and Moore.[46] A transmission electron microscopy study of STM tips sharpened by electrochemical etching followed by argon ion milling is reported in Biegelsen et al.[10] By virtue of removing residual contamination left from chemical etching, "Ion milling is a simple, reproducible technique to achieve high yield for STM operation." McCord and Pease[165] have used STM tips as micromechanical tool bits for machining away 20-nm-thick strips of an insulating film without damage to either the substrate or the tips. (As might be expected, the process also generates submicroscopic corkscrew shavings.)

NANOOPTICAL MICROSCOPY

STM tips may be used to make subwavelength-dimension apertures.[13,14] Such apertures, in combination with STMs to position them extremely close to an object to be viewed, may be used in Scanning Near-Field Optical Microscopes (SNOMs). Durig, Pohl and Rohner[54,55] describe these remarkable instruments which can provide images with 20-nm resolution! Other optical techniques that provide subwavelength imaging capabilities are the De Brabander et al.[39] approach and the Allen[2] approach; both involve some form of "video-enhanced contrast optical microscopy."

ATOMIC FORCE MICROSCOPES

An outgrowth of the STM, the *atomic force microscope* (AFM) is another very high-resolution, mechanically scanned microscope that uses nanoscale tip-to-substrate atomic forces rather than tunneling currents to "image" surface topography. Unlike STMs, AFMs can image insulators. The AFM was invented by Binnig, Quate and Gerber,[16] who reported that:

> The scanning tunneling microscope is proposed as a method to measure forces as small as 10^{-18} newtons. As one application to this concept, we introduce a new type of microscope capable of investigating surfaces of *insulators* on an atomic scale. The atomic force microscope is a combination of the principles of the scanning tunneling microscope and the stylus profilometer. It incorporates a probe that *does not damage the surface*. Our preliminary results in air demonstrate a lateral resolution of 30 Å and a vertical resolution of less than 1 Å ...

We are concerned in this paper with the measurement of ultra small forces on particles as small as single atoms. We propose to do this by monitoring the elastic deformation of various types of springs with the STM.

As with the STM, the AFM has undergone a rapid series of technical improvements. Atomic resolution images of graphite have been achieved by Binnig et al.[18] using an STM constructed using a silicon micromechanical lever as the deflection element. Atomic force profiling utilizing contact forces has been demonstrated by Yang, Miller and Bryant.[247] Atomic forces can also be measured directly with an STM using a mechanical technique developed by Tang, Bokor and Storz.[223] Marti et al.[161] have imaged an organic monolayer using an AFM and Marti, Drake and Hansma have achieved atomic resolution images of liquid covered surfaces with and AFM. Gould et al.[101] have even used an AFM to image amino acid crystals with subnanometer resolution.

The STM, AFM, and SNOM, as newly invented instruments, are expected to revolutionize our basic knowledge of atomic forces and atomic structure. In the past few years STMs have demonstrated that they can "image" single atoms at room temperature in either air or water, probe the forces acting between atoms and measure the density of states of molecular structures. Becker, Golovchenko and

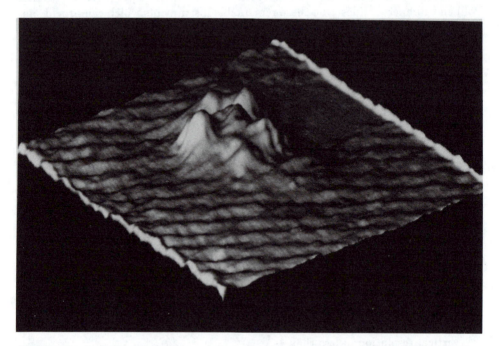

FIGURE 5   A STM image of an organic molecule pinned to a graphite substrate by the same STM. Photograph courtesy of John Foster, IBM Almaden Research Center, San Jose, California.

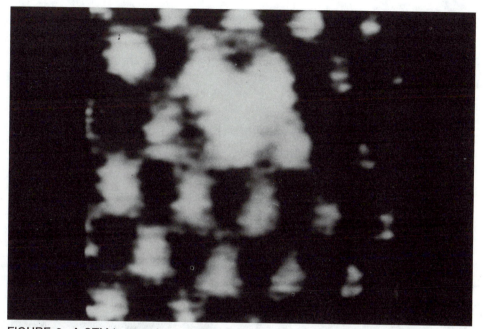

FIGURE 6 A STM image of an organic molecule that was modified (probably partially cleaved) by the same STM. Photograph courtesy of John Foster, IBM Almaden Research Center, San Jose, California.

Swartzentruber[5] reported using an STM to deposit a single atom on a surface! The achievement of the atomic placement capability suggested by Feynman and illustrated in Figure 10 seems feasible in the not-to-distant future. STM-related technology should enable researchers to harmlessly image and manipulate molecules, proteins, genes, viruses, bacteria, semiconductors, metals, dielectric nanoparticles.

Many additional papers on STMs and AFMs may be found in Feenstra.[67]

Now let's see some real STM images of atoms and molecules. Figure 2 shows a STM image of sulpher adatoms on an Mo(111) substrate at atmospheric pressure; it was made at Lawrence Livermore National Laboratory.[159] Figure 3 shows STM images of graphite and molybdenum disulfide at atmospheric pressure; it was made at Digital Instruments, using a commercially available STM. Note how *individual* atoms are clearly visible in these scans.

Figures 5 and 6 provide a dramatic demonstration of *actual molecular manipulation* using STMs at IBM's Almaden Research Center. Figure 6 shows a STM image of an organic molecule that was pinned to the surface of graphite using a voltage pulse to the STM tip. The image frame is 27 Å × 27 Å. Figure 6 shows a STM image of an organic molecule that was first pinned to the surface of graphite using a voltage pulse to the STM tip, and then subjected to a second voltage pulse.

The small size of the remaining fragment (about 4 Å) indicates that the molecule was probable cleaved by the second pulse. The image frame is 11 Å × 11 Å. This fascinating work was reported in Foster, Frommer, and Arnett.[86] Their work also holds great potential for computer memories. Quate[187] recently received a patent for the application of STMs to data storage.

FIGURE 7   NanoTechnology WorkStation: (a) overall view of (simplified) system, (b) graphic displays, (c) multi-tip analysis of a microtubule, and (d) STM-induced/ detected molecular conformation change.

NANOTECHNOLOGY WORKSTATIONS

One tool that would be useful is a "nanotechnology workstation," proposed in Schneiker and Hameroff.[202] The nanotechnology workstation concept combines multiple STMs, optical microscopes plus illumination and detector fiber optic waveguides (Figure 7(a) and (b)). Some of the possibilities for using such multi-tip STM capability are shown in Figures 7(c) and (d). There are many more applications for the nanotechnology workstation described below; these are also very important since they can lead to better understanding of the relevant chemistry and physics of nanoscale systems, and thus ultimately for artificial life. For convenience, both STMs used as FMs and STM-derived FMs will be denoted *STM/FMs* below.

The positioning systems for STMs are accurate to fractions of an Å and can be easily adapted to a wide variety of STM/FMs. For our purposes, the STM can be viewed as an ultraminiature robot finger that can both "see" and be used to directly and precisely manipulate nanostructures along the lines suggested by Feynman.

There are many possibilities for STM modifications and associated functions. Some possibilities are: (1) tip shape modifications—including scalpel, chisel, cylindrical, and other configurations—for scanning, scribing, etching, milling, and polishing operations or for electrical interfaces, electrochemical synthesis or machining; (2) structures attached to tips—including enzymes, synthetic catalysts, shape selective crown ethers, transducer molecules, etc.—for molecular recognition with species selective (and perhaps electrostatic or electromagnet assisted) pick, place, join, and cleave operations, or nano-environmental sensing; (3) multi-tip configurations—in parallel, radial, etc., configurations—for use as ultraminiature tweezers, jigs, and arcs or interface electrodes, or to generate rapidly rotating electric fields; and (4) tip materials modification—including insulating, semiconducting (silicon, ion implanted diamond), ferroelectric or ferromagnetic—for electrostatic, electromagnetic, magnetic, kHz-GHz acoustic (longitudinal, transverse, or torsional) and THz optical modulation in mono- or multi-polar configurations.

In addition to the above modifications, the STM/FMs or their tips could be augmented with a wide variety of sensors and transducers; the atomic force microscope of Binnig, Quate and Gerber[16] is an excellent example of the many possibilities here. The augmentation of STM/FMs with fiber optic interferometers (or comparable techniques) could provide extremely accurate real-time calibration of absolute and relative STM/FM tip positioning, thus overcoming the problems of electrical noise, creep, ageing and hysteresis inherent in present STM piezo-positioning systems. The technique given in Dietrich, Lanz and Moore[46] for making tips with uniform tip-to-base conical profiles, augmented by the techniques of Fink[76] for making perfect monoatomic tips, would be useful for STM/FMs using closely spaced multiple tips.

Using properly configured feedback control instrumentation, STM/FMs can operate as machine tools with effectively perfect lead screws and bearings, although of course their components include neither of these items. Augmented STM/FMs also can be used as sub-atomic resolution proximity detectors and coordinate measuring machines (on conducting surfaces) to monitor nearly perfect, superaccurate nanomachining operations (limited by the graininess of atoms and other materials science considerations). Many useful macroscopic mechanical structures and mechanisms may thus be duplicated at the submicron level, and many of these mechanisms may require no lubrication due to force/area scaling and very rapid heat dissipation.[72] For even smaller mechanisms, a switch to Feynman's mechanical chemistry approach would be needed to build up molecular devices in a series of joining and trimming operations.

Over 6 million chemical structures have been cataloged to date. Rather than mechanically build up structures atom by atom, it is likely that some of these existing chemicals would be used as elementary building blocks in a series of joining (Van der Waals, ionic, and covalent bonding) and trimming (bond cleavage) operations, using electrical and mechanical operations.

As an example of one of the building blocks available, Yamamoto[245] reports on molecular gears and Iwamura[127] has prepared a system that forms a chain of beveled molecular gears. For simple molecular gears, the rotation rate is thought to approach 10^9/sec. Yamamoto[245] describes compounds that "exist in conformations which are regarded as static meshed gears with a two-toothed and a three-toothed wheels and that some of them behave as dynamic gears..." In addition, Iwamura[127] reports that, "Recently we have prepared a doubly connected bevel gear system...[Transfer] of information from one end of the molecule to the other end could take place in large molecules via cooperativity of the torsional motions of the chain."

STM/FMs may execute many complex mechanical motions for driving such nanomechanisms. For instance, various three-dimensional tip motions (single straight lines, circles, spirals, helices, etc.) can be made at speeds presently limited mainly by mechanical resonances in the driving system of current STMs—another motivation for miniaturizing them considerably. Thus it would be possible, for example, to directly drive a molecular-scale mechanical positioning device and other nanomechanical mechanisms. Fluorescence techniques might be used to determine when an artificial receptor molecule attached to an STM tip has acquired or released a specific host molecule.

The capabilities of STM/FMs have been heretofore nonexistent; therefore, applications may abound in the near future as scientists in many disciplines become aware of this versatile and inexpensive technology. Certain techniques developed for electron microscopy might be adapted and used for STMs: (1) Furuya et al.[90] have used metal cluster labels with their scanning transmission microscope to implement a nanoscale distance measurement technique. (2) Panitz[177,178] has developed techniques for preparing and characterizing immunologically active, field emitter tips. His goal is to develop a single molecul detection capability.

NANOMILLING AND NANOLITHOGRAPHY

Dietrich, Lanz and Moore[46] describe how to use argon ion milling to produce very uniform STM points with a *sharpness of 2 nm and better*. A similar etch process can also produce knife edges. These have potential use as electrodes and nanotools as well.

McCord and Pease[163] have analyzed the STM's capabilities for generating intense electron beams and their application. They observe that the

> ...scanning tunneling microscope is a recent demonstration of an extreme case of this principle [of field emission]; the probe comes to within about 1 nm of the target and the resulting vacuum tunneling occurs over a distance of the same order. The potential difference between target and source is usually less than 1 V. In this paper we describe some preliminary calculations of an intermediate configuration in which the potential difference and the source-to-target distance are both increased. The resulting probe should be energetic enough for certain writing mechanisms and hence, might have application in microlithography or microstorage...
>
> We have the possibility of heating materials on a much finer lateral scale than had been previously thought possible...
>
> Another striking departure from the limits on beams formed by focusing optics is that the minimum beam radius should reduce in value as the electron energy is reduced. This conclusion, although surprising, is consistent with the very fine (< 1 nm) probes obtained experimentally in the scanning tunneling microscope.

Ringger[190] demonstrates that a nanometer pattern can be fabricated using an STM and concludes that the STM could yield a new tool for nanometer lithography. Further demonstrations of nanolithography are reported by McCord and Pease.[164,165]

NANOSTRUCTURE AND NANOPATTERN FORMATION

Structure and pattern formation are fundamental processes needed for artificial life. This section describes a few of the (non-STM) techniques that are available for forming nanostructures and nanopatterns.

Although Langmuir-Blodgett films have been major workhorses of molecular electronics research, especially due the work of Hans Kuhn[141] and his associates, researchers are also branching out into other systems. Netzer, Iscovici and Jacob[175] describe one such alternative:

Monolayer formation by adsorption offers certain important advantages as compared with the Langmuir-Blodgett method; adsorption is a spontaneous process leading to thermodynamically equilibrated film structures, there is no mechanical manipulation of the films, water is not indispensable for monolayer formation, monolayer composition and structure are usually dependent on the chemical nature and microscopic organization of the solid surface, covalent binding to the substrate and intralayer polymerization may take place simultaneously with the monolayer formation process, there are no restrictions regarding the macroscopic shape and size of the substrate ... Our present results demonstrate that a molecular spontaneous organization process combined with a suitable chemical triggering procedure may result into a controllable route to the synthesis of an artificial super molecular organizate.

Williams and Giordano[241] have "fabricated Au wires as small as 80 Å in diameter using a technique involving etched nuclear tracks in mica." Sacharoff, Westervelt and Bevk[195] have "... produced single ultrathin Pt wires with diameters as small as 80 angstroms ..." They "have also produced Pt yarn containing over 1.5 million individual Pt wires ..."

Craighead[34] notes, "An analytical electron microscope has been used to fabricate a variety of ultrasmall structures and to test the size limits on some basic concepts of high-resolution, electron-beam lithography ... Devices and structures have been made so small that the width of these structures was only tens of atoms."

Salisbury et al.[196] reports "that holes and lines of about 3-nm width may be cut directly in calcium fluoride by electron-beam lithography technique which has been designated sub-nanometer cutting and ruling by an intense beam of electrons (SCRIBE), in which no chemical development stage is required ... It is clearly possible to machine external shapes on a nanometer scale."

Fischer and Zingsheim[81] have developed submicron optical pattern transfer processes that use visible light for replicating submicron structures. Fischer and Zingsheim[77] further note that the resolution of contact imaging with light is limited by the distance between object and image and not by wavelength ... when the bleaching of a dye is inhibited by energy transfer to a metal in close proximity.

Garber et al.[89] note that,

Certain insoluble copolymers form characteristic liquid crystalline structures at water/copolymer interfaces. The formation of these structures ... correlates with biological activities as immunological adjuvants ... These casts appeared highly accurate ... resolved to less than 10 nm. This technique seems ideal for accurate replication of surfaces formed by nonmiscible liquids and should prove helpful in studies of other materials in aqueous suspension.

Ehrlich[62] notes that, "The behavior of individual atoms on solids has long been of interest for understanding the physical and chemical properties of surfaces. Now, through the use of the field ion microscope, it is possible to directly image single atoms on metal surfaces and to examine their properties quantitatively."

Koma, Sunouchi and Miyajima[137] describe a process in which epitaxial growth proceeds with Van der Waals forces, called Van der Waals epitaxy, which "seems to be one of the most hopeful techniques used to prepare a good quality of heterostructure with atomic order thickness...The present technique has opened a new way to prepare many kinds of heterostructures consisting of metal, semiconductor or insulator films with any thickness from subnanometer by using various transition metal dichalogenide materials."

Deckman et al.[40−43] have developed a number of nanoscale structuring techniques. One technique involved using a regular monolayer lattice of natural molecules as a lithographic pattern mask. Variations on this technique involve using reactive ion etching, chemical etching, and the use of amorphous semiconductor superlattice substrates. Accurately controllable slot widths may be fabricated in the 10 Å to 500 Å range. Douglas and Clark[50] describe a related technique for nanometer molecular lithography.

Control of electric currents and fields, magnetic fields, highly localized temperature gradients, etc., to modulate local surface growth processes are an alternative to atom-by-atom placement for nanoconstruction. The use of ultrastrong magnetic fields to grow higher quality crystal substrates for integrated electronic circuits is a macroscale analogy of such possibilities.

Howard et al.[116] show how advances in microfabrication technology are greatly advancing scientific instrumentation; Feynman Machines obviously have enormous scientific potential in this context. There are many synergistic possibilities. Just as STMs have been combined with SEMs (Gerber et al.[95]), they might later be combined with FEMs (Erlich[62]) or TEMs. The Japanese Research and Development Corporation is organizing a " NanoMechanism" research project in view of the scientific and engineering possibilities of NT.

NANOCOMPUTERS: MOLECULAR ELECTRONICS AND QUANTUM COMPUTING

Carter[22] organized the First International Workshop on Molecular Electronic Devices, held at the Naval Research Laboratories on March 23–24, 1981. He notes:

> Simple extrapolation suggests that in approximately two decades electronic switches will be the size of large molecules. It takes little imagination to recognize that the practical realization of this possibility will produce a

revolution in the areas of computation, technology, science, medicine, warfare, and lifestyle that will be more significant than any which occurred in the last fifty years.

Yates[248] organized another conference on chemical computing. Carter[25,26] organized two more conferences on molecular electronic devices. Conrad[32] presents several interesting and innovative possibilities for molecular computing, based on studies spanning nearly two decades.

Carter[23] describes three molecular fabrication techniques: (1) modular chemistry: a potentially very powerful generalized Merrifeld approach [which works by precisely controlled, sequential addition of specific, individual molecular groups onto a collection of growing molecular chains], (2) molecular epitaxial superstructuring and modulated structuring, and (3) Langmuir-Blodgett film structuring. "The point of view adapted here is that of building structures up from the molecular level rather than imposing structure from the outside." Additional techniques are proposed in Carter.[24] Robinson and Seeman[192] present a more recent proposal for constructing molecular-scale computer memories using self-assembling nucleic acid junctions.

The ability to use multi-tip STM/FMs for mechanical chemistry and as electrical interfaces might solve two formidable problems in the development of molecular electronic devices as envisioned by Carter[23]: (1) the synthesis of prototype devices, and (2) making individual, reliable, electrical connections to them for testing.

STM/FMs could be used for more conventional ultraprecise circuit or component trimming and repair operations. The extremely accurate positioning systems of STM/FMs can be exploited for minimal-scale wirebonding systems. Even though STM/FMs may not ultimately be used to mass produce such devices, they could be used to construct prototypes, help characterize test devices, and be used to optimize them. STM tip-induced sputtering[13,14] might be used to etch or mill ultrafine conductors for such devices.

Efficiently interfacing these and other such devices to the outside world in a manner that can effectively utilize their potential computing bandwidth presents some interesting problems in clocking, input/output, etc. The very-high-frequency "optical electronics" technology suggested by Javan[128,129] may find application here, and in other nanoapplications as well. Using micro- and nano-antennas coupled to laser radiation, it is possible to generate intense, localized, high frequency electric fields (modulated by the laser beam's intensity, phase, and polarization) which could control and power swarms of STM/FM constructed nanoscale devices.

During a recent talk on his quantum computing ideas, Feynman briefly speculated on a simple possibility for making nanocomputer components: use a STM tip to make tiny holes in very thin metal sheets, thus forming grids for tunneling "nano-vacuum tubes," perhaps around 3 to 10 nm in size.[75] Even more compact and potentially faster experimental configurations using several ultrasharp STM tips were mentioned in further speculations by Feynman.[73,75] Unfortunately, no one

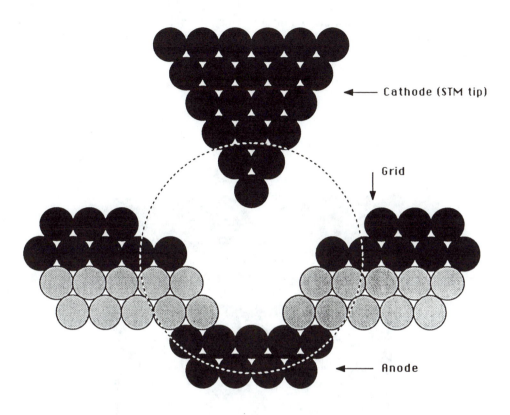

Cathode (STM tip)

Grid

Anode

FIGURE 8   A schematic drawing illustrating the concept of a "nano-vacuum tube." This is the near-minimum-scale case of the submicron triodes proposed by Shoulders, but using an STM tip as suggested by Feynman. A perfect monoatomic STM tip, similar to one later invented by Fink, is shown here.

has followed up on his ideas with either calculations or experiments. Analogous, but much larger (down to 100 nm scale) devices with calculated, *subpicosecond*-range, switching speeds have been proposed by Shoulders.[209] Although he considered even smaller, faster devices, limitations of electron beam micromachining technology at that time prevented further size reduction.[210]

Computing in general and quantum computing in particular was Feynman's "Holy Grail" of nanotechnology; when I told him about STMs, his first thoughts about it involved computing devices. Computing components each consisting of a few atoms might use quantized energy levels or quantum-mechanical spin effects.[72] Feynman[74] analyzes quantum computing and concludes that, "At any rate, it seems that the laws of physics present no barrier to reducing the size of computers until bits are the size of atoms, and quantum behavior holds dominant sway." *If* such devices could be designed and assembled, they could be *10,000 times faster* than conventional transistors.

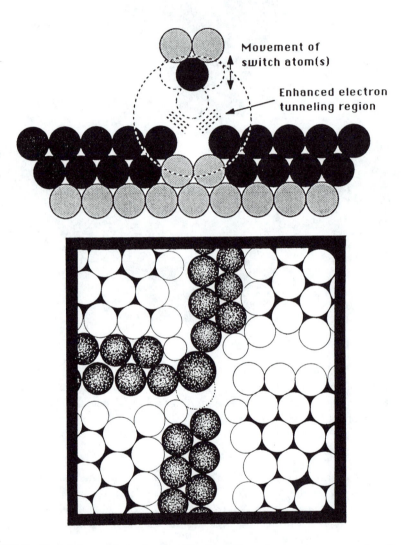

FIGURE 9 Modulation of tunnel gaps. The tunneling atom or group may be physically translated or rotated (by direct mechanical positioning, molecular conformation changes, electrostatic forces, etc.) or their orbitals may be otherwise modified (via interactions involving photons, solitons, electromagnetic fields, etc.).

The development of quantum computing holds fascinating possibilities for artificial life based on automata. Albert[1] states: "An automaton whose states are solutions of quantum-mechanical equations of motion is described, and the capacities of such an automaton to 'measure' and to 'know' and to 'predict' certain physical properties of the world are considered." He notes that its self-description "has no precedent and no analogue among classical automata."

Further developments in the field of systems-theoretic controllability of quantum-mechanical systems may provide some of the missing links needed to implement quantum-mechanical computers, and comparable sensors and effectors.[119,176] Erber and Putterman[61] examine the implications of the quantum theory of single atoms for algorithmic complexity theory and chaotic functions. Other researchers are studying a surprising new class of computation that directly depends on the nonclassical operating modes that quantum computers could potentially implement[157,158]; they could be *formally/computationally more powerful than universal Turing Machines*. See also Conrad and Rossler,[32] Beinoff,[6] Bennett et al.[8] and Deutsch[44,45] for related topics of interest. Bennett[7] discusses changes required in programming techniques for quantum computing.

Using STMs and AFMs for holding single atoms or molecules would be very interesting for quantum computing studies, and for developing modulated tunnel effect devices, as illustrated in Figure 9. The substrates shown can be replaced by more sophisticated molecular structures. Tunnel *gaps* will be an essential part of some types of nanodevices. A very large class of nanoswitches and nanosensors may ultimately be based on this simple principle.[203,204] This same principle may also be applied on a slightly larger size scale at higher voltages by using field emission rather than tunneling.

ARTIFICIAL LIFE-RELATED APPLICATIONS

Nondestructive STM interactions with biological material have immense potential, assuming certain technical obstacles are overcome.

Several groups have succeeded in imaging biomaterials in air such as protein-coated DNA and virus structures. The tunneling is thought to occur from tip onto biomolecular surface followed by low-resistance electron transport to the conducting substrate.[231] Simultaneous optical microscopy, as can occur with the nanotech workstation, may help this situation since photons have been shown to lower tunneling barriers, thus an appropriate choice of optical microscopy wavelengths may facilitate STM imaging by permitting non-damaging tunneling currents. Immunofluorescence techniques may be utilized to identify specific biomolecular targets. The STM probes may also be made (immuno)fluorescent to enhance their visibility.

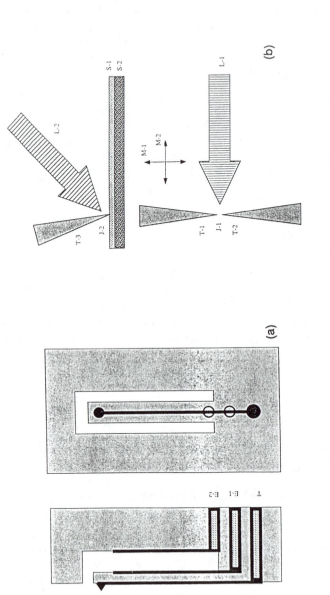

FIGURE 10 A-B (a) A micro-STM formed on a silicon substrate; thousands of these structures may be placed on a single silicon chip. The side and top views of a fine Z-axis STM drive are shown. Microlithography techniques are used to etch a cantilever beam that is deflected by the voltage across electrodes E-1 and E-2. T is the electrode that connects to the STM tip at the end of the cantilever. Non-STM applications for computer memories, nanolithography, and nanomachining are also possible. Thousands of these micro-actuators may be fabricated on a single chip for parallel operation. (b) tunnel junctions (with MEDs at tips) modulated at optical frequencies with lasers. L-1,L-2: laser beams, phase, intensity, polarization or frequency modulation or detection; T-1,T-2,T-3: STM tips used as nanoantennas; J-1,J-2: tunneling junctions; with vacuum, liquid, air or molecular device/layer filler; S-1,S-2: selected substrates; M-1,M-2: some mechanical junction width modulation modes (due to external force/sound/pressure, thermal expansion/contraction, etc.).

An alternative approach is to utilize an atomic force microscope (AFM) mode of operation for the nanotechnology workstation. In this case, a cantilever arrangement is adapted to, and mounted on, the STM which trails along just above the surface, or is held steady to observe mechanical dynamics of the material (i.e., protein conformational change). The movement of the cantilever is monitored by the STM, so that mapping and dynamics might be observed without direct tunneling through the biomaterial.

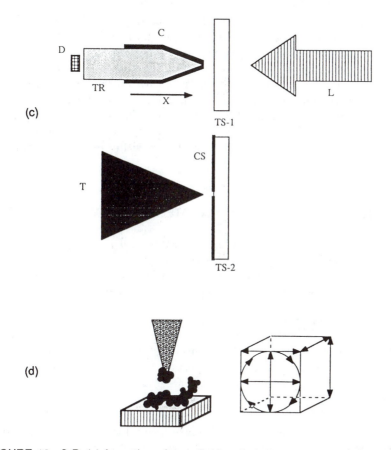

FIGURE 10 C-D (c) formation of near-field optical nano-apertures. L: light; D: detector; TS-1,TS-2: transparent substrates; TR: transparent rod; X: direction of movement until light detected; CS: conducting substrate; and T: STM tip. (d) molecular movement and construction with an STM. There are a wide range of STM tip motions that may be employed for pick and place operations. Multiple tips may be used where needed.

Biomaterials should be studied in a stable aqueous environment. Temperature, pH, ionic concentrations, availability of high-energy phosphate groups and numerous other parameters need to be closely regulated. Paul Hansma's studies at the University of California, Santa Barbara, have demonstrated that STM imaging can occur at ionic liquid-solid interfaces. By insulating STM probes to very near their tips, significant leakage of current to the ionic aqueous environment is apparently avoided.[216]

Potential applications of STM (and the nanotech workstation in particular) to biology and medicine—and hence artificial life—are abundant. The possibility of studying dynamic structural changes (via alterations in the AFM mode), detection of propagating phonons or solitons (using the multiple STM configurations, one tip can perturb and another detect perturbation at a second position on a macromolecule or polymer), spectroscopic analysis (i.e., DNA base-pair reading), and real-time imaging (via high-speed scanning) on living structures can expand the horizons of experimental biosciences. Further, a capability for nanoscale manipulation of biomaterials and organelles offers a host of imaginative possibilities. The most advanced operations require the development of atomically sharp, near perfectly conical STM tips; this objective is probably within reach of present-day nanomachining technology.

MICROTECHNOLOGY: A BRIDGE TO NANOTECHNOLOGY

Yates' conference[248] on chemical computing contains several discussions of the enormous technical difficulties facing developers of molecular computing systems. Additionally, Haddon[105] points out that the prospects for "conventional" technology present a much stronger and rapidly advancing challenge than is usually assumed:

> It is claimed that advances in lithography will soon allow feature sizes of [10 nm]...[It] is even claimed that 3-dimensional systems with stacked elements can be constructed—all by extension of *conventional* chip manufacturing techniques. It is therefore important to realize that the silicon semiconductor industry presents a moving target and has yet to achieve its full promise.

This paper goes on to describe the many difficult tasks and problems that must be solved in order to build a working molecular electronic or a biochip computer. These problems suggest that the most viable paths to general nanotechnology may simply result from pushing rather conventional technology toward its ultimate limits. For instance, many advances in nanometer *semiconductor* structure electronics are proceeding rapidly.[246] There is another reason why much future nanotechnology (and artificial life) might be partly semiconductor based: the semiconductor processing technology developed for integrated circuits is also being adapted for making silicon microstructures, including a wide variety of sensors and effectors.[181,182] Additional work on silicon micromechanics has been done by Csepregi[36] and Kaminsky.[134] Fan, Tai and Miller[66] have extended this technology to make micro-pin joints, microgears, microsprings and microcranks. Gabriel, Mehragany and Trimmer[87] describe additional micromechanical components. A recent conference held on the subject of microrobotics and teleoperators[125] was dominated by papers on silicon microstructures; the possibility and desirability of developing silicon microrobots, teleoperators and replicators was suggested previously in Schneiker.[197,198] The same scaling advances toward the nanoscale cited by Haddon above for silicon integrated circuits will also apply to these systems. Indeed, this technology can be used to mass produce micro-STMs and micro-AFMs (Figure 10). See Howard, Jackel and Skocpol[116] for recent work on semiconductor nanostructure fabrication.

MICRO VON NEUMANN MACHINES FOR NANOGENESIS

As anticipated by Shoulders,[206] the concept of replication can be extended into the artificial nanoworld. In contrast to earlier proposals that would require major advances in the chemistry of molecular structures, his proposal required mainly micro- or nano-milling and electroforming; *this is a tremendously important (i.e., practical) technological simplification.* The replication concept may be applied to the full range of potential nanotechnology applications we mentioned earlier in connection with the work of Feynman, Shoulders, von Hippel, Ettinger, Ellis and Laing.

The first forms of artificial life involving nanotechnology seem most likely to be based on distributed micro-systems augmented with nano-tools rather than nanoreplicators. Constructing artificial (nonbiological) nanoreplicators is probably a much more difficult task than developing constructing microreplicators for these reasons: (1) potentially greater design difficulty, (2) debugging difficulty due to complexity coupled with analytical equipment/inspection limitations, (3) reliability or repeatibility requirements, 4) materials handling limitations involving transfer, purity/contamination, stability and waste by-products of fabrication, packaging or testing, and (5) inability to simply bootstrap with parts made by silicon micromachining using the mass production, pattern-making capabilities from the integrated circuit industry.

Since microreplicators can be designed to make micro Feynman Machines, they can make nanotools and parts. So even without nanoreplicators, you can still effectively replicate arbitrary quantities of specific nanotools or nanoparts. As new nanotechnological devices are developed with top-down micro Feynman Machines, they can be replicated without nanoreplicators, i.e., *nanoreplicators are totally unnecessary* for the unlimited mass production desired for the widespread application of an evolving nanotechnology industrial base. (Invoking the scale-invariance property for artificial life, we see immediately that even if we desire replication, there is no reason whatsoever to require *nano*replication.) Indeed, with appropriate forms of self-assembly, or techniques suggested by Carter[23] and others, one may even dispense with replicators altogether. Likewise, with the hierarchical manufacturing processes envisioned by Feynman for generating huge numbers of computing elements, especially if based on another very-high-volume mass production technology (such as very-large-scale integrated circuit fabrication technology), the ultimate productivity possible would be so large as to also render replicators superfluous. Thus, the development of general nanotechnology does not hinge on the *prior* development of nanoreplicators.

Dyson[56] suggested that for practical reasons, early replicators may initially be a *symbiotic* combination of natural life and artificial automata, located in space:

> It will probably contain a small colony of microscopic plants which are able to utilize efficiently the feeble sunlight ... It will also contain a self-reproducing automaton ... One of the functions of the automaton will be to build a green-house out of local materials for its colony of plants. One of the functions of the plants will be to supply construction materials and fuel to the automaton ... Also the machine may be able to take care of the problem of maintenance by mixing biological and mechanical techniques.

The design of real-world nontrivial replicators is a very complex and difficult task. Von Neumann made many simplifying assumptions for his famous proof and noted that he may have avoided the most difficult part of the problem. It is important to simplify the problem. This can be done by: reducing the number of types of materials/parts needed, reducing the number of scale-dependent factors (path lengths for material/signal transport, replicator population), reducing the needed part tolerances and wear, increasing reliability (to eliminate/reduce repair), and design for automation (to reduce assembly complexity). It is also important to keep the numerous possibilities for simple hybrids in mind, for example, combining sets of 1–3 enzymes with 1–3 molecular electronic devices and 1–3 (silicon) nanostructures in a living cell. Other possibilities might start with the nonliving artificial cells developed by Chang[27] or the vesicle systems of Fendler.[71]

Dyson's[58] study of life's origins suggests that: (1) deferring the development of nanoreplicators, and (2) not requiring atomically perfect replication, may actually speed up the development of artificial life incorporating nanotechnology. Dyson believes that "replication and error-tolerance are naturally antagonistic principles" and that *precise* molecular replication was a secondary development of early life, preceded by the process of cellular reproduction by division "into two cells with approximately equal shares of their cellular constituents." He "considers the primal characteristics of life to be *homeostasis* rather than replication, ... the error-tolerance of the whole rather than the precision of the parts." In 1950, Crane[35] described some principles for the assembly of larger parts from smaller parts which could automatically generate many biological forms. He noted that if, at each level in the hierarchical assembly process, imperfect units were cast aside, then complex structures could be generated without requiring high accuracy at any step. In the 1970's, a process which increased the selectivity of enzymes involved in genetic translation, termed kinetic proofreading, was proposed and discovered.[104] Some or all of these principles may be applied to the development of artificial life.

One of the advantages of the STM/FM approach is that it is also applicable to larger structures, just on this side of the "optical ledge," where visible feedback is readily obtained. This makes it very valuable for interfacing—a key consideration for the efficient development, deployment and application of nanotechnology—and microreplicators. Interfacing is necessary to build on the enormous technical base of microtechnology and to capitalize on intermediate hybrid technology. This intermediate range of possibilities, at the borderline between microreplication and nanoreplication seems a desirable extrapolation of earlier work cited in Schneiker.[197,201] This certainly seems like the approach that would be the most feasible for first generation replicators; moreover, the preliminary development work could be initiated immediately.

Microreplicators have the advantage of being more easily able to use conventional casting, forging, milling, cutting, etching, joining, and molding operations. While these operations can be scaled to atomic dimensions, nano-phenomena such as surface diffusion, Van der Waals forces, etc., will make design, implementation, and debugging operations much more difficult for non-trivial systems.

Replicator, micromachine, and nanomachine component design might be greatly simplified using the mechanical analogues of the Mead-Conway VLSI electronics design rule method. This is another reason to *not* push *all* linear dimensions to their smallest limit for early nanotools.

The recent, dramatic increase in the maximum superconducting transition temperature to above the boiling point of liquid nitrogen (77° K) may soon make the development of superconducting microreplicators easier than other alternatives—especially if the transition temperature is raised to room temperature. Schneiker[201] lists and discusses the design and operational advantages of such systems, which follow from the great ease and efficiency with which magnetic fields may be generated, maintained, altered, repelled, and sensed.

CONCLUSION

We are living in a truly remarkable era where the old nanotechnology dream of atomically precise mechanical manipulation of matter has finally been achieved in some very special and very limited cases. The possible synergies between various emerging nanotechnologies to build on these and the many other developments described above hold enormous potentials for still further acceleration of nanotechnology research and development. Looking back at the last 30 year's worth of nanotechnology developments, we can see that *a staggeringly enormous amount of work still needs to be done* to turn even Feynman's and Shoulder's delightfully straight-forward nanotechnology visions into an everyday reality, let alone the Promethean visions of Ettinger[64,65] and Feinberg[68,69]. In the mean time, advancing the development of nanotechnology using Feynman machines and scanning tunneling engineering with the aim of developing Artificial Life is a superb scientific quest of human understanding and advancement.

POSTSCRIPT—STM/FM COMPETITION

At the end of his classic paper, Feynman[72] offered two prizes, one for writing a page of a book with the linear scale reduced by $1/25,000$ and the other for constructing an operating electric motor with dimensions 1/64 inch cubed). Both prizes have been awarded.[47] His goal: to provide an economic incentive to get people interested in doing actual lab work in this field. For the same reason, Schneiker[201] and Hameroff[108] have offered a series of $1,000 prizes for building a particularly useful type of Feynman Machine: miniaturized STMs. In addition, we want to eventually set up another annual memorial "Feynman Prize" for the most significant yearly achievements in nanotechnology—which will bring us ever closer to the sublime possibilities of artificial life. As Feynman urged,

>...have some fun! *Let's have a competition between laboratories.*

ACKNOWLEDGMENTS

The hand artwork is the craftsmanship of Paul Jablonka. HDS Systems supplied the hardware for the author's computer drafted drawings. Discussions with Joe Andrade, A. Aviram, A. Baratoff, Steve Bell, G. Binnig, Lang Brod, Forrest L. Carter, Eustace Dereniak, Virgil Elings, Richard P. Feynman, Hans-Werner Fink, Bill Ganoe, U. Fischer, John Foster, N. Garcia, J. K. Gimzewski, Stuart Hameroff, Paul K. Hansma, Paul Jablonka, Steve Jacobson, Ali Javan, Arthur Kantrowitz, Chris Langton, Martin Nonnenmacher, Howard Nebeck, Dieter Pohl, Cal Quate, H. Rohrer, Erhard Schreck, Alwin Scott, Roland Shack, Freeman Sheppard, Ken R. Shoulders, Wiegbert Siekhaus, Clayton Teague, G. Travaglini, and Mark Voelker were educational. They do not necessarily agree with any opinions expressed here. Research supported in part by the Department of Anesthesiology and Stuart Hameroff.

APPENDIX—SOME LONG-TERM NANOTECHNOLOGY POSSIBILITIES

The following possibilities involve nanotechnology and were *all* first anticipated *over two decades ago*. Picking up and moving *individual* atoms and molecules, building miniature telerobotic surgeons that operate on neurons, *unlimited* life span and *reversal* of ageing using robot-directed, *molecular-level repair*, boundless wealth, defenses against virtually all diseases and virtually unlimited *in situ evolution*, super compact *personal* computer databases that could easily store the Library of Congress, etc. Let's take a closer look at the historical context of these proposals.

In the prehistory of nanotechnology, the late science fiction author Robert Heinlein[111] envisioned the extensive use of life-size teleoperator hands, called "waldoes," complete with sensory feedback for full, remote-controlled telepresence. His hero, A.K.A. Waldo, used a series of these mechanical teleoperated hands, for building and operating a series of ever smaller sets of such mechanical hands. The smallest mechanical hands, "hardly an eighth of an inch across," were equipped with micro-surgical instruments and stereo "scanners." The smallest set of hands were used to "manipulate living nerve tissue, [to examine] its performance *in situ*" and for nerve surgery.

Feynman[72] reports that a friend, Albert R. Hibbs, suggests that you could put a "mechanical surgeon inside the blood vessel...Other small machines might be permanently incorporated inside the body..." Such machines could be very small, for as Feynman explicitly stressed in presenting the context of his ideas, "there is *plenty* of room at the bottom" for miniaturization. Feynman urged us to consider the possibility, in connection with biological cells, "...that we can manufacture an object that maneuvers at that level!" Ultimately, as Ettinger[64] originally suggested, cellular-level or even molecular-level repair might be be developed for life extension. Ettinger forecast that "...surgeon machines, working twenty four hours a day for decades or even centuries, will tenderly restore the frozen brains, cell by cell, or even molecule by molecule in critical areas..."

In 1962 and 1964, Ettinger[63,64] argued that continued scientific progress would logically one day make it possible to repair and reanimate properly frozen humans. Feinberg[68] also supported Ettinger's conclusions. Ettinger[64,65] cited prior references suggesting some possibilities for molecular-scale technology and suggested some other interesting possibilities.

In 1968, Taylor,[227] in examining the anticipated progress in biology, also concluded that death was not necessary. He cited the possibilities for genetic engineering and even genetic surgery: "The microsurgery of DNA may possibly be achieved by physical methods: fine beams of radiation (probably laser light or pulsed X-rays) may be used to slice through the DNA molecule at desired points..." He also cited predictions that bacteria would soon be programmed, and that, in combination with related developments, "such work opens up practical prospects at which the imagination boggles." He further notes that many scientists consider the creation of life feasible. Indeed, in 1965, *the synthesis of life was publicly proposed as a national goal* by the president of the American Chemical Society, Professor Charles Price. He pointed out that many new types of life might be made, not "mere imitations" of biology as we know it.

In 1967, Asimov[3] suggested the future possibility of "factories...where the working machinery consists of submicroscopic nucleic acids" and that a "repertoire of hundreds or thousands of complex enzymes" could be used to "bring about chemical reactions more conveniently than any methods now used" and also for "helping to construct life."

In 1970, Jeon et al.[130] carried out "the reassembly of *Amoeba proteus* from its major components: namely nucleus, cytoplasm, and cell membrane," taken from three different cells.

In 1970, Vol'kenshtein[234] noted that "the creation of a nonmacromolecular system which would act as a model for a living organisms is definitely possible" but could not arise by itself, and that the macromolecularity of present organisms is not essential, but due to their evolutionary origins. "Consequently, the *cybernetic nonmacromolecular machine*, which simulates life, could have been and can be created on earth only by man. Then it could perfect itself without limits."

Nemes'[174] discussion of self-reproducing machines includes the description of how to construct " an automatic lathe able to reproduce itself," a concept developed before Von Neumann's work on replication. The same principle could be applied to Feynman's[73] suggested nano-lathes.

In 1972, Danielli[37] described an array of possibilities for generating new life forms via "life-synthesis" and genetic engineering. He also noted that "macromolecular engineering" might enable the development of very powerful and compact macromolecular computer systems. In his Nobel lecture, Lehn[150] summarizes the wide range of advances in supramolecular design and engineering that have occurred since then and which "open perspectives toward the realization of molecular photonic, electronic, and ionic devices that would perform highly selective recognition, reaction, and transfer operations for signal and information processing at the molecular level."

In 1974, Halacy[106] noted "some rather inglorious ways" to use "the miracle of artificial life," including potential capabilities for *growing* diverse items ranging from *computers to airplanes.* Halacy also noted that "Leonard Engel wrote in 1962 that 'cold-war-minded scientists have in fact urged a crash program to guarantee a U.S. first' in creating living matter in the laboratory."

In 1974, Morowitz[170] suggested cooling cells to cryogenic temperatures in order to analyze and determine their structure. Artificial cells would likewise be constructed at such temperatures and then set in motion by thawing. He further reported that microsurgery experiments on amoebas "have been most dramatic. Cell fractions from four *different* animals can be injected into the eviscerated ghost of a fifth amoeba, and a living functioning organism results."

In 1974, Ettinger[65] proposed genetic engineering for making nanorobots: "Genetic engineering's most sensational impact will concern the modification of humans; but it will have other uses as well. Some of the 'robots' that will serve us will need to be nanominiaturized..." Ettinger[65] also cites the earlier ideas of White[239] for programmable, i.e., effectively computerized, cell repair machines:

It has been proposed that appropriate genetic information be introduced by means of artificially constructed virus particles into a congenitally defective cell for remedy; similar means may be used for the more general case of repair. Progress has been made in many relevant areas. The repair program must use means such as protein synthesis and metabolic pathways to diagnose and repair any damage... [Information] can be preserved by specifying that the repair program incorporate appropriate RNA tapes into itself...

In 1977, Darwin[38] further develops several notable advances on the theme of tissue repair and cellular repair machines. In 1978, Donaldson[48] presents an updated case for the viability of molecular level repair.

In addition to proposing nanorobotics, Ettinger also suggests a form of hybrid artificial life via the modification of existing organisms:

If we can design sufficiently complex behavior patterns into microscopically small organisms, there are obvious and endless possibilities, some of the most important in the medical area. Perhaps we can carry guardian and scavenger organisms in the blood, superior to the leukocytes and other agents of our human heritage, that will efficiently hunt down and clean out a wide variety of hostile or damaging invaders.

Ettinger also cited even earlier suggestions of future molecular computing, (i.e., nanocomputing) technology.

Ettinger notes the possibilities of artificial human evolution: "...in principle a machine can be made to do anything that is physically possible; and if we envision the human brain coupled to a machine or complex of machines—so that the machines are extensions of the person—then, with only modest reservations to be noted later, we can do anything, which means *we can be anything*."

In 1981, Donaldson[49] elaborated on his earlier discussion on how cryonically suspended human beings could be repaired, which he had discussed earlier in Donaldson.[48] He extended the earlier concepts of Ettinger and White, concluding that with such hybrid technology as *"micro-miniature biological-mechanical machines the size of viruses"* and related technology, "it seems unlikely that (to repair a single cell) there would be any difficulty at all in principle to carrying out *any imaginable repair.*" He calculated that about ten programmable cell-repair machines could be introduced into a cell to be repaired without causing too much mechanical disruption. He noted some other interesting possibilities for repair if for some reason more machinery was required than could be introduced into a given cell.

In 1985, Feinberg[70] described several possible approaches to nanoengineering and some related applications.

In 1986, Drexler[53] expands on Drexler[52] and contains a discussion of some potential developments, applications, and implications of nanotechnology; however, references to the real originators of many of the ideas that he discusses are not given.

Thompsen[230] reports on *optically* pumped X-ray lasers at Lawrence Livermore National Laboratory, which could become tremendously valuable for studying natural life in addition to helping "debug" first attempts at artificial life based on nanotechnology.

> The laser's wavelength is now around 10 nanometers. The experimenters would like to get down to 3 or 4 nanometers. At 10 nanometers scientists could do surface studies of nonliving materials, as well as holograms of living cell surfaces, but they are limited mainly to surfaces. Between 3 and 4 nanometers, imagery can go inside the cell. Water will transmit the X-rays—"Most of these things are swimming in water," Matthews reminds us—and images of both carbonaceous and calciferous structures in cells could be made. "With enough resolution you could even see the chains of carbon in the DNA," he says.

Even better resolution might be obtained with a gamma-ray laser (see Collins et al.[29]

One of the most prolific sources of very speculative and yet often plausible nanotechnology schemes used to appear at the end of the Ariadne column of the *New Scientist* magazine. It used to detail the latest exploits of the extraordinarily creative Daedalus, which frequently involved molecular mechanisms. In recent months, columns on the ingenious Daedalus have been sighted in the journal *Nature*. Unfortunately, lack of space precludes summarizing his numerous and extensive contributions.

As for the *very* long-term future of life, we will close on a positive note by citing Dyson's[57] conclusion that if the universe is open, *life may continue forever*. May you do the same.

REFERENCES

1. Albert, D. Z. (1983), "On Quantum-Mechanical Automata," *Physics Letters* **98A(5,6)**, 249–252.

2. Allen, R. D. (1985), "New Observations on Cell Architecture and Dynamics by Video-Enhanced Contrast Optical Microscopy," *Ann. Rev. Biophys. Chem.* **14**, 265–290.

3. Asimov, I. (1967), *Is Anyone There?* (New York: Ace Books).

4. Aviram, A. and M. A. Ratner (1974), "Molecular Rectifiers," *Chemical Physics Letters* **29(2)**, 277–283.

5. Becker, R. S., J. A. Golovchenko and B. S. Swartzentruber (1987), "Atomic-Scale Surface Modifications Using a Tunnelling Microscope," *Nature* **325**, 419–421.

6. Beinoff, P. (1980), "The Computer as a Physical System: A Microscopic Quantum Mechanical Hamiltonian Model of Computers as Represented by Turing Machines," *J. Stat. Phys.* **22**, 563–591.

7. Bennett, C. H. (1982), "The Thermodynamics of Computation–A Review," *Int. J. Theor. Phys.* **21(12)**, 905–940.

8. Bennett, C. H., G. Brassard, S. Breidbart and S. Wiesner (1983), "Quantum Cryptography, or Unforgeable Subway Tokens," *Advances in Cryptography. Proc. of Crypto* (New York: Plenum).

9. Betzig, E., A. Lewis, A. Harootunian, N. Isaacson and E. Kratschmer (1986), "Near-Field Scanning Optical Microscopy (NSOM), Development and Biological Applications," *Biophysics Journal* **49**, 269–279.

10. Biegelsen, D. K., F. A. Ponce, J. C. Tramontana, and S. M. Koch (1987), "Ion Milled Tips for Scanning Tunneling Microscopy," *Appl. Phys. Lett.* **50(11)**, 697–698.

11. Binnig, G., H. Rohrer, Ch. Gerber and E. Weibel (1982a), "Surface Studies by Scanning Tunneling Microscopy," *Phys. Rev. Letts.* **49**, 57–61.

12. Binnig, G., H. Rohrer, Ch. Gerber and E. Weibel (1982b), "Tunneling Through A Controllable Vacuum Gap," *Appl. Phys. Lett.* **40**, 178–179.

13. Binnig, G., Ch. Gerber, H. Rohrer and E. Weibel (1985), "Nano-Aperture," *IBM Tech. Disclosure Bul.* **27(8)**, 4893.

14. Binnig, G., Ch. Gerber, H. Rohrer and E. Weibel (1985), "Sputter Tip," *IBM Tech. Discl. Bul.* **27**, 4890.

15. Binnig, G., and H. Rohrer (1985), "Scanning Tunneling Microscopy," *Surface Science* **152**, 17–26.

16. Binnig, G., C. F. Quate and C. Gerber (1986), "Atomic Force Microscope," *Phys. Rev. Lett.* **56(9)**, 930–933.

17. Binnig, G., and D. P. E. Smith (1986), "Single-Tube Three-Dimensional Scanner for Scanning Tunneling Microscopy," *Rev. Sci. Instrum.* **57(8)**, 1688–1689.

18. Binnig, G. and H. Rohrer (1987), "Scanning Tunneling Microscopy—From Birth to Adolescence," *Rev. Mod. Phys.* **59(3)**, 615–625.

19. Bryant, A., D. P. E. Smith and C. F. Quate (1986), "Imaging in Real Time with the Tunneling Microscope," *Appl. Phys. Lett.* **48**, 832–834.

20. Carter, F. L. (1979), "Problems and Prospects of Future Electroactive Polymers and 'Molecular' Electronic Devices," *the NRL Program on Electroactive Polymers, First Annual Report*, Ed. L. D. Lockhart, Jr. *NRL Memorandum Report 3960*, 121.

21. Carter, F. L. (1980), "Further Considerations on 'Molecular' Electronic Devices," *Second Annual Report*, Ed. R. B. Fox, *NRL Memorandum Report 4335*, 35.

22. Carter, F. L., ed. (1982), *Molecular Electronic Devices* (New York: Marcel Dekker).

23. Carter, F. L. (1983), "Molecular Level Fabrication Techniques and Molecular Electronic Devices," *J. Vac. Sci. Technol. B* **1(4)**, 959–968.

24. Carter, F. L., (1986), "Chemistry and Microsturctures: Fabrication at the Molecular Size Level, *Superlattices and Microstructures* **2(2)**, 113–128.

25. Carter, F. L., ed. (1987), *Proceedings of the 2nd International Workshop on Molecular Electronic Devices (1983)* (New York: Marcel Dekker), in press.

26. Carter, F. L., ed. (1987), *Proceedings of the 3rd International Workshop on Molecular Electronic Devices (1986)* (Amsterdam: Elsevier North-Holland), in preparation.

27. Chang, T. M. S. (1972), *Artifical Cells* (Springfield, IL: Thomas Books).

28. Changeux, J.-P. (1965), "The Control of Biochemical Reactions," *Scientific American* **212(4)**, 36–45.

29. Collins, C. B., F. W. Lee, D. M. Shemwell and B. D. DePaola (1982), "The Coherent and Incoherent Pumping of a Gamma Ray Laser with Intense Optical Radiation," *J. Appl. Phys.* **53(7)**, 4645–4651.

30. Conrad, M. (1972), "Information Processing in Molecular Systems," *Currents in Modern Biology* **5**, 1–14.

31. Conrad, M. (1974), "Molecular Automata," *Physics and Mathematics of the Nervous System*, Eds. M. Conrad, W. Guttinger and M. Dal Cin (New York: Springer-Verlag), 419–430.

32. Conrad, M., and O. Rossler (1982), "Example of a System which is Computation Universal but not Effectively Programmable," *Bull. of Math. Biol.* **44(3)**, 443–447.

33. Conrad, M. (1986), "The Lure of Molecular Computers," *IEEE Spectrum* **23(10)**, 55–60.

34. Craighead, H. G. (1984), "10-nm Resolution Electron-Beam Lithography," *J. Appl. Phys.* **55(12)**, 4430–4435.

35. Crane, H. R. (1950), "Principles and Problems of Biological Growth," *The Scientific Monthly* **70(6)**, 376–389.

36. Csepregi, L. (1985), "Micromechanics: A Silicon Microfabrication Technology," *Microelectronic Engineering* **3**, 221–234.

37. Danielli, J. F. (1972), "Artificial Synthesis of New Life Forms," *Bull. Atomic Scientists* **28(10)**, 20–24.

38. Darwin, M. G. (1977), "The Anabolocyte: A Biological Approach to Repairing Cryoinjury," *Life Extension Magazine* **July-August**, 80–83.

39. De Brabander, M., R. Nuydens, G. Geuens, M. Moeremans, J. De Mey and C. Hopkins (1988), "Nanovid Ultramicroscopy: A New Non-Destructive Apporach Providing New Insights in Subcellular Motility, *Microtubules and Microtubule Inhibitors 1985*, Eds. M. De Brabander and J. De Mey (Amsterdam: Elsevier Science Publishers, North Holland), 187–196.

40. Deckman, H. W., and J. H. Dunsmuir (1982), "Natural Lithography," *Appl. Phys. Lett.* **41(4)**, 377–378.

41. Deckman, H. W., and J. H. Dunsmuir (1983), "Applications of Surface Textures Produced with Natural Lithography," *J. Vac. Sci. Technol. B.* **1(4)**, 1109–1112.

42. Deckman, H. W., J. H. Dunsmuir and B. Abeles (1985), "Microfabricated TEM Sections of Amorphous Superlattices," *J. Vac. Sci. Technol.* **A3(3)**, 950–954.

43. Deckman, H. W., B. Abeles, J. H. Dunsmuir and C. B. Roxlo (1987), "Microfabrication of Molecular Scale Microstructures," *Appl. Phys. Lett.* **0(9)**, 504–506.

44. Deutsch, D. (1982), "Is There a Fundamental Bound on the Rate at Which Information Can Be Processes?," *Phys. Rev. Lett.* **48(4)**, 286–288.

45. Deutsch, D. (1985), "Quantum Theory, the Church-Turing Principle and the Universal Quantum Computer," *Proc. R. Soc. Lond. A* **400**, 97–117.

46. Dietrich, H. P., M. Lanz and D. F. Moore (1984), "Ion Beam Machining of Very Sharp Points," *IBM Tech. Discl. Bul.* **27**, 3039–3040.

47. Dietrich, J. (1986), "Tiny Tale Gets Grand," *Engineering and Science* **January**, 24–26.

48. Donaldson, T. (1978). *A Brief Scientific Introduction to Cryonics*, reprinted by ALCOR Life Extension Foundation.

49. Donaldson, T. (1981), "How Will They Bring Us Back, 200 Years From Now?," *The Immortalist* **12(3)**, March, 5-10.

50. Douglas, K. and N. A. Clark (1986), "Nanometer Molecular Lithography," *Appl. Phys. Letts.* **48(10)**, 10 March, 676–678.

51. Drake, B., R. Sonnenfeld, J. Schneir, P. K. Hansma, G. Solugh and R. V. Coleman (1986), "A Tunneling Microscope for Operation in Air or Fluids," *Rev. Sci. Instr.* **57**, 441–445.

52. Drexler, K. E. (1981), "Molecular Engineering: An Approach to the Development of General Capabilities for Molecular Manipulation," *Proc. Natl. Acad. Sci. USA* **78(9)**, 5275–5278.

53. Drexler, K. E. (1986), *Engines of Creation* (Garden City, NY: Anchor Press/ Doubleday).

54. Durig, U., D. W. Pohl and F. Rohner (1986), "Near-Field Optical Scanning Microscopy," *J. Appl. Phys.* **59(10)**, 15 May, 3318–3327.

55. Durig, U., D. Pohl and F. Rohner (1986), "Near-Field Optical Scanning Microscopy with Tunnel-Distance Regulation," *IBM J. Res. Develop.* **30(5)**, Sept., 478–483.

56. Dyson, F. J. (1970), "The Twenty-First Century," *Vanuxem Lecture, delivered at Princeton University, 26 Feb. 1970*; mostly published in *The Key Reporter* **42(3)**, 1977.

57. Dyson, F. J. (1979), "Time Without End: Physics and Biology in an Open Universe," *Reviews of Modern Physics* **51(3)**, 447–460.

58. Dyson, F. J. (1985a), *Origins of Life* (Cambridge: Cambridge University Press).

59. Dyson, F. J. (1985b), personal communication.

60. Ellis, G. W. (1962), "Piezoelectric Micromanipulators," *Science* **138**, 84–91.

61. Erber, T., and S. Putterman (1985), Randomness in Quantum Mechanics—Nature's Ultimate Cryptogram," *Nature* **318**, 41–43.

62. Erlich, G. (1980), "Quantitative Examination of Individual Atomic Events on Solids," *J. Vac. Sci. Technol.* **7(1)**, 9–14.

63. Ettinger, R. C. W. (1962), *The Prospect of Immortality*, privately published.

64. Ettinger, R. C. W. (1964), *The Prospect of Immortality*. (New York: Doubleday).

65. Ettinger, R. C. W. (1972), *Man Into Superman*. (New York: St. Martin's).

66. Fan, L. S., Y. C. Tai and R. S. Miller (1987), "Pin Joints, Gears, Springs, Cranks, and Other Novel Micromechanical Structures," *Proc. Transducers '87 (The 4th International Conference on Solid-State Sensors and Actuators, June 1987, Tokyo, Japan)*, 849–852.

67. Feenstra, R. M., ed. (1988). *Proceedings of the Second International Conference on Scanning Tunneling Microscopy.* (New York: American Institute of Physics).

68. Feinberg, G. (1966), "Physics and Life Prolongation," *Physics Today* **November**, 45–48.

69. Feinberg, G. (1969), *The Prometheus Project; Mankind's Search for Long-Range Goals* (Garden City, NY: Anchor Books, Doubleday and Company, Inc.).

70. Feinberg, G. (1985), *Solid Clues: Quantum Physics, Molecular Biology, and the Future of Science* (New York: Touchstone Books, Simon and Schuster).

71. Fendler, J. H. (1984), "Polymerized Surfactant Vesicles: Novel Membrane Mimetic Systems," *Science* **223**, 886–894.

72. Feynman, R. P. (1960a), "There's Plenty of Room at the Bottom," *Engrg. and Sci. (Cal. Inst. of Tech.)* **Feb.**, 22–36; also in: *Miniaturization*, Ed. H. D. Gilbert (New York: Reinhold), 1961, 282–296, and Schneiker .[201]

73. Feynman, R. P. (1960b), "The Wonders that Await a Micro-Microscope," *Saturday Review* **43**, 45–47.

74. Feynman, R. P. (1984, 1985, 1986), personal communication.

75. Feynman, R. P. (1985), "Quantum Mechanical Computing," *a talk given to the Cal. Tech. Society of Physics Students, March 12.*

76. Feynman, R. P. (1985a), "Quantum Mechanical Computers," *Optics News* **Feb.**, 11–20.

77. Feynman, R. P. (1986a), "Quantum Mechanical Computers," *Foundations of Physics* **16(6)**, 507–531.

78. Fink, H.-W. (1986), "Mono-Atomic Tips for Scanning Tunneling Microscopy," *IBM J. Res. Develop.* **30(5)**, 460–465.

79. Fink, H.-W. (1988), *Physica Scripta*, in press.

80. Fischer, U. Ch. and H. P. Zingsheim (1981), "Submicroscopic Pattern Replication with Visible Light," *J. Vac. Sci. Technol.* **19(4)**, 881–885.

81. Fischer, U. Ch. and H. P. Zingsheim (1982), "Submicroscopic Contact Imaging with Visible Light by Energy Transfer," *Appl. Phys. Lett.* **40(3)**, 195–197.

82. Fischer, U. Ch. (1986), "Submicrometer Aperture in a Thin Metal Film as a Probe of Its Microenvironment through Enhanced Light Scattering and Fluorescence," *J. Opt. Soc. Am. B* **3(10)**, 1239–1244.

83. Fischer, U. Ch., U. T. Durig and D. W. Pohl (1988), "Near-field Optical Scanning Microscopy in Reflection," *Appl. Phys. Lett.* **52(4)**, 249–251.

84. Flanders, D. C. and A. E. White (1981), "Application of 100 Å Linewidth Structures Fabricated by Shadowing Techniques," *J. Vac. Sci. Technol.* **9(4)**, 892–896.

85. Forrest, D. R. (1985), "Nanotechnology," unpublished paper.

86. Foster, J. S., J. E. Frommer and P. C. Arnett (1988), "Molecular Manipulation Using a Tunneling Microscope," *Nature* **331**, 324–326.

87. Franks, A. (1987a), "Nanotechnology Opportunities," *J. Phys. E: Sci. Instrum.* **20**, 237.

88. Franks, A. (1987b), "Nanotechnology," *J. Phys. E: Sci. Instrum.* **20 Dec.**, 1442–1451.

89. Fuller, R. B., and J. Applewhite (1975), *Synergetics* (New York: Macmillan).

90. Furuya, F. R., L. L. Miller, J. F. Hainfeld, W. C. Christophel and P. W. Kenny (1988), "Use of $Ir_4(CO)_{11}$ To Measure the Lengths of Organic Molecules with a Scanning Transmission Electron Microscope," *J. Am. Chem. Soc.* **110(2)**, 641–643.

91. Gabriel, J. K., M. M. Mehragany and W. S. N. Trimmer (1987), "Micro Mechanical Components," *Modeling and Control of Robotic Manipulators and Manufacturing Process, Winter Annual Meeting of the American Society of Mechanical Engineers, December*, 397–401.

92. Gadzuk, J. W. (1987), "STM Not Developed in a Vacuum," *Physics Today* **1/11**.

93. Garber, M. A., W. McManus, B. Bennett, and R. L. Hunter (1985), "Protein Casting: A New Replication Technique for Observing Fluid Surface Structures," *J. Electron Microscopy Technique* **2**, 71–72.

94. García, N., ed. (1987), *STM '86, Proceedings of the First International Conference on Scanning Tunneling Microscopy, Santiago de Compostela, Spain, 14–18 July 1986* (Amsterdam: North Holland).

95. Gerber, C., G. Binnig, H. Fuchs, O. Marti and H. Rohrer (1985), "Scanning Tunneling Microscope Combined with a Scanning Electron Microscope," *Rev. Sci. Instrum.* **57(2)**, 221–223.

96. Gobrecht, J. and J. B. Pethica (1986), "The Potential of Mechanical Microlithography for Submicron Patterning," *Microelectronic Engineering* **5**, 471–475.

97. Godin, T. J. and R. Haydock (1986), "Quantum Circuit Theory," *Superlattices and Microstructures* **2(6)**, 597–600.

98. Golvochenko, J. A. (1986)., "The Tunneling Microscope: A New Look at the Atomic World," *Science* **232**, 48–53.

99. Gomer, R. (1986), "Possible Mechanisms of Atom Transfer in Scanning Tunneling Microscopy," *IBM J. Res. Develop.* **30(4)**, 428–430.

100. Gomez, J., L. Vazquez, A. M. Baro, N. Garcia, C. L. Perdriel, W. E. Triaca and A. J. Arvia (1986), "Surface Topography of (100)-Type Electro-Faceted Platinum from Scanning Tunneling Microscopy and Electrochemistry," *Nature* **323**, 612–614.

101. Gould, S., O. Marti, B. Drake, L. Hellemans, C. E. Bracker, P. K. Hansma, N. L. Keder, M. M. Eddy and G. D. Stucky (1988), "Molecular Resolution Images of Amino Acid Crystals with the Atomic Force Microscope," *Nature* **332**, 332–334.

102. Granqvist, C. G. and R. A. Buhrman (1976), "Ultrafine Metal Particles," *J. Appl. Phys.* **47(5)**, 2200–2219.

103. Griffiths, D. W. and M. L. Bender (1973), "Cycloamyloses as Catalysts. Introduction, *Advances in Catalysis*, Eds. D. D. Eley, H. Pines and P. B. Weisz (New York: Academic Press), 209–261.

104. Gueron, M. (1978), "Enhanced Selectivity of Enzymes by Kinetic Proof-reading," *Am. Scientist* **66(2)**, 202–208.

105. Haddon, R. C. and A. A. Lamola (1985), "The Molecular Electronic Device and the Biochip Computer: Present Status," *Proc. Nat. Acad. Sci. USA* **82**, 1874–1878.

106. Halacy, D. S., Jr. (1974), *Genetic Revolution, Shaping Life for Tomorrow* (New York: Harper and Row).

107. Hameroff, S. R. (1987), *Ultimate Computing: Biomolecular Consciousness and Nanotechnology* (Amsterdam: Elsevier North-Holland).

108. Hansma, P. K. and J. Tersoff (1987), "Scanning Tunneling Microscopy," *J. Appl. Phys.* **61(2)**, R1–R23.

109. Harootunian, A., E. Betzig, M. Isaacson and A. Lewis (1986), "Super-Resolution Fluorescence Near-Field Scanning Optical Microscopy," *Appl. Phys. Lett.* **49(11)**, 674–676.

110. Hartline, F. F. (1979), "Biological Applications for Voltage Sensitive Dyes," *Science* **203**, 992–994.

111. Heinlein, R. A. (1940), *Waldo and Magic, Inc.* (New York: Doubleday & Co.).

112. Hirschfeld, T. (1984), "Providing Innovative System Monitoring and Reliability Assessment Through Microengineering," *Energy and Technology Review (LLNL)* **Feb.**, 16–23.

113. Hirschfeld, T. (1985), "Instrumentation in the Next Decade," *Science* **230**, 286–291.

114. Howard, R. E. and D. E. Prober (1982), "Nanometer-Scale Fabrication Techniques," *VLSI Electronics Microstructure Science*, Ed. N. G. Einspruch (New York: Academic Press), vol. 5., 146–189.

115. Howard, R. E., P. F. Liao, W. J. Skoepol, L. D. Jackel and H. G. Craighead (1983), "Microfabrication as a Scientific Tool," *Science* **221**, 117–121.

116. Howard, R. E., L. D. Jackel and W. J. Skocpol (1985), "Nanostructures: Fabrication and Applications," *Microelectronic Engineering* **3**, 3–16.

117. Howells, Malcolm, J. Kirz, D. Saryre and G. Schmahl (1985), "Soft X-ray Microscopes," *Physics Today* **38(6)**, 22–32.

118. Huang, G. M., T. J. Tarn and J. W. Clark (1983), "On the Controllability of Quantum-Mechanical Systems," *J. Math. Phys.* **24**, 2608–2618.

119. Humphreys, C. J., I. G. Salisbury, S. D. Berger, R. S. Timsit and M. E. Mochel (1985), "Nanometre-Scale Electron Beam Lithography," *Inst. Phys. Conf. Ser. No. 78, Chapter 1, Paper presented at EMAG '85, Sept., 1–10.*

120. IBM (1985), "Optical Microscope Beyond the Diffraction Limit," *IBM Technical Disclosure Bulletin* **28(6)**, 2691–2693.

121. IBM (1985f), "Quantum Public Key Distribution System," *IBM Tech. Discl. Bulletin* **8(7)**, 3153–3163.

122. IBM (1986c), "Six-Axis Precision Positioning System," *IBM Tech. Discl. Bulletin* **(12)**, 5582–5583.

123. IBM (1986e), "Nanostructures for High Resolution Measurements of Magnetic Fields," *IBM Technical Disclosure Bulletin* **29(1)**, 312–314.

124. IEEE (1987), "Micro Robots and Teleoperators Workshop," *IEEE Proceedings, November* (New York: IEEE).

125. Isaacson, M. and A. Muray (1981), "Nanolithography Using In Situ Electro Beam Vaporization of Very Low Molecular Weight Resists," *Molecular Electronic Devices*, Ed. F. L. Carter (New York: Marcel Deker), 165–174.

126. Isaacson, M. and A. Murray (1981), "In situ Vaporization of Very Low Molecular Weight Resists Using 1/2 nm Diameter Electron Beams," *J. Vac. Sci. Technol.* **19(4)**, 1117–1120.

127. Iwamura, H. (1985), "Molecular Design of Correlated Internal Rotation," *J. Molecular Struct.* **126**, 401–412.

128. Javan, A. (1985), "Interview," *Lasers and Applications* **Oct.**, 52.

129. Javan, A. (1986), personal communication.

130. Jeon, K. W., I. J. Lorch, and J. F. Danielli (1970), "Reassembly of Living Cells from Dissociated Components," *Science* **167**, 1626–1627.

131. Johnstone, K. I. (1973), *Micromanipulation of Bacteria* (London: Churchill Livingstone).

132. Jones, R. V. (1967), "The Measurement and Control of Small Displacements," *32nd Parsons Memorial Lecture*, 325–336.

133. Kakuchi, M., M. Hikita and T. Tamamura (1986), "Amorphous Carbon Films as Resist Masks with High Reactive Ion Etching Resistance for Nanometer Lithography," *Appl. Phys. Letts.* **48(13)**, 835–837.

134. Kaminsky, G. (1985), "Micromachining of Silicon Mechanical Structures," *J. Vac. Sci. Technol. B* **3(4)**, 1015–1024.

135. Katzir, A. and F. Vranty (1984), "Use of PVC for Replecating Submicron Features for Microelectronic Devices," *Rev. Sci. Inst.* **55(4)**, 633–634.

136. Kiewit, D. A. (1973), "Microtool Fabrication by Etch Pit Replication," *Rev. Sci. Instrum.* **44(12)**, 1741–1742.

137. Koma, A., K. Sunouchi and T. Miyajima (1985), "Fabrication of Ultrathin Heterostructures with Van der Waals Epitaxy," *J. Vac. Sci. Tech. B* **3(2)**, 724.

138. Kratschmer, E., B. Whitehead, M. Isaacson and E. Wolf (1985)., "Nanometer Scale Metal Wire Fabrication," *Microelectronic Engineering* **3**, 25–32.

139. Kratschmer E., and M. Isaacson (1986), "Nanostructure Fabrication in Metals, Insulators, and Semiconductors using Self-Developing Metal Inorganic Resist," *J. Vac. Sci. Technol. B* **4(1)**, 361–364.

140. Krumhansl, James A. and Yoh-Han Pao (1979), "Microscience: An Overview," *Physics Today* **November**, 25–32.

141. Kuhn, H. (1983), "Functionalized Monolayer Assembly Manipulation," *Thin Solid Films* **99**, 1–16.

142. Kuhn, H. (1985), personal communication.

143. Kuki, A. and P. G. Wolynes (1987), "Electron Tunneling Paths in Proteins," *Science* **236**, 1647–1652.

144. Laing, R. (1974), "Some Forms of Replication in Artificial Molecular Machines," *Proceedings of the 1974 Conference on Biologically Motivated Automata Theory, McLean, Virginia, 6–8.*

145. Laing, R. (1975), "Some Alternative Reproductive Strategies in Artificial Molecular Machines," *J. Theor. Biol.* **54**, 63–84.

146. Laing, R. (1976), "Automaton Inspection," *J. Comp. Sys. Sciences* **13**, 172–183.

147. Laing, R. (1977), "Automaton Models of Reproduction by Self-Inspection," *J. Theor. Biol.* **66**, 437–456.

148. Laing, R. (1979), "Machines as Organisms: An Exploration of the Relevance of Recent Results," *BioSystems* **11**, 201–215.

149. Lang, N. D. (1986), "Electronic Structure and Tunneling Current for Chemisorbed Atoms," *IBM J. Res. Develop.* **30(4)**, 374–379.

150. Lehn, J. M. (1988), "Supramolecular Chemistry—Scope and Perspectives Molecules, Supermolecules, and Molecular Dvices (Nobel Lecture," *Angewandte Chemie* **27(1)**, 89–112.

151. Lewis, A., M. Isaacson, A. Harootunian and A. Muray (1984), "Development of a 500 Å Spatial Resolution Light Microscope," *Ultramicroscopy* **13**, 227–232.

152. Liberman, K. A. (1979a), "Analog-Digital Molecular Cell Computer," *BioSystems* **11**, 111–124.

153. Liberman, E. A. (1979b), "Biological Physics and the Physics of the Real World," *BioSystems* **11**, 323-327.

154. Likharev, K. K. (1987), "Single-Electron Transistors: Electrostatic Analogs of the DC Squids," *IEEE Transactions on Magnetics* **23(2)**, 1142–1145.

155. Lin, C. W., F.-R. F. Fan and A. J. Bard (1987), "High Resolution Photoelectrochemical Etching of n-GaAs with the Scanning Electrochemical and Tunneling Microscope," *Electrochemical Society* **134(4)**, 1038–1039.

156. Long, J. E. and T. J. Healy (1980), *Advanced Automation for Space Missions—Technical Summary: A Report of the 1980 NASA/American Society for Engineering Education Summer Study on the Feasibility of Using Machine Intelligence in Space Applications, Univ. Santa Clara, Sept.*

157. Maddox, J. (1985a), "Monitoring Quantum Jumps," *Nature* **314**, 493.

158. Maddox, J. (1985b), "Towards the Quantum Computer?," *Nature* **316**, 573.

159. Marchon, B., P. Bernhardt, M. E. Bussell, G. A. Somorjai, M. Salmeron and W. Siekhaus (1988), "Atomic Arrangement of Sulfur Adatoms on Mo(001) at Atmospheric Pressure: A Scanning Tunneling Microscopy Study," *Phys. Rev. Lett.*, in press.

160. Mark, H. F. (1966), *Giant Molecules* (New York: Time Inc.), 171–173.

161. Marti, O., H. O. Ribi, B. Drake, T. R. Albrecht, C. F. Quate and P. K. Hansma (1988), "Atomic Force Microscopy of an Organic Monolayer," *Science* **239**, 50–52.

162. McClaire, W. F. (1971), "Chemical Machines, Maxwell's Demon and Living Organisms," *J. Theor. Biol.* **30**, 1–34.

163. McCord, M. A., and R. F. W. Pease (1985), "High Resolution, Low-Voltage Probes From a Field Emission Source Close to the Target Plane," *J. Vac. Sci. Technol. B.* **3(1)**, 198–201.

164. McCord, M. A. and R. F. W. Pease (1986), "Lithography with the Scanning Tunneling Microscope," *J. Vac. Sci. Technol. B* **4(1)**, 86–88.

165. McCord, M. A. and R. F. W. Pease (1987), "Exposure of Calcium Fluoride Resist with the Scanning Tunneling Microscope," *J. Vac. Sci. Technol. B* **5(1)**, 430–433.

166. Miranda, J. R., N. Garcia, A. Baro, R. Garcia, J. L. Pena, H. Rohrer (1985), "Technological Applications of Scanning Tunneling Microscopy at Atmospheric Pressure," *Appl. Phys. Letts.* **47**, 367–369.

167. Mitchell, P. (1982), "Osmoenzymology: The Study of Molecular Machines," *Cell Function and Differentiation* (New York: Alan R. Liss), F.E.B.S. vol. 65, Part B, 399–408.

168. Mochel, M. E., C. J. Humphreys, J. A. Eades, J. M. Mochel and A. M. Petford (1983), "Electron Beam Writing on a 20-Å Scale in Metal β-Aluminas," *Appl. Phys. Lett.* **42(4)**, 392–394.

169. Morita, S., et al. (1987), "Construction of a Scanning Tunneling Microscope for Electrochemical Studies," *Japanese Journal of Applied Physics* **26(11)**, L1853–L1855.

170. Morowitz, H. J. (1974), "Manufacturing a Living Organism," *Hospital Practice* **9(11)**, 210–215.

171. Morrison, P. (1964), "A Thermodynamic Characterization of Self-Reproduction," *Review of Modern Physics* **36(2)**, 517–524.

172. Muralt, P., D. Pohl and W. Denk (1986), "Wide-Range, Low-Operating-Voltage, Bimorph STM: Application as Potentiometer," *IBM J. Res. Develop.* **30(5)**, 443–450.

173. NASA (1986), "Electrochemical Process Makes Fine Needles," *NASA Tech Briefs* **May/June**, 135.

174. Nemes, T. (1970), *Cybernetic Machines* (New York: Gordon and Breach Science Publishers).

175. Netzer, L., R. Iscovici, and J. Sangiv (1983), "Planned Multilayer Assemblies by Adsorption," *Molec. Cryst.* **93(1–4)**, 415–417.

176. Ong, C. K., G. M. Huang, T. J. Tarn and J. W. Clark (1984), "Invertibility of Quantum-Mechanical Control Systems," *Math. Sys. Theory* **17**, 335–350.

177. Panitz, J. A. (1986), "Immunologic Layer Formation on Metal Microelectrodes," *J. Colloid and Interface Science* **111(2)**, 516–528.

178. Panitz, J. A. (1984), "Biomolecular Adsorption and the Life Detector," *Journal De Physique* **C9, 12**, 285–291.

179. Pattee, H. H. (1961), "On The Origin of Macromolecular Sequences," *Biophys. J.* **1**, 683–710.

180. Peres, A. (1985), "Reversible Logic and Quantum Computers," *Phys. Rev. A* **2(6)**, 3266–3276.

181. Peterson, K. E. (1978), "Dynamic Micromechanics on Silicon: Techniques and Devices," *IEEE Trans. Electron Devices* **25**, 1241–1250.

182. Peterson, K. E. (1982), "Silicon as a Mechanical Material," *Proceedings of the IEEE* **70(5)**, 420–457.

183. Pethica, J. B. (1988), "Atomic-Scale Engineering," *Nature* **331**, 301.

184. Pohl, D. W. (1986), "Some Design Critieria in Scanning Tunneling Microscopy," *IBM J. Res. Develop.* **30(4)**, 417–427.

185. Pohl, D. W., W. Denk and M. Lanz (1984), "Optical Stethoscopy: Image Recording with Resolution," *Appl. Phys. Lett.* **44(7)**, 651–653.

186. Pohl, D. W., W. Denk and U. Duerig (1985), "Optical Stethoscopy: Imaging with $\lambda/20$ Resolution," *Proc. of the SPIE, 565, Micron and Submicron Integrated Circuit Metrology, August* (Bellingham, WA: SPIE), 56–61.

187. Quate, C. F. (1986), "Method and Means for Data Storage using Tunnel Current Data Readout," *US Patent 4,575,822, 11 March.*

188. Reiss, H. and C. Huang (1971), "Statistical Thermodynamic Formalism in the Solution of Information Theory Problems," *J. Stat. Phys.* **3**, 191–210.

189. Rhodes, C. K. (1985), "Multiphoton Ionization of Atoms," *Science* **229(4720)**, 1345–1351.

190. Ringger, M., H. R. Hidber, R. Schlögl, P. Oelhafen and H. J. Güntherodt (1985), "Nanometer Lithography With the Scanning Tunneling Microscope," *Appl. Phys. Lett.* **46(9)**, 832–834.

191. Ritter, E., R. J. Behm, G. Potschke and J. Wintterlin (1986), "Direct Observation of a Nucleation and Growth Process on an Atomic Scale," *STM '86, Proceedings of the First International Conference on Scanning Tunneling Microscopy, Santiago de Compostela, Spain, 14–18 July 1986*, Ed. N. García (Amsterdam: North Holland).

192. Robinson, B. H., and N. C. Seeman (1987), "The Design of a Biochip: A Self-Assembling Molecular-Scale Memory Device," *Protein Engineering* **1(4)**, 299–300.

193. Rosser, R. J. (1984), "X-Ray Microscopy at Imperial College," *X-Ray Microscopy*, Eds. Schmahl, G. and D. Rudolph (New York: Springer-Verlag).

194. Rothstein, J. (1963), "On Fundamental Limitations of Chemical and Bionic Information Storage Systems," *IEEE Tran. on Military Electronics* **2 and 3**, 205–208.

195. Sacharoff, A. C., R. M. Westervelt and J. Bevk (1985), "Fabrication of Ultrathin Drawn Pt Wires by an Extension of the Wollaston Process," *Rev. Sci. Instrum.* **56(7)**, 1344–1346.

196. Salisbury, I. G., R. S. Timsit, S. D. Berger and C. J. Hunphreys (1984), "Nanometer Scale Electron Beam Lithography in Inorganic Materials," *Appl. Phys. Letts.* **45(12)**, 1289–1291.

197. Schneiker, C. W. (1983), "Prospects and Applications for Genesis and Ultra Mass Production of Sub-Millimeter Machines, Devices, and Replicating Systems," distributed at 1983 Space Development Conference, Houston, Texas; revised and published in *Cryonics*.

198. Schneiker, C. W. (1984), "Some Research Topics for a New Space Development Programme," *J. Brittish Interplanetary Society* **37**, 190–192.

199. Schneiker, C. W. (1985), "NanoTechnology," unpublished manuscript; distributed at 1985 Space Development Conference, Washington, D.C.

200. Schneiker, C. W. (1986a), "NanoTechnology with STMs, Feynman Machines and von Neumann Machines," poster presented at 1986 Spain STM Conference by Stuart Hameroff, 14-16 July.

201. Schneiker, C. (1986b), *NanoTech: Science, Engineering and Related Topics* (Berkeley, CA: Applied AI); book available from Martin Brooks in paper or electronic form: Applied AI, Box 9002, Berkeley, CA 94709.

202. Schneiker, C. and S. R. Hameroff (1986), "NanoTechnology Workstation Based on Scanning Tunneling/Optical Microscopy: Applications to Molecular Scale Devices," *Proceedings of the Third International Workshop on Molecular Electronic Devices, 6–8 October, 1986, Arlington, Virginia*, Eds. F. L. Carter & H. Wohlgen (Elsevier: North-Holland), in press.

203. Schneiker, C. W. (1988a), "Scanning Tunneling Engineering," poster for Scanning Tunneling Microscopy '88 Conference, 4–8 July, 1988, Oxford, England.

204. Schneiker, C. W. (1988b), "Scanning Tunneling Engineering, Molecular Electronic Devices, and Nanotechnology," poster for Symposium on Molecular Electronics, 19–22 July, 1988, Santa Clara, California.

205. Schrage, H., J. Franke, F. Vogtle and E. Steckhan (1986), "Bicyclic Host Cavaties Derived from Triphenylamine—Guest Selectivity and Redox Properties," *Angew. Chem. Int. Ed. Engl.* **25**, 4, 336–338.

206. Shoulders, K. R. (1960), "On Microelectronic Components, Interconnections, and System Fabrication," *Proc. Western Joint Computer Conference* (Palo Alto, CA: National Press), 251–258.

207. Shoulders, K. R. (1961), "Microelectronics Using Electron-Beam-Activated Machining Techniques," *Advances in Computers*, Ed. Franz Alt (New York: Academic Press), 135–293.

208. Shoulders, K. R. (1962), "On Microelectronic Components, Interconnections, and System Fabrication," *Aspects of the Theory of Artificial Intelligence*, Ed. C. A. Muses (New York: Plenum Press), 217–235.

209. Shoulders, K. R. (1965), "Toward Complex Systems," *Microelectronics and Large Systems* (Washington, D.C.: Spartan Books), 97–128.

210. Shoulders, K. R. (1985, 1986), personal communication.

211. Silver, R. M., E. E. Ehrichs and A. L. de Lozanne (1987), "Direct Writing of Submicron Metallic Features with a Scanning Tunneling Microscope," *Appl. Phys. Lett.* **51**(4), 247-249.

212. Simpson, M., P. Smith and G. A. Dederski (1987), "Atomic Layer Epitaxy," *Surface Engineering* **3(4)**, 343–348.

213. Smith, C., H. Ahmed, M. J. Kelly and M. N. Wybourne (1985), "The Physics and Fabrication of Ultra-Thin Free Standing Wires," *Superlattices and Microstructures* **1(2)**, 153–154.

214. Smith, D. P. E., G. Binnig and C. F. Quate (1986), "Atomic Point-Contact Imaging," *Appl. Phys. Lett.* **49(18)**, 1166–1168.

215. Smith, D. P. E., M. D. Kirk and C. F. Quate (1987), "Molecular Images and Vibrational Spectroscopy of Sorbic Acid with the Scanning Tunneling Microscope," *J. Chem. Phys.* **86(11)**, 6034–6038.

216. Sonnenfeld, R. and P. K. Hansma (1986), "Atomic-Resolution Microscopy in Water," *Science* **232**, 211–213.

217. Sonnenfeld, R. and B. C. Schardt (1986), "Tunneling Microscopy in an Electrochemical Cell: Images of Ag Plating," *Appl. Phys. Lett.* **49(18)**, 1172–1174.

218. Srivastava, Y., A. Widom and M. H. Friedman (1985), "Microchips as Precision Quantum-Electrodynamic Probes," *Phys. Rev. Letts.* **55(21)**, 2246–2248.

219. Staufer, U., R. Wiesendanger, L. Eng, L. Rosenthaler, H. R. Hidber, H.-J. Güntherodt and N. Garcia (1987), "Nanometer Scale Structure Fabrication with the Scanning Tunneling Microscope," *Appl. Phys. Lett.* **51(4)**, 244-246.

220. Stefanides, E. J. (1986), "Focusing Ring: First Step Toward Scanning X-Ray Microscope," *Design News* **42(3)**, 102–106.

221. Stroscio, M. A. (1986a), "Comment on 'Microchips as Precision Quantum-Electrodynamic Probes,'" *Phys. Rev. Lett.* **56(19)**, 2107.

222. Stroscio, M. A. (1986b), "Quantum-Based Electronic Devices," *Superlattices and Microstructures* **2(1)**, 45–47.

223. Tang, S. L., J. Bokor, and R. H. Storz (1988), "Direct Force Measurement in Scanning Tunneling Microscopy," *Appl. Phys. Lett.* **42(3)**, 188–190.

224. Taniguchi, N. (1974), "On the Basic Concept of Nanotechnology," *Proc. Int. Conf. Prod. Eng. Tokyo, Part 2* (Tokoyo: JSPE), 18–23.

225. Taniguchi, N. (1983), "Current Status in, and Future Trends of, Ultraprecision Machining and Ultrafine Materials Processing," *Ann. CIRP* **32**, 1–10.

226. Taniguchi, N. (1985), "Atomic Bit Machining by Energy Beam Processes," *Prec. Eng.* **4**, 145–155.

227. Taylor, G. R. (1968), *The Biological Time Bomb* (New York: World Publishing Company).

228. Thomas, R. N., R. A. Wickstrom, D. K. Schroder and H. C. Nathanson (1974), "Fabrication and Some Applications of Large-Area Silicon Field Emission Arrays," *Solid-State Electronics* **17**, 155–163.

229. Thompson, W. A. and S. F. Hanrahan (1976), "Thermal Drive Apparatus for Direct Vacuum Tunneling Experiments," *Rev. Sci. Instrum.* **47(10)**, 1303.

230. Thomsen, D. E. (1986), "A Powerful Way to Make an X-Ray Laser," *Science News* **129(22)**, 348–349.

231. Travaglini, G., H. Rohrer, M. Amrein and H. Gross (1986), "Scanning Tunneling Microscopy on Biological Matter, in press.

232. Umentani, Y. (1978), "Principle of a Piezo-Electric Micro Manipulator With Tactile Sensibility," *Fourth International Conference on Industrial Robot Technology* (Bedford: International Fluidics Services Ltd).

233. Van de Walle, G. F. A., H. van Kempen and P. Wyder (1986), "Tip Structure Determination by Scanning Tunneling Microscopy," *Surf. Science* **167**, L219–L224.

234. Vol'kebshtein, M. V. (1970), *Molecules and Life* (New York: Plenum Press).

235. Von Foester, H. (1964), "Molecular Bionics," *Third Bionics Symposium* (Alexandria, VA: Defense Documentation Center), 161–190.

236. Von Hippel, A. R. (1962), "Molecular Designing of Materials," *Science* **138**, 91–108.

237. Von Neumann, J. (1966), *Theory of Self-Reproducing Automata*, Edited and completed by A. W. Burks (Urbana: University of Illinois Press).

238. Wang, C. P., R. L. Varwig and P. J. Ackman (1982), "Measurement and Control of Subangstrom Mirror Displacement by Acousto-Optical Technique," *Rev. Sci. Instrum.* **53(7)**, 963–966.

239. White, J. (1969), "Viral Induced Repair of Damaged Neurons with Preservation of Long-Term Information Content," paper read at the Second Annual Cryonics Conference, Ann Arbor, Michigan, April 11, 1969; reprints available from the Cryonics Society of Michigan.

240. Wiesner, J. B. (1961), "Electronics and Evolution," *Proceedings of the IRE*, May, 653–654.

241. Williams, W. D. and N. Giordano (1984), "Fabrication of 80 Angstrom Metal Wires," *Rev. Sci. Instr.* **55(3)**, 410–412.

242. Wohltjen, H. (1982), *Microfabrication Techniques: Current and Future. In: Molecular Electronic Devices*, Ed. F. L. Carter (New York: Marcel Deker), 231–243.

243. Wolf, E. L. (1984), *Principles of Electron Tunneling Spectroscopy* (New York: Oxford University Press; Oxford: Clarendon Press).

244. Wulff, Gunter (1985), "Molecular Recognition in Polymers Prepared by Imprinting with Templates," *Polymeric Reagents and Catalysis*, Ed. W. T. Ford (Washington, D.C.: American Chemical Society), 186–230.

245. Yamamoto, G. (1985), "Molecular Gears with Two-Toothed and Three-Toothed Wheels," *J. Molecular Struct.* **126**, 413–420.

246. Yamamura, Y., T. Fujisawa and S. Namba (1984), "Nanometer Structure Electronics, An Investigation of the Future of Microelectronics," *Proc. of the International Symposium on Nanometer Structure Electronics, April 16–18, Osaka University, Toyonaka, Japan* (Amsterdam: North Holland).

247. Yang, R., R. Miller, and P. J. Bryant (1987), "Atomic Force Profiling by Utilizing Contact Forces," *J. Appl. Physc.* **63(2)**, 570–572.

248. Yates, F. E. (1984), "Report on Conference on Chemically-Based Computer Designs," *Krump Institute for Medical Engineering Report CIME TR/84/1, University of California, Los Angeles, California 90024*.

249. Young, R. D., J. Ward and F. Scire (1971), "Observation of Metal-Vacuum-Metal Tunneling, Field Emission, and the Transition Region," *Phys. Rev. Lett.* **27**, 922–924.

250. Young, R., J. Ward and F. Scire (1972), "The Topografiner: An Instrument for Measuring Surface Microtopography," *Rev. Sci. Instr.* **43(7)**, 999–1011.

251. Zasadzinski, J. A. N., J. Schneir, J. Gurley, V. Elings, P. K. Hansma (1988), "Scanning Tunneling Microscopy of Freeze-Fracture Replicas of Biomembranes," *Science* **239** 1013–1015.

252. Zingsheim, H. P. (1976), *Ber. d. Bunsengesellsch. Phys. Chem.* **80.**, 1185.

253. Zingsheim, H. P. (1978), "Molecular Engineering Using Nanometer Surface Microstructures," *Proceedings: NSF Workshop on Opportunities for Microstructure Science, Engineering and Technology in Cooperation with the NRC Panel on Thin Film Microstructure Science and Technology. November 19–22, 1978, Airlie House, Airlie, Virginia* (Washington: National Science Foundation), 44–48.

K. ERIC DREXLER
Visiting Scholar, Stanford University. Box 60775, Palo Alto, California 94306

Biological and Nanomechanical Systems: Contrasts in Evolutionary Capacity

The field of nanotechnology includes the study of certain classes of artificial molecular machines and self-replicating systems. Its concern with molecular replicators relates it to the study of living systems, both natural and artificial. Consideration of proposed nanomachines and replicators shows that they lack certain basic characteristics that are essential to the evolutionary capacity of living things. This paper examines these characteristics and their evolutionary significance.

The first section below provides an overview of nanotechnology, comparing and contrasting it with biological systems. The next examines several distinctions in styles of development and function: diffusive vs. channeled transport, matching vs. positional assembly, topological vs. geometric structure, and adaptive vs. inert building blocks. These distinctions are used to define two overall styles of organization, *organic* and *mechanical*. The succeeding sections relate these styles to the evolutionary capacity (or incapacity) of biological and nanomechanical systems and then summarize conclusions regarding evolution, replicating systems, and proposals for artificial life.

Artificial Life, SFI Studies in the Sciences of Complexity,
Ed. C. Langton, Addison-Wesley Publishing Company, 1988   **501**

OVERVIEW OF NANOTECHNOLOGY

Nanotechnology is a projected technology based on a general ability to build objects to complex atomic specifications.[6] It takes its name from the nanometer scale of the structures it can produce; a cubic nanometer of material typically contains over a hundred atoms. Not all processes that make nanometer-scale products (which include simple molecules, ultrathin films, and submicron lines) are examples of nanotechnology, just as cigarettes and bubble pipes (making micron-scale smoke particles and soap films) are not tools of microtechnology. Nanotechnology implies atom-by-atom control of complex structures; microtechnology implies the fabrication of complex, microscopic structures without this control. (Nanotechnology will not be limited to small structures, however.[8])

The molecular machinery of life demonstrates functions that will be important in nanotechnology. Some enzymes assemble small reactive molecules to build larger molecules. Ribosomes are genetically programmed machine tools that assemble small reactive molecules in complex patterns to form large molecular machines.

Nanotechnology will be based on programmable machine tools with more general abilities—devices termed *assemblers*—which will enable the construction of a wide range of molecular structures. Ribosomes can build machines only of protein, but a molecular-scale robot arm, able to work with a wide range of reactive molecules, should be able to build molecular machines with almost any chemically reasonable structure.[6,7] Synthetic organic chemists make a wide range of molecular structures by mixing reactive molecules in solution. Characteristically, they cannot make very complex structures (say, a billion-atom molecule with the complexity of an integrated circuit), owing to the difficulty of controlling the site of a reaction on the surface of a large molecule. Diffusion bumps molecules together in all positions and orientations; reactions occur wherever they are chemically feasible. Assemblers will sidestep this limit by eliminating diffusion: they will position reactive molecules mechanically, making reactions occur only at the sites selected by the designer.

A well-established nanotechnology will likely make little use of biomolecules. The molecular machines envisioned for that era are surprisingly conventional, including gears and bearings,[10] electric motors (electrostatic, rather than electromagnetic[7]), and a full range of moving parts. Analysis indicates that digital logic systems based on molecular mechanical devices can be compact (fitting the capacity of a mainframe computer into a cubic micron) and can be reliable despite thermal noise.[9,11]

To visualize mechanical devices on this scale, it is important to recognize that molecules are objects, with size, shape, mass, strength, stiffness, and so forth. Large machines are made of parts with many atoms; nanomachines will be made of parts with few. Just as engineers prefer to work with light, rigid materials on a large scale, so nanoengineers will prefer such materials on a small scale. Thus, parts will typically contain patterns of atoms like those found in engineering plastics, ceramics, graphite, and diamond.

PATHS TO NANOTECHNOLOGY

The core idea of nanotechnology is to use molecular machines (assemblers) to build assemblers and other products. This appears reasonable, but circular. How might assemblers be built in the first place? Just as there was no principle preventing crude machine tools from building better machine tools during the development of macrotechnology, so there is no principle preventing crude molecular machines from building better molecular machines during the development of nanotechnology. Several paths lead toward this sort of spiraling advance of technology.

First-generation assemblers may be developed through protein engineering; biochemical analogies indicate that protein engineering (when sufficiently advanced) will enable the design and fabrication of complex, self-assembling molecular machines.[6,23,24,25,27] Likewise, first-generation assemblers may be developed through the synthesis of self-assembling sets of non-protein molecules.[14,16,17] Alternatively, advances in micromanipulation may enable the construction of first-generation assemblers through mechanically directed molecular assembly; reports of atomic rearrangement through field-induced evaporation from scanning tunneling microscope (STM) tips[1] and of highly localized chemical reactions induced by currents at an STM tip[13] are suggestive in this regard. In practice, development may well involve a combination of chemical, biochemical, and micromechanical techniques. However assemblers may first be built, later assemblers will be built using assemblers. The nature of nanotechnology and its capabilities will then be independent of the nature of proteins, conventional chemistry, and initial micromanipulation technologies.

Since multiple paths lead to molecular machines and nanotechnology, no one problem with development can block advance in this direction. With multiple paths, multiple research groups, and multiple chains of short-term rewards along each path, it is (in a competitive world) hard to imagine that nanotechnology will not eventually be realized. This adds to its interest as an object of study.

NANOREPLICATORS AND BIOREPLICATORS

If assemblers, guided by nanocomputers, can build almost anything, then with proper programming they should be able to build copies of themselves (and of the nanocomputer, and its instructions, and so forth). If assemblers are to process large quantities of material atom-by-atom, many will be needed; this makes pursuit of self-replicating systems a natural goal. The availability of atoms as prefabricated building blocks simplifies self-replication on this scale.

The following will compare and contrast living systems with systems (especially replicators) based on anticipated styles of nanomachinery. For present purposes, "living systems" are defined as systems based on cells, ranging from bacteria to blue whales (many observations will apply to viruses as well). For convenience, self-replicating systems of nanomachinery will here be termed *nanoreplicators*; living systems will occasionally be termed *bioreplicators*. (Note that this use of the term "replicator" is distinct from Dawkins' use,[2] in that it refers to the whole replicating system, genotype and phenotype, rather than to just its genetic material. In this context, a replicator in Dawkins' sense can be termed a "genetic replicator"; the distinction is vital, since only *genetic* replicators pass on mutations and evolve.)

The parallels between existing bioreplicators and proposed nanoreplicators are strong. Both rely on the use of molecular machines to position reactive molecules, thus directing the synthesis of complex systems, including more molecular machines. Assemblers are analogous to ribosomes; the systems that supply them with reactive molecules are analogous to metabolic enzyme systems. Both bioreplicators and nanoreplicators rely on digital control systems: the genetic system directs ribosomes; nanocomputers are expected to direct assemblers.[7,11] In a broad sense, each may be viewed as an instantiation of von Neumann's architecture for self-replicating systems.

Despite these parallels, the differences between existing bioreplicators and proposed nanoreplicators are great. Ribosomes get their parts, energy, and directions via diffusion, but assemblers in proposed nanoreplicators will get them via fixed channels. Ribosomes self-assemble via diffusion and matching of complementary parts, but assemblers will be made by operations analogous to manual construction. Cells and organisms have structures defined chiefly by patterns of containment and interconnection, but nanoreplicators will have structures defined by a specific geometry. Organisms grow, with their parts adapting to one another, but nanoreplicators will be constructed from parts of fixed structure. In summary, where bioreplicators have an "organic" style, proposed nanoreplicators will have a "mechanical" style, resembling factories more than they do living cells and organisms. The following sections will explore these differences in more detail, then argue that life has this "organic" style, not for reasons of technical efficiency, but because alternative "mechanical" systems could not arise through conventional evolution.

STYLES OF DEVELOPMENT, STRUCTURE, AND FUNCTION

Systems can differ in their means of transporting materials, energy, and information; in their means of assembling parts; in the definition of their structures; and in the adaptability of their parts. These differences distinguish different styles of development, structure, and function.

DIFFUSIVE VS. CHANNELED TRANSPORT

Bioreplicators make heavy use of diffusive transport for materials, energy and information. In living cells, metabolic substrates diffuse from enzyme to enzyme, as do energy-transmitting molecules, such as ATP. Small, diffusing molecules, such as cyclic AMP, serve as signals; diffusing RNA molecules carry whole blocks of organized, digital information.

Like factories, proposed nanoreplicators make heavy use of channeled transport systems. Examples of these include conveyor belts and pipes for moving materials, wires and drive shafts for moving energy, and cables for moving information. Compared to diffusive transport systems, channeled systems commonly have technical advantages in compactness, speed of transportation, and minimization of inventory.

Materials handling is important in manufacturing systems, including replicators. Typically, a part will go through several manufacturing operations, each performed by a distinct machine. This pattern is familiar both in factories and in cell metabolism (where the parts are molecules and the machines are enzymes); it is to be expected in nanoreplicators as well. In a diffusive system, every machine is effectively linked to every other—it can accept inputs from anywhere, and its outputs are available everywhere. In a prototypical channeled system, in contrast, every machine must be specifically linked (by conveyor belts or the equivalent) to its input-suppliers and output-consumers. Thus, in a channeled system, a new machine can do useful work only if aided by corresponding additions to the transportation system; in a diffusive system, a new machine can do useful work without such additions. As will be seen, this difference is of basic evolutionary importance.

The general pattern of diffusive transport in living cells has limitations and exceptions. Eukaryotic cells contain numerous membrane compartments, placing regional controls on diffusion; active molecules pump some materials across membranes against concentration gradients. These modifications do not suffice to make the transportation system channeled, however. It has recently been argued[26] that some enzymes seldom release their products to diffuse freely, but instead transfer them directly to the active site of the next enzyme in the metabolic pathway (on encountering that enzyme through a diffusive process). Since these enzymes *can* transfer their products diffusively, however, they are not subject to the limits of a truly channeled system. More significant is the presence of systems such as the fatty-acid synthetase complex, which holds a partially completed molecule on the end of a swinging arm, cycling it through different active sites on the complex to add a series of two-carbon units, building up a fatty-acid chain.[18] This system is effectively channeled; it is significant that systems of this sort appear rare in cells, despite their technical advantages in materials transport. These channeled islands are linked by a diffusive sea.

Diffusive systems have an advantage in reliability over simple one-path channeled systems. In a diffusive system, no vital channel can fail or be blocked by the failure of a processing-machine, since no such channel exists. A more complex channeled system can gain comparable reliability, however, by incorporating redundant paths and machines connected in a suitable network.

MATCHING VS. POSITIONAL ASSEMBLY

Bioreplicators make heavy use of spontaneous assembly based on diffusion and matching. Molecular parts (such as the RNA and protein molecules that make up ribosomes) diffuse and bump together in all possible positions and orientations. Those that have corresponding surfaces (matching patterns of bumps and hollows, positive and negative charge, hydrophobicity, etc.) pull together and stick, forming a specific structure.

Automated factories and proposed nanoreplicators, in contrast, make heavy use of positional assembly. Here, the prototype is a blind robot thrusting a pin into the expected location of a hole. There is no finding and matching of parts—if the hole is elsewhere, the operation fails. Positional assembly has potential advantages in speed, and in the lesser constraints it places on the structure of device interfaces (no need to induce self-assembly, and no need to guide it by providing unique interfaces for differing parts).

In a system made by a matching assembly process, an increase in the number of matching parts A and B leads naturally to an increase in the number of assemblies AB. In a positional assembly process, in contrast, new parts A must be placed in new positions to which parts B must be brought; new assemblies AB thus require corresponding changes in the assembly process. This is directly analogous to the requirement, in a channeled transport system, for new channels corresponding to new machines.

In a matching assembly process, a change in the size or shape of a part, if it does not disturb its interface to another part, will seldom disturb assembly. In a positional assembly process, however, a change in position constitutes a significant disturbance to the interface: for example, the insertion position of a screw on top of a carburetor will change if the carburetor grows taller, and the screw will miss the hole. Worse, any change in the height of any part on which the carburetor is mounted will cause the same problem. If the screw were to diffuse to the hole, then match and stick, such problems would not arise.

TOPOLOGICAL VS. GEOMETRIC STRUCTURES

Geometric structures characterize conventional machines and proposed nanoreplicators. Parts have definite sizes, shapes, and positions with respect to one another. The resulting fixed geometry lends itself to positional assembly and channeled transport systems.

The structures of cells and living organisms, however, are largely organized in a way that can be described as *topological*—characterized not so much by specific positions as by patterns of connectivity. The shape of a membrane compartment in a cell matters less than its continuity and the contents of the volume it defines. Likewise, the length of a muscle matters less than its attachment points. Diffusive transport and matching assembly, with their lack of position dependence, lend themselves to use in the assembly and functioning of topological structures.

Adding a part inside a densely organized geometric structure typically requires changes in the relative positions of many other parts, and hence corresponding adjustments in design. Adding a part inside a densely organized topological structure, in contrast, typically leaves topologies unchanged—room can be made by stretching and shifting other parts, with no change in their essential design.

ADAPTIVE VS. INERT PARTS

Closely related to the notion of topological and geometric structures is the notion of adaptive and inert parts. The prototype of an inert part is a rigid object with a flange having a special shape and a special pattern of bolt holes—it fits a corresponding part, or it doesn't; a change in the interface of one demands a compensating change in the interface of the other. The prototype of an adaptive part is a coat of spray paint—it fits the part it coats, with no delicate dependence on that part's size or shape. Rubber hoses are relatively adaptive; rubber gaskets are less so. Typical metal and ceramic parts, like the rigid parts of proposed nanomachines, are essentially inert.

Bioreplicators make extensive use of adaptive parts. Skin grows to cover an organism; it need not be redesigned when genes or environment give rise to a giant. Likewise, skulls grow to cover brains, muscles grow to match bone-lengths, and vascular systems grow to permeate tissues. The inherent adaptiveness of tissues and organs is demonstrated by healing in adults, and by the development of strangely connected but locally plausible organ systems in Siamese twins.

SUMMARY: O-STYLE VS. M-STYLE SYSTEMS

Living things are characterized by heavy use of diffusive transport, matching assembly, topological structures, and adaptive parts. Let us call systems that share these characteristics, whether living or not, *O-style systems* (O is mnemonic for organic).

Mechanical systems are characterized by heavy use of channeled transport, positional assembly, geometric structures, and inert parts. Let us call systems that share these characteristics *M-style systems* (M is mnemonic for mechanical).

The difference between O-style and M-style is not a hard distinction, but a matter of degree. The following will often speak of them as if they were distinct, but they form, at least in principle, a continuum. Molecular machines inside cells typically have M-style features; their parts are relatively geometric and inert. Automobiles contain hoses and coats of paint with a measure of O-style adaptiveness. Still, on the whole, cells (with their diffusive transport, matching assembly, topological structures, and adaptive parts) are strongly O-style while automobiles (with their channeled transport, lack of assembly operations, geometric structures, and inert parts) are strongly M-style. By this measure proposed nanoreplicators are far closer to cars than to cells and other living systems.

O-STYLE, M-STYLE, AND EVOLUTION

Each of the characteristics distinguishing O-style from M-style is of considerable importance to evolutionary capacity. In each case, the M-style characteristic introduces dependencies among parts such that typical changes in the structure of one are of no benefit (or do harm) without simultaneous, corresponding changes in the structure of others.

In a conventional evolutionary system, the genetic system does not somehow convert single mutations into properly corresponding changes in multiple parts. Further, selection pressures are applied after each mutation, with no favor extended to promising-but-harmful mutations while they await redemption in the form of a corresponding mutation elsewhere. In these circumstances, the characteristics of M-style systems effectively destroy their evolutionary capacity; the characteristics of O-style systems sustain it.

M-STYLE SYSTEMS AND EVOLUTION

Consider an integrated, strongly M-style box (perhaps a subsystem of a manufacturing system). It is built on a robotic assembly line and consists of a large number of rigid parts, some movable, mounted in a chassis. It includes motors, drive systems, and transport paths for workpieces and products. For the sake of familiarity, imagine that this box is a macromechanism made by machining and assembling metal parts, rather than a nanomechanism made by assembling reactive molecules; the issues raised are the same. For concreteness, imagine a system like the mechanism of a xerographic copier.

If this system is to serve as an example of (attempted) M-style evolution, we need some concept of a genetic system and associated embryology. For an M-style system, engineering practice suggests the following assumption: digital programs form the genetic system. They control machines that shape and assemble parts to make the box; this process constitutes the embryology. To make repeated structures, these digital programs might make repeated use of some segments of code. M-style positional assembly implies that these shaping and assembly operations involve moving tools to a series of specific three-dimensional coordinates (with respect to a local "workbench," say). Some mutation operations to programs will add or delete material to parts (changing their size and shape); others will change the coordinates at which a part is placed during assembly.

Certain trivial evolutionary changes are quite feasible in this system. Sections of individual parts that do not interact directly with other parts (or with assembly tools) can change shape with only local effects. A part might become thicker and stronger, and hence more reliable, or it might become thinner and lighter, and hence less costly. Plausible embryologies and selective pressures could lead to considerable optimization of part shapes.

Other trivial evolutionary changes run into difficulties. Some sections of individual parts form interfaces to neighboring parts. If a flange has a particular size and shape, its neighbor must correspond. A substantial "favorable" mutation in one (say, toward a larger, more robust configuration) would disrupt the interface in the absence of a (highly unlikely) simultaneous, corresponding mutation in the other. Only creeping changes in dimensions would be feasible, keeping each part's change within the other's tolerance at each step.

The trivial change of, say, lengthening a bracket similarly tends to disrupt positional assembly. Changing the size of a part on which other parts are mounted changes the position of the mounting-points. A substantial mutation of this sort, if it is to yield a functioning system, must be matched by a simultaneous mutation in the assembly coordinates of all the affected parts—and every added requirement for simultaneous mutation vastly lengthens the odds. This requirement is a direct consequence of positional assembly (and our choice of a simple genetic system and embryology). Again, only creeping change is possible—this time with each change falling within the tolerance of a potentially large number of other parts and assembly steps.

Non-trivial changes add parts or change system organization. Examples include inserting a gasket and connecting a tank to a pipe. The former forces a discrete change in the separation of two surfaces, precluding creeping change. The latter raises the specter of a useless section of pipe, running up to a tank with no opening, or (worse) a tank with a hole and no attached pipe. With positional assembly of non-adaptive parts, there seems no escape from the need for multiple, simultaneous, coordinated changes in such cases. In a viable system, however, mutations at any one site will be extremely rare; a simultaneous, matching mutation at another specific site will be astronomically rare. Several such mutations become effectively impossible.

A genuinely significant evolutionary change, for many purposes, would be one which lets our hypothetical box make a new product. This will, in general, require the addition of many parts, forming a new processing subsystem. In addition, parts will be required to form the channels linking the subsystem with sources of power, with preceding and following processing subsystems, and so forth. Finally, given geometric structures, simply opening enough room for the new subsystem will require wide-spread restructuring of other parts and systems. In short, even small changes rapidly approach impossibility, and the changes required to acquire new capabilities would be large.

It is easy to get some rough idea of the probabilities involved. In modern digital systems (which can incorporate error-correcting codes), an error rate of one bit in a billion is commonly considered high; error rates in fact can be made arbitrarily low through redundancy.[21] DNA replication (with error-correcting enzymes) can achieve bit-error rates as low as one in one hundred billion.[5] In an M-style system (macro- or nano-) designed for reliability, transmission of genetic information should be at least this accurate.

Attaching a new pipe to a tank requires several coordinated changes: making a hole, making a fitting, attaching the fitting, making a pipe, and attaching the pipe. If each of these five changes took as few as eight bits to specify, then 40 changed bits would be needed. Given a 10^{-9} probability of changing a single, specific bit in a generation, the probability of independently changing sixteen specific bits is 10^{-360}. If every hydrogen atom in the observable universe were a genome and had undergone one generation every nanosecond for 10 billion years, the probability of having seen this 40-bit change anywhere, at any time, would be less than one in 10^{270}. A simultaneous, coordinated change of this sort is effectively impossible.

This resembles bogus arguments raised against the feasibility of biological evolution. How do living things escape its application?

O-STYLE SYSTEMS AND EVOLUTION

Bioreplicators have patterns of development, structure, and function that enable evolution to proceed without coordinated, simultaneous genetic changes. Each O-style characteristic contributes to this result.

Because cells and organisms make widespread use of diffusive transport for energy, information, and molecular parts, the evolution of new processing entities (enzymes, glands) is facilitated. A genetic change that introduces an enzyme with a new function can have immediate favorable effects because diffusion automatically links the enzyme to all other enzymes, energy sources, and signal molecules in the same membrane compartment of the cell (and often beyond). No new channels need be built at the same time, because transport isn't channeled. What is more, no special space need be set aside for the enzyme, because device placement isn't geometric.

Changes in the number of parts—so difficult in a rigid M-style system—become easy. There are no strong geometric or transport constraints; this often allows the number of molecular parts in a cell to be a variable, statistical quantity. With many copies of a part, a mutation that changes the instructions for some copies is less likely to be fatal. Thus, diffusive transport facilitates quantitative redundancy, which facilitates qualitative evolutionary experimentation.

A matching assembly process (as in the formation of ribosomes, microtubules, and so forth) tolerates variations in system geometry and numbers of parts that would disrupt positional assembly. Further, the mechanical compliance of biomolecules, such as proteins, gives them a bit of adaptability, allowing small changes in the interface of one molecule to be tolerated by the matching process, giving time for a corresponding change in the facing molecule to occur.

At the level of multicellular organisms, the striking adaptability of tissues and organs ensures that basic requirements for viability, such as continuity of skin and vascularization of tissues, continue to be met despite changes in size and structure. If skin and vascular systems were inert parts, they would require compensating adjustments for such changes. We have seen the problems involved in a minor change in M-style plumbing, yet every individual has a different detailed vascular topology, without corresponding genetic gymnastics.

Thus, O-style systems are not described by calculations like the one above because they can undergo significant evolution *without requiring multiple, simultaneous, coordinated changes*. If one were to perform a similar calculation, allowing the 40 one-bit changes to occur separately, accumulating across generations, then the waiting time for the desired combination would fall from vastly longer than the age of the universe to a fraction of a second.

In short, the characteristics of O-style systems enable cumulative selection to operate; the nature and power of this mechanism have been well described by Richard Dawkins.[3] This mechanism, as Dawkins notes, does not enable all imaginable evolutionary steps, but only some. Among the prohibited steps are those that require multiple, simultaneous, coordinated changes. For example, vertebrate retinas have their neural wiring in front of their photosensors, reducing optical quality and necessitating a blind spot where the optic nerve passes through the sensor layer. Cephalopod retinas have the sensible structure, with the wiring behind. Why hasn't evolution flipped the vertebrate retina? Presumably because there is no small genetic change that would do the whole job, rather than just some damaging part of it; success would require multiple, simultaneous, coordinated changes. Likewise, all living things share essentially the same genetic code for translation between DNA sequences and amino acid sequences in proteins. Why hasn't the code changed in recent evolutionary time, perhaps to add a new amino acid? Presumably because to change the translation mechanism would (among other things) require the simultaneous recoding of the structures of many vital proteins—again, an evolutionarily prohibited step.

Today's O-style biological systems owe their existence to their ancestors' evolutionary flexibility. Since they have inherited that flexibility, they retain the capacity for further evolution; they can be said to have *evolved for evolvability*.[4] M-style systems—with their radically different patterns of development, structure, and function—have not done so and hence lack O-style flexibility. As we have seen, even O-style systems suffer from substantial constraints on their available evolutionary moves; it seems that M-style systems suffer from constraints that effectively eliminate significant evolutionary moves.

M-STYLE SYSTEMS THROUGH O-STYLE DESIGN

If M-style systems cannot evolve, how can they exist? The answer lies in the relationship between design and evolution.

The above argument for the evolutionary incapacity of M-style systems depended on a certain kind of genetic system and embryology operating in a certain kind of evolutionary environment. It assumed that mutations produced isolated changes in the shapes and positions of parts (however broad the consequences of those changes might be), and that selective pressures went to work immediately—in particular, that unworkable designs would fail and be lost, not kept and tinkered with. These assumptions regarding genetics are appropriate for an automatic manufacturing system patterned on present engineering practice, with computer programs playing the role of the genome. They are likewise appropriate for a similarly programmed, self-replicating manufacturing system, like a nanoreplicator. The assumptions regarding selective pressures are appropriate for a system in a situation analogous to the natural environment, as opposed to a development laboratory.

Design is an evolutionary process that operates on different genetic replicators. It works, not by mutating computer programs in a factory, but by mutating ideas in the mind of a designer (or, eventually, high-level representations in an AI design system). If introspection is any guide, ideas are not limited to channeled transport and positional assembly within the mind. Rather, they "diffuse," encountering each other in various patterns and combinations. Some "match" and stick, forming larger systems. These systems seem more topological than geometric, in that their patterns of connectivity are important, and they seldom seem to have anything analogous to a detailed position or alignment that can be globally disturbed by introducing a new piece in the structure. Finally, ideas are typically adaptive, taking a form that depends on their relationships to other ideas. Design concepts—particularly in their formative stages—have the sort of O-style flexibility that implemented machines typically lack.

(Note, incidentally, that the development of software from machine languages to modern AI languages has moved away from M-style toward O-style characteristics. The former are strongly "channeled" and "positional," but rule-based AI systems use mechanisms analogous to diffusion and matching—this enables the introduction of rules without wiring them into an elaborate and rigid control structure. A program based on these principles—EURISKO[20]—is one of the better examples of self-evolving software; Holland's classifier systems,[15] based on genetic algorithms, have these same abstract properties, as do proposed market-based software systems.[12,22] Attempts to make M-style programs evolve through mutation consistently failed.[19])

The evolutionary process of design has another fundamental advantage, independent of the O-style/M-style distinction: it operates under different selection pressures.[7] For a replicator in nature, selective pressures depend directly on function. If a system doesn't work, its failure has a direct, negative impact. In the design process, however, selection pressures differ. If a design doesn't work, it may still be retained because it is *promising*. A whole series of unworkable designs—some containing errors, others simply too sketchy to be implemented—can all be the genetic ancestors (in a memetic sense[2]) of a later design that is novel and workable. The freedom of design processes from the constant-workability constraint of ordinary evolution is a powerful advantage: it enables the introduction of huge numbers of simultaneous, coordinated changes in a single "generation" between working systems. This breaks down otherwise insurmountable barriers.

GENETIC DIFFUSION AND MATCHING

Genes in a population of sexually reproducing organisms diffuse, encountering each other in various combinations. Those combinations that "match," in the sense of being advantageous, tend to "stick" via differential reproduction of that pattern. The genes themselves typically code for parts of diffusive, matching-based systems, increasing the opportunities for recombination to produce useful results. Holland[15] notes some subtle advantages of genetic recombination in testing combinations of genes, but an elementary, quantitative effect is also of interest in efforts to model biological evolution.

A naive model of biological evolution treats it as similar to early experiments in machine learning[19] or to Dawkins' simple computer model of cumulative selection[3]: a generation is equated to a single trial in which one or a few mutant individuals are generated and compared to a single parent, selecting the best as parent of the next generation. The accumulation of mutations in an individual of any generation can, in this model, be seen as the sum of past favorable mutations to that same individual, with unfavorable mutations being discarded.

Consider a population with genetic diffusion. Consider a time span long enough to spread a favorable mutation (and eliminate an unfavorable mutation). The number of generations required to spread a favorable mutation is a selection-pressure-dependent multiple of the logarithm of the population size, in a well-mixed population. For times that are long compared to this time span, the accumulation of mutations in any individual of any generation is (roughly) *the sum of the favorable mutations to all past individuals in the population*, discarding all unfavorable mutations. If the population size is a million, then for the naive model above to have comparable quantitative results, the favorable-mutation rate would have to be multiplied by a factor of a million.

A million generations of a species with a million individuals can be expected to achieve a modest amount of evolution, in biological terms. Many species produce a generation a year, and it has been hundreds of millions of years since animals emerged onto land. A review of early machine-learning experiments describes millions of generations as "an immense number," but this equates to mere millions of trials. A million generations of a million-member species equates to a *trillion* trial-lifetimes. Even at one simulated lifetime per second, this many trials would take a computer roughly 30,000 years.

In considering the evolution of the protein machinery of modern cells, bacterial numbers and generation times are relevant, since bacteria dominated the biosphere for billions of years and eukaryotes are relatively recent. (Note that bacterial genes diffuse through a variety of viral and "sexual" recombination mechanisms.) A planet-wide monolayer of bacteria managing one generation per day would, in a billion years, make some 10^{38} trials. For comparison, this is the number of machine cycles that a trillion computers, each with a gigahertz CPU, would execute in about three billion years.

Genetic diffusion typically multiplies the rate of evolution by many orders of magnitude. Thus, the evolution of mechanisms for genetic diffusion is a prime example of evolving for greater evolutionary capacity. There is, however, no engineering reason to include such mechanisms in typical nanoreplicators.

STYLES AND GENOTYPE-PHENOTYPE MAPPINGS

One can describe the difference between O-style and M-style systems in a picture analogous to Sewall Wright's "genetic landscape"—an n-dimensional space, in which each point corresponds to a combination of genes (actually, a combination of gene-*frequencies*, in Wright's model) and the "height" at that point corresponds to the combination's fitness. In this picture, evolution tends to climb hills and may be blocked from a certain path by a deep enough valley (to sink too low is fatal). To move from Wright's model to one appropriate here, we define two points to be neighbors if they are separated by a single mutation. (This gives it the structure of Dawkins' "genetic space."[3])

O-style systems, because of their flexible organization, can vary in many ways without drastic, deleterious results. They have a relatively smooth mapping of genotypes to phenotypes. Accordingly, their genetic landscape is relatively smooth and continuous, enabling many long, uphill runs.

M-style systems, because of their brittle organizations, can survive few variations. Most significant genetic changes cause a mismatch in the patterns of inert parts and positional assembly procedures, and plunge the system into a deep valley. These systems have a relatively discontinuous mapping of genotypes to phenotypes, in functional terms. Accordingly, their genetic landscape is dissected into a host of tiny, isolated peaks; smooth changes in genetic space do not lead to smooth changes in results. This blocks cumulative selection by preventing small, beneficial steps.

One can imagine an even worse situation. A replicator (whether M-style or O-style) could have an encrypted genome. Each offspring replicator would receive a copy of the genome, then decrypt it in order to read its instructions. With a suitable choice of encryption algorithm, every bit in the genome would affect every bit in the decrypted result, and a single-bit mutation would lead to an effectively random output, flipping half the bits. To expect such an output to be a viable program for replication makes as much sense as programming a computer to perform as a text editor by loading its memory with bits from a random-number generator. (This is not a recommended software engineering practice.) For such a system, with an everywhere-discontinuous mapping of genotypes to phenotypes, the viable peaks in the genetic landscape would consist of isolated points, and cumulative selection would be utterly impossible.

NANOREPLICATORS AND EVOLUTION

Proposed nanoreplicators will be M-style systems, using channeled transport of materials, energy and information, positional assembly of parts, a geometric structure, and rigid, inert parts. Thus, for them to evolve would require multiple, simultaneous, coordinated changes that are impossibly unlikely to occur by accident. (Further, they will lack mechanisms for genetic diffusion.) While such nanoreplicators are amenable to design and modification by engineers (or presumably by AI systems), it seems they cannot undergo significant evolution.

One can imagine designing O-style nanoreplicators, perhaps patterned on bio-replicators, but there is reason to believe that their flexibility would be bought at the price of reduced technical efficiency and ease of design. One can imagine building M-style nanoreplicators controlled by O-style software, allowing evolution through what would amount to clever genotype-phenotype mappings, or through heuristically guided mutation (like that in EURISKO[20]). Achieving this would, however, entail a substantial research task beyond and independent of the task of developing a functional replicator. Further, it would lower the performance of the final devices by imposing computational overhead.

It seems that building a self-replicating molecular system based on nanomachinery does not entail building a system capable of evolution. Indeed, it seems that the latter would be a distinct and challenging goal.

NANOREPLICATORS, BIOREPLICATORS, AND EVOLUTION

The differences between O-style bioreplicators and M-style nanoreplicators are of more than academic interest. If nanotechnology in fact will be developed, then it is important to understand its relationship to familiar biological models and biological hazards.

Living systems are obvious models for nanoreplicators: they replicate, and they are based on molecular components. Indeed, the physical principles they demonstrate provide a firm basis for projecting the feasibility of many of the molecular operations needed in proposed nanoreplicators.

Beyond this, however, the models diverge: in structure and function, proposed nanoreplicators resemble factories more closely than they do cells. The difference between O-style and M-style systems is, in this case, the difference between evolving and nonevolving systems. What is more, living systems are *evolved* systems, while nanoreplicators will be designed: where the former are shaped to serve the goal of their own survival and replication in a natural environment, the latter will be shaped (whether well or poorly) to serve human goals, perhaps in an artificial environment. These differences greatly limit the utility of analogies between living systems and nanoreplicators. They likewise make genetic engineering a poor prototype for nanoengineering.

Genetic engineering today involves not design of replicators from scratch, but tinkering with the molecular machinery of existing bioreplicators. Since bioreplicators were not designed, they are not necessarily structured in a way that lends itself to understanding. Processes based on diffusion and matching allow complex nonlocal interactions that can be hard to trace; not having designed them or completely analyzed them, we still lack complete system specifications. These replicators can be crippled, but having evolved in nature, they resemble systems that can survive in nature. Typically, they are able to exchange genetic information with wild organisms, raising the possibility of the introduction of new, unconstrained replicators in the natural environment. Finally, having evolved to evolve, they have a capacity for further evolution—to serve their own survival, not human goals.

These concerns have inspired great caution regarding genetic engineering. They are substantially mitigated, of course, by the observation that nature has been tinkering with genes for a long time, and that engineered organisms are typically modified in ways that do nothing to help them survive in competition with their wild cousins.

Nanoengineering, in contrast, will involve building replicators from scratch. Because nanoreplicators will differ fundamentally from biological systems, there is reason to believe that novel and remarkably dangerous systems could be constructed—but are they likely to appear by accident?

Several facts make such accidents easy to avoid and difficult to cause. The most obvious and least fundamental of these is that, since these systems will be designed, their parts and structures will be known; moreover, with M-style organization, the relationships among their parts will be designed and fixed. More important, however, proposed nanoreplicators will be fundamentally alien to the biosphere, unrelated to anything that has evolved to survive in nature. For reasons of efficiency and technical simplicity, it will be natural to design nanoreplicators to function in environments found only in special chemical vats (providing, say, hydrogen peroxide as a source of energy and oxygen), placing a straightforward limit on their spread. Finally, engineering experience shows that, while the "capability" to fail (even explosively) can appear by accident, the capability to perform complex organized activities does not. Replicating in a natural environment, without the assumed special chemicals, would be such a complex activity.

The basic benefits of "free-living" replicators can be had without building such things. In a replicating system, a device A makes multiple copies of A, enabling exponential growth in a special environment. If A can in addition makes copies of B, which can make C (but not A or B) which can make D (but not A, B, or C), then many copies of D can be had by starting with a stream of copies of B. Devices B, C, and D can operate in any environment without raising the possibility of uncontrolled, exponential replication—though a copy of B can ultimately give rise to many copies of D, none of the devices in the chain can replicate. The ability to produce large amounts of D without replication reduces the incentive to build nanoreplicators that operate in natural environments.

Genetic engineering operates on the design level in modifying replicators that are themselves evolved to evolve. Nanoengineering will operate on the design level in constructing replicators that need have no ability to evolve; with such replicators, there will be a clean separation of the evolutionary mechanism (designs and designers) from the individual replicator's genetic mechanism (its embedded program). In light of general engineering practice and specific efficiency concerns, this seems the natural way to proceed. Developers would have to go out of their way to give M-style nanoreplicators an evolutionary capacity, at a substantial cost in design effort and software complexity.

In light of all this, it seems that by simply neglecting to solve some difficult problems, we need never come close to building nanoreplicators capable of runaway exponential growth, or capable of evolving into systems that pose that threat. Scenarios of massive destruction are a concern,[7] but accidents seem easy to avoid. The problem to focus on is not that of accidents, but of deliberate abuse.

CONCLUSION

When we speak of life, we speak of organic self-replicating systems with structures and behaviors that result from evolution and have a capacity for further evolution. M-style self-replicating systems will result from deliberate design, and (barring extraordinary efforts) will lack the capacity for further evolution. As a consequence, their behaviors can be expected to be stable and designed to serve human goals (however imperfectly) rather than being mutable and evolved to act as robust survival-and-replication systems. Although it is sometimes useful to consider them from the perspective of biological systems, M-style nanoreplicators will differ from organisms in such fundamental ways that it would be misleading to describe them as living things. It is entirely accurate to call them machines.

The pursuit of genuine artificial life will require special attention to O-style organization, or to conditions that can lead to its evolution. Work in artificial life will not automatically be furthered by the pursuit of useful nanomechanisms and self-replicating systems.

REFERENCES

1. Becker, R. S., J. A. Golovchenko, and B. S. Swartzentruber (1987), "Atomic-Scale Surface Modifications Using a Tunnelling Microscope," *Nature* **325**, 419–421.
2. Dawkins, R. (1976), *The Selfish Gene* (New York: Oxford Univ. Press).
3. Dawkins, R. (1987), *The Blind Watchmaker* (New York: Norton).
4. Dawkins, R. (1988), "The Evolution of Evolability," these proceedings.
5. Drake, J. (1969), "Comparative Rates of Spontaneous Mutation," *Nature* **221**, 1132.
6. Drexler, K. E. (1981), "Molecular Engineering: An Approach to the Development of General Capabilities for Molecular Manipulation," *Proc. Natl. Acad. Sci.* **78**, 5275–5278.
7. Drexler, K. E. (1986), *Engines of Creation* (New York: Doubleday).
8. Drexler, K. E. (1986), "Molecular Engineering: Assemblers and Future Space Hardware," *Proceedings of the 33rd Annual Meeting of the American Astronautical Society, Boulder, October, 1986.*
9. Drexler, K. E. (1987), "Molecular Machinery and Molecular Devices," *Molecular Electronic Devices II*, Ed. Forrest Carter (New York: Marcel Dekker).
10. Drexler, K. E. (1987), "Nanomachinery: Atomically Precise Gears and Bearings," *Proceedings of the IEEE Micro Robots and Teleoperators Workshop, Hyannis, November, 1987.*
11. Drexler, K. E. (1988), "Rod Logic and Thermal Noise in the Mechanical Nanocomputer," *Proceedings of the Third International Symposium on Molecular Electronic Devices*, Ed. Forrest Carter (Amsterdam: Elsevier).
12. Drexler, K. E., and M. S. Miller (1988), "Incentive Engineering for Computational Resource Management," *The Ecology of Computation*, Ed. Bernardo Huberman (Amsterdam: Elsevier).
13. Foster, J. S., J. E. Frommer, and P. C. Arnett (1988), "Molecular Manipulation Using a Tunnelling Microscope," *Nature* **331**, 324–326.
14. Hayward, R. C. (1983), "Abiotic Receptors," *Chem. Soc. Rev.* **12**, 285–308.
15. Holland, J. H. (1986), "Escaping Brittleness: The Possibilities of General Purpose Machine Learning Algorithms Applied to Parallel Rule-Based Systems," *Machine Learning: An Artificial Intelligence Approach*, vol. 2, Eds. R. S. Michalski, J. G. Carbonell, and T. M. Mitchell (Los Altos, CA: Kaufmann).
16. Kelly, T. R., and M. P. Maguire (1987), "A Receptor for the Oriented Binding of Uric Acid Type Molecules," *J. Am. Chem. Soc.* **109**, 6549–6551.
17. Lehn, J.-M. (1985), "Supramolecular Chemistry: Receptors, Catalysts, and Carriers," *Science* **227**, 849–856.
18. Lehninger, A. L. (1975), *Biochemistry* (New York: Worth).
19. Lenat, D. B. (1983), "The Role of Heuristics in Learning by Discovery: Three Case Studies," *Machine Learning*, Eds. R. S. Michalski, J. G. Carbonell, and T. M. Mitchell (Palo Alto, CA: Tioga).

20. Lenat, D. B., and J. S. Brown (1984), "Why AM and EURISKO Appear to Work," *Artificial Intelligence* **23**, 269–294.
21. McEliece, R. (1985), "The Reliability of Computer Memory," *Scientific American* **248**, 88–92.
22. Miller, M. S., and K. E. Drexler (1988), "Comparative Ecology: A Computational Perspective" and "Markets and Computation: Agoric Open Systems," *The Ecology of Computation*, Ed. Bernardo Huberman (Amsterdam: Elsevier).
23. Pabo, C. O., and E. G. Suchanek (1986), "Computer-Aided Model-Building Strategies for Protein Design," *Biochemistry* **25**, 5987–5991.
24. Ponder, J. W., and F. M. Richards (1987), "Tertiary Templates for Proteins," *J. Mol. Biol.* **193**, 775–791.
25. Rastetter, W. H. (1983), "Enzyme Engineering," *Appl. Biochem. and Biotech.* **8**, 423–436.
26. Srivastava, D. K., and S. A. Bernhard (1986), "Metabolite Transfer via Enzyme-Enzyme Complexes," *Science* **234**, 1081–1086.
27. Ulmer, K. M. (1983), "Protein Engineering," *Science* **219**, 666–671.

Stuart Hameroff,† Steen Rasmussen‡ and Bengt Månsson*

†Advanced Biotechnology Laboratory, Department of Anesthesiology, University of Arizona College of Medicine, Tucson, Arizona 85724, ‡Physics Laboratory III and Center for Modelling, Nonlinear Dynamics and Irreversible Thermodynamics, Technical University of Denmark, 309C, DK-2800 Lyngby, Denmark, and *Physical Resource Theory, Chalmers University of Technology, S-412 96 Goteborg, Sweden

Molecular Automata in Microtubules: Basic Computational Logic of the Living State?

1. INTRODUCTION

The study of "Artificial Life" includes a number of approaches to understanding the "molecular logic of the living state."[27] Among these are the study of self-organizing processes, dynamic simulation of life–like behavior, computer programs which evolve and compete, robotics, origin of life, and computer based cellular automata. Further, the study of Artificial Life may serve to emulate and illuminate possible mechanisms of molecular logic in biological structures. Defined as a "unique set of relationships characterizing the nature, function, and interactions of biomolecules,"[29] life's molecular logic remains elusive and Artificial Life researchers generally seek to reproduce this logic independent of the particular biological "wetware" involved. However, certain classes of biomolecules may have appropriate structure and functional performance to warrant consideration as sites for life's molecular logic. For example, DNA and RNA store hereditary data which, when dynamically implemented, manifest genetic information and lead to the full variety found in the many forms of life. Other classes of biomolecules are directly involved

Artificial Life, SFI Studies in the Sciences of Complexity,
Ed. C. Langton, Addison-Wesley Publishing Company, 1988

in real-time biological information processing in which genetically determined factors interact with input from the environment to determine dynamic behavior. Specifically, the cytoskeleton is a class of protein polymers which comprises intelligent parallel networks within the interiors of living eukaryotic cells. Among these cytoskeletal polymers, microtubules are the most conspicuous.

The cytoskeleton was originally thought to provide merely structural "bonelike" support within cells, but is now known to dynamically organize living cellular activities. Complex activities of microtubules and other cytoskeletal elements (i.e., axoplasmic transport, mitosis, cell growth, shape and differentiation, locomotion, synapse modulation, etc.) have prompted belief that the cytoskeleton is the cell's nervous system. A series of models of dynamic cytoskeletal information processing have been proposed.[1,3,5,8,19,22,26,36] Among these are microtubule cellular automata[20,21,41] in which individual automata "cells" are microtubule protein subunits whose polymerized form is a hexagonal lattice wrapped in a cylinder. In these models, based as much as possible on known microtubule biochemistry, each subunit within the cylindrical lattice may exist in two or more conformational states, depending on lattice neighbor conformational states at discrete time intervals. Fröhlich's model of coherent protein oscillations can provide a cooperative "clocking" mechanism in microtubules for discrete generations or time steps for cellular automata.[13-15] Automata behavior within microtubules and other cytoskeletal structures could explain their capacity for intelligent organization—at least one form of the "molecular logic of the living state."

In this paper we investigate molecular-level automata based on known microtubule structure and feasible nearest neighbor interaction "rules" derived from geometrically determined dipole coupling among microtubule polymer subunits. Study of microtubule automata is of interest because of their unique geometry (parallel, interconnected, cylindrical, hexagonal lattice automata), resemblance to spin-glass systems, possibility of "nesting" in hierarchical automata (e.g., the brain) and because they may represent true biological mechanisms. Our simulations of microtubule automata demonstrate categories of behavior including robust gliders, blinkers and emergent patterns which selectively propagate. These behaviors appear capable of selective transport of molecules and organelles, molecular information processing and the orchestration of intracellular activities. Thus microtubule automata are examples of "Artificial Life" which may represent molecular logic in life's basic computing units.

CYTOPLASM AND THE CYTOSKELETON

Living organisms are collective assemblies of cells which contain collective assemblies of organized material called protoplasm. In turn, protoplasm consists of membranes, organelles, nuclei and the bulk interior medium of living cells: cytoplasm.

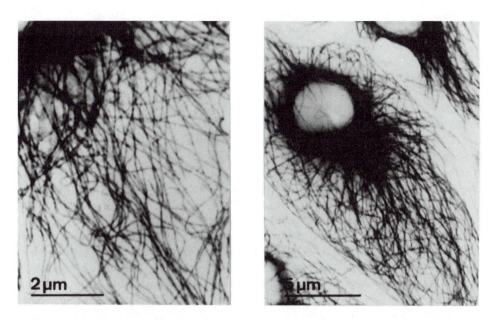

FIGURE 1   Tissue culture cells illustrating microtubules (MT) by dark staining immuno-fluorescence. Dense arrays of MT near nucleus emanate from centrioles. With permission from Geuens, Gundersen, Nuydens, Cornellisen, Bulinski, and DeBrabander.[17]

Dynamic rearrangements of cytoplasm within eukaryotic cells account for their changing shape, repositioning of internal organelles, and in many cases, movement from one place to another. In brain nerve cells, cytoplasmic rearrangements lead to formation, maintenance and modulation of synaptic connections, the cornerstone of contemporary "neural net" theories of learning and memory.

The nature of cytoplasm has been intensely studied since the advent of light microscopy in the last century. Many conflicting theories and descriptions emerged and characterized living cytoplasm as containing or consisting of "reticular threads," "alveolar foam," "watery soup," or "gel: an elastic intermeshing of linear crystalline units giving elasticity and rigidity to a fluid while allowing it to flow."[7] Initially, development of the electron microscope through the 1960's did not illuminate the substructure of cytoplasm. Portions of cells which were optically empty by light microscopy persisted in being empty in electron micrographs. The cell was perceived to be a "bag of watery enzymes." In some cases, however, fibrillar structures seen with light microscopy appeared as fine tubular filaments with the electron microscope. These tubular filaments comprised the internal structure of motile organelles such as cilia, flagella, centrioles and basal bodies, and were prominent in the mitotic apparatus of dividing cells. Tubular filaments were also observed throughout the cytoplasm of cells including the axons and dendrites of neurons. Ironically, the fixative then used in electron microscopy, osmium tetroxide, had been dissolving filamentous elements so that their presence was observed only sporadically. Later in the early 1970's with the advent of glutaraldehyde fixation for electron microscopy, delicate tubular structures were found to be present in virtually all cell types and they came to be called microtubules (Figure 1). Subsequent characterization of other cytoskeletal elements such as actin, intermediate filaments, and the structure of centrioles led to the recognition that cell interiors were comprised of dynamic networks of connecting filaments and brought the cytoskeleton out of the closet.

In the context of evolution, the cytoskeleton plays an interesting and perhaps crucial role. Several billion years ago, life on earth underwent a large nonlinear evolutionary jump when simple, immobile bacteria (prokaryotes) "suddenly" developed into a rich variety of mobile organisms (eukaryotes) over a surprisingly brief time span. One theory which explains this evolutionary leap is that of a symbiotic association of several life forms. Such symbiosis (a "mutually beneficial association") was proposed by Marishkowski in 1905, Wallen in 1922, and refined by Margulis in 1975.[30,33] Margulis proposed that prokaryotic cells underwent a sequence of symbiotic events leading to the first eukaryotes. The first step, according to Margulis, was during the period of adaptation to oxygen breathing when a non-oxygen breathing prokaryote engulfed an oxygen breathing prokaryote which became the ancestor of the mitochondria, capable of converting oxygen to chemical energy packets (i.e., ATP). The next symbiotic event, according to Margulis, was ingestion of a spirochete: a motile organism which traveled by whip-like beating of its tail-like flagellum composed of contractile cytoskeletal filaments. Ingestion and retention of flagellae and their intracellular anchors, basal bodies, are thought to have led to cilia, centrioles, and microtubules: cytoskeletal structural and organizational elements which brought the capabilities for intelligent cell movement and cytoplasmic compartmentalization and organization. Multiple cilia anchored to basal bodies and extending outward through cell membranes enabled single cell organisms such as paramecium to swim about in their aqueous medium, greatly expanding their ability to find food, avoid predators, and extend their horizons. In other stationary cells, cilia could propel the environmental medium past the organism. Within the cytoplasm, cytoskeletal structures such as centrioles, basal bodies and microtubules organized, oriented, and transported organelles and materials including the separation of chromosomes.[31] Cells became diversified as the cytoskeleton took on functions akin to mechanical scaffolding, conveyor lattice, and the cell's own nervous system. The resultant "eukaryotic" cell was as different from prokaryotes as a main-frame from an abacus.

CENTRIOLES AND MICROTUBULES

Cilia, flagella, basal bodies, and centrioles are assemblies of microtubules, themselves complex cylindrical assemblies of protein subunits, and are ubiquitous

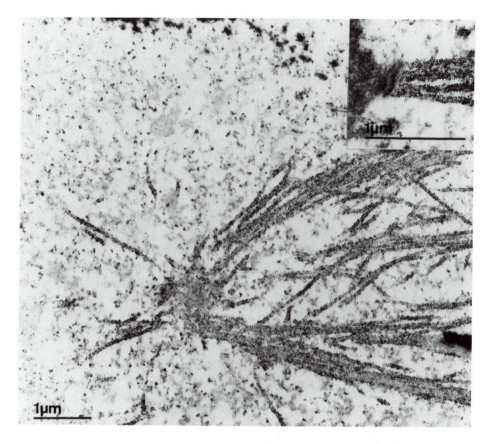

FIGURE 2 Microtubules (MT) labelled with tyrosine tubulin immunogold in tissue culture cell during mitosis. Mitotic spindle MT emanate from centrioles at left. Insert upper right: MT attaching to chromosomes. From Geuens, Gundersen, Nuydens, Cornellisen, Bulinski, and DeBrabander.[17]

throughout eukaryotic biology. Centrioles are the specific apparatus within living cells which trigger and guide major reorganizations of cellular structure occurring during mitosis, growth and differentiation. In centrioles, nine pairs or triplets of microtubules are arranged in self-replicating super-cylinders, which are always found in pairs oriented perpendicular to each other. To initiate cell division, two centriole pairs migrate to opposite poles of the cell from where microtubule "mitotic spindles" separate chromosomes and establish orientation and architecture for the next generation cells[10] (Figure 2). Centrioles are the cell's navigators, gravity sensors and the focal point of the cytoskeleton. Their mystery and aesthetic elegance have created an enigmatic aura about these marvelous organelles. In Wheatley's[46] book

"Centrioles: A Central Enigma in Cell Biology," B.R. Brinkley refers to centrioles as the "center of the cytoplasmic universe."

In addition to specialized cytoskeletal organelles like centrioles, bulk cytoplasm consists of networks of individual microtubules, arrayed in parallel and interconnected by filamentous strands (Figure 3). Other interconnecting networks of smaller filamentous proteins (actin, intermediate filaments, etc.) intersperse with microtubules to form a dynamic gel whose activities (i.e., mitosis, growth and differentiation, locomotion, food ingestion or phagocytosis, synapse modulation, dendritic spine formation, cytoplasmic movement, etc.) define the living state. It seems likely, even necessary, that some cytoskeletal mechanism for information processing serves to organize cytoplasmic behavior. Of the various filamentous structures which comprise the cytoskeleton, microtubules (MT) are the most widely observed, best characterized and perhaps best suited for dynamic information processing (Figures 4–6). Most MT are assemblies of 13 longitudinal protofilaments which are each a series of polar, subunit dimer proteins known as tubulin. Each tubulin dimer consists of two slightly different classes of 55 kilodalton monomers (α and β tubulin) which are 4 nanometers (nm) in diameter.[12] MT assembly and disassembly are dynamic, complex processes which depend on calcium ion concentration and other factors. Oriented by centrioles, MT polymerization determines the architecture and form of cells which can quickly change by MT depolymerization and reassembly in another direction.[25] GTP, an energy providing analog of ATP, binds to polymerizing

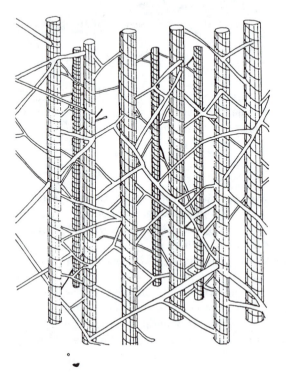

FIGURE 3 Artist's impression of parallel microtubule (MT) network within a cell. Straight cylinders are microtubules (MT), 25 nanometers in diameter. Hatched areas on MT represent tubulin subunit dimers. Branching interconnections are microtubule associated proteins (MAPs) and/or other cytoskeletal filaments. Actin and intermediate filaments are not shown.
By Jamie Bowman Hameroff.

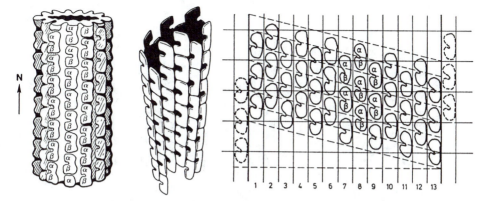

FIGURE 4 Microtubule (MT) structure and computer simulation unfolding of MT lattice. Left: MT structure by X-ray diffraction analysis.[2] Each subunit is a dimer with an α and β monomer. Middle: stylized tubule showing spiral offset of 1-1/2 dimers (3 monomers) per row. Right: MT lattice flattened for computer simulation. α and β monomers within a single neighborhood of central dimer and surrounding hexagonal dimers are labelled.

tubulin. GTP hydrolysis energy is delivered to assembled MT, although the precise utilization of the energy is not understood. One possible utilization is the production of coherent lattice excitations as proposed by Fröhlich.[13–15]

Hexagonal packing of MT subunit dimers is "twisted" resulting in slightly different subunit neighbor relationships. MT outer diameters are 25–30 nm, inner diameters are 14–15 nm and functional MT lengths may range from hundreds of nm to micrometers and meters in some mammalian neurons.[35] When viewed in cross section by electron microscopy, MT outer surfaces are surrounded by a "clear zone" of several nm which apparently represents an electronegative field due to excess electrons in tubulin and may also organize cytoplasmic water and enzymes.[42] MT, as well as their individual dimers, have dipoles with negative charges localized toward alpha monomers.[8] Thus MT are "electrets": oriented assemblies of dipoles which are predicted to have piezoelectric properties.[4,33] Contractile or enzymatic proteins (microtubule associated proteins: "MAPs") may be attached to MT at specific dimer sites; MAP attachments result in various helical patterns on MT surfaces.[6] Proteins, organelles, calcium ions and other materials are transported by MT and their contractile MAPs at rates from one to 400 millimeters per day.[37] The contractile MAPs use ATP as their energy source, but the methods of control and orchestration of these tiny "arms" are not understood.

MT and cytoskeletal function demonstrate the importance of cytoskeletal activities to intelligent behavior at the cellular level. These activities and the lattice polymer structure of MT have suggested capabilities for information processing.

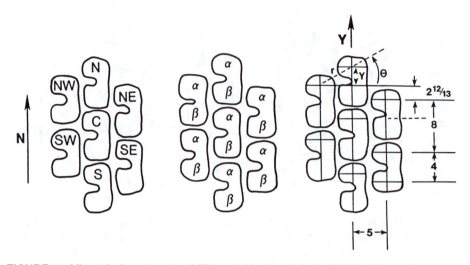

FIGURE 5   Microtubule automata (MTA) neighborhood. Left: Neighborhood dimers oriented to longitudinal MT axis with α ends to the north (C=center, N=north, NE= northeast, SE=southeast, S=south, SW=southwest, NW=northwest). Center: α and β monomers within each dimer are labelled. Right: Distances (in nanometers) and orientation among lattice neighbors. Interaction forces are calculated using y=r sin θ.

At least a dozen author groups have published models of MT/cytoskeletal information processing which liken cytoskeletal function to computer related technologies.[19] Among these is a model of cellular automata behavior resulting from coherent dipole excitations among MT subunits.

COHERENT PROTEIN DIPOLE EXCITATIONS

Proteins are vibrant, dynamic structures in physiological conditions. A variety of recent techniques (nuclear magnetic resonance, X-ray diffraction, fluorescence depolarization, infrared spectroscopy, Raman and Brillouin laser scattering) have shown that proteins and their component parts undergo conformational motions over a range of time scales from femtoseconds (10^{-15} sec) to many minutes. The most functionally significant conformational vibrations appear to be roughly in the range of 10^{-9} to 10^{-11} sec, or from nanoseconds to ten picoseconds. Biologically relevant motions of globular proteins in this time scale are thought to include "collective elastic body modes, coupled atom fluctuations, solitons and other nonlinear motions, and coherent excitations."[23,24]

Collective conformational states near the nanosecond time domain have been woven into a theory of coherent protein excitations by Herbert Fröhlich. A major contributor to early theory of superconductivity, Fröhlich turned to the study of biology in the late 1960's and came to several conclusions with profound implications. One is that changes in protein conformation are triggered by charge redistributions such as dipole oscillations within hydrophobic regions of proteins.[14] Another Fröhlich concept[15] is that a set of proteins connected in a common physical structure and voltage gradient field such as within a membrane or polymer electret like a microtubule would be excited coherently if biochemical energy such as ATP or GTP were supplied. Coherent excitation frequencies on the order of 10^9 to 10^{11} Hz are deduced by Fröhlich, who cites as evidence sharp windows of sensitivity to electromagnetic energy in this region by a variety of biological systems.[18] Other aspects of Fröhlich's model include "metastable states" (longer-lived conformational state patterns stabilized by local factors) and polarization waves, traveling regions of conformational states out of phase with the majority of coherently excited states. Solitons may be an equivalent description of these phenomena.[42,46] Fröhlich's model of coherence can explain long-range cooperative effects by which proteins and nucleic acids in biological systems may communicate. The major component of his theory suggests that random supply of energy to a system of nonlinearly coupled dipoles can lead to coherent excitation of a single vibrational mode, provided the energy exceeds a critical threshold. Important biological consequences may be expected from such coherent excitations and long-range cooperation.

Coherent excitations and cooperative coupling among tubulin dimers within microtubules may serve as a "clocking" mechanism to generate discrete "generations" for cellular automata behavior in MT. A crude calculation, assuming the clocking mechanism to be a sound wave ($V_{sound} \cong 10^3$m/s), yields a clocking frequency of approximately 10^{11} Hz for one wave across the MT diameter ($\cong$ 25nm). Thus the orientation, state, or phase of any tubulin dimer at any given excitation period would depend, according to Fröhlich's model, on factors which include electrostatic dipole neighbor interactions, conformational states, binding of water, ions, or microtubule associated proteins (MAPs), bridges to other MT, energy-providing phosphate nucleotides (i.e., GTP) and associated proteins, and/or intrinsic genetically determined subunit factors. Net effects of these influences could alter phases of particular subunits in coherent biomolecular arrays, possibly resulting in dynamic pattern polarization waves, metastable states and long-range order. For our microtubule automata neighbor rules, we have focused solely on electrostatic dipole interactions among coherently excited subunits within MT.

CELLULAR AUTOMATA IN MICROTUBULES

Cellular automata behavior including dynamic patterns and capabilities for information processing depend on a lattice whose subunits can exist in two or more states at discrete time steps, and neighbor interaction rules which determine those states among lattice subunits.[11,16,45,47,49,50,51] Coherent excitations could effect discretetime steps among MT lattice neighbors. Fröhlich's model predicts coherent excitations within a range of 10^{-9} to 10^{-11} seconds which matches the conclusions of Karplus and McCammon[23,24] that biologically relevant protein motions occur in this time domain. We will nominally refer to the clocking period, or generation time as "nanosecond" (10^{-9} sec), although shorter times (i.e., 10^{-11} sec) imply faster automata.

The term cellular automata undoubtedly arose because of the conception of biological cells as indivisible subunits. However, biological cells are complex dynamic entities whose behavior is a collective effect of component systems including the cytoskeleton. "Cellular" automata therefore may be an ironic misnomer and "cells" in our automata model are component subunits of microtubules: tubulin dimers. The structure of MT and our derivation of automata neighborhoods and neighbor dynamics are illustrated in Figures 4–6. The MT cylinder has a circumference of 13 units and the pitch pattern of the leftward helix is 1-1/2 dimers (three tubulin monomers—Figure 4). We consider 7-member neighborhoods of tubulin dimers: a central dimer surrounded by a twisted hexagon of 6 neighbor dimers. Taking the cylinder axis of the MT to define the "y" axis, the dimer neighborhood is defined as follows: "C" is the center dimer and "N" (north) is the nearest neighbor dimer in the positive direction of the y-axis (Figure 5). Similarly, dimers labeled "NE" (northeast), "SE" (southeast), "S" (south), "SW" (southwest), and "NW"(northwest) are appropriately oriented around the "C" dimer. Within a dimer the α-monomer is north of its β partner. Along the longitudinal MT "y" or "north" axis, monomer centers in each longitudinal row are 4 nm apart, and the distance between longitudinal rows is 5 nm. The helical twist yielding an offset along the y axis for NW, SW, NE, and SE neighbors lead to neighbor dimer distances and dipole interaction forces which distinctly differ among the six neighbors surrounding the C dimer. These dipole interaction forces, coupled to tubulin conformational states, are considered as neighbor rules in our model of microtubule automata. The differing forces from the six neighbors, along with the "wraparound" cylindrical structure are geometrical properties of MT which contribute to interesting automata behavior.

The basis for neighbor interactive forces is in dipole coupling among MT dimers. Each dimer may be viewed as having a mobile electron shared by the two monomers. At each time step, the electron's average position is considered to be oriented either more toward the alpha monomer ("alpha state") or more toward the beta monomer ("beta state"—Figure 6). MT net dipoles are negative toward the alpha ends and oscillations need not necessarily change direction totally to effect a conformational change. Fröhlich's theory suggests that each dimer's conformational state, coupled to its dipole orientation, will "update" every nanosecond. The electrostatic forces exerted on each dimer's mobile electron by the mobile electrons of its nearest neighbor dimers would depend on the dipole orientations of the neighbor dimers and may serve as the basis for automata "rules." The magnitude of the electrostatic force exerted by any one neighbor is:

$$f = \frac{e^2}{4\pi\varepsilon r^2} \tag{1}$$

where ε is the average permittivity in MT (determined by the permittivity of free space times the MT dielectric constant, typically 10 for proteins[29]), e is the charge on an electron and r is the distance between the electrons. The distance r between electrons differs among neighbor pairs because of the MT lattice screw symmetry and dipole state of each neighbor. Net positive neighbor forces will induce center dimer alpha states, and net negative forces will induce center dimer beta states at each time step. We have assumed that only the y-component of the interaction forces are effective, and neglected any net force around the circumference.

Considering the net forces due to MT lattice geometry, the resulting force acting on an electron in a central dimer can then be calculated as:

$$f_{\text{net}} = \frac{e^2}{4\pi\varepsilon} \sum_{i=1}^{6} \frac{y_i}{r_i^3} \tag{2}$$

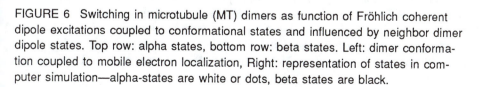

FIGURE 6 Switching in microtubule (MT) dimers as function of Fröhlich coherent dipole excitations coupled to conformational states and influenced by neighbor dimer dipole states. Top row: alpha states, bottom row: beta states. Left: dimer conformation coupled to mobile electron localization, Right: representation of states in computer simulation—alpha-states are white or dots, beta states are black.

From Eq. (2) it is seen that only the relative magnitudes of neighbor forces are necessary for the automata simulations. Distances between pairs of neighbor dimer electron sites, symmetry considerations, and relative neighbor dipole coupling forces yielding automata "rules" are shown in Tables 1–3 in the Appendix. Calculated summation of 128 possible 7-member neighborhoods are shown in Tables 4 and 5 in the Appendix. Relative forces in Appendix Tables 3–5 may be multiplied times 2.3×10^{-14} Newtons to obtain absolute values. Accordingly, the typical absolute value for electron-electron interactions among MT dimer neighbors are of the order of 10^{-13} Newtons.

To simulate MT automata, we represent MT structure as a two-dimensional grid in which the cylindrical MT has been fileted open, flattened, and placed horizontal with the "north" and "y" direction now pointing to the right (Figure 7). The grid consists of MT subunit dimer loci which can exist in either an "alpha state" (blank with dot—Figure 6) or "beta state" (solid black—Figure 6) at each time step or generation. To run the simulation, each dimer in an MT grid is treated as the "C" dimer for each time step. We have studied 2 different boundary conditions and grid formats: 1) a 40-dimer-long MT in which the "north" and "south" ends are connected, producing a toroidal surface (Figures 7, 9, 11, & 13-15) and 2) a 125-dimer-long, open-ended MT in which states at opposite ends are independent (Figures 10 & 12).

The extent to which each dimer is influenced by net neighbor forces acting upon its mobile electron may be represented by a "threshold" parameter. The higher the threshold, the greater are the summated neighbor forces necessary to induce a transition. For example, a threshold of ± 9.000 means that net neighbor forces greater than $9.000 \times 2.3 \times 10^{-14}$ Newtons will induce an alpha state, and negative forces of less than $-9.000 \times 2.3 \times 10^{-14}$ Newtons will induce a beta state. We have investigated the effects of varying thresholds on MT automata behavior as well as the effects of asymmetrical thresholds in which alpha and beta transition thresholds are different. Biological factors which could equate to threshold include temperature, acid/base balance, voltage gradients, ionic concentration, genetically determined variability in individual dimers, and binding of molecules including MAPs and/or drugs to dimer subunits. In nerve cells, traveling membrane depolarizations could induce transient waves of lowered threshold along parallel-arrayed MT. Such coupling could provide a hierarchical nesting mechanism between neuronal level activities ("neural nets") and cytoskeletal automata.

In the MT automata of Smith, Watt, and Hameroff[41] and Hameroff, Smith, and Watt,[20,21] boundary conditions were chosen in which reflection of automata patterns occurred at each end of simulated tubules. With this boundary condition and a constant low threshold, patterns of alpha/beta states developed which included traveling wave fronts, oscillatory standing waves, traveling kinks, diamonds and triangles. These traveling structures "interacted with the background, could be destroyed, absorbed by, or bounce off other traveling structures and disturb the background, reorganizing as they go." Our current study sought to corroborate these

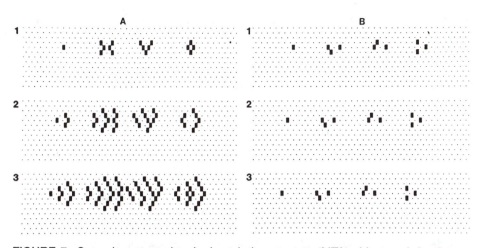

FIGURE 7 Some important virtual microtubule automata (MTA) objects existing for two different threshold values. (a) Left: Objects in top panel are labelled from left to right, dot glider, spider glider, triangle glider, and diamond blinker. Gliders are objects which move without losing their shape, and blinkers are stationary oscillating objects. The two lower left-hand panels show the next two generations of these β objects on an α background. The thresholds are $\pm$ 1.000. Note that the objects leave a "wake", which in time leads to a traveling wave structure (see also Figure 9). (b) Right: the first three generations of some other gliders existing in the system with thresholds around ±9.000. Only the dot glider of the objects from (a) survives in this system. Here we see different triangle gliders, which are elements in bus gliders developing from random seeds in MTA (also see Figure 13). Here at a threshold of $\pm$ 9.000, gliders travel without creating any wake.

preliminary findings, investigate different boundary conditions, catalogue effects of altered thresholds, and describe behavior of "virtual" automata which develop in MT.

SIMULATIONS AND RESULTS

For our MT automata simulations we have studied two kinds of boundary conditions which differ from the reflection boundaries previously used. In both cases, the East and West boundaries are contiguous (with a three-monomer offset) to simulate the cylindrical MT geometry. One of the two new "North-South" boundary models we have studied is a torus, in which the two ends of an MT segment which is 40 dimers in length are connected to each other and information can flow around the closed system toroidal surface. The other boundary condition is an open ended MT which is 125 dimers in length in which the states of the dimers of the extreme South and North ends of the MT are independent. We have investigated cases where these states all are set at random at each time step, as well as where the North-end dimers are all in one state. Thus there is a flow of information into and out of the MT. In both boundary-condition models, varying the threshold significantly changes the behavior of the MT automata patterns.

In toroidal boundary conditions, six qualitatively different behaviors are observed which depend on threshold conditions and randomly chosen starting patterns. In Figures 7a and 7b, toroidal microtubule automata ("MTA") start with an α background and β "seeds": this means that at the initiation of a sequence of automata time-steps, most MT grid subunits are in α states, with a few randomly selected, isolated β states. From top to bottom Figure 7a shows evolution of some β "seed" objects at a low threshold ($\pm$ 1.000). We refer to these MTA objects which originate in the top row of Figure 7a (from left to right) as: dot glider, spider glider, triangle glider, and diamond blinker. Gliders are objects which move without losing

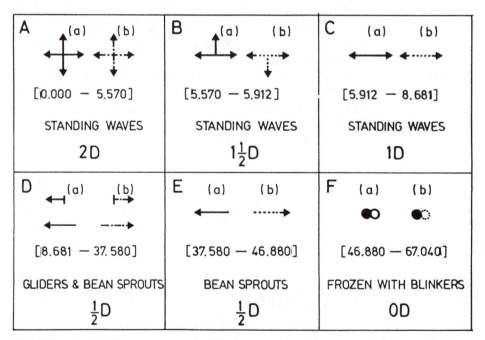

FIGURE 8 Dependence of toroidal MTA dynamics on symmetrical dipole coupling threshold. The upper items in each sector (A–F) indicate the dominant global behavior for (a) β seeds induced on an α background, and (b) α seeds induced on a β background. Below the dominant behaviors are the threshold intervals, what we call this type of dynamic phenomena, and the dimensionality of the growth patterns.

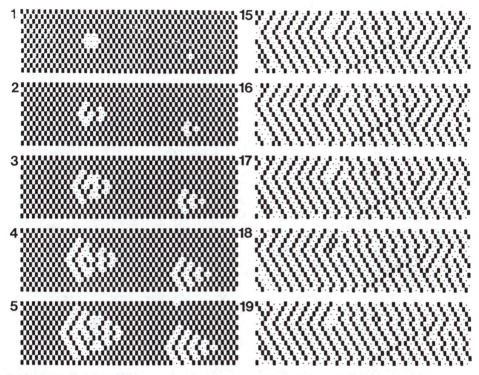

FIGURE 9   Toroidal MTA with thresholds ±1.000. Numbers at left indicate generation. Initial pattern is a β background with α seeds which develop into ripple patterns or traveling waves. Last generations (15-19) consist of traveling waves containing virtual structures as diamond blinkers, triangle gliders and spider gliders. The spider glider is the most robust virtual structure in MTA.

their shape, and blinkers are stationary objects which oscillate every other generation. The middle and lower sections of Figure 7a show the next two generations of these β seed objects leaving "wakes", altered background patterns which lead to traveling and/or standing wave structures (see also Figure 9). In Figure 7b, three generations of similar seeds are shown with a higher threshold of ± 9.000. These objects travel as gliders, but without the wakes and traveling waves observed with the lower threshold in Figure 7a. The dot glider behavior exemplifies the differing effects of the 2 thresholds. The other triangle gliders are elements of "bus gliders" which can develop from random seeds (Figure 13). MTA gliders generally travel at a velocity of one dimer per generation. Since each dimer is 8 nm in length, this glider velocity translates to 8 to 800 meters per second over a range of time-steps from 10^{-9} to 10^{-11} seconds. Perhaps coincidentally, this velocity range correlates with that of traveling nerve action potentials.

Figure 8 summarizes the dependence of toroidal MTA behavior on threshold. The upper portion of each sector of the diagram schematically indicates the dominant global behavior for (a) β seeds on an α background, and (b) α seeds on a β

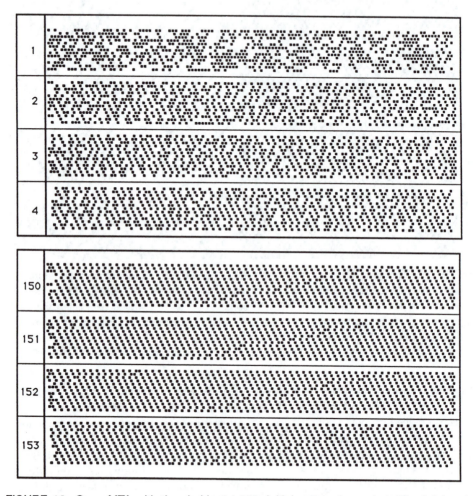

FIGURE 10 Open MTA with thresholds ±1.000. Initial pattern is random. The left-hand end is random at each time step and the right hand end is kept in β states at all times. By generations 150–153, the system self-organizes into a crystal-like pattern with dislocations. β spiders move through the crystal and change the dislocations. Near the left-hand end there is a glancing collision of β and α spider gliders. Although the wave-fronts can lie in two directions, the SW-NE direction is always favored (see also Figure 9).

background. Below the schematic behaviors are shown the corresponding threshold intervals followed by our qualitative description of the behavior. At the bottom of each sector, the dimensionality of the growth behavior is indicated.

In Figure 9 the dynamics of a toroidal MTA at a symmetric threshold ± 1.000 is shown; Figure 10 shows an "open" MTA at the same threshold. Compared to higher-threshold MTA, these are more sensitive automata in that the net forces of all configurations (except the pure α and the pure β configuration where net forces are zero) exceed the threshold. In Figure 9 the development from an initial configuration with a β background and a few α "seeds" is seen. These seeds grow in all four directions and an asymmetric ripple pattern is generated which evolves to traveling waves on the torus. The wave fronts are initially oriented both SW-NE and SE-NW, but in later generations (18 and 19) they are mostly SW-NE oriented. Also observed are blinkers, stationary objects which alternate between two patterns, and gliders, objects which travel at a velocity of one dimer per time step. Both α and β gliders exist simultaneously in the system: the α gliders always traveling south and the β gliders always traveling north. Generally, interactions among the traveling gliders and stationary blinkers may be viewed as calculations and endow MTA with potential computational capabilities.

In the open MTA depicted in Figure 10, the "South"-end (left-hand) dimers are given random states at each time step, and the "North"-end (right-hand) dimers are kept in the β state at all times. The initial configuration of the open MTA is random [probability (α state) = probability (β state) = 0.5]. The most interesting feature of this system is its self-organization from the random configuration into an ordered "crystal like" pattern with a SW-NE orientation. Proceeding from the South end, the ordering is "catalyzed" by the gliders induced by the randomization of the end states. These gliders move North until they hit an irregularity. They then annihilate together with parts of the irregularity resulting in an increase of the ordered area. Randomizing the right-hand end introduces α gliders. The ordering then proceeds from both ends, but ordering is less complete since head-on collision of α and β gliders result in blinkers. Once the crystal-like traveling wave pattern has been formed, β gliders will travel along the MT without any change. This can be viewed as transmission of signals, albeit rather trivial ones. All kinds of seeds initiate the creation of this traveling wave pattern for thresholds ≤ 5.570, but only very few conserve their shape. These structures can be considered as kinds of virtual automata[27,28] which depend on certain threshold values and global MTA dynamics. At higher thresholds, traveling wave structures disappear and at even higher thresholds, gliders disappear as well. Thus traveling waves and gliders are examples of emergent MTA properties.

Increasing the threshold further leads to new features. In Figure 11 (threshold ± 5.750), the initial configuration is again α seed on β background. The wave pattern now develops only in three directions forming a wedge-shaped structure. The wavefronts are all in the SW-NE orientation rather than SE-NW, a preference which is also apparent in Figures 9 and 10. It appears to be an effect of the "left-handed" helical structure of MT.

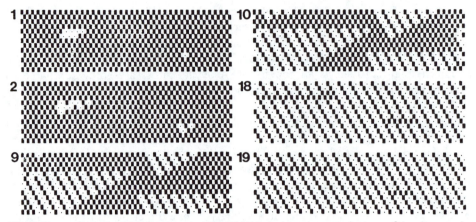

FIGURE 11   Toroidal MTA at thresholds ±5.750. Induced α seeds on a β background yield a crystal-like structure with the preferred SW–NE direction (Figures 9 and 10). Note the evolution of typical wedge patterns for these threshold values due to the three growth directions for the α seeds. Mirrored wedges develop from β seeds on an α background.

With higher thresholds, the wavefronts disappear, and a variety of glider structures exist. In Figure 12 (threshold ± 7.400), an open MTA with a random initial configuration gradually evolves into a system where the entire pattern moves north without any structural changes. This may be described as a "giant glider" onto which new parts are added at the south end, and old ones disappear at the north end. In terms of information transmission, this may be more useful than the system with threshold ± 1.000, since a large number of different signals are faithfully transmitted along the MTA. Furthermore, not all signals are allowed into the MTA "transmission line." Only certain of the $2^{13} = 8192$ possible configurations introduced at the South-end "layer" are accepted (the selection depends upon the state of the first few "layers"). Such selection may be viewed as filtering or manipulation of information, a form of computation.

Figures 7 and 13 show another virtual structure, the "bus glider," for thresholds around ± 9.000. For a very broad class of initial configurations, the system organizes the β seeds into a bus, which then goes South (to get back, you have to take the α but on a β background). All β seeds move South after a transient "wiggle," but only a limited number preserve their shape. Most of the significant structures, like the bus gliders shown in Figures 7 and 13, are extremely stable to perturbations within certain threshold intervals. As shown in Figures 7 and 13, a threshold of ± 9.000 allows passage of certain glider structures without leaving any lasting wake or track behind. Sequences of gliders could be used to transmit "messages" with MTA always ready to read a new message soon after the previous one. Encoding of such messages may be both spatial (location of glider on MTA) and temporal (time sequence of gliders). Isolated gliders existing for thresholds around ± 9.000 also seem ideal for regulation of certain biological functions such as sequential contraction of MAPs and movement of molecules along microtubules.[19]

At higher symmetric threshold values ($\geq$ 37.580), even the dot glider cannot exist any more. The typical dynamics for MTA in the range up to 46.880 is "bean sprout growth" from any induced seed (Figure 14). Bean sprouts are linear patterns

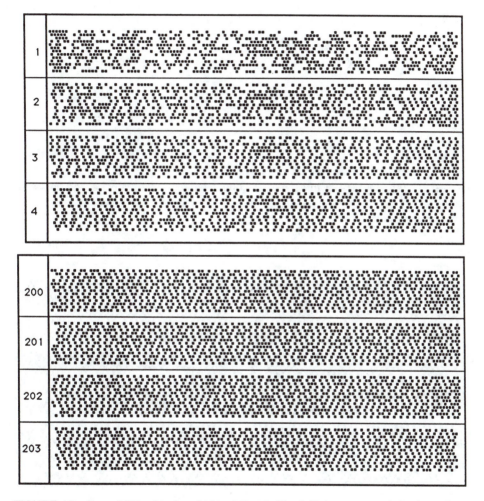

FIGURE 12 Open MTA with thresholds ±7.400. The initial pattern and the boundary states are as in Figure 10. Comparison of Figures 200 and 201 shows that the entire pattern moves north along the MTA without any changes. There is an autonomous selection of configurations at the south input end.

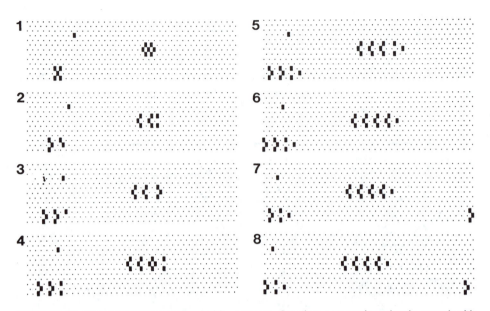

FIGURE 13 Toroidal MTA at thresholds ±9.000. Starting pattern is α background with β seeds which "wiggle" and expand bidirectionally, organizing themselves into bus gliders moving south.

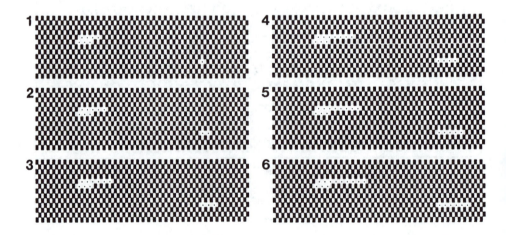

FIGURE 14 Toroidal MTA at thresholds ±40.000. Starting pattern is β background with α seeds which only move in the north direction (bean sprout growth), β seeds on an α background result in south-growing bean sprouts.

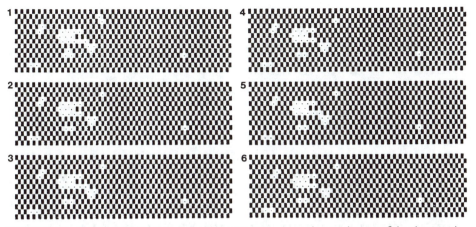

FIGURE 15   Toroidal MTA at thresholds ±47.000. α seeds starting on β background are nearly frozen. Only blinker activity is evident (note alternating connection between large α seed and smaller α seed to northeast).

which add one dimer each time step. As the threshold is increased further, all patterns freeze. At thresholds which are not too high (i.e., ± 47.000), blinkers may exist in the frozen structures (Figure 15). For symmetric thresholds at ± 62.020, no dynamics exist any more. The dynamical properties for MTA over the entire threshold range are summarized in Figure 8.

The capability to freeze a pattern may usefully function as a memory mechanism for MTA. However the frozen configuration heavily depends on the dynamics of the freezing process, e.g., the relation between pattern updating time (clocking frequency) and the velocity of the freezing process, and further on whether the threshold is symmetrical or not. This situation is analogous in many ways to spin-glass systems. A slow "freezing" (i.e., increase in threshold) always results in a north-south striped pattern, due to the bean sprout growth interval. But a more rapid freezing process allows for conservation of a variety of memory structures. It might even be possible to imitate and explain patterns of attachment of MT associated proteins ("MAPs"[19]) by modulating the "freezing" process. More detailed comparisons of MTA and spin-glass systems will be discussed elsewhere.

Thus far we have only discussed MTA dynamics for symmetrical threshold values. However, MT are electrets with net dipoles on the entire macromolecule, and like-oriented dipoles on each subunit; the α ends are negative relative to the β plus ends.[4,8,33] Thus electrons are most likely to be found in the α end of the tubulin dimers, suggesting an asymmetrical threshold for MTA in which escape from the α state is more difficult. Asymmetrical thresholds offer new dynamical properties. Figure 16 shows toroidal MTA with an α to β threshold of -20.000, and a β to α threshold of +2.000. α seeds form gliders which travel in opposite directions simultaneously. Such a phenomenon could account for observations of concurrent bilateral transport along single MT,[37,38] as well as bilateral information exchange.

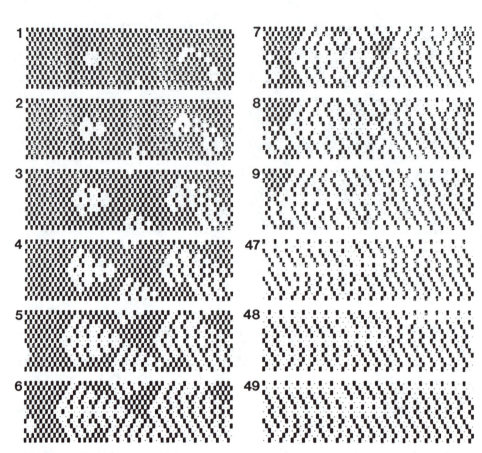

FIGURE 16 Toroidal MTA with asymmetrical thresholds. For α to β the threshold is -20.000. For β to α the threshold is 2.000. Some α seeds are induced on a β background. These seeds evolve into two α gliders traveling in opposite directions. The north-going glider is guided by a black spider, the south-going glider is guided by a white spider. Such an emergent property may account for the observed phenomena of organelles moving simultaneously in opposite directions along MT.

DISCUSSION

We have demonstrated automata behavior in the lattice structure of cytoskeletal microtubules (MT) based on calculated dipole coupling among MT subunits. We have observed automata behavior in MT of both a general nature (i.e., "virtual automata" gliders and blinkers) similar to behavior in other automata systems, and of a specific nature (i.e., wedge patterns, NE-SW orientations) derived from the unique MT geometry. Further, we have catalogued a variety of MT automata (MTA) behavior dependent on a threshold parameter, a measure of dipole coupling sensitivity among coherently oscillating MT subunits. These threshold dependent behaviors include gliders (dot, spider, bus, and giant gliders), traveling and standing wave patterns, blinkers, linearly growing patterns ("bean sprouts"), memory wakes, bidirectional gliders, and frozen patterns. These MTA behaviors may be capable of information processing and computation and MT appear to be well designed nanoscale automata computers if Fröhlich's coherent excitations or some comparable mechanisms do occur. Biological activities including guidance and movement of single-cell organisms, transport of molecules within cells, and cognitive functions of the human brain may utilize MTA behavior. Specific automata behavior within the cytoskeleton can regulate processes like axoplasmic transport, growth, cellular movement and specific functioning while serving as each cell's "on-board computer." In nerve cell axons and dendrites, massively parallel arrays of interconnected MT and neurofilaments could serve as specialized information circuits. MT automata gliders (8 to 800 meters/sec) may exist as traveling depolarization waves, solitons, or localized electrostatic fields which bind and transport molecules. In nerve cells, traveling gliders may correlate with traveling nerve membrane action potentials. Nerve membrane depolarization may interact with MT automata by transiently altering dipole coupling thresholds via ionic fluxes, voltage gradients or direct connection to the cytoskeleton. This coupling can lead to a view of the brain/mind as a hierarchy of nested automata with a previously overlooked basal dimension. For example, artificial intelligencer/roboticist Hans Moravec[36] has calculated the computing power of the brain based on the classical "neural net" assumption that each neuron–neuron synaptic connection is a fundamental binary switch. He assumes 40 billion neurons which can change state hundreds of times per second, resulting in roughly 4×10^{12} "bits" per second. However, rather than a simple switch, each neuronal synapse is extremely complex and dynamically controlled and modulated by the cytoskeleton. (Further, single-cell organisms like amoeba and paramecium perform complex tasks without benefit of synapse, brain, or nervous system.) Considering the cytoskeleton as a sub-dimension in a hierarchy of nested automata increases the calculated information capacity of the brain/mind immensely. Assuming parallel MT spaced about 100 nanometers apart and the volume of brain which is neuronal cytoplasm to be about 40 percent leads to approximately 10^{14} MT subunits in a human brain (ignoring neurofilaments and other cytoskeletal structures). If each subunit can change state every 10^{-9} to 10^{-11} seconds (as per Fröhlich's theory), the brain's MT would have an information capacity of roughly 10^{23} to 10^{25} bits per second! This would allow for massive parallelism and redundancy as interacting conformational state patterns compute at a sub-level of the brain's hierarchy of parallel information processing systems. The branching patterns of neural axons and dendrites have been likened to trees. MT and related filaments may represent a cytoskeletal forest within those trees.

Our future work will include further MTA investigations in which families of MTA communicate via MAPs and interconnecting filaments allowing for parallel computation and decision making inside a neuron. Modeling a "neural net" in which interneural connection strengths are modulated by intraneural MT automata may more accurately emulate brain function. Experimentally we are pursuing dynamic patterns of tubulin dimer conformational states in MT, as proposed by the MTA model, using technologies like the scanning tunneling microscope.[39] Just as the genetic code was deciphered, we hope to decrypt what may be real-time information codes in the cytoskeletons of eukaryotic cells. By so doing, we can perhaps communicate with and program cytoskeletal structures to perform tasks including nanoscale surgery for a variety of medical problems, or the self-assembly of large-scale cognitive arrays. These would truly be new frontiers of "Artificial Life."

APPENDIX

MT AUTOMATA NEIGHBORHOOD CHARACTERIZATION

We consider local neighborhoods within MT hexagonal lattices to be a central dimer subunit surrounded by its six nearest neighbor dimers (N, NE, SE, S, SW, NW). Tables 1, 2, and 3 show distances between pairs of neighbor dimer electron sites, symmetry considerations, and relative forces for neighbor interactions, respectively. Each of the seven neighbor dimers can be in either an α state, or β state at each excitation period. Thus there are $2^7 = 128$ possible neighborhood patterns, each of which has a net summation of forces influencing the central dimer's state at the subsequent excitation period. Positive forces induce α states and negative forces induce β states, provided the net forces exceed a transition threshold. To identify and characterize the 128 neighborhood patterns, we sought to "label" each using a binary pattern (Figure 17). (In Figures 17 and 18 the MT neighborhoods are stylized and do not indicate the spiral offset of the MT lattice). A particular neighborhood configuration can be numerically characterized in the following "binary" way (Figure 17). An α state is 1, a β state is 0. The SW dimer binary value is multiplied times 2 raised to the zero power (equal to one), the SE dimer in binary value is multiplied times 2 raised to the first power (equal to two), the S dimer binary value is multiplied times 2 raised to the second power (four) and so on in the sequence of NW, NE, N, and C. The values for the 7 dimers are summed to give a number from 0 to 127, corresponding to the $2^7 = 128$ possible configurations of a 7-member neighborhood. In Table 4 the net forces for the 128 possible neighborhood configurations are shown in increasing order of absolute magnitude.

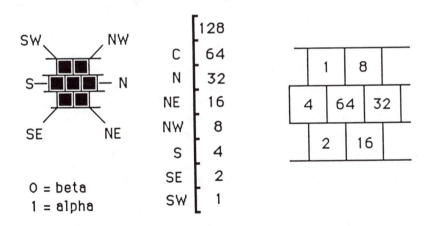

FIGURE 17 Binary characterization of MTA neighborhood patterns 0-127. For explanation, see Appendix text.

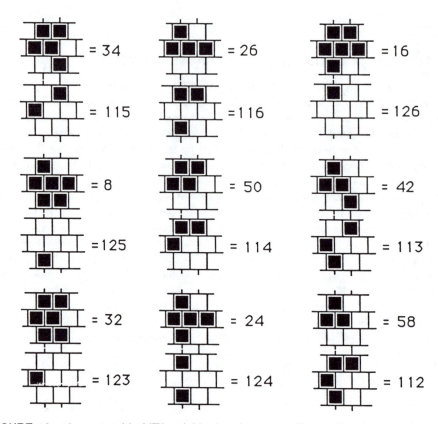

FIGURE 18   18 most stable MTA neighborhood patterns. For explanation, see Appendi
text.

The lower numbers (absolute values) indicate relatively stable neighborhood patterns whose central dimer state will change only at threshold values which are
low. The neighborhoods with high net forces will be unstable and tend to change
their central dimer state except at thresholds of high values. Figure 18 shows the
18 most stable neighborhood patterns (leaving out patterns where the net force
acts to conserve the state of the center dimers) and their numerical representation by binary code. Net forces for neighborhoods 34 and 115, for example, are
1.1453 and −1.1453. Thus the central β in neighborhood 34 will change to α only
at thresholds below 1.1453, and the central α in neighborhood 115 will change to β
only at negative thresholds below (less negative than) −1.1453. Thus these neighborhoods have relatively stable central dimer states. The remaining neighborhood
pairs in Figure 18 have increasing net force values, and thus have progressively
less stable central dimer states. The net forces for these neighborhood pairs are:
$26, 116 \pm 4.254$; $16, 126 = \pm 5.570$; $8, 125 = \pm 5.912$; $50, 114 \pm 7.023$; $42, 113 = \pm 7.365$;
$32, 123 = \pm 8.681$; $24, 124 = \pm 11.48$; and $58, 112 = \pm 12.94$. These neighborhood

patterns and their binary representations allow description and comparison of MTA and further illustrate the potential for pure information representation within MT automata.

TABLE 1 Distances (nm) between pairs of MT neighborhood dimer electron sites

Neighbor Position	Central Dimer α		Central Dimer β	
	Neighbor α	Neighbor β	Neighbor α	Neighbor β
North	$+8$	$+4$	$+12$	$+8$
Northeast	$+3\frac{1}{13}$	$-\frac{12}{13}$	$+7\frac{1}{13}$	$+3\frac{1}{13}$
Southeast	$-4\frac{12}{13}$	$-8\frac{12}{13}$	$-\frac{12}{13}$	$-4\frac{12}{13}$
South	-8	-12	-4	-8
Southwest	$-3\frac{1}{13}$	$-7\frac{1}{13}$	$+\frac{12}{13}$	$-3\frac{1}{13}$
Northwest	$+4\frac{12}{13}$	$+\frac{12}{13}$	$+8\frac{12}{13}$	$+4\frac{12}{13}$

TABLE 2 Relations between Dimer Interactions

Due to symmetry, the following relations between dimer interactions must be fulfilled: f = force, $(\alpha, N\text{-}\beta)$ = Center in α state and North in β state, etc.

$f(\alpha,N\text{-}\alpha)$	$=$	$-f(\alpha,S\text{-}\alpha)$	$=$	$f(\beta,N\text{-}\beta)$	$=$	$-f(s,S\text{-}\beta)$
$f(\alpha,NE\text{-}\alpha)$	$=$	$-f(\alpha,SW\text{-}\alpha)$	$=$	$f(\beta,NE\text{-}\beta)$	$=$	$-f(\beta,SW\text{-}\beta)$
$f(\alpha,NW\text{-}\alpha)$	$=$	$-f(\alpha,SW\text{-}\alpha)$	$=$	$f(\beta,NW\text{-}\beta)$	$=$	$-f(\beta,SE\text{-}\beta)$
$f(\alpha,NE\text{-}\beta)$	$=$	$-f(\alpha,NW\text{-}\beta)$	$=$	$f(\beta,SE\text{-}\alpha)$	$=$	$-f(\beta,SW\text{-}\alpha)$
$f(\alpha,N\text{-}\beta)$	$=$	$-f(\beta,S\text{-}\alpha)$				
$f(\alpha,S\text{-}\beta)$	$=$	$-f(\beta,N\text{-}\alpha)$				
$f(\alpha,SE\text{-}\beta)$	$=$	$-f(\beta,NW\text{-}\alpha)$				
$f(\alpha,SW\text{-}\beta)$	$=$	$-f(\beta,NE\text{-}\alpha)$				

TABLE 3 Relative Forces $[(= (y/-r^3) \times 1000)]$ for Neighbor Configurations.[1]

Neighbor Position	Central Dimer α Neighbor α	Neighbor β	Central Dimer β Neighbor α	Neighbor β
North	+15.625	+62.500	+ 6.944	+15.625
Northeast	+15.205	− 7.022	+ 9.635	+15.205
Southeast	−14.250	− 8.338	− 7.022	−14.250
South	−15.625	− 6.944	−62.500	−15.625
Southwest	−15.205	− 9.635	+ 7.022	−15.205
Northwest	+14.250	+ 7.022	+ 8.338	+14.250

[1] Net forces (Tables 4 & 5) are summation of six neighbors.

TABLE 4 Net Interaction Forces ($\times 2.3 \times 10^{-14}$ Newtons) for the 128 Different Neighborhood Configurations in MTA.[1]

0	0.000	32	8.681	64	−37.58	96	9.292
1	−22.23	33	−13.55	65	−32.01	97	14.86
2	− 7.228	34	1.453	66	−31.67	98	15.20
3	−29.45	35	−20.77	67	−26.10	99	20.77
4	46.88	36	55.56	68	−28.90	100	17.97
5	24.65	37	33.33	69	−23.33	101	23.54
6	39.65	38	48.33	70	−22.99	102	23.88
7	17.42	39	26.10	71	−17.42	103	29.45
8	5.912	40	14.59	72	−44.81	104	2.064
9	−16.31	41	− 7.634	73	−39.24	105	7.634
10	− 1.316	42	7.365	74	−38.90	106	7.976
11	−23.54	43	−14.86	75	−33.33	107	13.55
12	52.79	44	61.47	76	−36.13	108	10.74
13	30.56	45	39.24	77	−30.56	109	16.31
14	45.56	46	54.24	78	−30.22	110	16.66
15	23.33	47	32.01	79	−24.65	111	22.23
16	5.570	48	14.25	80	−59.81	112	−12.94
17	−16.66	49	− 7.976	81	−54.24	113	− 7.365
18	− 1.658	50	7.023	82	−53.90	114	− 7.023
19	−23.88	51	−15.20	83	−48.33	115	− 1.453
20	52.45	52	61.13	84	−51.13	116	− 4.254
21	30.22	53	38.90	85	−45.56	117	1.316
22	45.22	54	53.90	86	−45.22	118	1.658
23	22.99	55	31.67	87	−39.65	119	7.228
24	11.48	56	20.16	88	−67.04	120	−20.16
25	−10.74	57	− 2.064	89	−61.47	121	−14.59
26	4.254	58	12.93	90	−61.13	122	−14.25
27	−17.97	59	− 9.292	91	−55.56	123	− 8.681
28	58.36	60	67.04	92	−58.36	124	−11.48
29	36.13	61	44.81	93	−52.79	125	− 5.912
30	51.13	62	59.81	94	−52.44	126	− 5.570
31	28.90	63	37.58	95	−46.88	127	0.000

[1] A positive force, if larger than threshold, makes C go into α state; similarly, a negative force makes C go into β state. For characterization of the 128 neighborhoods, see Appendix.

TABLE 5 Net Interaction Fforces ($\times 2.3 \times 10^{-14}$ Newtons) for the 128 Different Neighborhood Configurations in MTA in Pairs of Increasing Net Forces.

0	0.000	127	0.000	101	23.54	11	−23.54
117	1.316	10	−1.316	102	23.88	19	−23.88
34	1.453	115	−1.453	5	24.65	79	−24.65
118	1.658	18	−1.658	39	26.10	67	−26.10
104	2.064	57	−2.064	31	28.90	68	−28.90
26	4.254	116	−4.254	103	29.45	3	−29.45
16	5.570	126	−5.570	21	30.22	78	−30.22
8	5.912	125	−5.912	13	30.56	77	−30.56
50	7.023	114	−7.023	55	31.67	66	−31.67
119	7.228	2	−7.228	47	32.01	65	−32.01
42	7.365	113	−7.365	37	33.33	75	−33.33
105	7.634	41	−7.634	29	36.13	76	−36.13
106	7.976	49	−7.976	63	37.58	64	−37.58
32	8.681	123	−8.681	53	38.90	74	−38.90
96	9.292	59	−9.292	45	39.24	73	−39.24
108	10.74	25	−10.74	6	39.65	87	−39.65
24	11.48	124	−11.48	61	44.81	72	−44.81
58	12.94	112	−12.94	22	45.22	86	−45.22
107	13.55	33	−13.55	14	45.56	85	−45.56
48	14.25	122	−14.25	4	46.88	95	−46.88
40	14.59	121	−14.59	38	48.33	83	−48.33
97	14.86	43	−14.86	30	51.13	84	−51.13
98	15.20	51	−15.20	20	52.45	94	−52.45
109	16.31	9	−16.31	12	52.79	93	−52.79
110	16.66	17	−16.66	54	53.90	82	−53.90
7	17.42	71	−17.42	46	54.24	81	−54.24
100	17.97	27	−17.97	36	55.56	91	−55.56
56	20.16	120	−20.16	28	58.36	92	−58.36
99	20.77	35	−20.77	62	59.81	80	−59.81
111	22.23	1	−22.23	52	61.13	90	−61.13
23	22.99	70	−22.99	44	61.47	89	−61.47
15	23.33	69	−23.33	60	67.04	88	−67.04

REFERENCES

1. Albrecht-Buehler, G. (1985), "Is the Cytoplasm Intelligent Too?", *Cell and Muscle Motility* **6**, 1–21.
2. Amos, L. A., and A. Klug (1974), "Arrangement of Subunits in Flagellar Microtubules," *J. Cell Sci.* **14**, 523–550.
3. Atema, J. (1973), "Microtubule Theory of Sensory Transduction," *J. Theor. Biol.* **38**, 181–190.
4. Athenstaedt, H. (1974), "Pyroelectric and Piezoelectric Properties of Vertebrates," *Ann. NY Acad. Sci.* **238**, 68–93.
5. Barnett, M. P. (1988), "Molecular Systems to Process Analog and Digital Data Associatively," *Proceedings of Third International Symposium on Molecular Electronic Devices, October 6–8, 1986, Arlington, Virginia*, Ed. F. L. Carter and H. Wohltjen (Elsevier: North Holland), in press.
6. Burns, R. B. (1978), "Spatial Organization of the Microtubule Associated Proteins of Reassembled Brain Microtubules," *J. Ultrastruct. Res.* **65**, 73–82.
7. Burnside, B. (1974), "The Form and Arrangement of Microtubules: An Historical, Primarily Morphological Review," *Ann. NY Acad. Sci.* **253**, 14–26.
8. DeBrabander, M. (1982), "A Model for the Microtubule Organizing Activity of the Centrosomes and Kinetochores in Mammalian Cells," *Cell Biol. Intern. Rep.* **6(10)**, 901–915.
9. DeBrabander, M., and J. DeMey (1985), *Microtubules and Microtubule Inhibitors* (Amsterdam: Elsevier).
10. DeBrabander, M., G. Geuens, J. DeMey, and M. Joniav (1986), "The Organized Assembly and Function of the Microtubule System Throughout the Cell Cycle," *Cell Movement and Neoplasia*, Ed. M. DeBrabander (Oxford: Pergamon Press).
11. Dewdney, A. K. (1985), "Computer Recreations," *Sci. Am.* **252**, 18–30.
12. Dustin, P. (1984), *Microtubules* (Berlin: Springer–Verlag), 2nd rev. ed., 442.
13. Fröhlich, H. (1986), "Coherent Excitations in Active Biological Systems," *Modern Bioelectrochemistry*, Eds. F. Gutmann and H. Keyzer (New York: Plenum Press), 241–261.
14. Fröhlich, H. (1975), "The Extraordinary Dielectric Properties of Biological Materials and the Action of Enzymes," *Proc. Natl. Acad. Sci.* **72(11)**, 4211–4215.
15. Fröhlich, H. (1970), "Long Range Coherence and the Actions of Enzymes," *Nature* **228**, 1093.
16. Gardner, M. (1970), "Mathematical Games: The Fantastic Combination of John Conway's New Solitaire Game 'Life,'" *Sci. Am.* **223**, 120–123.
17. Geuens, G., G. G. Gundersen, R. Nuydens, F. Cornelissen, J. C. Bulinski, and M. DeBrabander (1986), "Ultrastructural Co-Localization of Tyrosinated and Nontyrosinated Alpha Tubulin in Interphase and Mitotic Cells," *J. Cell Biol.* **103(5)**, 1883–1893.

18. Grundler, W., and F. Keilmann (1983), "Sharp Resonances in Yeast Growth Prove Nonthermal Sensitivity to Microwaves," *Phys. Rev. Letts.* **51**, 1214–1216.

19. Hameroff, S.R. (1987), *Ultimate Computing: Biomolecular Consciousness and Nanotechnology* (Elsevier: North-Holland).

20. Hameroff, S. R., S. A. Smith, and R. C. Watt (1984), "Nonlinear Electrodynamics in Cytoskeletal Protein Lattices," *Nonlinear Electrodynamics in Biological Systems*, Eds. W.R. Adey and A.F. Lawrence (New York: Plenum Press), 567–583.

21. Hameroff, S. R., S. A. Smith, and R. C. Watt (1986), "Automaton Model of Dynamic Organization in Microtubules," *Ann. NY Acad. Sci.* **466**, 949–952.

22. Hameroff, S. R., and R. C.Watt (1982), "Information Processing in Microtubules," *J. Theor. Biol.* **98**, 549–561.

23. Karplus, M., and J. A. McCammon (1979), "Protein Structural Fluctuations During a Period of 100 ps," *Nature* **277**, 578.

24. Karplus, M., and J. A. McCammon (1983), "Protein Ion Channels, Gates, Receptors," *Dynamics of Proteins: Elements and Function, Ann. Rev. Biochem.*, Ed. J. King (Menlo Park: Benjamin/Cummings), vol. 53, 263–300.

25. Kirschner, M., and T. Mitchison (1986), "Beyond Self Assembly: From Microtubules to Morphogenesis," *Cell* **45**, 329–342.

26. Koruga, D. (1984), "Microtubule Screw Symmetry: Packing of Spheres as a Latent Bioinformation Code," *Ann. NY Acad. Sci.* **466**, 953–955.

27. Langton, C. G. (1986), "Studying Artificial Life with Cellular Automata," *Physica* **10D(22)**, 120–149.

28. Langton, C. G. (1987), "Virtual State Machines and Cellular Automata" *Complex Systems* **1**, 257–271.

29. Lehninger, A. L. (1982), *Principles of Biochemistry* (New York: Worth).

30. Lindeburg, M. R. (1982), *Engineer In Training Review Manual* (San Carlos, CA: Professional Publications).

31. Margulis, L. (1975), *Origin of Eukaryotic Cells* (New Haven: Yale University Press).

32. Margulis, L. and D. Sagan (1986), "Strange Fruit on the Tree of Life," *The Sciences* **May/June**, 38–45.

33. Margulis, L., L. To, and D. Chase (1978), "Microtubules in Prokaryotes," *Science* **200**, 1118–1124.

34. Mascarenhas, S. (1974), "The Electret Effect in Bone and Biopolymers and the Bound Water Problem," *Ann. NY Acad. Sci.* **238**, 36–52.

35. Matsumoto, G., and H. Sakai (1979), "Microtubules Inside the Plasma Membrane of Squid Giant Axons and Their Possible Physiological Function," *J. Membr. Biol.* **50**, 1–14.

36. Moravec, H. (1987), *Mind Children* (San Francisco: University of California Press); also see Moravec's contribution to these proceedings.

37. Ochs, S. (1982), *Axoplasmic Transport and Its Relation to Other Nerve Functions* (New York: Wiley Interscience).

38. Roth, L. E., D. J. Pihlaja, and Y. Shigenaka (1970), "Microtubules in the Heliazoan Axopodium. I. The Gradion Hypothesis of Allosterism in Structural Proteins," *J. Ultrastr. Res.* **30**, 7–37.

39. Schnapp, B. J., R. D. Vale, M. P. Sheetz and T. S. Reese (1985), "Single Microtubules From Squid Axoplasm Support Directional Movement of Organelles," *Cell* **40**, 455–462.

40. Schnapp, B. J., R. D. Vale, M. P. Sheetz and T. S. Reese (1986), "Microtubules and the Mechanism of Directed Organelle Movement," *Ann. NY Acad. Sci.* **466**, 909–918.

41. Schneiker, C. W., and S. R. Hameroff (in press), "Nanotechnology Workstations Based on Scanning Tunneling/Optical Microscopy: Applications to Molecular Scale Devices," *Proceedings of the Third International Symposium on Molecular Electronic Devices, Oct. 6–8, 1986, Arlington, Virginia*, Eds. F.L. Carter, and H. Wohltjen (Elsevier: North–Holland).

42. Scott, A. C. (1984), "Solitons and Bioenergetics," *Nonlinear Electrodynamics in Biological Systems*, Eds. W. R. Adey, and A. F. Lawrence (New York: Plenum Press).

43. Smith, S.A., R.C. Watt, and S.R. Hameroff (1984), "Cellular Automata in Cytoskeletal Lattices," *Physica* **10D**, 168–174.

44. Stebbings, H., and C. Hunt (1982), "The Nature of the Clear Zone Around Microtubules," *Cell Tissue Res.* **227**, 609–617.

45. Toffoli, T. (1984), "Cellular Automata as an Alternative to (Rather Than an Approximation of) Differential Equations in Modeling Physics," *Physica* **10D**, 117–127.

46. Tuszynski, J. A., R. Paul, R. Chatterjee, and S. R. Sreenivasa (1984), "Relationship Between Fröhlich and Davydov Models of Biological Order," *Phys. Rev. A* **30(5)**, 2666–2675.

47. Von Neumann, J. (1966), *Theory of Self-Reproducing Automata*, Ed. A.W. Burks (Urbana: University of Illinois Press).

48. Wheatley, D. N. (1982), *The Centriole: A Central Enigma of Cell Biology* (Amsterdam: Elsevier).

49. Wolfram, S. (1984), "Cellular Automata as Models of Complexity," *Nature* **311**, 419–424.

50. Wolfram, S. (1984), "Universality and Complexity in Cellular Automata," *Physica* **10D**, 1–35.

51. Wright, R. (1985), "The On/Off Universe. The Information Age," *The Sciences* **May/June**, 7–9.

Authors cited but not listed are Lehninger, 1982, and Ochs, 1982.

Valentino Braitenberg
Max Planck Institute of Biological Cybernetics, Tübingen, F.R.G.

Some Types of Movements

Automatic vehicles on four wheels are lifelike in their front-to-back polariza-tion (sensors in front, motors in the back), in their up-down polarization reflecting their contact with the ground (or water surface), and in the bilateral symmetry of their body with paired sensors and motors. There are animals so constituted mainly among the small aquatic species. Larger animals moving in the water, in the air or on the surface of the earth have articulated appendages which provide the motor system with an increased number of degrees of freedom. The nerve cen-ters which govern their coordinated activity are large and poorly understood. They pose fascinating problems of interpretation, not so much, I suspect, because of their intrinsic complexity, but mainly because we do not know in what terms the states and changes of the motor system are represented in the brain. The cerebellum is probably the part of the vertebrate brain in which the "wiring" is best known, but the basic operation it performs is still obscure. Its action is undoubtedly important in motor coordination. When it is disturbed, the impression one gains is that of movements which reach their goals in a qualitative way, but with difficulty, as if the physics of the moving parts were disregarded. Possibly the intact cerebellum com-putes the equations of mechanics, adding "smoothness" to the movements dictated in a rough qualitative way by the cerebral cortex. But does the cerebellum "know" Newton's laws and Hamiltonian mechanics? I doubt it. Are there other ways of computing movements? Are there shortcuts which provide the desired effects with-out cumbersome calculations? These are questions relevant to future theories of

Artificial Life, SFI Studies in the Sciences of Complexity,
Ed. C. Langton, Addison-Wesley Publishing Company, 1988   **555**

the cerebellum, and already touched upon by some.[6] I want to contribute a few general considerations on movements which are perhaps preparatory to an understanding of the physics of movement as the cerebellum understands it. Some of these thoughts emerged from conversations I had with Dr. Gin McCollum during her stay in Tübingen in 1986. Sloppy physics, however, is not her responsibility.

MOVEMENT BY REDEFINITION OF THE SYSTEM

Linear and angular momentum are conserved. A rocket emitting a jet of gas, a gun recoiling, a man catching a heavy ball move only in so far as the thing moving is redefined with respect to the original condition: the rocket without the gas, the gun without the bullet (and the gunpowder), the man plus the ball. It must be remembered that living beings are essentially of one piece. They do incorporate and expel matter, but the dynamics of this is very rarely used for propulsion: the jet of larval libellulas, or that of the octopus escaping. Catching a heavy thing, resisting the jump of an aggressor are, however, of this kind: the body has to deal with a change in linear and/or angular momentum.

ROTATION VS. EXPANSION

Generally, skeletal muscles are attached to bones in such a way that the result of their contraction is rotation around a joint (visceral muscles are different). Thus torque, moment of inertia, and angular momentum are expressions which occur naturally in the descriptions of such movements, and terms akin to these physical definitions most likely are used in the neurological computation of movement. Not only do the various segments of a jointed arm rotate around the axes of the joints, but the arm itself, when lengthening or shortening, may impart angular momentum to the system body plus arm as a whole. The angular momenta, and the torques produced by the segments of a limb often cancel. Arms and legs are composed of two main segments each (upper arm and forearm, thigh and lower leg) which contribute to extension and flexion with opposite and roughly equal angular momenta. Extension of the two arms in opposite directions produces no net angular momentum in the body as a whole, nor does symmetrical extension or flexion of both legs. Such movements can be described in terms of contraction and expansion like some of the movements of mollusks.

Contraction and expansion change the moments of inertia around some of the axes of the body. Moments of inertia must be of prime importance in the control of movements if torques are what the muscles produce. The physics of jointed limbs is complicated because rotation in one joint generally changes the moment of inertia involved in the rotation of another joint. Think of an arm positioned with a right angle at the elbow. Pendular movements of the upper arm and forearm cannot be both harmonic when performed at the same time, since the change of the moment of inertia due to movement of the forearm interferes with the pendular movement of the arm around the shoulder (Figure 1).

MOVEMENTS WITHOUT SUPPORT

A space ship at rest in absolute space is condemned to stay put with its center of mass, although its shell may be seen to move for short distances when there are changes of its inner configuration: the crew rushing from one side of the cabin to the other. Can it suddenly start to rotate? It can, with the crew marching around in the opposite direction to preserve angular momentum. Can it rotate around a certain angle and then stop, with exactly the same inner configuration at the beginning and at the end of the movement? It can, if the dance of the crew stops after having gone around an integer number of times. Could an animal, suspended in free space, change its orientation by rotating around an axis? This would seem to be more complicated, since there are no wheels inside the animal which may rotate and then stop regaining the original configuration: everything is connected in the animal body; any rotation would necessarily result in a twist. Still, by combining contraction, expansion and bending an animal could move part of its mass around in a circle to produce a net rotation of its body with exactly the same configuration as before. Think of a weightless man in space holding a heavy mass close to his body (Figure 2a).

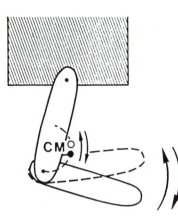

FIGURE 1   The center of mass of the arm (CM) shifts with changes of the elbow angle. Pendular motion of the forearm interferes with pendular motion in the shoulder joint.

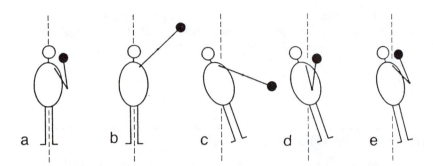

FIGURE 2   The sequence a to e shows how a man suspended in space can change his position.

He may extend the arm holding the mass (b), then swing the arm around (c), imparting equal and opposite torques to the arm and to the body, then bring the mass again close to his body (d) and finally rotate the arm back to its original position (e). The moment of inertia of the arm after the flexion is smaller than that of the extended arm, while the moment of inertia of the body stays the same. The torque which brings the arm back to its original position produces rotation of the body by an angle less than that of its (opposite) rotation during the swing of the extended arm. The net result is a rotation of the whole system with an unchanged internal configuration. It is said that cats land on their feet thanks to rotations of their tail by which they correct deviations from the vertical position. Extension, rotation and contraction are the only way a skater can put himself back in the upright position when he threatens to fall, since he cannot exert any torques against the practically frictionless contact of his legs with the ground. Most likely this kind of correction is also applied with normal upright posture even when friction against the ground is available. The complicated sequence of contractions in the muscles of the whole body which follows a jerk of the platform on which the subject is standing[2,5] may reflect (in part) such dynamic strategies.

NAHVI'S PRINCIPLE

Mahmood Nahvi learned this observing crane operators on construction sites. A crane holding a load suspended from a long cable, if it were rotated in one swing to the position where the cargo is to be grounded would produce long-lasting pendular movements of the load, making it impossible for the workers to get at it. The strategy used by the crane operators is the following (Figure 3): They rotate the crane quickly to a position half way between the starting position and the target position and stop it there. The load swings to where it is going to end up and loses its kinetic energy at the end of the pendular movement. At that time the crane is quickly swung around the rest of the way, the load is thereby lowered, losing its gravitational energy as well, and can be safely delivered.

It was Nahvi's contention[4] that with all the elastic energy stored in tendons and in the muscles themselves between the contracting elements and the load, every contraction would be followed by oscillations, unless the temporal pattern of the contraction were preprogrammed to compensate for the stored energy, quite analogously to the strategy of the crane operators. Nahvi thought that the generality of this principle in motor control, and the large amount of stored information and/or computation which it involves, well deserves a computer of the size and complexity of the cerebellum.

The strategy used by the crane operator may be a clue to the understanding of the triphasic, sometimes quadriphasic pattern of muscular contractions observed in many rapid movements of the limb,[8] of the head[3] and of the whole body.[2] A contraction of the agonist muscle is followed by a contraction of the antagonist, stopping the movement before it reaches the goal, which then again is followed by a contraction of the agonist, completing the movement. This is strongly reminiscent of Nahvi's principle.

CENTERS OF MASS

One piece of information which must at all times be present in the control apparatus governing a system of jointed masses is the position of the center of mass of the

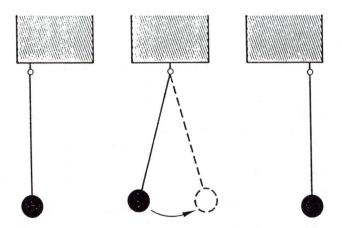

FIGURE 3   The strategy used by the crane operator.

body. The center of mass shifts with changes of the configuration of the body, e.g., extension or flexion of limbs, when they are not balanced around it, as they sometimes are. It may be located outside of the body as defined by its outer surface, e.g., in a person bending over or sitting on a chair. The position of the center of mass is important because the geometrical relation between it and the vector of displacement of partial masses of the body decides whether the result is rotation one way or the opposite way or just linear displacement of the body, or no movement at all.

In cases where the movements of a (relatively light) arm can be regarded in isolation, being approximately independent from the concomitant movements of a (relatively massy) trunk, the center of mass of the arm itself becomes important. Except for the fully extended position, the center of mass is outside the arm, on the bisectrix of the elbow angle (assuming the upper arm and forearm as having the same shape and mass, Figure 1). The positions of the center of mass correspond one-to-one to those of the tip of the extremity.

It has been shown[7] that circles drawn repetitively in the frontal plane, like on a blackboard, with the extended arm, turn out quite satisfactory, whereas circles drawn on a sagittal plane (i.e., parallel to the plane of mirror symmetry of the body) are strongly distorted. In the latter case, flexion of the elbow is involved, the center of mass moving in and out along the bisectrix of the elbow. The distortion observed may be related to a tendency of the center of mass to move on a circle which would result in a distorted circle at the tip of the extremity holding the drawing implement (Figure 4). In the case of the circle drawn with the extended arm, the center of mass does move on a circle, a movement which may be simply compounded out of two pendular harmonic motions of the arm, at right angles to each other. This is apparently easy to perform.

THROWING AND JUMPING

Throwing an object implies imparting it acceleration toward the velocity with which it leaves the hand. The hand accompanies the object, exerting force on it for a certain time, during which it accelerates. In order to keep in contact, the hand has to accelerate too. In certain cases, e.g., when throwing with a flexion of the forearm (Figure 5a) starting with an extended arm, the accelerated motion is apparently accomplished entirely through an accelerated contraction of the muscle. In other cases (Figure 5b), the geometry of the joint and of the tendon help, as when throwing by an extension of the forearm starting from a bent elbow (which is in fact more efficient). There, in the course of the movement, the lever arm on the side of the muscle, although anatomically constant, becomes actually shorter because its projection on a line perpendicular to the direction of the muscle shortens. Thus the contraction of the muscle at a constant speed results in an accelerated rotation of the forearm in the elbow joint.

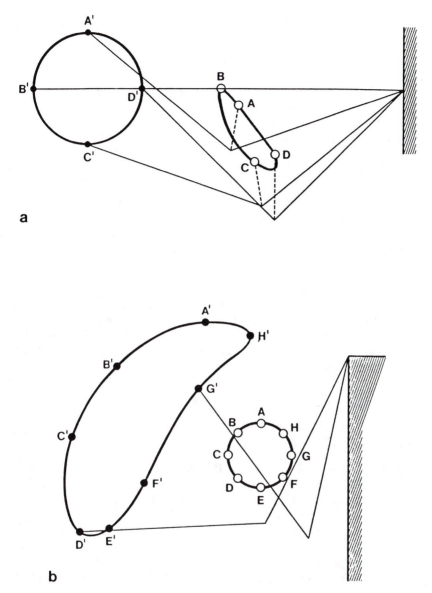

FIGURE 4 The tendency of the center of mass of the arm to move on a circle makes it difficult to draw a circle. a: If the hand is constrained to move on a circle, the center of mass describes the awkward figure A, B, C, D. b: Conversely, if the center of mass is allowed to move on a circle A to H, the figure described by the tip of the arm is A' to H'.

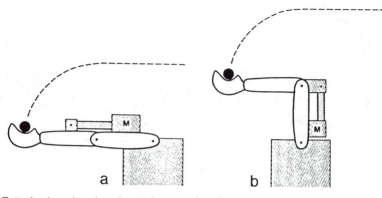

FIGURE 5   In throwing, i.e., imparting acceleration to a mass, the geometry of the joint is important. Going from flexion to extension (b) is more advantageous than going from extension to flexion (a).

The situation is even more dramatic in jumping, which is really "throwing" the entire body. Here the geometry of the knee joint (Figure 6) is definitely advantageous, with the femur rolling on the plane of the tibia, making the gain of the joint (degrees of rotation per unit of shortening of the muscle) progressively greater between flexion and extension.

HARMONIC OSCILLATIONS

In very rapid oscillatory movements of the forearm, as occur sometimes in musical performance, e.g., in the bowing of string instruments, the movement becomes almost perfectly sinusoidal.[1] When the voluntary movement is stopped, the arm continues to oscillate for a few cycles at roughly the same frequency, implying that there is a spring-like operation involved in the movement. What the nervous system apparently does in this case, is to adjust the elasticity of the joint by setting the "tonus" of the muscles to a certain value, appropriate for the desired frequency, and to feed the resulting spring-like oscillation with energy at the correct phase.

"GAITS" OF A JOINTED ARM

Limbs are composed of discrete masses connected by a number of joints, each having between 1 and 3 degrees of freedom. The movements of such a system thus

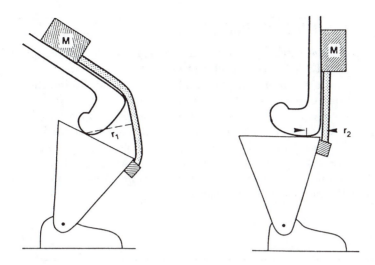

FIGURE 6   The knee joint: variable leverage.

can be classified in a natural way, by noting the qualitative relationships between the rotations in the various joints. The human arm can be schematized as three segments (upper arm, forearm and hand) moving in three joints (shoulder, elbow and wrist). When the movement is confined to one plane, as it is in the bowing arm of a violinist, each of the joints has one degree of freedom, the shoulder losing its second degree of freedom of rotation and the rotation of the forearm around its long axis being eliminated. Consider a periodic movement of the arm such as up and down bowing. Each joint may contribute to the movement, or oppose it by moving in the opposite direction, or not move at all. Thus the number of qualitatively different movements of the arm is $3^3 = 27$. If movement of the bow in one direction is to result, we must eliminate the 2^3 cases in which none of the joints moves in that direction, all of them either being at rest or moving in the opposite direction. Thus the useful patterns of movement for any one direction are $3^3 - 2^3 = 19$. In theory, any of these patterns for the up-bowing may be combined with any of the patterns for the down-bowing, bringing the number of "gaits" for the up-and-down movement to 361, although most of the time, but by no means always, symmetrical patterns are used for the up and down strokes. A number of these gaits are used, more or less consciously, by the violinist, in order to produce varying musical effects. The teacher may use such expressions such as "with a stiff wrist" or "movement from the shoulder" or "elastic operation of the wrist" to induce one or the other movement.

OTHER KINDS OF MOVEMENT

Besides throwing and catching, pushing and pulling are also interactions with foreign bodies, where the communal center of mass of the actor and the object acted upon certainly plays an important role in the control of the system. Here the internal representation of events in the outside world, which we usually do not associate with the motor system, obviously must become part of the motor control.

A very special kind of movement, reserved to primates and particularly prominent in humans, is the detailed operation of the fingers in performances such as peeling an orange, extracting a worm from a hole, drawing, handwriting, typing or playing the piano. Here the interest is more in the realm of information and control, rather than in the problems of biomechanics.

CONCLUSION

It was my intention to show that the movements in space of articulated bodies pose problems of a different kind from those treated in elementary or even not so elementary textbooks of mechanics. The account which I gave was naive on purpose. It is my contention that the brain deals with movements in a naive way. If we hope to understand the role of the cerebellum and of other centers concerned with movement, we must lower ourselves to that level.

ACKNOWLEDGMENTS

Claudia Martin-Schubert's help with the drawings and Margarete Ghasroldashti's patience in typing is most gratefully acknowledged.

REFERENCES

1. Braitenberg, V. (1988), "The Cerebellum and the Physics of Movement: Some Speculations," *Cerebellum and Neuronal Plasticity*, Ed. M. Glickstein (New York: Plenum).
2. Diener, H. C., J. Dichgans, F. Bootz, and M. Bacher (1984), "Early Stabilization of Human Posture after a Sudden Disturbance: Influence of Rate and Amplitude of Displacement," *Exp. Brain Res.* **56**, 126–134.
3. Hannaford, B., and L. Stark (1985), "Roles of the Elements of the Triphasic Control Signal," *Experimental Neurology* **90 (3)**, 619–634.
4. Nahvi, M. J., and M. R. Hashemi (1984), "A Synthetic Motor Control System; Possible Parallels with Transformations in Cerebellar Cortex," *Cerebellar Functions*, Eds. J. R. Bloedel, J. Dichgans and W. Precht (Berlin: Springer Verlag).
5. Nashner, L. M., and G. McCollum (1985), "The Organization of Human Postural Movements: A Formal Basis and Experimental Synthesis," *Behav. Brain Sci.* **8**, 135–172.
6. Pellionisz, A., and S. R. Lli (1980), "Tensorial Approach to the Geometry of Brain Function. Cerebellar Coordination via Metric Tensor," *Neuroscience* **5**, 1125–1136.
7. Soechting, J. F., F. Lacquaniti, and C. A. Terzuolo (1986), "Coordination of Arm Movements in Three-Dimensional Space. Sensorimotor Mapping during Drawing Movement," *Neuroscience* **17(2)**, 295–311.
8. Wadmann, W. J., J. J. Denier van der Gon, R. H. Geuze, and C. R. Mol (1979), "Control of Fast Goal-Directed Arm Movements," *Journal of Human Movement Studies* **5**, 3–17.

Annotated Bibliography

This bibliography consists of two sections:

Section-1: Annotated Bibliography on Artificial Life.
Chris Langton and workshop participants

Section-2: Annotated, Categorical Bibliography on L-Systems.
Aristid Lindenmayer and Przemyslaw Prusinkiewicz

SECTION 1:
ANNOTATED BIBLIOGRAPHY IN ALPHABETICAL ORDER

As stated in the preface, the first workshop on Artificial Life grew largely out of my frustration with the fragmented nature of the literature on biological modeling. This bibliography is a first attempt to collect a wide spectrum of references together in one location.

The following collection is unavoidably biased, incomplete, and otherwise inadequate in a number of respects. First, the majority of references have come from my own personal bibliography, which necessarily reflects my own eclectic interests.

Thus, there is a heavy bias in favor of models of biological phenomena based on formal automata and language theory.

Second, the sheer volume of literature on biological modeling would require several man-years of effort to catalogue and categorize, let alone annotate, even if there was a clear and consistent way to find all of it, and it is probably growing at more than a shelf-foot per month. I am sure that I have missed many works which others would consider fundamental. As this has primarily been a one man effort, gaping holes are regrettable but unavoidable. I did include a plea in the workshop announcement package that participants bring with them their personal reference lists, with the result that about 10 people contributed their favorite citations. I'd like to thank Ed Carlson, Michael Conrad, Stuart Hameroff, Marek Lugowski, Nancy Norris, Clifford Pickover, Milan Zeleny, and the other workshop participants who contributed reference lists. I'd especially like to thank Aristid Lindenmayer and Przemyslaw Prusinkiewicz for contributing the annotated and categorized bibliography on L-systems, which constitutes Section 2 of this bibliography.

Third, a bibliography of nearly 500 annotated entries is approaching the upper limit for listing in alphabetical order. The ideal bibliography would be both annotated and broken down into categories with inter-category cross-citations. However, at this point in the development of the field of Artificial Life, it is very difficult—and probably unwise—to attempt to impose a rigid taxonomic framework on top of such an amorphous and rapidly evolving area of research. To use a biological analogy, the field of Artificial Life *as a coherent, unified discipline* is in an embryonic form. The embryo consists of differentiated cells, to be sure, but they are primarily stem cells, and what they themselves will give rise to is as yet anybody's guess. The ultimate phenotype of Artificial Life itself is quite unpredictable.

As a result, this bibliography should be treated as an additional article, to be read through as one would read through the other articles in these proceedings, keeping in mind the subjective bias and limited vision of the author.

In order to compensate for the inadequacies of this bibliography, it is being ported to an on-line bibliographic data base with an intelligent front end which will allow Internet access for searching, downloading, and updating citations. Citations will be able to be retrieved, added, or updated in either BiBTEX or Refer format. We expect that the bibliography will be available on-line early in 1989. For information on accessing the bibliography via Internet, contact Richard K. Belew, CSE Dept., UCSD, La Jolla, CA, 92093. Electronic mail should be addressed to `rik%cs@ucsd.edu`.

For the purposes of this proceedings, I have not included all of the references in section 2 in section 1, nor the long bibliography on nanotechnology in Schneiker's article in these proceedings.

I think of the bibliography as one of the most important results of the workshop, and I think it is important that the research community at large helps to correct some of its inadequacies and biases by contributing further citations, whether by use of the on-line facility or by mailing references directly to me. Ideally, contributions should be accompanied by a brief annotation and a list of index terms.

AAAS88   AAAS. *DÆDALUS: Special Issue on Artificial Intelligence.* AAAS, Canton, Mass., Winter 1988.

Amounts largely to an in-print debate between mainstream AI and the new connectionist school. Many of the issues apply equally well to Artificial Life.

Aladyev74   V. Aladyev. "Survey of Research in the Theory of Homogeneous Structures." *Mathematical Biosciences,* 22, 1974.

Survey of primarily Russian work on cellular automata and related structures.

Alberts83   B. Alberts, D. Bray, J. Lewis, M. Raff, K. Roberts, and J.D. Watson. *Molecular Biology of the Cell.* Garland, New York, 1983.

A thorough introduction to the mechanics of the cell. See also Watson87.

Albrecht78   G. Albrecht-Buehler. "The Tracks of Moving Cells." *Scientific American,* 238(4):69–76, April 1978.

Description of similarities in the behavior of the daughter cells resulting from a cell-division.

Albus82   J.S. Albus. *Brains, Behavior, and Robotics.* Byte Books, Peterborough, NH, 1982.

Neurologically astute model for simulating intelligence.

Aleksander83   I. Aleksander and P. Burnett. *Reinventing Man: The Robot becomes Reality.* Holt, Rinehart and Winston, New York, 1983.

An excellent book on the fundamental principles involved in robotics. Especially good coverage of the social implications of robotics.

Amoroso71   S. Amoroso and G. Cooper. "Tesselation Structures for Reproduction of Arbitrary Patterns." *Journal of Computer and System Sciences,* 5(5), 1971.

Automata theory relating to pattern reproduction.

Anderson83   P. W. Anderson. "Suggested Model for Prebiotic Evolution: The Use of Chaos." *Proc. Natl. Acad. Sci,* 80:3386–3390, 1983.

A "spin-glass" model for the origin of life.

Aoki84　I. Aoki. "Internal Dynamics of Fish Schools in Relation to Inter-Fish Distance." *Bulletin of the Japanese Society of Scientific Fisheries*, 50(5):751–758, 1984.

Dynamic properties of fluctuations in nearest neighbor distance are determined by using power spectra.

Apter70　M. J. Apter. *The Computer Simulation of Behavior.* Hutchinson University Library Press, London, 1970.

Application of computers to psychology. Many issues important to artificial life are raised here, and the author even speculates about artificial life at the end of the book.

Arbib66a　M. A. Arbib. "Simple Self-Reproducing Universal Automata." *Information and Control*, 9:177–189, 1966.

Arbib's novel contribution to the pool of self-reproducing structures. This is a CA-like system in which a more realistic approximation to a chemical-bond is achieved than in von Neumann's model.

Arbib66b　M. A. Arbib. "Self-Reproducing Automata—Some Implications for Theoretical Biology." In C. H. Waddington, editor, *Towards a Theoretical Biology*, pages 204–226, Edinburgh University Press, 1966.

Arbib67　M. Arbib. "Automata Theory and Development: Part I." *J. Theor. Biol.*, 14:131–156, 1967.

Automata theory as it relates to the process of development.

Arbib69　M. Arbib. *Theories of Abstract Automata.* Prentice-Hall, Englewood, CA, 1969.

Formal automata theory with an eye towards modeling biological phenomena.

Ashby47　W. R. Ashby. "The Nervous System as Physical Machine, with Special Reference to the Origin of Adaptive Behavior." *Mind*, 56:44ff, 1947.

Ashby48　W. R. Ashby. "The Homeostat." *Electronic Engineering*, 20:380ff, 1948.

Early cybernetic work on a goal-seeking or "teleological" system.

Ashby52　W. R. Ashby. *Design for a Brain.* Wiley & Sons, New York, 1952.

A classic book representing "the path not taken" by AI.

Atkins84   P. W. Atkins. *The Second Law.* Scientific American Library, W. H. Freeman, New York, 1984.

An excellent layman's account of the Second Law of Thermodynamics. Contains a chapter devoted to the thermodynamics of life with a cellular automaton model of the generation of complex structures.

Axelrod81   R. Axelrod and W.D. Hamilton. "The Evolution of Cooperation." *Science,* 211:1390–1396, 1981.

A classic paper on the evolution of altruistic behavior which won Science magazine's "best paper of the year" award.

Axelrod84   R. Axelrod. *The Evolution of Cooperation.* Basic Books, New York, 1984.

A fine treatment of the selective value of cooperation. Includes a chapter on the iterated prisoner's dilemma and the results of a computer contest organized by Axelrod to test strategies for dealing with the dilemma.

Babloyantz86   A. Babloyantz. *Molecules, Dynamics, and Life: An Introduction to the Self-Organization of Matter.* Wiley Interscience, New York, 1986.

Thermodynamic analysis of "how inert matter can acquire self-organizing properties once thought unique to living things."

Backus78   J. Backus. "Can Programming Be Liberated from the von Neumann Style?" *Comm. ACM,* 8:613–641, 1978.

Classic analysis of the primary problem with serial programming —the dreaded "von Neumann bottleneck"—and suggestions for a programming style that avoids it.

Bagley67   J.D. Bagley. *The Behavior of Adaptive Systems which Employ Genetic and Correlation Algorithms.* Ph.D. thesis, The University of Michigan, 1967.

BaiLin84   H. Bai-Lin, editor. *Chaos.* World Scientific, Singapore, 1984.

A collection of reprints of early papers on chaos. Includes papers by Lorenz, May, Feigenbaum, and etc.

Barricelli62   N.A. Barricelli. "Numerical Testing of Evolution Theories: I & II." *Acta Biotheoretica,* 16, 1962.

An early computer model of evolution featuring a cellular automaton.

Barto75   A.G. Barto. *Cellular Automata as Models of Natural Systems*. Ph.D. thesis, The University of Michigan, 1975.

Investigation of when it is and is not appropriate to use cellular automata when modeling natural systems.

Bays87   C. Bays. "Candidates for the Game of Life in Three Dimensions." *Complex Systems*, 1(3):373–400, 1987.

3D CA rules with dynamics similar to Conway's game of Life See Gardner70 & Berlekamp82. See also Dewdney87b.

Beeler73   M. Beeler. *Patterson's Worm*. AI Memo 290, MIT AI Laboratory, 1973.

Description of simple system for exploring optimal exploration pattern for a simple organism "feeding" on a lattice. See also Gardner73.

Benes80   J. Benes. "Whither Cybernetics: Past Achievements and Future Prospects." *Kybernetes*, 9(4):283–288, 1980.

Recent survey of the field of cybernetics.

Berlekamp82   E. Berlekamp, J.H. Conway, and R. Guy. *Winning Ways for Your Mathematical Plays*. Academic Press, New York, 1982.

In two volumes. Volume II contains the proof that the game of LIFE is computation universal and describes the construction of a self-reproducing configuration.

Bernal67   J.D. Bernal. *The Origin of Life*. World Publishing, New York, 1967.

Bernard62   E.E. Bernard and M.R. Kare, editors. *Biological Prototypes and Synthetic Systems*. Plenum, New York, 1962.

Proceedings of the Second Annual Bionics Symposium held at Cornell University, August 30-September 1, 1961. From the cybernetics era. Includes articles by Foerster on "Bio-Logic," Lofgren on self-repair, and McCulloch on "Biomimesis."

Bethke81   A.D. Bethke. *Genetic Algorithms as Function Optimizers*. Ph.D. thesis, The University of Michigan, 1981.

A nice application of the GA that illustrates the power of recombination and "implicit-parallelism."

Blum51   H.F. Blum. *Time's Arrow and Evolution*. Princeton University Press, Princeton, 1951.

Thermodynamics of life and its origin. Third edition published in 1968.

Boden77   M. Boden. *Artificial Intelligence and Natural Man*. Basic Books, New York, 1977.

History and philosophy of Artificial Intelligence.

Booker82   L. Booker. *Intelligent Behavior as an Adaptation to the Task Environment*. Ph.D. thesis, The University of Michigan, 1982.

The genetic algorithm is applied to simulated organisms based on Holland's classifier systems (Holland86a) to adapt their behavioral responses to complex environments.

Booker88   L. Booker, D.E. Goldberg, and J.H. Holland. "Classifier Systems and Genetic Algorithms." *Artificial Intelligence*, 1988—in press.

Tutorial overview.

Bosworth72   J. Bosworth, N. Foo, and B.P. Zeigler. *Comparison of Genetic Algorithms with Conjugate Gradient Methods*. Technical Report T.R. no. 012520-5-T, The University of Michigan, 1972.

Boyd85   R. Boyd and P.J. Richerson. *Culture and the Evolutionary Process*. University of Chicago Press, Chicago, 1985.

A good text on the extra-genetic process of cultural evolution.

Braitenberg84   V. Braitenberg. *Vehicles: Experiments in Synthetic Psychology*. MIT Press, Cambridge, 1984.

A classic book on the synthesis of complex behavior from the interactions of simple components.

Braitley72   P. Bratley and J. Millo. "Computer Recreations: Self-Reproducing Programs." *Software-Practice and Experience*, 2:397–400, 1972.

Breder51   C.M. Breder. "Studies in the Structure of the Fish School." *Bull. Am. Museum Nat. Hist.*, 98:7ff, 1951.

Study of schooling phenomena and their sensitivity to environmental factors.

Breder59   C.M. Breder. "Studies in the Social Grouping of Fishes." *Bull. Am. Museum Nat. Hist.*, 117:399ff, 1959.

Special forms, environmental influences, structural details, and the evolution of fish groups are discussed.

Bremmerman62   H.J. Bremmerman. "Optimization through Evolution and Recombination." In M.C. Yovits, G.T. Jacobi, and G.D. Goldstein, editors, *Self-Organizing Systems*, page 93ff, Spartan Books, Washington, DC, 1962.

One of a number of pioneering papers that initiated the evolutionary approach to optimization.

Breslow82   R. Breslow. "Artificial Enzymes." *Science*, 218:532–537, 1982.

Description of syntheticly engineered molecules with catalytic properties.

Broadwell87   P. Broadwell. "Plasm: A Fish Sample."

Computer simulation of an aquarium in which the "fish" respond to each other according to local rules for behavior. Demonstrated at the Artificial Life Workshop, Sept. 21-25, 1987, Los Alamos.

Brown87   D. Brown. "Competition of Cellular Automata Rules." *Complex Systems*, 1(1):169–180, 1987.

Different rules in a heterogeneous CA compete with their neighbors for "territory" in the lattice (see Edelman87).

Bryant77   P.J. Bryant, S.V. Bryant, and V. French. "Biological Regeneration and Pattern Formation." *Scientific American*, 237(1), July 1977.

Discusses the role of cellular interactions in development.

Burger80   J. Burger, D. Brill, and F. Machi. "Self-Reproducing Programs." *Byte*, 5, 1980.

Burks61   A.W. Burks. "Computation, Behavior, and Structure in Fixed and Growing Automata." *Behavioral Science*, 6:5–22, 1961.

Automata theory suited for the modeling of natural growth and development.

Burks66   A.W. Burks. "Automata Models of Self-Reproduction." In *Information Processing*, pages 121–123, Information Processing Society of Japan, 1966.

Burks70   A.W. Burks. *Essays on Cellular Automata.* University of Illinois
            Press, Urbana, 1970.

            *A classic collection of papers on cellular automata intended as
            a companion to von Neumann's Theory of Self Reproducing Au-
            tomata (vonNeumann66). Includes papers by Burks, Thatcher,
            Moore, Myhill, Ulam, and Holland.*

Burks73   A.W. Burks. "Logic, Computers, and Men." *Proceedings and Ad-
            dresses of the American Philosophical Association,* 46:39–57,
            1973.

Burks74   A.W. Burks. "Cellular Automata and Natural Systems." In W.D. Kei-
            del, W. Händler, and M. Spreng, editors, *Cybernetics and Bionics,*
            pages 190–204, R. Oldenbourg, Munich, 1974.

            *Cellular Automata as a useful bridge between natural systems
            and formal computational systems.*

Burks75   A.W. Burks. "Logic, Biology, and Automata—Some Historical Reflec-
            tions." *Int. J. Man-Machine Studies,* 7:297–312, 1975.

            *"[A] historical study of the relation of logic (including automata
            theory and inductive logic) to the biological sciences, broadly
            conceived."*

Burks81   A.W. Burks. *Programming and Structure Changes in Parallel
            Computers.* Technical Report, Logic of Computers Group, The Uni-
            versity of Michigan, 1981.

            *Discussion of the merits of self-restructuring, parallel comput-
            ers. This was the keynote address at the Conference on Parallel
            Computers, held in Nürnberg Germany, June 1981.*

Butler78   J.T. Butler. "Analysis of Cellular Automata Growth Models." In *Pro-
            ceedings of 1978 ACM Computer Science Conference, Detroit
            Michigan, Feb. 1978,* Association for Computing Machinery, New
            York, 1978.

Cairns-Smith66   A.G. Cairns-Smith. "The Origin of Life and the Nature of the Primitive
            Gene." *J. Theor. Biol.,* 10:53–88, 1966.

            Early article on his life-from-clays hypothesis.

Cairns-Smith68   A.G. Cairns-Smith. "An Approach to a Blueprint for a Primitive Organ-
            ism." In C.H. Waddington, editor, *Towards a Theoretical Biology,*
            pages 57–66, Edinburgh University Press, 1968.

            An early presentation of the "life-from-clays" hypothesis.

Cairns-Smith73 A.G. Cairns-Smith. *The Life Puzzle: On Crystals and Organisms and on the Possibility of a Crystal as an Ancestor*. University of Toronto Press, Toronto, 1973.

Cairns-Smith82 A.G. Cairns-Smith. *Genetic Takeover and the Mineral Origins of Life*. Cambridge University Press, Cambridge, 1982.

A full exposition of Cairns-Smith's theory that life originated with replicating clays.

Cairns-Smith85a A.G. Cairns-Smith. *Seven Clues to the Origin of Life*. Cambridge University Press, Cambridge, 1985.

A layman's version of Genetic Takeover (Cairns-Smith82).

Cairns-Smith85b A.G. Cairns-Smith. "The First Organisms." *Scientific American*, 252(6):90–100, June 1985.

Another readable exposition of life-from-clays and genetic takeover.

Calvin69 M. Calvin. *Chemical Evolution: Molecular Evolution towards the Origin of Living Things on the Earth and Elsewhere*. Oxford University Press, London, 1969.

Caraco82 T. Caraco and M.C. Bayham. "Some Geometric Aspects of House Sparrow Flocks." *Anim. Behav.*, 30:990–996, 1982.

Analysis of the 3D structure of a bird-flock.

Cardelli85 L. Cardelli. *Crabs: The Bitmap Terror*. Technical Report, Bell Laboratories, 1985.

An "unofficial" Bell-Labs report detailing the escapades of tiny bitmap "creatures" which chew up window systems and interact with one another on screen.

Casjens85 S. Casjens, editor. *Virus Structure and Assembly*. Jones and Bartlett, Boston, 1985.

Describes self-assembly of virus.

Caspar62 D.L. Caspar and A. Klug. "Physical Principles in the Construction of Regular Viruses." *Cold Spring Harbor Symp. Quant. Biol.*, 27:1–24, 1962.

Virus self-assembly.

Caspar63   D.L. Caspar. "Assembly and Stability of the Tobacco Mosaic Virus Particles." *Adv. Prot. Chem.*, 18:37–121, 1963.

Virus self-assembly.

Caspar76   D.L. Caspar. "Switching in the Self-Control of Self-Assembly." In R. Markham and R.W. Horne, editors, *Structure-Function Relationship of Proteins*, pages 85–98, Elsevier/North-Holland, New York, 1976.

Describes the role of "conformational switching" to control the sequence of self-assembly of biological components. See the article by Goel and Thompson in these proceedings for a model incorporating conformational switching.

Cavicchio70   D.J. Cavicchio. *Adaptive Search using Simulated Evolution.* Ph.D. thesis, The University of Michigan, 1970.

Adaptive evolution in a population of devices (pattern recognition programs) is discussed and criteria for effective reproductive plans are presented.

Chaitin70   G. J. Chaitin. "To a Mathematical Definition of 'Life.'" *ACM SICACT News*, 4:12–18, 1970.

A definition of life based on the notion that a living organism is its own simplest description.

Chaitin79   G. J. Chaitin. "Toward a Mathematical Definition of 'Life.'" In R.D. Levine and M. Tribus, editors, *The Maximum Entropy Formalism*, pages 477–498, MIT Press, 1979.

Application of algorithmic complexity theory to the quantitative measurement of organization in living organisms and other complex systems.

Chapuis58   A. Chapuis and E. Droz. *Automata: A Historical and Technological Study.* B.T. Batsford Ltd, London, 1958.

A thorough, scholarly study of the history of clockwork automata, full of pictures and technical details.

Clark59   F. Clark and R.L.M. Synge, editors. *The Origin of Life on Earth.* Pergamon Press, London, 1959.

Codd68   E.F. Codd. *Cellular Automata.* Academic Press, New York, 1968.

Codd's Ph.D. thesis in which he details an 8-state, self-reproducing, universal computer/constructor.

Conrad70a   M. Conrad. *Computer Experiments on the Evolution of Coadaptation in a Primitive Ecosystem.* Ph.D. thesis, Stanford University, 1970.

This was the first in a series of studies of evolution in artificial ecosystems. Further papers in the sequence are listed below.

Conrad70b   M. Conrad and H.H. Pattee. "Evolution Experiments with an Artificial Ecosystem." *J. Theor. Biol.*, 28:393–409, 1970.

Conrad74   M. Conrad. "Molecular Automata." In M. Conrad, W. Guttinger, and M. Dal Cin, editors, *Physics and Mathematics of the Nervous System*, pages 419–430, Springer-Verlag, Berlin, 1974.

Discusses possibility for implementing molecular-scale computations.

Conrad77   M. Conrad. "Evolutionary Adaptability of Biological Macromolecules." *J. Molec. Evol.*, 10:87–91, 1977.

This paper discusses the idea that the amenability of a biological organization to evolution can hitchike along with the advantageous traits whose appearance it facilitates.

Conrad79   M. Conrad. "Bootstrapping on the Adaptive Landscape." *BioSystems*, 11:167–182, 1979.

Another discussion of the evolutionary self-facilitation of evolvability.

Conrad81   M. Conrad. "Algorithmic Specification as a Technique for Computing with Informal Biological Models." *BioSystems*, 13:303–320, 1981.

Algorithms as proper subjects for biological experiments.

Conrad83   M. Conrad. *Adaptability.* Plenum Press, New York, 1983.

This book gives an extensive analysis and discussion of the concept of evolvability and the evolutionary self-facilitation principle—how the structure of biological systems becomes organized for more effective evolution through the process of evolution.

Conrad85   M. Conrad and M. Strizich. "EVOLVE II: A Computer Model of an Evolving Ecosystem." *BioSystems*, 17:245–258, 1985.

Continuation of the artificial ecosystem series. For EVOLVE III see 344 or Rizki85.

Crick81   F. Crick. *Life Itself (Its Origin and Nature)*. Simon and Schuster, New York, 1981.

Crick's views on the nature and origins of life. Includes a discussion of his "Directed Panspermia" hypothesis of an extraterrestrial origin for life.

Darwin   C. Darwin. *On the Origin of Species*. John Murray, London, 1859.

The first edition. No bibliography on life would be complete without this entry!

Davis65   M. Davis, editor. *The Undecidable*. Raven Press, New York, 1965.

Collection of articles on formally undecidable propositions. Contains translations of some of Gödel's famous papers.

DavisL87   L. Davis, editor. *Genetic Algorithms and Simulated Annealing*. Pitman, London, 1987.

A collection of articles describing current results using genetic and/or annealed search.

DavisJM75   J.M. Davis. "Socially Induced Flight Reactions in Pigeons." *Anim. Behav.*, 23:597–601, 1975.

Description of the "contagion of flight."

DavisJM80   J.M. Davis. "The Coordinated Aerobatics of Dunlin Flocks." *Anim. Behav.*, 28:668–673, 1980.

Dynamic behavior of bird-flocks.

Dawkins76   R. Dawkins. *The Selfish Gene*. Oxford University Press, Oxford, 1976.

Classic book on the gene as the unit of selection. Final chapter proposes "memes"—the cultural equivalent of genes—as another form of self-interested replicator in nature.

Dawkins83   R. Dawkins. *The Extended Phenotype: The Gene as a Unit of Selection*. Oxford University Press, Oxford, 1983.

Extension of the ideas presented in "The Selfish Gene."

Dawkins86   R. Dawkins. *The Blind Watchmaker*. W.W. Norton, New York, 1986.

A thorough and careful explication of the nature of Darwinian evolution. The paperback edition contains an order form for the Blind Watchmaker program for the Apple Macintosh computer, described in Dawkins' contribution to these proceedings.

DeJong75   K.A. DeJong. *Analysis of Behavior of a Class of Genetic Adaptive Systems*. Ph.D. thesis, The University of Michigan, 1975.

DeJong80   K.A. DeJong. "Adaptive System Design: A Genetic Approach." *IEEE Transactions on Systems, Man, and Cybernetics*, SMC-10(9):566–574, 1980.

Dennett78   D.C. Dennett. *Brainstorms: Philosophical Essays on Mind and Psychology*. Bradford Books/MIT Press, Cambridge, 1978.

   A very nice book on philosophical issues regarding mind and computation.

Derrida86a   B. Derrida and Y. Pomeau. "Random Networks of Automata: A Simple Annealed Approximation." *Europhys. Lett.*, 1(2):45–49, 1986.

   This and the following article extend and analyze Kauffman's random boolean nets (Kauffman69).

Derrida86b   B. Derrida and H. Flyvbjerg. "Multi-Valley Structure in Kauffman's Model: Analogy with Spin Glasses: Letter to the Editor." *J. Phys. A: Math.Gen.*, 19:L1003–L1008, 1986.

Descartes87   R. Descartes. *Méditations on First Philosophy*. Cambridge University Press, Cambridge, 1987.

   Classic treatise in which Descartes compares the human body to a mechanical clock. This edition also has selections from the objections with replies. Translated by John Cottingham with an introduction by Bernard Williams.

Dewdney84a   A.K. Dewdney. "Computer Recreations: In the Game Called Core War Hostile Programs Engage in a Battle of Bits." *Scientific American*, 250(5):14–22, May 1984.

   Description of a computer game in which short pieces of code reproduce and compete for occupation of memory.

Dewdney84b   A.K. Dewdney. "Computer Recreations: Sharks and Fsh Sage an Ecological War on the Toroidal Planet Wa-tor." *Scientific American*, 251(6):14–22, December 1984.

   Description of program for simulating predator-prey dynamics.

Dewdney84c   A.K. Dewdney. *The Planiverse: Computer Contact with a Two-Dimensional World*. Poseidon Press, 1984.

   Entertaining and enlightening book on what life would be like in a 2D universe.

Dewdney85a A.K. Dewdney. "Computer Recreations: A Core War Bestiary of Viruses, Worms and Other Threats to Computer Memories." *Scientific American*, 252(3):14–23, March 1985.

Review of self-propagating programs.

Dewdney85b A.K. Dewdney. "Computer Recreations: A Circuitous Odyssey from Robotropolis to the Electronic Gates of Silicon Valley." *Scientific American*, 253(1):14–19, July 1985.

Description of a program for wiring up simple simulated robots using actuators, sensors, motors, and etc. in order to perform simple task in a simulated environment. Should be extended to allow interactions between robots.

Dewdney85c A.K. Dewdney. "Computer Recreations: Exploring the Field of Genetic Algorithms in a Primordial Computer Sea Full of Flibs." *Scientific American*, 253(5):21–32, November 1985.

Description of a program for experimenting with genetic algorithms.

Dewdney87a A.K. Dewdney. "Computer Recreations: A Program Called MICE Nibbles its Way to Victory at the First Core War Tournament." *Scientific American*, 256(1):14–20, January 1987.

Results of the first international competition.

Dewdney87b A.K. Dewdney. "Computer Recreations: The Game of Life Acquires Some Successors in Three Dimensions." *Scientific American*, 256(2):16–24, February 1987.

Description of Carter Bays' 3D extensions of Conway's game of LIFE. See also Bays87.

Dewdney87c A.K. Dewdney. "Computer Recreations: Braitenberg Memoirs— Vehicles for Probing Behavior Roam a Dark Plain Marked by Lights." *Scientific American*, 256(3):16–24, 1987.

Discussion of programming techniques for implementing Braitenberg's vehicles (Braitenberg84).

Dixon83 R. Dixon. "The Mathematics and Computer Graphics of Spirals in Plants." *Leonardo*, 16:86–90, 1983.

Dobzhansky72   K. Dobzhansky. "Darwinian Evolution and the Problem of Extraterrestrial Life." *Perspect. Biol. Med.*, 15:157–175, 1972.

Discussion of the impossibility of predicting the future course of evolution due to its sensitive and unpredictable dependence on "accidental" genetic and environmental events.

Drexler81   K.E. Drexler. "Molecular Engineering: An Approach to the Development of General Capabilities for Molecular Manipulation." *Proc. Nat. Acad. Sci. USA*, 78(9):5275–5278, 1981.

An early characterization of nanotechnology.

Drexler82   K.E. Drexler. "Mightier Machines from Tiny Atoms May Someday Grow." *Smithsonian*, 145–154, November 1982.

Short overview of nanotechnology.

Drexler86   E. Drexler. *Engines of Creation: Nano-Technology.* Anchor-Press/Doubleday, New York, 1986.

A popular treatment of nanotechnology.

Drexler87   K.E. Drexler. "Molecular Machinery and Molecular Devices." In F. Carter, editor, *Molecular Electronic Devices II*, Marcel Dekker, New York, 1987.

Dreyfus79   H. Dreyfus. *What Computers Can't Do: The Limits on Artificial Intelligence.* Harper & Row, New York, 1979.

Classic book arguing against the traditional AI approach to machine intelligence.

Dyson81   F. Dyson. *Disturbing the Universe.* Harper Colophon, New York, 1981.

Freeman Dyson's thoughts about the nature of life, the universe and everything.

Dyson85a   F. Dyson. *Origins of Life.* Cambridge University Press, Cambridge, 1985.

Dyson's theory that non-replicating, metabolizing fauna must have preceeded replicating organisms. Includes a brief survey and analysis of other origin theories.

Dyson85b   F. Dyson. "Space Butterflies and Other Speculations." *Science 85*, November 1985.

Proposal for simple, self-reproducing machines to explore the planets and stars.

Edelman82   G.M. Edelman. "Selective Networks Capable of Representative Trans-
            formations, Limited Generalizations, and Associative Memory." *Proc.
            Nat. Acad. Sci.*, 79:2091–2095, 1982.

            Description of the "Darwin" machine.

Edelman87   G.M. Edelman. *Neural Darwinism: The Theory of Neuronal
            Group Selection.* Basic Books, New York, 1987.

            *Presentation of Edelman's theory that high-level functions com-
            pete for neuronal support.*

Eigen71   M. Eigen. "Self-Organization of Matter and the Evolution of Biological
            Macromolecules." *Naturwissenschaften*, 10, 1971.

            *Theoretical discussion of evolution, natural selection among
            molecules, and the origin of life.*

Eigen73   M. Eigen. "The Origin of Biological Information." In J. Mehra, editor,
            The Physicist's Conception of Nature, pages 594–632, D. Reidel
            Publishing Co., 1973.

Eigen79   M. Eigen and P. Schuster. *The Hypercycle: A Principle of Natural
            Self-Organization.* Springer-Verlag, Berlin, 1979.

            *Theory of the emergence of hierarchies of auto-catalytic
            networks.*

Eigen81   M. Eigen, W. Gardiner, P. Schuster, and R. Winkler-Oswatitsch. "The
            Origin of Genetic Information." *Scientific American*, 244(4):88–118,
            April 1981.

            *A good, general introduction to Eigen's theories on quasi-species,
            hyper-cycles, and the origin of life.*

Eigen82a   M. Eigen and R. Winkler. *The Laws of the Game: How the Princi-
            ples of Nature Govern Chance.* Harper Colophon, New York, 1982.

            *A wide ranging description of how chance and low-level rules
            lead to complex behavior.*

Eigen82b   M. Eigen and P. Schuster. "Stages of Emerging Life—Five Principles
            of Early Organization." *J. Molec. Evol.*, 19:47–61, 1982.

            *The five are: formation of stereoregular heteropolymers, selec-
            tion through self-replication, evolution of quasispecies, emer-
            gence of cooperative catalytic hypercycles, and evaluation of
            translation products through compartmentalization.*

Elsasser81   W.M. Elsasser. "A Form of Logic Suited for Biology." *Prog. Theor. Biol.*, 6:23–61, 1981.

One presentation of the notion that the underlying "bio-logic" of living systems is independent of the particular physical law that governs the behavior of the material primitives of the system. Argues for a kind of "creative-selection" which is different from Darwinian selection.

Epstein83   I.R. Epstein, K. Kustin, P. De Kepper, and M. Orban. "Oscillating Chemical Reactions." *Scientific American*, 248(3):112–123, March 1983.

Chemical examples of the way in which simple, local reactions can lead to complex global behavior.

Fahlman79   S.E. Fahlman. *NETL: A System for Representing and Using Real-World Knowledge.* MIT Press, Cambridge, 1979.

Fahlman's implementation of a semantic network—a distributed knowledge representation scheme.

Farley77   J. Farley. *The Spontaneous Generation Controversy from Descartes to Oparin.* John Hopkins University Press, Baltimore, 1977.

Farmer84   J.D. Farmer, T. Toffoli, and S. Wolfram, editors. *Cellular Automata: Proceedings of an Interdisciplinary Workshop at Los Alamos, New Mexico, March 7-11, 1983*, North-Holland, Amsterdam, 1984.

A collection of papers on the theory and applications of cellular automata.

Farmer86a   J.D. Farmer, S.A. Kauffman, and N.H. Packard. "Autocatalytic Replication of Polymers." *Physica D*, 22:50–67, 1986.

A model of the emergence of autocatalytic networks using the concept of a reaction graph.

Farmer86b   J.D. Farmer, N.H. Packard, and A.S. Perelson. "The Immune System, Adaptation & Learning." *Physica D*, 22:187–204, 1986.

A dynamic model of the immune system based on Jerne's network hypothesis. Comparisons are drawn with Holland's Classifier Systems (Holland86a,Holland86b).

Farmer86c   J.D. Farmer, A. Lapedes, N.H. Packard, and B. Wendroff. *Evolution, Games and Learning*. North-Holland, Amsterdam, 1986.

Proceedings of the Fifth Annual International Conference of the Center for Nonlinear Studies, Los Alamos, New Mexico, May 20-24, 1985. Also Published as Physica D, Vol. 22, 1986.

Fasan70   E. Fasan. *Relations with Alien Intelligences*. Springer-Verlag, Berlin, 1970.

Written during the height of CETI enthusiasm (Communication with Extra-Terrestrial Intelligence—see Shklovskii66). Similiar considerations apply to intelligences or life-forms of our own manufacture.

Feigenbaum63   E.A. Feigenbaum and J. Feldman, editors. *Computers and Thought*. McGraw Hill, New York, 1963.

A description of the state of the art of AI as of 1960.

Feinberg80   G. Feinberg and R. Shapiro. *Life Beyond the Earth*. William Morrow, New York, 1980.

Discussion of likelihood and nature of life in the solar system and beyond.

Fisher58   R. Fisher. *Genetical Theory of Natural Selection—2nd edition*. Dover, New York, 1958.

A reprint of Fisher's classic 1929 mathematical analysis of the process of evolution by natural selection.

Fodor81   J.A. Fodor. "The Mind Body Problem." *Scientific American*, 244(1): 114–123, January 1981.

Hardware of the brain vs. behavior of that hardware.

Fogel66   L.J. Fogel, A.J. Owens, and M.J. Walsh. *Artificial Intelligence through Simulated Evolution*. John Wiley & Sons, New York, 1966.

This work has been criticized as involving overly simple evolution of finite automata, but it was of pioneering significance at the time. For a review see Lindsay68.

FoxRF88   R.F. Fox. *Energy and the Evolution of Life*. W.H. Freeman, New York, 1988.

A solid exposition of the relevance of energy flow to models of life and its origins.

FoxSW59   S.W. Fox, K. Harada, and J. Kendrick. "Production of Spherules from Synthetic Proteinoid and Hot Water." *Science*, 129:1221ff, 1959.

Fox et al's original paper on proteinoid spheres.

FoxSW60   S.W. Fox. "How Did Life Begin?" *Science*, 132:200ff, 1960.

Fox's theory of the origin of life based on proteinoid spheres.

FoxSW65   S.W. Fox, editor. *The Origins of Pre-Biological Systems and Molecular Structures.* Academic Press, 1965.

FoxSW77   S.W. Fox and K. Dose. *Molecular Evolution and the Origin of Life.* Marcel Dekker, New York, 1977.

Good review of the work of Sidney Fox.

FoxSW86   S.W. Fox, editor. *Selforganization.* Adenine Press, Guilderland, New York, 1986.

Proceedings of the Liberty Fund conference on self-organization held in Key Biscayne, Florida, 1984.

Foy74   J. Foy. *A Computer Simulation of Impulse Conduction in Cardiac Muscle.* Ph.D. thesis, The University of Michigan, 1974.

Simulation of the generation of periodic pulse-waves in an excitable, refractory medium, and analysis of perturbations which lead to a transition from the periodic to a chaotic regime reminiscent of fibrilation.

Frantz72   D.R. Frantz. *Non-Linearities in Genetic Adaptive Search.* Ph.D. thesis, The University of Michigan, 1972.

Freitas77a   R. Freitas. "The Legal Rights of Extraterrestrials." *Analog*, 97:54–67, 1977.

Much of this discussion could be applied to artificial terrestrial life as well.

Freitas77b   R. Freitas. "Metalaw and Interstellar Relations." *Mercury*, 6:15–17, 1977.

Relations with other life-forms.

Freitas80   R. Freitas. "A Self-Reproducing Interstellar Probe." *J. of the British Interplanetary Soc.*, 33:251–264, 1980.

Detailed plans for the construction of a self-reproducing fleet of interstellar probes. Includes a discussion of such questions as whether it is morally right for a reproducing star-probe to enter a foreign solar system and convert part of that system's mass and energy for its own use.

Freitas83a   R. Freitas. "Roboclone." *Omni*, 5:44–47, 1983.

Popular account of self-reproducing machines and their possible applications.

Freitas83b   R. Freitas. "Building Athens without the Slaves." *Technology Illustrated*, 3:16–20, 1983.

Self-reproducing machines could "expand human horizons dramatically, increase productivity, and cause an explosion of wealth before the end of this century."

Gardner70   M. Gardner. "Mathematical Games: The Fantastic Combinations of John Conway's New Solitaire Game 'Life.'" *Scientific American*, 223(4):120–123, October 1970.

The original description of Conway's game of LIFE.

Gardner71   M. Gardner. "Mathematical Games: On Cellular Automata, Self-Reproduction, the Garden of Eden and the Game of 'Life.'" *Scientific American*, 224(2):112–117, February 1971.

Follow-up article on LIFE and other CA rules.

Gardner73   M. Gardner. "Mathematical Games: Fantastic Patterns Traced by Programmed "Worms.'" *Scientific American*, 229(5):116–123, November 1973.

Description of Patterson's Worm system. Also see Beeler73.

Gardner78   M. Gardner. "Mathematical Games: A Mathematical Zoo of Astounding Critters, Imaginary and Otherwise." *Scientific American*, 238(6): 18–28, June 1978.

Organisms with mathematically interesting structure.

Gardner83   M. Gardner. *Wheels, Life, and Other Amusements.* W.H. Freeman, New York, 1983.

The last three chapters are devoted to Conway's game of LIFE.

Gatlin72   L. L. Gatlin. *Information Theory and the Living System.* Columbia University Press, New York, 1972.

An interesting analysis of the information-theoretic structure of DNA, together with a theory about the selection pressure on the DNA-to-protein channel. This work deserves to be followed up now that vastly more DNA sequence information is available.

Gebstadter79   E.B. Gebstadter. *Copper, Silver, Gold: An Indestructible Metallic Alloy.* Acidic Books, Perth, 1979.

This self-referential reference refers to a self-referential imaginary book (Gebstadter79) which is, however, cited elsewhere (Hofstadter79). Could this be a self-reproducing, self-referential reference?

Gleick87a   J. Gleick. "Artificial life: Can Computers Discern the Soul?" *The New York Times,* Thursday, Sept. 29, 1987.

Science Section: good review of the first Artificial Life Workshop, despite the title.

Gleick87b   J. Gleick. *Chaos.* Viking, New York, 1987.

A very good layman's introduction to the theory of chaotic dynamical systems and the people who have pioneered the understanding of the role of chaos in physics, chemistry, and biology.

Godel31   K. Gödel. "Uber Formal Unentscheidbare Satze der *Principia Matematica* und Verwandter Systeme I." *Monatshefte fur Matematik und Physik,* 38:173–98, 1931.

Gödel's famous proof of the formal undecidability of certain mathematical truths. Translation by Elliot Mendelsohn printed in Davis65.

Goel74   N.S. Goel. "Cooperative Processes in Biological Systems." *Prog. Theor. Biol.,* 2:213–302, 1974.

Goel78   N.S. Goel. "Computer Simulation of Engulfment and Other Movements of Embryonic Tissues." *J. Theor. Biol.,* 71:103–140, 1978.

Goel86   N.S. Goel and R.L. Thompson. "Organization of Biological Systems: Some Principles and Models." *Int. Rev. Cytol.,* 103:1–88, 1986.

Goel87   N.S. Goel and R.L. Thompson. "Microcomputer Modeling of Biological Systems." *The World & I,* 2:162–173, 1987.

Goel88a   N.S. Goel and R.L. Thompson. *Computer Simulation of Self-Organization in Biological Systems.* Croom Helm, London, 1988.

A review of computer models of self-organization in biological systems. Chapters cover computer models for protein folding, self-assembly of virus, aggregation of cells, protein biosynthesis, and the evolution of macromolecular machinery. Numerous citations.

Goel88b   N.S. Goel and R.L. Thompson. "Movable Finite Automata (MFA) Models for Biological Systems II: Protein Biosynthesis." *J. Theor. Biol.,* 1988—in press.

See Thompson88 for part I, as well as Goel and Thompson's contribution to these proceedings.

Goldberg83   D.E. Goldberg. *Computer-Aided Gas Pipeline Operation using Genetic Algorithms and Rule Learning.* Ph.D. thesis, The University of Michigan, 1983.

Example of the emergence of a default-hierarchy in one of the first applications of classifier systems and genetic algorithms to a real-world optimization problem.

Goldberg88   D.E. Goldberg and J.H. Holland. "Genetic Algorithms and Machine Learning: Introduction to the Special Issue on Genetic Algorithms." *Machine Learning,* 3, 1988—in press.

Collection of articles detailing current research using GA's.

Goldberg89   D.E. Goldberg. *Genetic Algorithms in Search, Optimization, and Machine Learning.* Addison-Wesley, Reading, Mass., 1989.

An introductory textbook and guide to current research. Contains examples, exercises, and sample programs as well as proofs of most of the major GA theorems on crossover and implicit parallelism.

Goodman71   E.E. Goodman, R. Weinberg, and R. Laing. "A Cell Space Embedding of Simulated Living Cells." *Bio-Medical Computing,* 2:121–136, 1971.

Goodwin68   B. Goodwin. "The Division of Cells and the Fusion of Ideas." In C. H. Waddington, editor, *Towards a Theoretical Biology,* pages 134–139, Edinburgh University Press, 1968.

Grefenstette85   J.J. Grefenstette, editor. *Proceedings of an International Confer-ence on Genetic Algorithms and their Applications,* Carnegie-Mellon University, July 24-26, 1985, Lawerence Earlbaum Assoc., Hillsdale, NJ, 1985.

              Proceedings of the first workshop on genetic algorithms.

Grefenstette87   J.J. Grefenstette, editor. *Genetic Algorithms and their Applica-tions: Proceedings of the Second International Conference on Genetic Algorithms,* MIT, July 28-31, 1987, Lawerence Earlbaum Assoc., Hillsdale, NJ, 1987.

Guttman66   B.S. Guttman. "A Resolution of Rosen's Paradox for Self-Reproducing Automata." *Bulletin of Mathematical Biophysics,* 28, 1966.

              A response to Rosen59.

Haken77a   H. Haken. *Synergetics—An Introduction: Nonequilibrium Phase Transitions and Self-Organization in Physics, Chemistry, and Biology.* Springer-Verlag, Berlin, 1977.

              This is the first volume in the Springer Series on "Synergetics": Haken's name for the study of complex dynamics in nonequilib-rium systems.

Haken83   H. Haken. *Advanced Synergetics: Instability Hierarchies of Self-Organizing Systems and Devices.* Springer-Verlag, Berlin, 1983.

              More on chaos and structure in nonequilibrium systems.

Hameroff87   S.R. Hameroff. *Ultimate Computing: Biomolecular Conscious-ness and Nanotechnology.* Elsevier, New York, 1987.

              Review of nanotechnology and the possibility for molecular-scale computation.

Hamilton64   W. Hamilton. "The Genetical Evolution of Social Behavior." *J. Theor. Biol.,* 7:1–31, 1964.

              Classic paper on the evolution of altruistic behavior.

Harary72   F. Harary. *Graph Theory.* Addison-Wesley, Reading Mass., 1972.

              The standard text on graph theory—a useful tool for describing and analyzing interaction "networks."

HayL80   L. Hay. "Self-Reproducing Programs." *Creative Computing,* 6:134–136, July 1980.

Hebb49   D.O. Hebb. *The Organization of Behavior*. Wiley & Sons, New York, 1949.

Hebb's "cell-assembly" model, in which aggregates of neurons form the physiological basis of concepts.

HermanGT69   G.T. Herman. "Computing Ability of a Developmental Model for Filamentous Organisms." *J. Theor. Biol.*, 25:421–435, 1969.

Analysis of computational capacity of L-systems (Lindenmayer68).

HermanGT70   G.T. Herman. "Role of Environments in Developmental Models." *J. Theor. Biol.*, 29:329–342, 1970.

HermanGT73   G.T. Herman. "On Universal Computer Constructors." *Information Processing Letters*, 2:61–64, 1973.

Proof that computation/construction universality is not sufficient to eliminate cases of trivial self-reproduction.

HermanGT75   G.T. Herman and G. Rozenberg. *Developmental Systems and Languages*. North-Holland, Amsterdam, 1975.

The first and a very useful textbook on L-systems.

Hewitt77   C. Hewitt. "Viewing Control Structures as Patterns of Passing Sessages." *Artificial Intelligence*, 8:323–364, 1977.

Influential description of a method for implementing distributed control structures: the ACTOR paradigm.

Hewitt80   C. Hewitt. "The Apiary Network Architecture for Knowledgeable Systems." In *Proceedings of Lisp Conference*, Stanford University, 1980.

Description of a hardware architecture to support the ACTOR paradigm.

Hillis82   W.D. Hillis. "New Computer Architectures and Their Relationship to Physics, or Why Computer Science is No Good." *International Journal of Theoretical Physics*, 21(3/4):255–262, 1982.

Why computer science should begin to look at physics.

Hillis84   W.D. Hillis. "The Connection Machine: A Computer Architecture Based on Cellular Automata." *Physica D*, 10:213–228, 1984.

An early description of the connection machine concept. This article contains many interesting comments on the motivations for massively parallel processors. Reprint of MIT AI-Lab Memo 646 (1981).

Hillis87   W.D. Hillis. *The Connection Machine*. MIT Press, Cambridge, 1987.

Danny Hillis' Ph.D. thesis describing the massively parallel (65,000 processor!) machine that is now being produced by his company: Thinking Machines Corp. Hillis refers to the human brain as "the competition."

Hillis88   W.D. Hillis. "Intelligence as an emergent behavior; or, the songs of Eden." *DÆDALUS*, 117(1):175–189, Winter 1988.

A metaphorical tale on the mutually beneficial co-evolution of brains and minds, the moral being that both are based on self-interested replicators in some sense. See also Dawkin's description of "memes" in Dawkins76.

Hinton81   G. Hinton and J.A. Anderson, editors. *Parallel Models of Associative Memory*. Lawrence Earlbaum and Associates, Hillsdale, NJ, 1981.

A collection of papers on distributed representations for memory.

Hofstadter79   D.R. Hofstadter. *Gödel, Escher, Bach: An Eternal Golden Braid*. Basic Books, New York, 1979.

Hofstadter's Pulitzer Prize winning treatise on the self-referential nature of life, language, intelligence, and computation. See especially Chapters X, XI, XII, XIV, XVI, XVII, and XX. Also, the "Ant Fugue" which follows Chapter X.

Hofstadter82   D.R. Hofstadter. *Artificial Intelligence: Subcognition as Computation*. Technical Report 132, Indiana University Computer Science Dept., 1982.

A well-written exposition on the difference between the top-down, computer-program metaphor adopted by traditional AI and the bottom-up, emergent-phenomena approach partially embraced by the new "connectionist" school. Reprinted as Chapter 26 of Hofstadter85.

Hofstadter85   D.R. Hofstadter. *Metamagical Themas: Questing for the Essence of Mind and Pattern*. Basic Books, New York, 1985.

A collection of Hofstadter's "Metamagical Themas" columns from Scientific American, elaborated upon with post-scripts several additional articles. See especially sections I, V, and VI.

Hogeweg80   P. Hogeweg. "Locally Synchronized Developmental Systems, Conceptual Advantages of the Discrete Event Formalism." *Int. J. General Systems*, 6:57–73, 1980.

Hogeweg81   P. Hogeweg and B. Hesper. "Two Predators and a Prey in a Patchy Environment: An Application of MICMAC Modeling." *J. Theor. Biol.*, 93:411–432, 1981.

A system for the simulation of ecological dynamics.

Hogeweg83   P. Hogeweg and B. Hesper. "The Ontogeny of the Interaction Structure in Bumble Bee Colonies: A MIRROR Model." *Behav. Ecol. Sociobiol.*, 12:271–283, 1983.

A later system for the simulation of ecological dynamics.

Hogeweg85   P. Hogeweg and B. Hesper. "Socioinformatic Processes, a MIRROR Modeling Methodology." *J. Theor. Biol.*, 113:311–330, 1985.

Hogeweg88   P. Hogeweg. "Cellular Automata as a Paradigm for Ecological Modelling." *Applied Math. & Computation*, in press, 1988.

Holland62   J.H. Holland. "Outline for a Logical Theory of Adaptive Systems." *JACM*, 9:297–314, 1962.

Early work on genetic algorithms. Reprinted in Burks70.

Holland67   J.H. Holland. "Nonlinear Environments Permitting Efficient Adaptation." In *Computer and Information Sciences II*, Academic Press, New York, 1967.

Considerations of effects of complexity of environment on the process of evolution.

Holland73   J.H. Holland. "Genetic Algorithms and the Optimal Allocation of Trials." *SIAM J. Comput.*, 2(2):88–105, 1973.

A mathematical analysis of genetic algorithms and their "implicit parallelism."

Holland75a   J. H. Holland. *Adaptation in Natural and Artificial Systems.* University of Michigan Press, Ann Arbor, 1975.

An early treatment of the genetic algorithm and classifier systems.

Holland76a   J. H. Holland. "Studies of the Spontaneous Emergence of Self-Repli-
             cating Systems using Cellular Automata and Formal Grammars." In
             A. Lindenmayer and G. Rozenberg, editors, *Automata, Languages,
             Development*, pages 385–404, North-Holland, 1976.
             *A study demonstrating the necessity of stable sub-configurations
             for the emergence of reproducing entities in any reasonable time.*

Holland76b   J. H. Holland. "Adaptation." In R. Rosen and F.M. Snell, editors,
             Progress in Theoretical Biology IV, pages 263–293, Academic
             Press, New York, 1976.

Holland86a   J.H. Holland. "Escaping Brittleness: The Possibilities of General Pur-
             pose Learning Algorithms Applied to Parallel Rule-Based Systems."
             In R.S. Mishalski, J.G. Carbonell, and T.M. Mitchell, editors, *Machine
             Learning II*, pages 593–623, Kaufman, 1986.
             An overview of classifier systems and the genetic algorithm.

Holland86b   J.H. Holland, K.J. Holyoak, R.E. Nisbett, and P.R. Thagard. *Induc-
             tion: Processes of Inference, Learning, and Discovery*. MIT Press,
             Cambridge, 1986.
             *Classifier systems and genetic algorithms viewed within the
             larger context of induction in general.*

HoiLe77   L. Hoi. "On Machines as Living Things." *Acta Cybernetica*, 3(4),
             1977.
             *Description of a cellular-automaton-like system in which the in-
             dividual "cells" can move around like molecules in a fluid (c.f.
             the paper by Lugowski in these proceedings).*

Hollstien71 R.B. Hollstien. *Artificial Genetic Adaptation in Computer Control
             Systems*. Ph.D. thesis, The University of Michigan, 1971.

Hopcroft79   J.E. Hopcroft and J.D. Ullman. *Introduction to Automata Theory,
             Languages, and Computation*. Addison-Wesley, Menlo Park, 1979.
             The classic text book on the formal theory of computation.

Hopfield82   J.J. Hopfield. "Neural Networks and Physical Systems with Emergent
             Collective Computational Abilities." *Proc. Nat. Acad. Sci.*, 79:2554–
             2558, 1982.
             *A proposal for a distributed system supporting emergent compu-
             tation.*

Huberman88   B. Huberman, editor. *The Ecology of Computation*. Elsevier, New York, 1988.

A collection of papers on distributed computations viewed as "ecosystems" of interacting computational agents.

Iyengar84   S.S. Iyengar. *Computer Modeling of Complex Biological Systems*. CRC Press, Boca Raton, 1984.

Methods for programming primarily mathematical models of biological systems.

Jacobs72   W. Jacobs. "How a Bug's Mind Works." In H.W. Robinson and D.E. Knight, editors, *Cybernetics, Artificial Intelligence, and Ecology*, Spartan Books, 1972.

Describes the structure and behavior of PERCY, a simulated bug capable of learning. Emphasis is on adaptation in simple behavior generators. See the articles by Coderre and Travers in these proceedings.

Jacobson58   H. J. Jacobson. "On Models of Reproduction." *American Scientist*, 46(3):255–284, September 1958.

Description of a self-reproducing train set!

Jean84   R.V. Jean. *Mathematical Approach to Pattern and Form in Plant Growth*. John Wiley & Sons, New York, 1984.

Joyce84   G.F. Joyce and R.D. Tschirgi. "Automata Simulation of the Selection Process." *BioSystems*, 17:65–72, 1984.

Kampis87   G. Kampis and V. Csanyi. "Replication in Abstract and Natural Systems." *BioSystems*, 20:143–152, 1987.

Argues against validity of automaton models of self-reproduction.

Karplus86   M. Karplus and J.A. McMammon. "The Dynamics of Proteins." *Scientific American*, 254(4):42–51, April, 1986.

Computer simulation of the dynamic structure of proteins.

Kauffman69   S.A. Kauffman. "Metabolic Stability and Epigenesis in Randomly Constructed Genetic Nets." *J. Theor. Biol.*, 22:437–467, 1969.

Analysis of the dynamics of random, complex networks of Boolean automata. Such nets have come to be referred to under the general title of "The Kauffman Model," and constitute an active area of research, e.g., Derrida86a, Derrida86b, & Weisbuch86.

Kauffman84   S.A. Kauffman. "Emergent Properties in Random Complex Automata." *Physica D*, 10, 1984.

Reviews the surprisingly ordered dynamics of randomly assembled Boolean networks of automata, and discusses the emergence of powerful subautomata called "forcing-structures," which come to dominate the dynamics.

Kauffman86a   S.A. Kauffman. "Autocatalytic Sets of Proteins." *J. Theor. Biol.*, 119:1–24, 1986.

Discussion of the possibilities for the emergence of autocatalytic sets of proteins and their relevance to the origin of life.

Kauffman86b   S.A. Kauffman and R. G. Smith. "Adaptive Automata Based on Darwinian Selection." *Physica D*, 22:68–82, 1986.

Discussion of issues plus results of experimental selection for automata whose dynamical attractors match a predetermined target pattern.

Kauffman87a   S.A. Kauffman. "Developmental Logic and Its Evolution." *BioEssays*, 6(2):82–87, 1987.

The emergence of small, stable attractors in random, complex boolean networks as models of cell-types and the selective evolution of those cell-types.

Kauffman87b   S.A. Kauffman and S. Levin. "Towards a General Theory of Adaptive Walks on Rugged Landscapes." *J. Theor. Biol.*, 128:11–45, 1987.

Combinatorial optimization, whether in biological evolution or in the synaptic-weight-space of neural-net models, confronts a rugged fitness landscape. This article takes a step towards studying the statistical structure of such landscapes.

Kauffman89   S.A. Kauffman. *Origins of Order: Self-Organization and Selection in Evolution.* Oxford University Press, Oxford, 1989—in press.

Discusses the balance struck between self-organized properties in complex systems and natural selection, with implications for evolution. Topics include: fitness landscapes, evolution of catalytic activities, origin of life, dynamics of genetic regulatory-networks, and morphogenesis.

Kemeny55   J. G. Kemeny. "Man Viewed as a Machine." *Scientific American,* 192(4), April 1955.

Kohonen77   T. Kohonen. *Associative Memory.* Springer-Verlag, Berlin, 1977.

Kolata84   G. Kolata. "Esoteric Math has Practical Result." *Science*, 225:494–495, 1984.

Generation of the detailed structure of leaves and other natural objects by means of a fixed-point algorithm.

Kornfield81   W.A. Kornfield and C. Hewitt. "The Scientific Community Metaphor." *IEEE Transactions on Systems, Man, and Cybernetics*, SMC-11:24–33, 1981.

Analysis of the problem-solving capabilities of scientific communities and discussion of a parallel, distributed programming methodology that incorporates these capabilities.

Laing69   R. Laing. *Formalisms for Living Systems (Part I)*. Technical Report T.R. no. 08226-8-T, University of Michigan, 1969.

Discussion and comparison of several different modeling strategies for living systems based on formal computational systems.

Laing70   R. Laing. *Asexual and Sexual Reproduction Expressed in the von Neumann Cellular System (Formalisms for Living Systems (Part I) Sections 9.2–9.5)*. Technical Report, University of Michigan, 1970.

An extension to the preceeding paper introducing the capacity for sexual reproduction to the von Neumann model.

Laing71   R. Laing. "Formalisms for Biology: A Hierarchy of Developmental Processes." *Intern. J. Neuroscience*, 2:219–232, 1971.

Demonstration of the applicability of formal grammar theory and automata theory to the modeling of biological development.

Laing72   R. Laing. "Artificial Organisms and Autonomous Cell Rules." *Journal of Cybernetics*, 2(1):38–49, 1972.

A formal system for modeling features of biological development.

Laing73   R. Laing. "The Capabilities of Some Species of Artificial Organisms." *Journal of Cybernetics*, 3(2):16–25, 1973.

Description of the behavioral potential of artificial organisms composed of freely interacting elementary components enclosed in an expandable envelope.

Laing75a   R. Laing. "Artificial Molecular Machines: A Rapproachment between Kinematic and Tessellation Automata." In *Proceedings of the International Symposium on Uniformly Structured Automata and Logic*, Tokyo, August 1975.

Artificial molecules consisting of chains of Wang program statements are demonstrated to be computation-universal and are shown to be capable of constructing copies of themselves and each other.

Laing75b   R. Laing. "Some Alternative Reproduction Strategies in Artificial Molecular Machines." *J. Theor. Biol.*, 54:63–84, 1975.

A more accessible version of the Artificial Molecular Machine report.

Laing76   R. Laing. "Automaton Introspection." *Journal of Computer and System Sciences*, 13:172–183, 1976.

Discussion of the manner in which an automaton can determine its own structure.

Laing77   R. Laing. "Automaton Models of Reproduction by Self-Inspection." *J. Theor. Biol.*, 66:437–456, 1977.

Introduces a self-reproducing automaton which can reproduce without a description of itself. Rather, the machine generates its own description by self-inspection. Unlike von Neumann's system, this system will support Lamarckian inheritance.

Laing79   R. Laing. "Machines as Organisms: An Exploration of the Relevance of Recent Results." *BioSystems*, 11, 1979.

A good overview of Laing's work on artificial organisms, with commentaries and his responses.

Laing82   R. Laing. "Replicating Systems Concepts: Self-Replicating Lunar Factory and Demonstration." In R. Freitas and W.P. Gilbreath, editors, *Advanced Automation for Space Missions*, pages 189–335, NASA Conf. Pub. 2255, 1982.

A detailed report commissioned by NASA on the feasability of constructing self-replicating lunar factories.

Langton84   C.G. Langton. "Self-Reproduction in Cellular Automata." *Physica D*, 10(1-2):135–144, 1984.

Details of a simple self-reproducing CA configuration.

Langton86   C.G. Langton. "Studying Artificial Life with Cellular Automata." *Physica D*, 22:120–149, 1986.

   A preliminary investigation of the potential of CA for supporting life.

Langton87   C.G. Langton. "Virtual State Machines in Cellular Automata." *Complex Systems*, 1:257–271, 1987.

   Discussion of the importance of propagating structures in CA. Contains example of a Turing machine implemented on the model of protein synthesis using propagating structures.

Langton88a   C.G. Langton and K. Kelley. "Toward Artificial Life." *Whole Earth Review*, 58:74–79, 1988.

   Review of the first Artificial Life workshop at Los Alamos. Includes pictures and descriptions of many of the presentations.

Leduc11   Stephane Leduc. *The Mechanism of Life.* Rebman, London, 1911.

   A controversial work on biogenesis. For a review, see the article by Zeleny et al. in these proceedings.

LeeCY63   C.Y. Lee. "A Turing Machine which Prints Its Own Code Script." In *Mathematical Theory of Automata*, pages 155–164, Polytechnic Press, 1963.

Lehninger82   A.L. Lehninger. *Principles of Biochemistry.* Worth, New York, 1982.

   A classic textbook. The introduction defines biology as the attempt to uncover "the molecular logic of the living state."

Letvin59   J.Y. Letvin, H.R. Maturana, W.S. McCulloch, and W.H. Pitts. "What the Frog's Eye Tells the Frog's Brain." *Proc. I.R.E.*, 47:1940ff, 1959.

   Classic paper identifying functional groups of neurons which act as feature detectors.

Lindenmayer68   A. Lindenmayer. "Mathematical Models for Cellular Interactions in Development, I & II." *Journal of Theoretical Biology*, 18:280–315, 1968.

   Lindenmayer's original articles on L-Systems.

Lindenmayer71   A. Lindenmayer. "Developmental Systems without Cellular Interactions: Their Languages and Grammars." *J. Theor. Biol.*, 30:455–484, 1971.

   Description of context-free L-systems.

Lindenmayer76   A. Lindenmayer and G. Rozenberg, editors. *Automata, Languages, Development*, North-Holland, Amsterdam, 1976.

Early collection of papers on L-Systems.

Lindenmayer77   A. Lindenmayer. "Theories and Observations of Developmental Biology." In R.E. Butts and J. Hintikka, editors, *Foundational Problems in Special Sciences*, pages 103–118, D. Reidel, 1977.

On the role of mathematical theories in biology.

Lindenmayer82   A. Lindenmayer. "Developmental Algorithms: Lineage Versus Interactive Control Mechanisms." In S. Subtelny and P.B. Green, editors, *Developmental Order: Its Origin and Regulation*, pages 219–245, A.R. Liss, New York, 1982.

Discussion of context-free vs. context-sensitive developmental controls.

Lindsay68   R.K. Lindsay. "Artificial Evolution of Intelligence." *Contemporary Psychology*, 13(3):113–116, 1968.

A critical review of Fogel66 including an interesting discussion of the problem.

Lodish79   H.F. Lodish and J.E. Rothman. "The Assembly of Cell Membranes." *Scientific American*, 240(1):38–53, January 1979.

Details of low-level self-assembly of cell membranes.

Lofgren62   L. Lofgren. "Kinematic and Tesselation Models of Self-Repair." In E.E. Bernard and M.R. Kare, editors, *Biological Prototypes and Synthetic Systems*, Plenum, 1962.

Formal analysis of self-repairing automata.

Lovelock72   J.E. Lovelock. "Gaia as Seen through the Atmosphere." *Atmosphere and Environment*, 6:579–580, 1972.

Lovelock79   J.E. Lovelock. *Gaia: A New Look at Life on Earth*. Oxford University Press, Oxford, 1979.

Presentation of the theory that the Earth itself must be seen as a living organism.

Lovelock83   J.E. Lovelock. "Daisy World: A Cybernetic Proof of the Gaia Hypothesis." *The CoEvolution Quarterly*, 38:66–72, 1983.

Lumsden81   C.J. Lumsden and E.O. Wilson. *Genes, Mind, and Culture: The Coevolutionary Process.* Harvard University Press, Cambridge, 1981.

Sociobiology extended to human culture. See WilsonEO75.

MacDonald83   N. MacDonald. *Trees and Networks in Biological Models.* John Wiley, Chichester, 1983.

Mathematical properties of trees and networks and applications to biology. Includes two chapters on L-systems.

MacLaren78   L. MacLaren. *A Production System Architecture Based on Biological Examples.* Ph.D. thesis, University of Washington, 1978.

Mallove80   E.F. Mallove, R.L. Forward, Z. Paprotny, and J. Lehmann. "Interstellar Travel and Communication: A Bibliography." *J Brit. Interplanetary Soc.*, 33:201–248, 1980.

A compilation of over 2500 references on the topic of interstellar transportation and communication, broken down into categories including: origin of extra-solarlife, exotic biochemistries, alien life morphologies, possible alien artifacts, the "zoo" hypothesis(!), and so forth.

Mandlebrot78   B. Mandelbrot. *Fractals: Form, Chance, and Dimension.* W.H. Freeman, New York, 1978.

This and the following book detail in elegant form the manner in which extremely complex sturctures can emerge from relatively simple recursive rules.

Mandelbrot83   B. Mandelbrot. *The Fractal Geometry of Nature.* W.H. Freeman, New York, 1983.

Margolus84   N. Margolus. "Physics-Like Models of Computation." *Physica D*, 10:81–95, 1984.

Discussion of reversible cellular automata illustrated by an implementation of Fredkin's Billiard-Ball model of computation.

Margulis70   L. Margulis. *Origin of Eucaryotic Cells.* Yale University Press, New Haven, 1970.

Presents the thesis that eukaryotic cells are the result of the development of a symbiotic relationship between early prokaryotic cells.

Margulis81   L. Margulis. *Symbosis in Cell Evolution.* W.H. Freeman, San Francisco, 1981.

Further evidence for the symbiosis theory presented in Margulis70.

Margulis86   L. Margulis and D. Sagan. *Microcosmos: Four Billion Years of Evolution from Our Microbial Ancestors.* Summet Books, New York, 1986.

Marquand68   J. Marquand. *Life: its Nature, Origins, and Distribution (C.S.P. no. 17).* Oliver and Boyd, Edinburgh, 1968.

Martinez79   H.M. Martinez. "An Automaton Analogue of Unicellularity." *BioSystems,* 11:133–162, 1979.

An attempt to formalize the evolution of cellular organization.

Matela85   R.J. Matela and R. Ransom. "A Topological Model of Cell Division: Structure of the Computer Program." *BioSystems,* 18:65–78, 1985.

Maturana75   H.R. Maturana. "The Organization of the Living." *Int. J. Man-Mach. Stud.,* 7:313ff, 1975.

An early exposition of Autopoiesis: a theory of "self-producing" structures.

May76   R.M. May. "Simple Mathematical Models with Very Complicated Dynamics." *Nature,* 261:459ff, 1976.

Excellent early review of simple chaotic dynamical systems with consequences for ecological modeling. Reprinted in BaiLin84.

Maynard-Smith75   J. Maynard-Smith. *The Theory of Evolution—3rd edition.* Penguin, New York, 1975.

A classic.

Maynard-Smith82   J. Maynard-Smith. *Evolution and the Theory of Games.* Cambridge University Press, Cambridge, 1982.

Application of von Neumann's theory of games to the modeling of evolution. Another classic.

Maynard-Smith86   J. Maynard-Smith. *The Problems of Biology.* Oxford University Press, Oxford, 1986.

Excellent lay introduction to the fundamental problems of biology.

McCulloch43 W.S. McCulloch and W. Pitts. "A Logical Calculus of the Idea Immanent in Nervous Activity." *Bulletin of Mathematical Biophysics*, 5:115–133, 1943.

The original paper on McCulloch-Pitts neurons.

Meinhardt82 H. Meinhardt. *Models of Biological Pattern Formation.* Academic Press, New York, 1982.

MillerJG78 J.G. Miller. *Living Systems.* McGraw Hill, New York, 1978.

A detailed, system-theoretic approach to life.

MillerSM74 S.M. Miller and L.E. Orgel. *The Origins of Life on Earth.* Prentice-Hall, Englewood Cliffs, NJ, 1974.

Theory of the origin of life based on the results of the classic Miller/Urey/Orgel experiments on the pre-biotic synthesis of organic compounds.

Minsky67 M. Minsky. *Computation, Finite and Infinite Machines.* Prentice-Hall, Englewood, CA, 1967.

An introduction to automata theory written at a time when the emulation of natural systems was still a guiding principle.

Minsky69 M. Minsky and S. Papert. *Perceptrons: An Introduction to Computational Geometry.* MIT Press, Cambridge, 1969.

A treatise on the capabilities and limitations of one class of perceptron models. This book is widely—but unfairly—considered responsible for halting early AI research on connectionist/neural-net models of intelligence.

Minsky86 M. Minsky. *The Society of Mind.* Simon and Schuster, New York, 1986.

A loosley-coupled theory of intelligence based on the model of a distributed society of interacting "experts."

Monod71 J. Monod. *Chance and Necessity.* Alfred A. Knopf, New York, 1971.

A classic monograph on "necessary" vs. "accidental" features of living organisms.

Moore56   E.F. Moore. "Artificial Living Plants." *Scientific American*, 195(4): 118–125, October 1956.

Proposal for a species of self-reproducing machines which would mine and refine raw materials to construct offspring, and which would then be harvested for the materials from which they constructed themselves.

Moore64a   E.F. Moore. "Mathematics in the Biological Sciences." *Scientific American*, 211(3):148–164, September 1964.

Moore predicts that the most significant applications of mathematics in the biological sciences will stem from the analysis of problems in the theory of computing machines.

Moore64b   E.F. Moore. "The Firing Squad Synchronization Problem." In E.F. Moore, editor, *Sequential Machines—Selected Papers*, pages 213–214, Addison-Wesley, 1964.

This is a classic problem in distributed control: how to get a line of locally-communicating agents to all do something at the same time, using only local message passing.

Moore70   E.F. Moore. "Machine Models of Self-Reproduction." In A.W. Burks, editor, *Essays on Cellular Automata*, pages 187–203, University of Illinois Press, 1970.

Moore's proof of the existence of "Garden of Eden" configurations in cellular automata: configurations which cannot occur under the action of a specific CA rule. Moore also provides a way around Rosen's paradox (Rosen59).

Moravec88   H. Moravec. *Mind Children: The Future of Robot and Human Intelligence.* Harvard University Press, Cambridge, 1988.

A wide ranging treatise on the past, present, and future capabilities of robots by the director of Carnegie-Mellon's Robotics Institute. Moravec predicts that robots will replace the human race within the next century!

Morowitz59   H.J. Morowitz. "A Model of Reproduction." *American Scientist*, 47(2):261–263, 1959.

A simple extension to Jacobson's self-reproducing train set (Jacobson58).

Mortimer70   J.A. Mortimer. *A Cellular Model for Mammalian Cerebeller Cortex*. Technical Report LCG-107, The University of Michigan, 1970.

"*An automata-theoretic model of mammalian cortex is developed to investigate the role of the four principle neuronal populations in determining the spatio-temporal response of the cortex to natural and electrical inputs.*"

Mosqueira83   F. G. Mosqueira. "A Simple Competition Model Involving Selection." *Bull. Math. Biol.*, 45(1):51–67, 1983.

Myhill70   J. Myhill. "The Abstract Theory of Self-Reproduction." In A.W. Burks, editor, *Essays on Cellular Automata*, pages 206–218, University of Illinois Press, 1970.

Nakano78   K. Nakano. "Models of Self-Repairing Systems." In *Proceedings of the 4th International Joint Conference on Pattern Recognition, held in Kyoto, Japan, 1978*, pages 374–378, Kyoto University, Kyoto, Japan, 1978.

A cellular model in which a perturbed pattern of cells returns to the original pattern.

Newell63   A. Newell and H.A. Simon. "GPS: A Program that Simulates Human Thought." In E.A. Feigenbaum and J. Feldman, editors, *Computers and Thought*, McGraw-Hill, New York, 1963.

Probably the most influential paper establishing the "physical symbol system" computational paradigm as the underlying model of intelligence used by AI.

Newell80   A. Newell. "Physical Symbol Systems." *Cognitive Science*, 4:135–183, 1980.

A more recent presentation of the physical symbol system hypothesis. For the opposite view, see Hofstadter82.

NicolisG77   G. Nicolis and I. Prigogine. *Self-Organization in Nonequilibrium Systems*. John Wiley & Sons, New York, 1977.

Classic treatise on the dynamics of dissipative systems.

NicolisJS86   J.S. Nicolis. *Dynamics of Hierarchical Systems*. Springer-Verlag, Berlin, 1986.

Aproaches to studying the dynamics and structures of hierarchical systems composed of many interacting parts.

Niesert87 U. Niesert. "How Many Genes to Start with? A Computer Simulation about the Origin of Life." *Origins of Life*, 17:155–169, 1987.

Nijhout81 H. F. Nijhout. The Color Patterns of Butterflies and Moths. *Scientific American*, 245(5), November 1981.

Description of the way in which the wide variety of complex wing patterns of butterflies and moths emerge from the actions of a few simple rules.

Niklas86 K.J. Niklas. "Computer-Simulated Plant Evolution." *Scientific American*, 254(3):78–86, March 1986.

A model which traces the fossil record of the evolution of plant structure.

Oparin38 A.I. Oparin. *The Origin of Life*. Macmillan, New York, 1938.

Oparin's coacervate hypothesis on the origin of life. 2nd edition published by Dover in 1967.

Oppenheimer86 P. Oppenheimer. "Real Time Design and Animation of Fractal Plants and Trees." *Computer Graphics*, 20(4), August 1986.

Orgel73 L.E. Orgel. *The Origins of Life: Molecules and Natural Selection*. Wiley, New York, 1973.

Ostrand71 T.J. Ostrand. "Pattern Reproduction in Tesselation Automata of Arbitrary Dimension." *Journal of Computer and System Sciences*, 5(6), 1971.

Pabo86 C.O. Pabo and E.G. Suchanek. "Computer-Aided Model-Building Strategies for Protein Design." *Biochemistry*, 25:5987–5991, 1986.

Pagels88 H.R. Pagels. *The Dreams of Reason: The Computer and the Rise of the Sciences of Complexity*. Simon and Schuster, New York, 1988.

An excellent book reviewing the emerging synthesis of nonlinear science and bottom-up computational approaches in the study of complex systems.

Palmer85 E.M. Palmer. *Graphical Evolution*. John-Wiley & Sons, New York, 1985.

An introduction to the theory of random graphs—an important tool in modeling networks of interactions.

Papert80   S. Papert. *Mindstorms: Children, Computers, and Powerful Ideas*. Basic Books, New York, 1980.

Papert's description of his LOGO language and its applications in children's education.

PartridgeBL80a   B.L. Partridge. "The Effect of School Size on the Structure and Dynamics of Minnow Schools." *Anim. Behav.*, 28:68–77, 1980.

PartridgeBL80b   B.L. Partridge, T. Pitcher, J.M. Cullen, and J. Wilson. The Three-Dimensional Structure of Fish Schools. *Behavioral Ecology and Sociobiology*, 6:277–288, 1980.

Partridge84   D. Partridge, P.D. Lopez, and V.S. Johnston. "Computer Programs as Theories in Biology." *J. Theor. Biol.*, 108:539–564, 1984.

Discussion of the notion that a program itself constitutes a theory of the system whose behavior it emulates. The inadequacy of traditional programming languages for this role are identified, a more adequate methodology is proposed, and examples of its use are presented.

Pattee66   H. H. Pattee, E.A. Edelsack, L. Fein, and A.B. Callahan, editors. *Natural Automata and Useful Simulations*, Macmillan, London, 1966.

Proceedings of the Symposium on Fundamental Biological Models, held at the Stanford Biophysics Laboratory, June 17, 1965. This symposium was dedicated to "the process of modeling itself as an instrument for gaining understanding of the behavior of living matter."

Pattee85   H.H. Pattee. "Universal Principles of Measurement and Language Function in Evolving Systems." In J. Casti and A. Karlqvist, editors, *Complexity, Language, and Life: Mathematical Approaches*, pages 168–281, Springer-Verlag, Berlin, 1985.

Pattee88   H.H. Pattee. "Instabilities and Information in Biological Self-Organization." In F.E. Yates, editor, *Self-Organizing Systems: The Emergence of Order*, Plenum, New York, 1988.

Peacocke83   A. R. Peacocke. *An Introduction to the Physical Chemistry of Biological Organization*. Clarendon Press, Oxford, 1983.

Excellent introduction to the thermodynamics of living systems. Contains an extensive bibliography on biological organization.

Penrose58a L.S. Penrose and R. Penrose. "A Self-Reproducing Analogue." *Nature*, 179, 1957.

Penrose's first self-reproduction model, which is really more like crystal growth.

Penrose58b L.S. Penrose. "Mechanics of Self-Reproduction." *Ann. Human Genetics*, 23:59–72, 1958.

Penrose59a L.S. Penrose. "Self-Reproducing Machines." *Scientific American*, 200(6):105–113, June 1959.

Self-reproducing machines constructed out of simple mechanical blocks which can hook onto one another to form chains.

Penrose59b L.S. Penrose. "Automatic Mechanical Self-Reproduction." *New Biology*, 28:92–117, 1959.

Penrose60 L.S. Penrose. "Developments in the Theory of Self-Replication." *New Biology*, 31:57–66, 1960.

Pickover87a C. Pickover. "Biomorphs: Computer Displays of Biological Forms Generated from Mathematical Feedback Loops." *Computer Graphics Forum*, 5(4):313–316, 1987.

A computer graphics algorithm is used to create complicated forms resembling invertebrate organisms.

Pickover87b C. Pickover. "Mathematics and Beauty IV: Computer Graphics and Wild Monopodial Tendril Plant Growth." *Computer Graphics World*, 10(7):143–145, 1987.

Algorithms for the artistic rendering of plant and coral-like forms are described.

Pitcher73 T.J. Pitcher. "The Three-Dimensional Structure of Schools in the Minnow." *Anim. Behav.*, 21:673–686, 1973.

Plum72 T. Plum. *Simulations of a Cell-Assembly Model.* Ph.D. thesis, The University of Michigan, 1972.

A simulation based on Hebb's cell-assembly model, in which clusters of neurons perform computational tasks in the absence of any global controller.

Pollard65 E.C. Pollard. "The Fine Structure of the Bacterial Cell and the Possibility of its Artificial Synthesis." *American Scientist*, 53, 1965.

Pontecorvo58   G. Pontecorvo. "Self-Reproduction and All That." *Symposium of the Society for Experimental Biology*, 12:1–5, 1958.

Poundstone85   W. Poundstone. *The Recursive Universe*. William Morrow, New York, 1985.

Review of discoveries made in Conway's game of LIFE, interspersed with speculations on its implications for understanding the universe.

Preston84   K. Preston and M.J.B. Duff. *Modern Cellular Automata Theory and Applications*. Plenum Press, New York, 1984.

A good book on CA's, with applications primarily in the domain of image processing. Some discussion of LIFE-like CA rules, e.g. in hexagonal lattices.

Priese76   L. Priese. "On a Simple Combinatorial Structure Sufficient for Nontrivial Self-Reproduction." *J. Cybern.*, 6(1-2):101–137, 1976.

Self-reproduction in a reversible system.

Quillian68   M.R. Quillian. "Semantic memory." In M. Minsky, editor, *Semantic Information Processing*, pages 227–270, MIT Press, Cambridge, 1968.

An early discussion of semantic networks—dynamic distributed knowledge representation networks.

Ransom81   R. Ransom. *Computers and Embryos: Models in Developmental Biology*. John Wiley, Chichester, 1981.

Rastetter83   W.H. Rastetter. "Enzyme Engineering." *Appl. Biochem. and Biotech.*, 8:423–436, 1983.

Recombinant DNA technology makes possible the production of modified protein catalysts.

Reed67   J. Reed. "Simulation of Biological Evolution and Machine Learning." *J. Theoret. Biol.*, 17:319–342, 1967.

An early attempt to apply evolution to machine learning in the context of a card-game.

Reeke88   G. Reeke and G. Edelman. "Real Brains and Artificial Intelligence." *Daedalus*, 117:143–174, 1988.

Critique of traditional AI and a description of Edelman's Neuronal Group Selection hypothesis (Edelman87).

Reichardt78   J. Reichardt. *Robots: Fact, Fiction, and Prediction.* Penguin Books, London, 1978.

Contains a detailed timeline of the historical roots of automata and robots.

Reynolds87   C.W. Reynolds. "Flocks, Herds, and Schools: A Distributed Behavioral Model." *Computer Graphics: Proceedings of SIGGRAPH '87*, 21(4):25–34, July 1987.

Description of Reynold's bottom-up system for the study of flocking behavior.

Rizki85   M.M. Rizki and M. Conrad. "EVOLVE III: A Discrete Events Model of an Evolutionary Ecosystem." *BioSystems*, 18:121–133, 1985.

A continuation in the series of artificial ecosystem models initiated by Conrad (see Conrad 70a).

Rizki86   M.M. Rizki and M. Conrad. "Computing the Theory of Evolution." *Physica D*, 22:83–99, 1986.

A discrete event simulation of an evolving artificial ecosystem. In this model, the fitness of an "organism" is an emergent property of the ecosystem.

Rosen59   R. Rosen. "On a Logical Paradox Implicit in the Notion of a Self-Reproducing Automaton." *Bull. Math. Biophysics*, 21:387–394, 1959.

A paradox based on a logical "Catch-22." For responses see Guttman 66, Moore 70.

Rosen66   R. Rosen. "A Note on Replication in (M,R) Systems." *Bulletin of Mathematical Biophysics*, 28:149–151, 1966.

Rosen72   R. Rosen, editor. *Foundations of Mathematical Biology.* Academic Press, New York, 1972-73.

An edited collection in three volumes: I) Subcellular Systems, II) Cellular Systems, and III) Supercellular Systems.

Rosenberg67   R. Rosenberg. *Simulation of Genetic Populations with Biochemical Properties.* Ph.D. thesis, University of Michigan, 1967.

An early study of adaptation in which the adaptation takes place in the enzymatic control over a set of artificial chemical interactions. This system exhibits a genuine genotype/phenotype distinction.

Rosenblatt62   F. Rosenblatt. *Principles of Neurodynamics: Perceptrons and the Theory of Brain Mechanisms.* Spartan Books, Washington, 1962.

Early work on what would now be referred to as a "connectionist" model.

Rothstein79a   J. Rothstein. "Generalized Entropy, Boundary Conditions, and Biology." In R.D. Levine and M. Tribus, editors, *The Maximum Entropy Formalism,* pages 423–468, MIT Press, Cambridge, 1979.

Argues that novel features in living systems are a consequence of complexity—rather than specifically of biology—and that other complex systems should exhibit similar novel features.

Rothstein79b   J. Rothstein. "Generalized Life." *Cosmic Search,* 1:35–37,44–46, 1979.

Organisms viewed as self-replicating, computer-controlled heat engines able to play survival games.

Rothstein82   J. Rothstein. "Physics of Selective Systems: Computation and Biology." *Int. J. Theor. Physics,* 21(3-4):327–350, 1982.

"The statistical thermodynamics of systems displaying selective behavior is used to discuss some important ultimate physical limitations of computers and biological systems."

Rothstein85   J. Rothstein. "On the scientific validity and utility of the living earth concept and its further generalization." In *Proceedings of the symposium: Is the Earth a Living Organism? University of Massachusetts, Amherst, August 1-6 1985,* Northeast Audubon Center, Sharon, Connecticut, 06069, 1985.

A critical look at the Gaia ("Earth as living organism") hypothesis (Lovelock79).

Royal81   The Royal Society. "Theories of Biological Pattern Formation." *Philosophical Transactions of the Royal Society of London,* B295:425ff, 1981.

Rozenberg74   G. Rozenberg and A. Salomaa. *L-Systems: Lecture Notes in Computer Science no. 15.* Springer-Verlag, Berlin, 1974.

Rozenberg86   G. Rozenberg and A. Salomaa. *The Book of L.* Springer-Verlag, Berlin, 1986.

Collection of papers on L-systems.

Rozonoer79a   L.I. Rozonoer and E.I. Sedykh. "Evolution Mechanisms of Self-Repro-ducing Systems I." *Automation and Remote Control*, 40(2):243–251, 1979.

This and the following two articles treat evolution in a system consisting of a large number of self-reproducing entities.

Rozonoer79b   L.I. Rozonoer and E.I. Sedykh. "Evolution Mechanisms of Self-Repro-ducing Systems II." *Automation and Remote Control*, 40(3):419–429, 1979.

Rozonoer79c   L.I. Rozonoer and E.I. Sedykh. "Evolution Mechanisms of Self-Repro-ducing Systems III." *Automation and Remote Control*, 40(5):741–749, 1979.

SaganC   C. Sagan. "Life." Encyclopedia Britannica, 15th *ed.* Macropaedia, Vol. 10.

A very nice article on definitions of life.

Samuel59   A.L. Samuel. "Some Studies in Machine Learning using the Game of Checkers." *IBM J. Res. Dev.*, 3:210–229, 1959.

Samuel's original paper on machine learning using adaptation.

Samuel67   A.L. Samuel. "Some Studies in Machine Learning using the Game of Checkers II—Recent Progress." *IBM J. Res. Dev.*, 11:601–617, 1967.

Sampson76   J.R. Sampson. *Adaptive Information Processing*. Springer-Verlag, Berlin, 1976.

A collection of topics which constituted a one-semester under-graduate course on adaptive information processing. Topics in-clude the formal theory of automata, biological information pro-cessing, and artificial intelligence.

Sampson84   J.R. Sampson. *Biological Information Processing: Current The-ory and Computer Simulation*. Wiley & Sons, New York, 1984.

A book which grew out of—and updates—Sampson's previous book Sampson76.

Savageau76   M.A. Savageau. *Biochemical Systems Analysis*. Addison-Wesley, Reading, MA, 1976.

Methods for the analysis of catalyzed reaction networks.

Schrandt60   R.G. Schrandt and S.M. Ulam. "On Patterns of Growth of Figures in Two Dimensions." *Notices of the American Mathematical Society,* 7:642ff, 1960.

Early studies of pattern development in cellular automata, including competitions between two kinds of patterns. Reprinted in Burks 70.

Schrodinger44   E. Schrödinger. *What is Life?* Cambridge University Press, Cambridge, 1944.

A classic treatise on the nature of life.

Schuster86   P. Schuster. "Dynamics of Molecular Evolution." *Physica D,* 22:100–119, 1986.

Schwartz83   J.T. Schwartz. *A Taxonomic Table of Parallel Computers, Based on 55 Designs.* Technical Report, Courant Institute, New York University, 1983.

An excellent comparison of a wide variety of actual and proposed parallel architectures.

Scott86   A. Scott. *The Creation of Life: Past, Future, Alien.* Blackwell, Oxford, 1986.

Popular review of theories of the origin of life, together with an examination of the possibilities for the creation of novel forms of life.

Searle80   J. Searle. "Minds, Brains, and Programs." *Behavioral & Brain Sciences,* 3:417–458, 1980.

The classic "Chineese room" argument against AI programs being considered "intelligent."

SegelLA80   L.A. Segel, editor. *Mathematical Models in Molecular and Cellular Biology.* Cambridge University Press, Cambridge, 1980.

SegelLA84   L.A. Segel. *Modeling Dynamic Phenomena in Molecular and Cellular Biology.* Cambridge University Press, Cambridge, 1984.

A general approach to the mathematical modeling of dynamic biological phenomena. Emphasizes role of computers in modeling.

Shannon49   C.E. Shannon and W. Weaver. *A Mathematical Theory of Communication.* University of Illinois Press, Urbana, Illinois, 1949.

The classic introduction to Shannon's information theory, which generalizes Boltzman's measure of thermodynamic entropy S. Shannon's entropy measure H is used by many researchers attempting to quantify the notion of biological organization, e.g., see Gatlin 72.

Shannon56a   C.E. Shannon. "A Universal Turing Machine with Two Internal States." In C.E. Shannon and J. McCarthy, editors, *Automata Studies*, Princeton University Press, Princeton, 1956.

A proof that a Universal Turing Machine can be constructed using only two internal states.

Shannon56b   C.E. Shannon and J. McCarthy. *Automata Studies.* Princeton University Press, Princeton, 1956.

An early collection of papers on automata theory including articles by many of the early pioneers in the application of automata theory to the study of natural systems: Shannon, von Neumann, Ashby, Minsky, Moore, McCarthy, Kleene, and others.

Shapiro86   R. Shapiro. *Origins: A Skeptic's Guide to the Creation of Life on Earth.* Summit Books, New York, 1986.

A very readable survey of theories of the origin of life together with analysis and criticism.

Shaw62   E. Shaw. "The Schooling of Fishes." *Scientific American*, 206:128–138, June 1962.

Study of the development of schooling behavior in young fish.

Shklovskii66   I. Shklovskii and C. Sagan. *Intelligent Life in the Universe.* Holden-Day, San Francisco, 1966.

The case for life elsewhere in the universe.

Shoch82   J.F. Shoch and J.A. Hupp. "Computing Practices: The 'Worm' Programs—Early Experience with a Distributed Computation." *Comm. ACM*, 25(3):172–180, 1982.

Simon69   Herbert A. Simon. *The Sciences of the Artificial.* M.I.T. Press, Boston, 1969.

An excellent monograph on the nature of complexity and the study of artificial systems.

Simons83   G. Simons. *Are Computers Alive? Evolution and New Life Forms.* Birkhauser, Boston, 1983.

A must read. See the following description.

Simons85   G. Simons. *The Biology of Computer Life.* Birkhauser, Boston, 1985.

These two books by Simons discuss the computer's potential for exercising free will and feeling emotions. Although the emphasis is more on traditional AI and computational methodologies, everything discussed in these books is central to the study of Artificial Life.

Skarda87   C.A. Skarda and W.J. Freeman. "How Brains Make Chaos in Order to Make Sense of the World." *Behavioral and Brain Sciences,* 10:161–195, 1987.

SmithA68   A.R. Smith III. "Simple Computation-Universal Cellular Spaces and Self-Reproduction." In *Ninth Annual Symposium on Switching and Automata Theory,* pages 269–277, IEEE, 1968.

SmithA71   A.R. Smith III. "Simple Computation-Universal Cellular Spaces." *JACM,* 18(3):339–353, 1971.

Proof that 1D cellular automata are capable of supporting universal computation.

SmithA76   A.R. Smith III. "Introduction to and Survey of Polyautomata Theory." In A. Lindenmayer and G. Rozenberg, editors, *Automata, Languages, Development,* pages 405–422, North-Holland, 1976.

A good survey of work on cellular-automaton-like systems up to 1975.

SmithA84a   A.R. Smith III. "Plants, Fractals, and Formal Languages." *Computer Graphics,* 18(3), 1984.

Applications of formal grammars for generating graphic fractals — "graftals" — in the modeling of plants.

SmithA84b   A. R. Smith III. *Graftal Formalism Notes.* Technical Report 114, Computer Graphics Department, Lucasfilm Ltd., 1984.

Formal results in Smith's "graftal" system.

Smith80   S.F. Smith. *A Learning System Based on Genetic Adaptive Algorithms.* Ph.D. thesis, University of Pittsburgh, 1980.

Spiegelman67   S. Spiegelman. "An *In-Vitro* Analysis of a Replicating Molecule." *American Scientist*, 55:3–68, 1967.

Description of "Spiegelman's Monster."

Stahl63   W.R. Stahl and H.E. Goheen. "Molecular Algorithms." *Journal of Theoretical Biology*, 5:266–287, 1963.

"Turing Machines are used to model 'algorithmic enzymes' which transform biochemicals represented as letter strings." Several dozen enzymes collected into a primitive "cell" exhibit a kind of "logical homeostasis."

Stahl64   W.R. Stahl, R.W. Coffin, and H.E. Goheen. "Simulation of Biological Cells by Systems Composed of String-Processing Finite Automata." In *AFIPS Conference Proceedings Vol 25—1964 Spring Joint Computer Conference*, pages 89–102, 1964.

Strings representing DNA, RNA, proteins and other biochemicals are processed by "algorithmic enzymes."

Stahl65a   W.R. Stahl. "Algorithmically Unsolvable Problems for a Cell Automaton." *J. Theoret. Biol.*, 8:371–394, 1965.

A Turing-Machine-based simulation of a cell is presented and it is demonstrated that if the cell were to attempt hereditary adaptation it would be faced with certain unsolvable problems.

Stahl65b   W.R. Stahl. "Self-Reproducing Automata." In *Perspectives in Biology and Medicine*, University of Chicago Press, 1965.

Overview of self-reproducing automata and discussion of his automata-based simulation of cellular self-reproduction.

Stahl67   W. R. Stahl. "A Computer Model of Cellular Self-Reproduction." *J. Theoret. Biol.*, 14:187–205, 1967.

A Turing-Machine-based simulation of a cell with 46 genes which is capable of repeated self-reproduction and differentiation.

Stanley86   H.E. Stanley and N. Ostrowsky, editors. *On Growth and Form*. Martinus Nijhoff Publishers, Boston, 1986.

Proceedings of a summer school held in Corsica, France in 1985 on the wide variety of complicated forms and patterns that emerge in systems whose fundamental components are essentially engaged in random walks.

Stevens74   P.S. Stevens. *Patterns in Nature*. Little, Brown and Co., Boston, 1974.

Sutton81   R.S. Sutton and A.G. Barto. "Toward a Modern Theory of Adaptive Networks: Expectation and Prediction." *Psychological Review*, 88:135–170, 1981.

Thatcher63   J. Thatcher. "The Construction of a Self-Describing Turing Machine." In *Proceedings of the Symposium on Mathematical Theory of Automata, New York, April 1962*, Polytechnic Press, New York, 1963.

Thatcher70a   J. Thatcher. "Self-Describing Turing Machines and Self-Reproducing Cellular Automata." In A.W. Burks, editor, *Essays on Cellular Automata*, pages 103–131, University of Illinois Press, 1970.

Thatcher70b   J. Thatcher. "Universality in the von Neumann Cellular Model." In A.W. Burks, editor, *Essays on Cellular Automata*, pages 132–186, University of Illinois Press, 1970.

Thom75   R. Thom. *Structural Stability and Morphogenesis*. W.A. Benjamin, Reading, Mass., 1975.

The presentation of "Catastrophe Theory" by its inventor.

Thomas74   L. Thomas. *The Lives of a Cell*. Viking, New York, 1974.

Philosophical, poetic, thought-provoking essays on biology.

Thompson42   d'Arcy. W. Thompson. *Growth and Form*. Macmillan, New York, 1942.

Influential text on morphogenesis. Shows how structural changes during development can be modeled as mapping transformations.

Thompson85   R.L. Thompson and N.S. Goel. "A Simulation of t4 Bacteriophage Assembly and Operation." *BioSystems*, 18:23–45, 1985.

Thompson88   R.L. Thompson and N.S. Goel. "Movable Finite Automata (MFA) Models for Biological Systems I: Bacteriophage Assembly and Operation." *J. Theor. Biol.*, In press, 1988.

See also Goel88b as well as Goel and Thompsons' contribution to these proceedings.

Toffoli77   T. Toffoli. *Cellular Automata Mechanics*. Ph.D. thesis, The University of Michigan, 1977.

Toffoli's demonstration of reversible universal computation.

Toffoli80 T. Toffoli. "Reversible Computing." In De Bakker and Van Leeuwen, editors, *Automata, Languages, and Programming*, pages 632–644, Springer-Verlag, 1980.

Toffoli84a T. Toffoli. "CAM6: A High-Performance Cellular-Automaton Machine." *Physica D*, 10, 1984.
Description of the architecture of a hardware CA machine.

Toffoli84b T. Toffoli. "Cellular Automata as an Alternative to (Rather than an Approximation of) Differential Equations in Modeling Physics." *Physica D*, 10, 1984.
An insightful discussion of the reasons for and against modeling physical systems with differential equations.

Toffoli87 T. Toffoli and N. Margolus. *Cellular Automata Machines*. MIT Press, Cambridge, 1987.
An excellent book on the application of cellular automata to modeling physical systems.

Toombs67 R. Toombs and N.A. Barricelli. "Simulation of Biological Evolution and Machine Learning." *J. Theoret. Biol.*, 17:319–342, 1967.

Turing37a A.M. Turing. "On Computable Numbers, with an Application to the Entscheidungsproblem." *Proceedings of the London Mathematical Society*, Series 2(42):230–265, 1936-1937.
Turing's famous demonstration of the formal limits on computation based on a proof that the halting problem is undecidable. Reprinted in Davis65.

Turing37b A.M. Turing. "A Correction." *Proceedings of the London Mathematical Society*, Series 2(43):544–546, 1937.
A correction to the previous paper.

Turing50 A.M. Turing. "Can a Machine Think?" *Mind*, 433–460, October 1950.
The paper in which Turing introduced the "Turing test" for machine intelligence. Reprinted in Feigenbaum63.

Turing52 A.M. Turing. "The Chemical Basis of Morphogenesis." *Phil. Trans. Roy. Soc. of London*, Series B: Biological Sciences(237):37–72, 1952.
A reaction-diffusion model for development.

Tsypkin71   Y.Z. Tsypkin. *Adaptation and Learning in Automatic Systems.* Academic Press, New York, 1971.

UlamSM62   S.M. Ulam. "On Some Mathematical Problems Connected with Patterns of Growth of Figures." *Proceedings of Symposia in Applied Mathematics*, 14:215–224, 1962.

An early study on pattern development in cellular automata by the man who suggested CA's to von Neumann. Reprinted in Burks 70.

Ulmer83   K.M. Ulmer. "Protein Engineering." *Science*, 219:666–671, 1983.

Urey52   H.C. Urey. "The Early Chemical History of the Earth and the Origin of Life." *Proc. Nat. Acad. Sci*, 38:351ff, 1958.

Extensions to Oparin's coacervate hypothesis for the origin of life.

Valdez80   F. Valdez and R. Freitas. "Comparison of Reproducing and Nonreproducing Starprobe Strategies for Galactic Exploration." *J. British Interplanetary Soc.*, 33:402–408, 1980.

VanOver80   R. Van Over. *Sun Songs: Creation Myths from Around the World.* New American Library, New York, 1980.

A very nice survey of creation myths.

Varela74   F.J. Varela, H.R. Maturana, and R. Uribe. "Autopoiesis: The Organization of Living Systems." *BioSystems*, 5(4):187–196, 1974.

One of the first presentations of the concept of autopoiesis.

Vartanian60   A. Vartanian. *La Mettrie's L'Homme Machine: A Study in the Origins of an Idea.* Princeton University Press, Princeton, 1960.

The classic 18th century treatise on man as machine. Translated into English with an introductory monograph and notes.

Vichniac84   G.Y. Vichniac. "Simulating Physics with Cellular Automata." *Physica D*, 10:96–116, 1984.

Bottom-up approach to modeling physical phenomena.

Vitanyi73   P.M.B. Vitayni. "Sexually Reproducing Cellular Automata." *Math. Biosci.*, 18:23–54, 1973.

vonNeumann45   J. von Neumann. *First Draft of a Report on the EDVAC*. Technical Report, University of Pennsylvania, 1945.

The report that got von Neumann's name associated with the serial, stored-program, general purpose, digital architecture upon which 99.99% of all computers today are based.

vonNeumann51   J. von Neumann. "The General and Logical Theory of Automata." In *Cerebral Mechanisms in Behavior*, pages 1–41, Wiley, New York, 1941.

vonNuemann56   J. von Neumann. "Probabilistic Logics and the Synthesis of Reliable Organisms from Unreliable Components." In C. Shannon and J. McCarthy, editors, *Automata Studies*, Princeton University Press, Princeton, 1956.

vonNeumann58   J. von Neumann. *The Computer and the Brain*. Yale University Press, New Haven, 1958.

A classic book comparing the architecture of computers with the architecture of the brain.

vonNeumann66   J. von Neumann. *Theory of Self-Reproducing Automata*. University of Illinois Press, Urbana, 1966.

Von Neumann's work on self-reproducing automata, completed and edited after his death by Arthur Burks. Also includes transcripts of von Nuemann's 1949 University of Illinois lectures on the "Theory and Organization of Complicated Automata."

vonTiesen80   G. von Tiesenhausen and W. Darbo. *Self-Replicating Systems— A Systems Engineering Approach*. Technical Report, NASA TM-78304, 1980.

Waddington68   C. H. Waddington, editor. *Towards a Theoretical Biology: Vols. 1-4*. Edinburgh University Press, Edinburgh, 1968, 1969, 1970, 1972.

A Classic. Proceedings of an enlightened series of meetings organized by Waddington just as the "dark-ages" were setting in. These meetings were attended by many of the early pioneers in Artificial Life.

Waksman69   A. Waksman. "A Model of Replication." *JACM*, 16:178–188, 1969.

WalterC69a   C. Walter. "Stability of Controlled Biological Systems." *J. Theor. Biol.*, 23:23–28, 1969.

WalterC69b   C. Walter. "The Absolute Stability of Certain Types of Controlled Biological Systems." *J. Theor. Biol.*, 23:39–52, 1969.

Walter50   W.G. Walter. "An Imitation of Life." *Scientific American*, 182(5):42–45, May 1950.

Reports experiments with two autonomous, electronic "turtles"— Elmer & Elsie—equipped with various sensors and simple control circuitry which wander around their environment and interact with one another.

Walter51   W.G. Walter. "A Machine that Learns." *Scientific American*, 185(2):60–63, August 1951.

Further experiments with Elmer & Elsie.

Watson87   J.D. Watson, N.H. Hopkins, J.W. Roberts, J.A. Steitz, and A.M. Weiner. *Molecular Biology of the Gene (4th Edition), Vols. I & II.* Benjamin/Cummings, Menlo Park, 1987.

The complete reference to current knowledge about the mechanisms of life. See also Alberts83.

Weinberg70a   R. Weinberg. *Computer Simulation of a Living Cell.* Ph.D. thesis, The University of Michigan, 1970.

Computer simulation of Escherichia Coli.

Weinberg70b   R. Weinberg. *Computer Simulation of a Primitive, Evolving Eco-System.* Technical Report T.R. no. 03296-6-T, The University of Michigan, 1970.

Application of the genetic algorithm to simulated populations of bacteria. The 40 loci on the genetic instruction string are indexed in order to make use of the inversion operator.

Weinberg71   R. Weinberg, B.P. Zeigler, and R.A. Laing. "Computer Simulation of a Living Cell: Metabolic Control System Experiments." *Journal of Cybernetics*, 1(2):34–48, 1971.

Weisbuch86   G. Weisbuch. "Mini Review: Networks of Automata and Biological Organization." *J. Theoret. Biol.*, 121:255–267, 1986.

Argues that "a conceptual breakthrough has been obtained by comparing the properties of organization of living systems to the dynamical properties of random or complex networks of automata."

West70   M.W. West and C. Ponnamperuma. "Bibliography on the Origin of Life." *Space Life Sci.*, 2:225–295, 1970.

A bibliography of some 1600 references to work on the origin of life. Supplement published in the same journal V. 3, pp 293-304 (1972).

Whyte69   L.L. Whyte, A.G. Wilson, and D. Wilson, editors. *Hierarchical Structures*. Elsevier, New York, 1969.

Wiener48   N. Wiener. *Cybernetics, or Control and Communication in the Animal and the Machine*. John Wiley, New York, 1948.

Wiener's classic book on cybernetics. Second edition with additions published in 1961.

Wigner61   E.P. Wigner. "The Probability of the Existence of a Self-Reproducing Unit." In *The Logic of Personal Knowledge*, pages 231–248, The Free Press, 1961.

WilsonEO71   E.O. Wilson. *The Insect Societies*. Belknap/Harvard University Press, Cambridge, 1971.

Social insects provide some of the most accessible natural examples of the manner in which complex behavior emerges from the collective behavior of relatively simple entities. This book provides a detailed overview of the social insects and includes an extensive bibliography.

WilsonEO75   E.O. Wilson. *Sociobiology: The New Synthesis*. Belknap/Harvard University Press, Cambridge, 1975.

The classic treatise on the genetic determination of social behavior. Sociobiology has started an enormous controversy within the social sciences concerning the relative importance of nature (genetics) vs. nurture (culture) in the determination of behavior.

WilsonEO85   E. O. Wilson. "The Sociogenesis of Insect Colonies." *Science*, 288(4707):1489–1495, 1985.

Sociogenesis is "the process by which colony members undergo changes in caste, behavior, and physical location incident to colonial development." Wilson argues that "some of the processes can also be usefully compared with morphogenesis at the levels of cells and tissues."

WilsonSW85   S.W. Wilson. "Knowledge Growth in an Artificial Animal." In J.J. Grefenstette, editor, *Proceedings of an International Conference on Genetic Algorithms and their Applications*, Carnegie-Mellon University, July 24-26 1985.

Winfree80   A.T. Winfree. *The Geometry of Biological Time*. Springer-Verlag, Berlin, 1980.

Dynamics of biological oscillators.

Winfree87   A. T. Winfree. *When Time Breaks Down*. Princeton University Press, Princeton, 1987.

Dysfunction in biological oscillators.

Wolfram83   S. Wolfram. "Stastical Mechanics of Cellular Automata." *Rev. Modern Physics*, 55:601–644, 1983.

Important paper largely responsible for the resurgence of interest in cellular automata.

Wolfram84a   S. Wolfram. "Cellular Automata as Models of Complexity." *Nature*, 311(4):419–424, 1984.

A well written account of the manner in which complex dynamics can emerge from simple components.

Wolfram84b   S. Wolfram. "Universality and Complexity in Cellular Automata." *Physica D*, 10:1–35, 1984.

Identifies four qualitative classes of CA dynamics.

Wolfram86   S. Wolfram, editor. *Theory and Applications of Cellular Automata*. World Scientific, Singapore, 1986.

Collection of papers on CA's. Contains an extensive bibliography.

Wolpert68   L. Wolpert. "The French Flag Problem: A Contribution to the Discussion on Pattern Development and Recognition." In C.H. Waddington, editor, *Towards a Theoretical Biology*, vol. I, pages 125–133, Edinburgh University Press, 1968.

Wolpert78   L. Wolpert. "Pattern Formation in Biological Development." *Scientific American*, 239(4):154–164, October 1978.

This and the preceding paper discuss the manner in which the context in which cells find themselves affects their development.

Wood80   W.A. Wood. "Bacteriophage T4 Morphogenesis as a Model for Assembly of Subcellular Structure.' *Quart. Rev. Biol.*, 55:353–367, 1980.

Woods75   W.A. Woods. "What's in a Link?" In E.G. Bobrow and A.M. Collins, editors, *Representation and Understanding: Studies in Cognitive Science*, pages 35–82, Academic Press, 1975.

Nice discussion of semantic networks.

Zeigler70   B.P. Zeigler and R. Weinberg. "System Theoretic Analysis of Models: Computer Simulation of a Living Cell." *J. Theor. Biol.*, 29:35–56, 1970.

Application of the concepts of homomorphism and aggregation to the modeling of a living cell: Escherichia Coli.

Zeigler76   B.P. Zeigler. *Theory of Modeling and Simulation.* Wiley & Sons, New York, 1976.

Presents the "Discrete Event" modeling scheme.

Zeleny77a   M. Zeleny. "Self-Organization of Living Systems: A Formal Model of Autopoiesis." *Int. J. General Systems*, 4:13–28, 1977.

A formal computer implementation and extension of the original Varela-Maturana-Uribe model of "autopoiesis"—the theory of "self-producing" systems: i.e., systems whose high-level functions include the maintainence of the low-level organization which supports the high-level functions.

Zeleny77b   M. Zeleny. "Organization as an Organism." In *Proceedings of the SGSR Annual Meeting, Denver Colo., 1977*, pages 262–270, Society for General Systems Research, Washington, D.C., 1977.

Basic notion of social autopoiesis and an institutional example.

Zeleny78   M. Zeleny. "APL-Autopoiesis: Experiments in Self-Organization of Complexity." *Progress in Cybernetics and Systems Research*, III, 1978.

Zeleny81a   M. Zeleny, editor. *Autopoiesis: A Theory of Living Organization.* Elsevier, New York, 1981.

A multi-author discussion of autopoiesis. A comprehensive summary of "self-producing" systems and their modeling.

Zeleny81b M. Zeleny. "Autopoiesis Today." *Cybernetics Forum: Special Issue devoted to Autopoiesis*, 10(2/3):3–6, 1981.

This volume—edited by Zeleny—contains reprints of classical papers on autopoiesis, including the works of Maturana, Varela, Uribe, and Zeleny. Also contains a comprehensive annotated bibliography.

SECTION 2:
AN ANNOTATED BIBLIOGRAPHY OF
PLANT MODELING AND GROWTH SIMULATION BY
ARISTID LINDENMAYER AND PRZEMYSLAW PRUSINKIEWICZ

1. General
2. Surveys
3. Theory of L-systems
    - 3.1 Books
    - 3.2. Bibliographies
    - 3.3. Papers
4. Geometrical interpretation of L-systems
5. L-systems in developmental analysis
6. Multidimensional L-systems
7. Synthesis of realistic plant images
    - 7.1. Synthesis methods based on L-systems
    - 7.2. Other synthesis methods
    - 7.3. Organ models
8. Related botanical papers
    - 8.1. Plant architecture
    - 8.2. Phylotaxis
    - 8.3. Inflorescences
    - 8.4. Leaves
    - 8.5. Population dynamics and modular organisms.

1. GENERAL

R. V. Jean. *Mathematical Approach to Pattern and Form in Plant Growth.* John Wiley and Sons, New York, 1984.

A. Lindenmayer. "Theories and Observations of Developmental Biology." In: R. E. Butts and J. Hintikka, editors, *Foundational Problems in Special Sciences.* D. Reidel Publ. Co., Dordrecht-Holland, pages 103–118, 1977.

On the role of mathematical theories in biology.

B. B. Mandelbrot. *The Fractal Geometry of Nature.* W. H. Freeman, San Francisco, 1982.

P. S. Stevens. *Patterns in Nature.* Little, Brown and Co., Boston, 1974.

d'Arcy Thompson. *On Growth and Form.* At the University Press, Cambridge, 1952.

2. SURVEYS

A. R. Smith. "Formal Geometric Languages for Natural Phenomena." In: *SIGGRAPH '87 Course Notes on the Modeling of Natural Phenomena,* 1987.

Personal reflections on the role of rewriting concepts in geometry, specifically as applied to the modeling of plants. Emphasizes the role of L-systems. Contains details on programs Gene and mktree, and an annotated graftal bibliography.

A. Fournier. "Prolegomenon." In: *SIGGRAPH '87 Course Notes on the Modeling of Natural Phenomena,* 1987.

A general introduction to the field of natural phenomena modeling, with an emphasis on fractals and the modeling of biological structures.

The following papers survey the development of ideas related to the L-systems in the biological context.

A. Lindenmayer. "Developmental Algorithms for Multicellular Organisms: A Survey of L-Systems." *Journal of Theoretical Biology* 54: 3-22, 1975.

A. Lindenmayer. "Algorithms for Plant Morphogenesis." In: R. Sattler, editor: *Theoretical Plant Morphology.* Leiden University Press, The Hague, pages 37–81, 1978.

A. Lindenmayer. "Developmental Algorithms: Lineage Versus Interactive Control Mechanisms." In: S. Subtelny and P. B. Green, editors: *Developmental Order: Its Origin and Regulation.* Alan R. Liss, Inc., New York, pages 219–245, 1982.

A. Lindenmayer. "Models for Multicellular Development: Characterization, Inference and Complexity of L-Systems." In: A. Kelmenová and J.Kelmen, editors: *Trends, Techniques and Problems in Theoretical Computer Science,* Lecture Notes in Computer Science 281:138–168, Springer-Verlag, Berlin, 1987.

N. Macdonald. *Trees and Networks in Biological Models.* J. Wiley, New York, 1983.

> *Chapters 18 and 19 present a good introduction to L-systems from the biological perspective.*

3. THEORY OF L-SYSTEMS

3.1. BOOKS.

G. T. Herman and G. Rozenberg. *Developmental Systems and Languages.* North-Holland, Amsterdam, 1975.

> *The first and a very useful textbook on L-systems.*

A. Lindenmayer and G. Rozenberg, editors. *Automata, Languages, Development.* North-Holland, Amsterdam, 1976.

G. Rozenberg and A. Salomaa. *The Mathematical Theory of L-Systems.* Academic Press, New York, 1980.

> *A more recent monograph of L-systems. Emphasis is put on 0L-systems.*

G. Rozenberg and A. Salomaa, editors. *The Book of L.* Springer-Verlag, Berlin, 1986.

A. Salomaa. *Formal Languages.* Academic Press, New York, 1973.

> *Contains a very lucid introduction to L-systems.*

P. M. B. Vitányi. *Lindenmayer Systems: Structure, Languages and Growth Functions.* Mathematical Centre, Amsterdam, 1980.

3.2. BIBLIOGRAPHIES.

G. Rozenberg, M. Penttonen and A. Salomaa. "Bibliography of L Systems." *Theoretical Computer Science* 5:339–354, 1977.

G. Rozenberg, J. Mäenpää, and A. Salomaa. "Supplementary Bibliography of L Systems." *Bull. Europ. Assoc. Theor. Comp. Sci. (Leiden)* 13: 64–79, 1981.

3.3. PAPERS.

A. Lindenmayer. "Mathematical Models for Cellular Interaction in Development, Parts I and II." *Journal of Theoretical Biology* 18:280–315, 1968.

> *The original paper on L-systems.*

P. Eichhorst and W. J. Savitch. "Growth Functions of Stochastic Lindenmayer Systems." *Information and Control* 45:217–228, 1980.

T. Yokomori. "Stochastic Characterizations of E0L Languages." *Information and Control* 45:26–33, 1980.

> *The above two papers introduce a definition of stochastic L-systems which is useful when adding specimen-to-specimen variation to models (cf. Nishida [1980], Prusinkiewicz [1987]):*

H. Jürgensen and A. Lindenmayer. "Inference Algorithms for Developmental Systems with Cell Lineages." *Bulletin of Mathematical Biology* 49, Nr. 1:93–123, 1987.

> *The problem is to find a D0L-system generating an observed sequence of tree structures.*

D. Wood. "Time-Delayed 0L Languages and Sequences." *Information Sciences* 8:271–281, 1975

D. Wood. "Generalized Time-Delayed 0L Languages." *Information Sciences* 12:151–155, 1977.

K. Culik, II and D. Wood. "Speed-Varying 0L Systems." *Information Sciences* 14:161–170, 1978.

> *The above three papers introduce the concept of time-delayed L-systems. It may be worthwhile to look at them again in the light of simulation of continuous growth.*

N. D. Jones and S. Skyum. "Complexity of Some Problems Concerning L-Systems." *Mathematical Systems Theory* 13:29-43, 1979.

> *A review of results on computation and complexity of L-systems.*

P. M. B. Vitányi. "Development, growth and time." In G. Rozenberg and A. Salomaa [Eds.]: *The Book of L.* Springer-Verlag, Berlin, pages 431–444, 1986.

> *An attempt to obtain the sigmoidal growth curve using L-systems; the concept of slowing down the progress of time at a certain age seems artificial.*

4. GEOMETRICAL INTERPRETATIONS OF L-SYSTEMS

L-systems were conceived for the purpose of describing development of plants, but originally this description was confined to the topological level. In order to present simulated plant development using computer-generated images it is necessary to specify a geometrical interpretation of L-systems as well. The first approaches—by Hogeweg and Hesper [1974], and Frijters and Lindenmayer [1974]—used the same branching angle throughout the entire structure. More flexible approaches are listed below.

A. L. Szilard and R. E. Quinton. "An Interpretation for DOL Systems by Computer Graphics." *The Science Terrapin* 4:8–13, 1979.

> *The original paper introducing the concept of using the symbols generated by an L-system to specify directions and angles between line segments.*

R. Siromoney and K. G. Subramanian. "Space-Filling Curves and Infinite Graphs." In H. Ehrig, M. Nagl and G. Rozenberg (Eds.): *Graph Grammars and Their Application to Computer Science; Second International Workshop.* Lecture Notes in Computer Science 153, Springer-Verlag, Berlin, pages 380-391, 1983.

> *Chain coding is used to graphically interpret strings generated by L-systems.*

H. Freeman. "On Encoding Arbitrary Geometric Configurations." *IRE Trans. Electron. Comput.* 10:260–268, 1961.

> *The classic paper on chain coding.*

P. Prusinkiewicz. "Graphical Applications of L-Systems." *Proceedings of Graphics Inter-face '86 — Vision Interface '86*:247–253, 1986.

> *Turtle geometry is used as the basis for interpreting strings gen-erated by L-systems. In contrast to chain coding which relies on absolute directions, in turtle geometry directions are specified relative to the previous ones.*

H. Abelson and A. A. diSessa. *Turtle Geometry.* M.I.T. Press, Cambridge, 1982.

> *The fundamental book on turtle geometry.*

H. A. Maurer, G. Rozenberg, and E. Welzl. "Using String Languages to Describe Picture Languages." *Information and Control* 54:155–185, 1982.

I. H. Sudborough and E. Welzl. "Complexity and Decidability for Chain Code Picture Lan-guages." *Theoretical Computer Science* 36:173–202, 1985.

> *The above two papers formally analyze the properties of pictures generated by Chomsky grammars under the chain interpretation. It would be interesting to apply a similar approach to L-systems.*

F. M. Dekking. *Recurrent Sets: A Fractal Formalism.* Delft University of Technology, Report 82-32, 1982.

F. M. Dekking. "Recurrent Sets." *Advances in Mathematics* 44, Nr. 1:78–104, 1982.

> *Two papers on generating fractals using string languages; a fairly formal approach.*

A. R. Smith. *Graftal Formalism Notes.* Technical Memo Nr. 114, Lucasfilm Computer Division, San Rafael, CA, 1984.

5. L-SYSTEMS IN DEVELOPMENTAL ANALYSIS

The subsequent papers present a progression of L-system-based techniques from a simple model of a context-free linear structure (blue-green bacteria) to a complex context-sensitive branching structure (*Mycelis muralis*).

R. Baker and G. T. Herman. "Simulation of Organisms using a Developmental Model, Parts I and II." *Int. J. of Bio-Medical Computing* 3:201–215 and 251–267, 1972.

G. T. Herman and W. H. Liu. "The Daughter of CELIA, the French Flag, and the Firing Squad." *Simulation* 21:33–41, 1973.

A. Lindenmayer. "Adding Continuous Components to L-Systems." In: G. Rozenberg and A. Salomaa, editors: *L Systems*, Lecture Notes in Computer Science 15, Springer-Verlag, Berlin, pages 53-68, 1974.

> *The above three papers present details of the first program for simulating plant development with L-systems, called CELIA.*

D. Frijters and A. Lindenmayer. "A Model for the Growth and Flowering of *Aster novae-angliae* on the Basis of Table (1, 0) L-Systems." In: G. Rozenberg and A. Salomaa, editors: *L Systems*, Lecture Notes in Computer Science 15, Springer-Verlag, Berlin, pages 24–52, 1974.

> *One of the first papers including computer-generated plots of plants generated using L-systems.*

D. Frijters. "Principles of Simulation of Inflorescence Development." *Annals of Botany* 42:549–560, 1978.

> *Delays, sequences of stages, developmental switches implemented using table 0L-systems and CELIA-style numerical attributes of segments are discussed, but all models presented are interactionless.*

D. Frijters. "Mechanisms of Developmental Integration of *Aster novae-angliae* L. and *Hieracium murorum* L." *Annals of Botany* 42:561–575, 1978.

> *This paper discusses mechanisms which control plant development, including delays, interaction and developmental switches, but does not specify them in the form of ready-to-use L-systems. However, many mathematical relationships between parts of the analyzed plants are given.*

A. Lindenmayer. "Positional and Temporal Control Mechanisms in Inflorescence Development." In: P. W. Barlow and D. J. Carr , editors: *Positional Controls in Plant Development*. University Press, Cambridge, 1984.

J. M. Janssen and A. Lindenmayer. "Models for the Control of Branch Positions and Flowering Sequences of Capitula in *Mycelis muralis* (L.) Dumont (Compositae)." *New Phytologist* 105:191–220, 1987.

> *The second paper is a refinement of the first one; both use Mycelis muralis as an example.*

Some other papers using L-systems to analyze biological development are listed below.

D. Frijters and A. Lindenmayer. "Developmental Descriptions of Branching Patterns with Paracladial Relationships." In: A. Lindenmayer and G. Rozenberg, editors: *Automata, Languages, Development*. North-Holland, Amsterdam, pages 57–73, 1976.

> *Paracladial relations describe self-similarities in branching structures on the topological level. It would be interesting to investigate the relationship between paracladial relations and geometric self-similarity as found in fractals.*

A. Lindenmayer. "Paracladial Relationships in Leaves." *Ber. Deutsch Bot. Ges. Bd.* 90:287–301, 1977.

> *Self-similarity of leaf structures is described in terms of D0L-systems. A simple L-system describing leaf topology of Delphinium ajacis is given, but addition of geometry is not a trivial problem in this case.*

T. Nishida. "KOL-Systems Simulating Almost but Not Exactly the Same Development—The Case of Japanese Cypress." *Memoirs Fac. Sci., Kyoto University, Ser. Bio.* 8:97–122, 1980.

> *Development of shoots of Japanese cypress is expressed in terms of stochastic L-systems, directly applicable to the production of realistic computer-generated images.*

D. F. Robinson. "A Notation for the Growth of Inflorescences." *New Phytologist* 103:587–596, 1986.

> *The proposed notation is not the same as L-systems, but a translation between the two is straightforward. Out of several detailed examples, only Lychnis coronaria has been modeled by us using computer graphics so far.*

---

6. MULTIDIMENSIONAL L-SYSTEMS

A. Habel and H.-J. Kreowski. "On Context-Free Graph Languages Generated by Edge Replacement." In: H. Ehrig, M. Nagl and G. Rozenberg, editors: *Graph Grammars and Their Application to Computer Science; Second International Workshop*, Lecture Notes in Computer Science 153, Springer-Verlag, Berlin, pages 143–158, 1983

> *Graph-rewriting systems can be divided into two categories, those which replace nodes and those which replace edges. Edge-rewriting systems have direct biological applications. This paper surveys the theoretical background of edge rewriting.*

Two approaches to parallel graph rewriting are known as *map L-systems* and *double-wall L-systems*. The following papers describe map L-systems or ideas leading to them.

K. Culik II and A. Lindenmayer. "Parallel Graph Generating and Graph Recurrence Systems for Multicellular Development." *Int. J. General Systems* 3:53–66, 1976.

A. Lindenmayer and K. Culik II. "Growing Cellular Systems: Generation of Graphs by Parallel Rewriting." *Int. J. General Systems* 5:45–55, 1979.

K. Culik II and D. Wood. "A Mathematical Investigation of Propagating Graph 0L-Systems." *Information and Control* 43:50–82, 1979.

Three papers on parallel graph generation which precede the concept of map L-systems.

A. Lindenmayer and G. Rozenberg. "Parallel Generation of Maps: Developmental Systems for Cell Layers." In: V. Claus, H. Ehrig and G. Rozenberg (Eds.): *Graph Grammars and Their Application to Computer Science; First International Workshop*, Lecture Notes in Computer Science 73, Springer-Verlag, Berlin, pages 301–316, 1979.

The original paper on map L-systems.

P. L. J. Siero, G. Rozenberg, and A. Lindenmayer. "Cell Division Patterns: Syntactical Description and Implementation." *Computer Graphics and Image Processing* 18:329–346, 1982.

A. Lindenmayer and M. de Does. "Algorithms for the Generation and Drawing of Maps Representing Cell Clones." In: H. Ehrig, M. Nagl and G. Rozenberg (Eds.): *Graph Grammars and Their Application to Computer Science; Second International Workshop*, Lecture Notes in Computer Science 153, Springer-Verlag, Berlin, pages 39–57, 1983.

The above papers contain detailed examples of how map L-systems work. There also introduce graphical interpretations of maps.

A. Lindenmayer. "Models for Plant Tissue Development with Cell Division Orientation Regulated by Preprophase Bands of Microtubules." *Differentiation* 26:1-10, 1984.

Example of a three-dimensional L-system, or cellwork. Includes a calculation of edge lengths after subdivision.

A. Lindenmayer. "An Introduction to Parallel Map Generating Systems." In: H. Ehrig, M. Nagl, A. Rosenfeld and G. Rozenberg, editors: *Graph Grammars and Their Application to Computer Science; Third International Workshop*, Lecture Notes in Computer Science 291, Springer-Verlag, Berlin, pages 27–40, 1987.

A different approach to define cellular structures was developed by J. Lück and H. B. Lück under the name of double-wall L-systems.

J. Lück and H. B. Lück. "Proposition d'une Typologie de l'Organisation Cellulaire des Tissus Végétaux." In: Hervé Le Guardier and Thiebut Moulin, editors: *Actes du Premier Séminaire de l'Ecole de Biologie Théorique du CNRS*, Ecole Nationale Superieure de Techniques Avancées, Paris, pages 335–371, 1981.

J. Lück and H. B. Lück. "Sur la Structure de l'Organisation Tissulaire et Son Incidence sur la Morphogenèse." In: Hervé Le Guardier, editors: *Actes du Deuxième Séminaire de l'Ecole de Biologie Théorique du CNRS*. Publications de l'Université de Rouen, Abbaye de Solignac, pages 385–397, 1982

J. Lück and H. Lück. "Generation of 3-Dimensional Plant Bodies by Double Wall Map and Stereomap Systems." In: H. Ehrig, M. Nagl and G. Rozenberg, editors: *Graph-Grammars and Their Application to Computer Science; Second International Workshop*, Lecture Notes in Computer Science 153, Springer-Verlag, Berlin, pages 219–231, 1983.

J. Lück and H. B. Lück. "Comparative Plant Morphogenesis Founded on Map and Stereomap Generating Systems." In: J. Demongeot, E. Goles and M. Tchuente, editors: *Dynamical Systems and Cellular Automata*. Academic Press, London, pages 111–121, 1985.

J. Lück and H. B. Lück. "Un Mécanisme Générateur d'Hélices Phyllotaxiques." In: G. Benchetrit and J. Demongeot, editors: *Actes du IVe Séminaire de l'Ecole de Biologie Théorique*, Editions du CNRS, Paris, pages 317–330, 1985.

H. B. Lück and J. Lück. "Unconventional Leaves (An Application of Map 0L-Systems to Biology)." In: G. Rozenberg and A. Salomaa, editors: *The Book of L*. Springer-Verlag, Berlin - Heidelberg - New York - Tokyo, pages 275–289, 1986.

J. Lück and H. B. Lück. "From 0L and IL Map Systems to Indeterminate and Determinate Growth in Plant Morphogenesis." In: H. Ehrig, M. Nagl, A. Rosenfeld and G. Rozenberg, editors: *Graph Grammars and Their Application to Computer Science; Third International Workshop*, Lecture Notes in Computer Science **291**, Springer-Verlag, Berlin, pages 393–410, 1987.

J. Lück, A. Lindenmayer, and H. B. Lück. "Models for Cell Tetrads and Clones in Meristematic Cell Layers." *Botanical Gazette*, 1988, in press.

7. SYNTHESIS OF REALISTIC PLANT IMAGES
7.1. SYNTHESIS METHODS BASED ON L-SYSTEMS.

P. Hogeweg and B. Hesper. "A Model Study on Biomorphological Description." *Pattern Recognition* 6:165–179, 1974.

> *The techniques of A. R. Smith were inspired by this paper.*

A. R. Smith. "About the Cover: Reconfigurable Machines." *Computer* 11, Nr. 7:3–4, 1978.

> *The first realistic images of computer-generated plants.*

A. R. Smith. "Plants, Fractals, and Formal Languages." *Computer Graphics* 8, Nr. 3:1–10, 1984

> *A classic; the first paper to recognize the role of formal languages in computer graphics. Coins the term "database amplification". Contains many realistic plant images.*

A. R. Smith. *Grammars for Generating the Complexity of Reality.* Video tape, Lucasfilm/PIXAR, 1985.

> *A video illustration of the above paper.*

P. Prusinkiewicz. "Applications of L-Systems to Computer Imagery." In: H. Ehrig, M. Nagl, A. Rosenfeld and G. Rozenberg, editors: *Graph Grammars and Their Application to Computer Science; Third International Workshop*, Lecture Notes in Computer Science **291**, Springer-Verlag, Berlin, pages 534–548, 1987.

> *The turtle interpretation of L-systems is generalized to three dimensions. Stochastic L-systems are used to model specimen-to-specimen variation within a species, and bicubic patches are integrated with the L-system-based model. L-systems generating several plant images are specified in detail.*

T. Beyer and M. Friedell. "Generative Scene Modelling." *Proceedings of EUROGRAPH-ICS '87*, pages 151–158 and 571, 1987.

> *The authors claim to have invented a "new theory of scene modelling" while ignoring previous work in the field, including the 1984 paper by A. R. Smith. The extension of L-systems to graphs with cycles seems to be different from the map L-systems of Lindenmayer and the double-wall L-systems of the Lücks, although the description is too superficial to determine what was really done.*

7.2. OTHER SYNTHESIS METHODS.

M. Aono and T. L. Kunii. "Botanical Tree Image Generation." *IEEE Computer Graphics and Applications* 4, Nr. 5:10–34, 1984.

> *Topology is defined by recursive algorithms; variation between trees is achieved primarily by manipulating branching angles. Based on the analysis of tree architecture by Honda.*

M. Aono and T. L. Kunii. *Botanical Tree Image Generation*, Video tape, IBM Japan, Tokyo, 1985.

J. Bloomenthal. "Modeling the Mighty Maple." *Computer Graphics* 19, Nr. 3:305–311, 1985.

> *An exercise in realism; uses splines to model branches, carefully models branching points, and texture-maps leaves and bark.*

J. Bloomenthal. *Polygonization of Implicit Surfaces.* Report CSL-87-2, Xerox Corporation, Palo Alto, CA, 1987.

> *A method for modeling branching points in trees with wide branches.*

G. Eyrolles. *Synthèse d'Images Figuratives d'Arbres par des Méthodes Combinatoires.* Ph. D. Thesis, Université de Bordeaux I, 1986.

> *Realistic two-dimensional tree silhouettes are generated using a stochastic technique based on Horton-Strahler analysis (a classic method for analysing river systems.)*

Y. Kawaguchi. "A Morphological Study of the Form of Nature." *Computer Graphics* 16, Nr. 3:223–232, 1982.

> *Generation of abstract branching structures.*

P. Oppenheimer. "Real Time Design and Animation of Fractal Plants and Trees." *Computer Graphics* 20, Nr. 4:55–64, 1986.

W. T. Reeves and R. Blau. "Approximate and Probabilistic Algorithms for Shading and Rendering Structured Particle Systems." *Computer Graphics* **19**, Nr. 3:313–322, 1985.

7.3. ORGAN MODELS.

P. Lienhardt. *Modélisation et Évolution de Surfaces Libres.* Ph. D. Thesis, Université Louis Pasteur, Strasbourg, 1987.

P. Lienhardt, and J. Francon. *Synthèse d'Images de Feuilles Végétales.* Technical Report R-87-1, Département d'informatique, Université Louis Pasteur, Strasbourg, 1987

8. RELATED BOTANICAL PAPERS
8.1. PLANT ARCHITECTURE.

Two fundamental books on trees.

F. Hallé, R. A. A. Oldeman, and P. B. Tomlinson. *Tropical Trees and Forests: An Architectural Analysis.* Springer-Verlag, Berlin, 1978.

M. H. Zimmerman and C. L. Brown. *Trees—Structure and Function.* Springer-Verlag, Berlin, 1971.

The series of papers authored or co-authored by Honda provides increasingly refined models of trees. While the approach based on L-systems starts from the topology of branching structures and complements it with geometrical details, Honda starts from the branching geometry. However, as both models get increasingly accurate, the same types of problems, such as interaction and growth control mechanisms, are addressed.

H. Honda. "Description of the Form of Trees by the Parameters of the Tree-Like Body: Effects of the Branching Angle and the Branch Length on the Shape of the Tree-Like Body." *J. Theoretical Biology* 31:331–338, 1971.

Tree structures are defined by simple recursive algorithms. Emphasis is put on geometry. Special attention is given to the definition of branching angles.

J. B. Fisher and H. Honda. "Computer Simulation of Branching Pattern and Geometry in *Terminalia* (Combretaceae), a Tropical Tree." *Botanical Gazette* **138**, Nr. 4:377–384, 1977.

H. Honda, P. B. Tomlinson, and J. B. Fisher. "Two Geometrical Models of Branching of Botanical Trees." *Annals of Botany* 49:1–11, 1982.

Extensions and improvements of the previous paper.

H. Honda and J. B. Fisher. "Tree Branch Angle: Maximizing Effective Leaf Area." *Science* 199:888–890, 1978.

Observed branch angles in Terminalia *are shown to be similar to theoretically optimal values that produce the maximum effective leaf area.*

H. Honda and J. B. Fisher. "Ratio of Tree Branch Lengths: The Equitable Distribution of Leaf Clusters on Branches." *Proceedings of the National Academy of Sciences USA* 76, Nr. 8:3875–3879, 1979.

Continuation of the previous paper; observed branch angles in Terminalia *also produce the most equitable distribution of leaf clusters.*

H. Honda, P. B. Tomlinson, and J. B. Fisher. "Computer Simulation of Branch Interaction and Regulation by Unequal Flow Rates in Botanical Trees." *Amer. J. Botany* 68:569–585, 1981.

R. Borchert and H. Honda. "Control of Development in the Bifurcating Branch System of *Tabebuia rosea*: A Computer Simulation." *Botanical Gazette* 145, Nr. 2:184–195, 1984.

Branch interaction is applied to control the number of branches of various orders. Yet another specification of branching geometry is introduced. The conclusion states: Our results suggest that any realistic simulation of the development of the branch system in botanical trees requires at least four sets of rules addressing (1) geometry of the branch system, (2) reduction of branch numbers at higher orders through reduction of branching in lateral branches, (3) the establishment of apical control through branch interaction, and (4) control of growth of the branch system as a whole by a sigmoid growth function. *It would be very interesting to express these results in terms of L-systems, and compare them with the* Mycelis *model.*

There are also other papers which analyze plant architecture in a quantitative way, thus providing useful data for simulation.

A. D. Bell, D. Roberts, and A. Smith. "Branching Patterns: The Simulation of Plant Architecture." *Journal of Theoretical Biology* 81:351–375, 1979.

Simulation of two-dimensional branching patterns of rhizomatous plants.

B. Descoings. "Les Types Morphologiques et Biomorphologiques des Especes Graminoides dans les Formations Herbeuses Tropicales." *Naturalia Monspeliensia, sér. Bot* 25:23–25, 1975.

H. Ford. "Investigating the Ecological and Evolutionary Significance of Plant Growth Form using Stochastic Simulation." *Annals of Botany* 59:487–494, 1987.

W. R. Remphery, B. R. Neal, and T. A. Steeves. "The Morphology and Growth of *Arctostaphylos uva-ursi* (Bearberry): An Architectural Analysis". *Canadian Journal of Botany* 61, Nr. 9:2430-2450, 1983.

W. R. Remphery, B. R. Neal, and T. A. Steeves. "The Morphology and Growth of *Arctostaphylos uva-ursi* (Bearberry): An Architectural Model Simulating Colonizing Growth". *Canadian Journal of Botany* 61, Nr. 9:2451-2458, 1983.

E. Renshaw. "Computer Simulation of Sitka Spruce: Spatial Branching Models for Canopy Growth and Root Structure." *IMA J. of Math. Applied in Medicine and Biology* 2:183–200, 1985.

B. H. Smith. "The Optimal Design of a Herbaceous Body." *The American Naturalist* 123:197–211, 1984.

O. Yu and M. Gnonot. "Recherches sur le Tallage Chez le Dactyle (*Dactylis glomerata* L.). I. Etude Experimentale de l'Effet de l'Azote sur le Tallage. II. Réflexions sur la Formalisation et la Modélisation des Processus de Tallage." *Acta Oecologica, Oecol. Plantarum* 2:351–365, and 3:39–47, 1981 and 1982.

The following papers analyze mechanical properties of tree structures.

C. D. Murray. "A Relationship between Circumference and Weight in Trees and Its Bearing on Branching Angles." *Journal of General Physiology* 10:725-739, 1927.

T. A. McMahon. "The Mechanical Design of Trees." *Scientific American*:93-102, 1975.

T. A. McMahon and R. E. Kronauer. "Tree Structures: Deducing the Principle of Mechanical Design." *Journal of Theoretical Biology* 59:443–466, 1976.

W. W. Armstrong. "The Dynamics of Tree Linkages with a Fixed Root Link and Limited Range of Rotation." *Actes du Colloque Internationale l'Imaginaire Numérique '86*:16-21, 1986.

8.2. PHYLLOTAXIS.

R. O. Erickson. "Tubular Packing of Spheres in Biological Fine Structure." *Science* 181:705-716, 1973

R. O. Erickson. "The Geometry of Phyllotaxis." In: J. E. Dale and F. L. Milthrope, editors: *The Growth and Functioning of Leaves.* University Press, Cambridge, pages 53–88, 1983.

An important comprehensive reference.

P. H. Hellendoorn and A. Lindenmayer. "Phyllotaxis in *Bryophyllum tubiflorum*: Morphogenetic Studies and Computer Simulations." *Acta Biol. Neerl.* 23, Nr. 4:473–492, 1974.

A. H. Veen and A. Lindenmayer. "Diffusion Mechanism for Phyllotaxis: Theoretical Physico-Chemical and Computer Study." *Plant Physiol.* 60:127–139, 1977.

W. L. Kilmer. "On Growing Pine Cones and Other Fibonacci Fruits—McCulloch's Localized Algorithm." *Mathematical Biosciences* 11:53-57, 1971.

A. M. Mathai and T. A. Davis. "Constructing the Sunflower Head." *Mathematical Biosciences* 20:117-133, 1974.

Some nice flower models are produced, but the underlying mathematical reasoning is not clear.

G. J. Mitchison. "Phyllotaxis and the Fibonacci Series." *Science* 196:270–275, 1977.

F. J. Richards. "Phyllotaxis: Its Quantitative Expression and Relation to Growth in the Apex." *Philos. Trans. Royal Society London*, Ser. B, 235:509–564, 1959.

8.3. INFLORESCENCES.

The first group of publications is devoted to the description and classification of inflorescences.

W. Troll. *Die Infloreszenzen*, Vol. I. Gustav Fischer Verlag, Stuttgart, 1964.

D. Müller-Doblies and U. Müller-Doblies. "Cautious Improvement of a Descriptive Terminology of Inflorescences." *Monocot Newsletter* 4, Institut für Biologie, Technical University of Berlin (West), 13 pages, 1987

F. Weberling "Typology of Inflorescences." *J. Linn. Soc. (Bot.)* 59, Nr. 378:215–222, 1965.

The second group of interest focuses on mechanisms which control inflorescence development and flowering sequences.

M. V. S. Raju, G. J. Jones, and G. F. Ledingham. " Floret Anthesis and Pollination in Wild Oats *(Avena fatua)*." *Canadian Journal of Botany* 63:2187-2195, 1984.

> *Contains data on the flowering sequence of the wild oats.*

Y. Sell. "Action de l'Apex Caulinaire et Maturation Florale dans le Cadre de la Floraison Descendante." *Flora* 169:15-22, 1980.

> *The role of apical dominance in inflorescence development.*

Y. Sell. "Physiological and Phylogenetic Significance of the Direction of Flowering in Inflorescence Complexes." *Flora* 169:282-294, 1980

> *From the summary:* Downward flowering of the lateral shoots of a herbaceous plant is here suggested to be the expression of antagonistic long-range interactions between various organs.

8.4. LEAVES.

W. K. Silk and R. O. Erickson. "Kinematics of Plant Growth." *Journal of Theoretical Biology* 76:481-501, 1979.

R. O. Erickson and W. K. Silk. "The Kinematics of Plant Growth." *Scientific American* 242, Nr. 5:134–151, 1980.

H. Scholten and A. Lindenmayer. "A Mathematical Model for the Laminar Development of Simple Leaves." In: W. van Cotthem, editor: *Morphologie—Anatomie und Systematik der Pflanzen, 5. Symposium.* Verlag Waegeman, Ninove, Belgium, pages 29–37, 1981.

> *An analytical description of growth is expressed in terms of growth fields and applied to leaf blades.*

B. Jeune. "Sur le Déterminisme de la Forme de Feuilles de Dicotylédones." *Adansonia* 2:83–94, 1978.

B. Jeune. "Modèle de Développement pour des Feuilles Basipètes de Dicotylédons." *Adansonia* 8:301–323, 1986.

> *A case study of the development of very young leaves.*

S. Kürbs. "Vergleichend-entwicklungsgeschichtliche Studien an Ranunculaceen-Fieder-blättern." Part I and II. *Bot. Jahrb. Syst.* 93 Nr. 1:130–167 and Nr. 3:325-371, 1973.

> *Analyzes leaves of the dissected and divided type.*

The following papers contain detailed pictures and discussion of venation patterns.

F. Bougnon, H. Dulieu, and M.-F. Turlier. "Morphologie Végétale — Rapports entre les Directions Fondamentales de Croissance dans l'Ébauche et la Nervation Foliaires." *C. R. Acad. Sc. Paris, Sér. D* 268:48–50, 1969.

N. Lersten. "Histogenesis of Leaf Venation in *Trifolium Wormskioldii* (Leguminosae)" *American Journal of Botany* 52, Nr. 8:767–774, 1965.

J. Philpott. "Blade Tissue Organization of Foliage Leaves of Some Carolina Shrub-Bog Species as Compared with Their Appalachian Mountain Affinities." *The Botanical Gazette* 118, Nr. 2, 1956.

T. R. Pray. "Foliar Venation of Angiosperms, Part I – IV." *American Journal of Botany* 41: 663–670; 42:18–27, 611–618 and 698–706, 1954 and 1955.

8.5. POPULATION DYNAMICS AND MODULAR ORGANISMS.

F. A. Bazzaz and J. L. Harper. "Demographic Analysis of the Growth of *Linum usitaissimum.*" *New Phytologist* 78:193–207, 1977.

A. D. Bell. "The Simulation of Branching Patterns in Modular Organisms." *Philos. Trans. Royal Society London* Ser. B, 313:143–169, 1986.

F. Hallé. "Modular Growth in Seed Plants." *Philos. Trans. Royal Society London* Ser. B, 313:77–87, 1986.

J. L. Harper and A. D. Bell. "The Population Dynamics of Growth Form in Organisms with Modular Growth." In: R. M. Anderson, B. O. Turner and L. R. Taylor, editors: *Population Dynamics*, Blackwell, Oxford, pages 29–52, 1979.

L. Maillette. "Structural Dynamics of Silver Birch. I. The Fates of Buds. II. A Matrix Model of the Bud Population." *J. Appl. Ecol.* 19:203–218 and 219–238, 1982.

J. R. Porter. "A Modular Approach to Analysis of Plant Growth. I. Theory and Principles. II. Methods and Results." *New Phytologist* 94:183–190 and 191–200, 1983.

Index

A

a posteriori fitness function, 145
a priori fitness function, 145
abacus sort, 365
abstractions, 352
acropetal signal, 234
actor, 439
adaptation, 275, 311, 426
    extrinsic, 141
    intrinsic, 141
adaptive behavior, 341
adaptive mechanisms, 351
adaptive memory, 351
adaptive parts, 507
adaptive processes, 355
AFM, 461
Agar, 422, 427
agents, 405, 407, 427-430
    programming with, 427-430
aggregate interactions, 344
Agre, P., 439
algae, 223
algorithm, 11
    genetic, 19, 157-158, 162, 164
algorithmic enzymes, 19
algorithms, 205
    genetic, 35, 253, 426
    recursive, 205
ambient indeterminacy, 349
analog computers, 170
analog models, 319
analysis, 39, 41
ANI, 439
animal behavior, 421
animal construction kit, 421
animal programs, 277
animal species
    designing, 277
animals, 438
animation, 4, 30, 422, 438
    actor, 439
    flocking, 438
    self-scripting, 438

Arbib, Michael, 16
Argument from Design, 43
Aristotle, 68
artificial, 3
artificial animals, 275, 279, 398
artificial bugs, 143
artificial cell, 127
artificial ecology, 143, 275
artificial environment, 278
artificial genes
    isolating effects of, 264
Artificial Intelligence, 21, 31, 38, 55, 63-64,
171, 174, 421-422
    for Typogenetics, 387
    formalists vs. realists, 64
    needed for artificial life, 387
    Platonic ontology and, 64
Artificial Intuition, 262
Artificial Life, 38, 341, 521-522, 544
    in a Typogenetical world, 387
    RAM, 275-293
    role of environment, 278
Artificial Life Toolkit, 397
artificial life-forms, 251-252
    mutants, 252
artificial molecular machines, 16
artificial organisms, 49, 54
    capabilities, 51
    ascertain componentry, 52
    characteristics of, 49
    consequences of action, 52
    homeostatic behavior, 52
    reproduction, 51
    rights of, 57
    self-repair, 51
    shaped by chance, 54
artificial physics, 66, 71, 122
artificial selection, 203
ASAS, 438
assemblers, 502
assembly, 505
    matching vs. positional, 505
atomic bit machining, 445

atomic force microscope, 461
autocatlytic emergence
    expected time of, 99
autocatalytic sets, 82
automata theory, 13
automata, 8
    cellular, 28, 80, 105-108, 222, 341,
    351, 521-530
    molecular, 521
    nested, 522, 532, 543
autopoiesis, 131
    processes, 133
Aviram, A., 456

B

Babbage, Charles, 193
basipetal signal, 234
behavior
    animal, 421
    hierarchical structures, 430
    local determination of, 42
behavior generation, 5
behavior generation problem, 5
behavioral PTYPE, 30
Belousov-Zhabotinsky reaction, 125, 108
Benard convection cells, 125
bilateral symmetry, 206
Binnig, G., 448
biochips, 444
bioinformatic models, 308
bioinformatics, 298
biological automata, 21
biological geometries, 341
    for computing, 344
biomorphs, 201-216, 436
bioreplicators, 504-505, 507
Blind Watchmaker, 43, 201, 436
blue-green bacteria, 223
Boids, 30
Boole, George, 388
bottom-up environments, 38
bottom-up modeling, 3
bottom-up specifications, 31, 42
boundary conditions, 349
    as algorithm specification, 349
bracketed L-systems, 226
brain/mind, 543
BrainWorks, 422
Braitenberg, Valentino, 401-402, 422, 436,
555
Braitenberg's vehicles, 401-402

branching growth, 26
branching structure, 225
branching trees, 415
Brenner, Sydney, 218
Brooks, R. A., 434, 439
Bruno, 301
bumblebees, 308
Burks, A. W., 56-57, 388

C

Cairns-Smith, A. G., 107, 136, 169
Carter, Forrest L., 443, 453, 456
catalytic species
    diffusion of, 83
causal links, 82
cell repair machines, 483
cell-space model, 50
cells, 158-164
    daughter, 159, 162
cellular automata, 28, 80, 105-108, 222, 341,
351, 521-530
    theory, 105
cellular automaton, 13
    Conway's game of LIFE, 20
cellular automaton models, 320
centrioles, 525
cerebellum, 555
Changeux, J.-P., 455
channeled transport, 504
chaotic system, 349
    enumerating, 349
Chapman, D., 439
chemical synthesis, 449
chemical turbulence, 115
children
    education of, 397
Chomsky, 242
Chomsky language classes, 222
    correspondence to L-systems, 243
Church, 10
classifier, 436
clay theory
    for the origin of life, 107
clays, 107
clock
    socially regulated, 308
clockwork technology, 7
clumps, 357
Codd, E. F., 16
Coderre, Bill, 407
cognitive activity, 67

cognitive activity (cont'd.)
  theory of, 67
cognitive science, 422
coherent excitation, 528-529, 543
ComMet, 341, 353
competition, 480
complex systems
  roles of computers in studying, 38
complexity, 81, 244, 259
composite structure, 341
computability in practice, 12
computability in principle, 12
computation, 188, 191, 349, 354
  as animation, 361
  fluid-like local, 354
    geometric programmability, 361
    reminiscent of fluids, 349
Computational Metabolism, 341, 353
  hierarchies, 341
  implemented instances of, 341
  information, 343
  semantics, 351
computational ethology, 436
computational evolution, 436
computational logic, 521, 537-538
computational metabolism, 341
computational paradigm, 38
computer animation, 30
  as testable theories, 276
computers
  analog, 170
  artificial intelligence, 171
  digital, 171
  general purpose, 11
  growth of processing power, 193
  role in studying complex systems, 38
computing, 341
  biological geometries for, 344
  nonsymbolic, 351
  via infections, 356
conformational changes, 333-334
conformational switching, 323
Conrad, Michael, 19, 71, 456
constrained embryology, 204
context free, 26
context sensitive, 27
context-sensitive rules, 31
control mechanisms, 9
Conway, John, 20
Conway's game of LIFE, 20

cooperative gene structures, 82
  emergence of, 95
Core Wars, 20
crossover, 35-36
cultural evolution, 167-168
cumulative selection, 209, 219, 511
cybernetics, 15, 170-171
cytoskeleton, 522, 524-526, 530, 543-544

D

Darwin, Charles, 43, 453
database amplification, 260, 266
dataflow systems, 408
dataflow trees, 417
Dawkins, Richard, 201, 436, 503, 511
deception, 433
decision trees, 415
delay neurons, 424
densest sphere packings, 350
Department of Defense
  Advanced Research Project
    Agency, 175
Descartes, 106
determinism, 349, 365
development, 157-158, 160-165
  interactionless, 223
  with interaction, 224
developmental models, 221
Devol, George, 173
Dewdney, A. K., 20
differentiation, 33
diffusion, 504
  hormone, 224
  of catalytic species, 83
diffusive transport, 504
digital computers, 171
Dijkstra, E., 363
dipoles, 527, 529-530, 541
discrete mathematics, 342
distributed control of behavior, 21
Donaldson, T., 453
Drexler, K. E., 501
drive energy, 435
Dutch Flag problem, 363
  n-dimensional, 363
dynamical systems, 66, 353
  computer study of, 66
Dyson, F. J., 478

E

echinoderms, 215

ecological simulation, 275
educational computing, 397-398, 400, 405
elongation cycle, 332
embedded computation, 30
embryology, 202-203
embryology (cont'd.)
    constrained, 204
    separation from genetics, 203
    Weismann's principle, 203
emergence, 71-73, 201, 341-343, 398-403
    autocatalytic, 99
    of cooperative gene structures, 95
emergent behavior, 2, 71, 439
    levels of, 71
emergent fitness function, 141, 37
emergent properties, 209
emergent structures, 312
emotions, 183
environment, 278
    bottom-up, 38
    top-down, 38
enzyme, 22
error replication, 82, 95
error-tolerance, 478
ethology, 421-422
    computational, 436
Ettinger, R. C. W., 453
eukaryotes
    theory of symbiotic origin, 524
EURISKO, 515
evolution, 33-34, 79, 100, 143, 157, 201, 275, 311, 341-342, 426, 436, 501, 508-510, 513
    computational, 436
    cultural, 168
    M-style, 508
    O-style, 508, 510
    parallel, 100
    simulated, 158, 165
    without kinetics, 90-94
evolutionarily pregnant, 218
evolutionary potential, 218
evolutionary process, 84-85, 87-88, 90
evolutionary watersheds, 217
evolvability, 201, 216
evolved for evolvability, 511
EVOLVEIII, 436
evolvors, 218
expert systems, 31, 172
experts, 409
explicit information, 343

extrinsic adaptation, 141

F
Farmer, Doyne, 82, 101
feedback, 15
feedback structures, 94
Feynman, Richard P., 443-499
Feynman machines, 448
Feynman Prize, 480
filaments, 222
finite automata, 13
firing-squad problem, 236
fitness function, 253
    a posteriori, 145
    a priori, 145
    emergent, 141
Flibs, 20
flocking, 30, 32, 438
fluid-like local computation, 354
formal limits of machine behaviors, 12
formal system, 356
Forrest, D. R., 453
foxes and rabbits, 288
fractal, 257
frame problem, 439
Franks, A., 444
Fröhlich, H., 527, 532, 543
Fuller, Buckminister, 454

G
Garden-of-Eden, 384
Gardner, Martin, 20
general purpose computers, 11
generalized genotype, 22
generalized permutations, 351
generalized phenotype, 22
genes, 22
genetic algorithm, 19, 35, 157-164, 253, 426
genetic diffusion, 512
genetic engineering, 253, 515-516
genetic landscape, 514
genetic operators, 35
genetic system, 217
genetic takeover theory, 169
genetics
    separation from embryology, 203
genotype, 6, 22, 157-158, 162, 253-254
    distinction from phenotype, 22
geometric interpretation, 264
geometric structures, 506
geometry, 354

giant component, 88
Gibson, Robert, 283
gliders, 20, 533
global interactions, 351
Goel, N. S., 133, 317
Goethe, 385
Goldman, S. R., 275
graftals, 264
graphic programming, 437
Greenberg-Hastings model, 108-113
Griffin, D. R., 182
growth rules, 159, 206
GTYPE, 22
    unpredictability of PTYPE from, 23
Gödel, K., 10, 53

H
Halting Problem, 12, 53
Hameroff, Stuart, 521
Hamming distance, 83
Hartman, Hyman, 105
Hebb synapses, 431
Hebb, 431
Heinlein, R. A., 448
herbaceous plants, 229
herding, 30
heterarchy, 436
heterostructure, 361
hierarchical behavior, 430
hierarchically structured centers, 435
hierarchies, 341, 352
    ComMet, 341
higher-level selection, 218
hill-climbing, 426
Hofstadter, Douglas R., 354, 369-370
Hogeweg, Pauline, 74, 297
Holland, John, 19, 35-37, 436, 512-513
homeostatic behavior, 52
hormone diffusion, 224
Hufford, K. D., 125
HyperCard, 438
hypotheses
    Physical Symbol System, 67, 69

I
impermeable membrane, 357
implicit information, 343
imprinting, 431
individual-oriented models, 302
inert parts, 507

inflorescence, 229-231, 236
    cymose, 231
information, 351
    explicit, 343
    implicit, 343
    spatially extended concept of, 351
informational fluid, 351
inheritance, Lamarckian, 16
initial spatial configurations, 349
innate releasing mechanisms, 435
intentionality, 434
interactionless development, 223
INTERLISP-D, 305
interneurons, 424
intrinsic adaptation, 141
intuition
    artificial, 262
inversion, 35

J
Jacks, 7
Jacobson, Homer, 18
Jefferson, D. R., 275
Johns Hopkins University Beast, 171

K
K-lines, 431
Kahn, K., 439
Kauffman, Stuart, 82, 101
kinematic model, 13, 50
kinetics, 82
    evolution without, 90-94
kinks, 119
Kleene, 10
Klir, G. J., 125
knowledge-based systems, 408
Kolmogorov, A. N., 245
Kuhn, H., 455

L
L-systems, 19, 25, 157, 221-246
    bracketed, 226
    complexity measure, 244
    correspondence to Chomsky language classes, 243
    differentiation for, 33
    map, 222, 239
Laing, Richard, 16, 49, 387, 456
Laing's molecular machines, 16, 387
Lamarckian inheritance, 16
Langton, Christopher, 1, 63

learning, 350, 431-434, 437
    Hebb synapses, 431
    K-lines, 431
    R-trees, 431
    script-agent, 432
    scripts, 431
Leduc, Stephane, 125-128, 130-135
LEGO/Logo, 397-405
lek formation, 283
level-bands, 432-433
Lieberman, E. A., 457
Liesegang patterns, 125
life in silico, 39
life in vitro, 38
Lindenmayer, Aristid, 19, 25, 221
linear growth, 26
linear systems, 41
    vs. nonlinear systems, 41
liquid computer, 344
liquid-in-motion dynamic, 346
local description, 301
local determination of behavior, 42
local indeterminacy, 349
local interaction, 352
local structures, 31
local timing regime, 301
locality, 425
localized recognition, 342
logic
    Typogenetics as design logic, 370
Logo, 227, 397-398, 400-401, 422, 438
Logo turtle, 425
Lorenz, K. Z., 435
Lugowski, M. W., 341
lunar factory, 58
    self-replicating, 58

M
M-style systems & evolution, 508
M-style systems, 507
machine behaviors
    formal limits of, 12
machines
    cell repair, 483
    Von Neumann for nanogenesis, 477
Macintosh, 438
Mandelbrot, B., 72
Månsson, Bengt, 521
map L-Systems, 239
mappings, 514
Margulis, L., 524

massively parallel computers, 356
matching vs. positional assembly, 505
Maturana, H. R., 131
Mayr, E., 292
McCarthy, John, 171, 175
McCulloch, Warren S., 19, 105, 423
McCulloch-Pitts neurons, 423
measurement theory, 74
memory, 51
    experiential, 51
meta-Typogenetics, 383
metabolism, 349
micromanipulators, 454
microreplicators, 477
microscopes
    atomic force, 461
    scanning tunneling, 446, 458, 503
microtubules, 522-530, 539, 542
Miller-Urey experiment, 107
miniaturization, 196
Minsky, Marvin, 19, 171, 415, 417, 429, 432
Minsky's Society of Mind model, 408
MIRROR modeling, 297-301
MIRROR models, 302-305
model of metaphor, 355
model-building, 435
modeling, 63, 436
    filaments, 222
    frame-based approach, 239
    mechanisms, 232-233
    organs, 237
    plants, 221
    realizations, 63
    simulations, 63
    types of, 319
models, 6
    analog, 319
    animal behavior, 407
    bioinformatic, 308
    biological using RAM, 282
    cell-space, 50
    cellular automata, 108
    cellular automaton, 320
    evolutionary process, 84-90
    Greenberg-Hastings, 108-113
    individual-oriented, 302
    kinematic, 50
    Minsky, 408
    MIRROR, 297, 301-305
    movable finite automata, 317-318,
        324, 332, 335

models (cont'd.)
 output-oriented, 298
 physically realistic, 319
 predator-prey, 288
 reaction-diffusion-equation
 based, 320
 self-assembly, 323
 self-organization, 319
 Society of Mind, 408, 415, 417, 429,
 431
 structure-oriented, 298
 Turing's for morphogenesis, 108
 variable structure, 302-303
 world, 184
molecular automata, 521
molecular electronic devices, 444
molecular engineering, 444
molecular logic, 521-522
molecular machines, 16, 444, 455
molecular replicators, 444
molecular-level repair, 481
Moravec, Hans, 167, 437, 543
morphogenesis, 22, 25, 33, 135
morphogens, 108
Morris, h. C., 369
mosquito control, 284
motor neurons, 425
movable finite automata models, 317-318,
324, 332, 335
movement
 artificial, 555-556
 oscillatory, 562
 without support, 557
Moynihan, Martin, 179
multi-layered dynamical descriptions, 353
multi-level feedback, 31
mutation, 35, 201, 426
mutual exclusion, 351
Myhill, J., 53-54

N

Nahvi, Mahmood, 558-559
Nahvi's principle, 558
nanocomputers, 444, 503
nanocomputing, 456
nanogenesis, 477
nanolithography, 459
nanomachines, 444
nanoreplicators, 444, 456, 477, 504-515
nanorobots, 483
nanostructure, 467

nanotechnology, 443-445, 501-503
 history of, 448
 K. R. Shoulders, 450
 workstation, 465
NASA, 58
natural selection, 34, 275, 343
neighborhood, 344
nested automata, 522, 532, 543
neurons, 422-424
 delay, 424
 McCulloch-Pitts, 423
 motor, 425
 sensorimotor, 424
Newell, A., 67, 171
Niklas, K. J., 436
nonlinear interactions, 31
nonlinear systems, 41
 vs. linear systems, 41
nonsymbolic computing, 351
nucleotides, 85

O

O-style systems & evolution, 508, 510
O-style systems, 507
object-oriented paradigm, 279
open-system formalisms, 355
Oppenheimer, Peter, 251
optical circuits, 195
organic forms, 135
orientation, 426
origin of life, 79, 107
osmosis, 130, 133
osmotic growth, 126, 128, 131, 135
osmotic precipitation, 126
output-oriented models, 298

P

Packard, Norman, 82, 101, 141
Paley, William, 43
Papert, Seymour, 19, 400
paradigm systems, 299-301
paradox
 Russellian, 385
paradox of self-reference, 385
parts
 adaptive, 507
 inert, 507
Pasteur, Louis, 126
Pattee, H. H., 63
pattern recognition, 350
pattern-in-time, 349

Pearson, J., 74
Penrose, L. S., 17
perceptrons, 185
persisting internal state, 351
Petworld, 436
Pfeffer, 130
phase beauty, 230
phenotype, 6, 22, 158, 162, 253-254
    distinction from genotype, 22
Physical Symbol System Hypothesis, 67,
69
physics of information, 342
picotechnology, 445
Pitts, W., 19, 423
pixel-peppering, 204
planning, 439
plants, 222
    herbaceous, 229
plurifunctionality, 353
Post, 10
power set, 36
predator-prey model, 288
Prigogine, I., 127, 72
principles
    Nahvi's, 558
    superposition, 41
probabilistic choice, 349
    as control structure, 349
problems
    behavior generation, 5
    syntactic inference, 244
    Turing's halting problem, 12
procedure, 10
process control, 9, 351
production rule, 429
production system program, 159
program, 10
    open vs. closed, 292
programinals, 282
Programmable Brick, 405
programmable controllers, 10
programming
    by example, 437
    graphic, 437
    metaphor, 438
    methodology, 437
    with agents, 427
programs as animals, 275
    kinetic, 479
proofs
    Typogenetics, 383

proofs, Typogenetics (cont'd.)
    meta-logical, 383
    reductio strategy, 383
    special derivation format, 379
propagating state, 351
    without global iterations, 351
    without process control, 351
propagating structures, 20
propagating wavefronts, 351
propagation, 370
    logic for investigating, 370
    varieties of, Typogenetics, 382-383
    emergent, 209
    self-structuring, 311
protein biosynthesis, 329, 334
protein conformation, 529
Prusinkiewicz, P., 221
PTYPE, 22
    behavioral, 30
    unpredictability from GTYPE, 23

Q
quantum computers, 444
quasi-species, 83

R
R-trees, 431
racemes, 230, 233
RAM, 275-293, 436
    biological models, 282
random graph, 90-91
random mutation, 350
randomness, 265
Rasmussen, Steen, 79, 521
Ratner, M. A., 456
reaction-diffusion, 105
    equation-based
    models, 320
reaction-diffusion systems, 106-108, 119
realistic models, 319
realization, 65, 68
recursion, 205, 208, 258
recursive algorithm, 205, 208, 258
recursively generated objects, 25
regress paradox
    and Typogenetics, 382-383
replicating machines, 456
replication rate, 98
replicator, 503
replicators, 202, 516
    bioreplicators, 504-507

replicators (cont'd.)
    nanoreplicators, 504-505, 514-515
representation, 434
reproduction, 426
Resnick, Mitchel, 397
retina, 187
reversible computation, 349
Reynolds, Craig W., 30, 438
ribosomes, 502
robotics, 172-173, 175-177, 180
Rohrer, H., 448
Rosenblatt, Frank, 19, 185
Rothstein, J., 456
routing, 351
rules, 157-164
    deletion, 161
    for expression, 158
    growth, 159, 162-164
    inhibitory, 161
    pyramid, 164
Russell-like paradox, 385, 369

S
SAIL, 175
Samuel, Arthur L., 18, 184
Sapir-Whorf hypothesis, 272
scanning tunneling engineering, 448
scanning tunneling microscope, 446, 458, 503
schema, 36
schema space, 36
Schneiker, Conrad, 443
schooling, 30
Scire, J., 457
script-agent, 432
scripts, 431, 438
Searle, John, 55-57, 63
second-order machines, 11
segmental gradients, 212
segmentation, 210
selection
    artificial, 203
    cumulative, 209, 219, 511
    higher-level, 218
    single-step, 219
    species, 219
self-assembly, 323
self-complementary, 381
self-copying, 381
self-maintenance, 126
self-organization, 80-81, 125, 319, 355

self-organization (cont'd.)
    processes, 81
self-organizing manufacturing methods, 355
self-replicating lunar factory, 58
self-replication, 69, 82, 95, 202, 380
    in Typogenetics, 382-383
    natural vs. artificial, 382-383
    RNA, 82
    von Neumann's description, 69
self-reproduction, 14, 16, 29, 33, 50, 58
    factory or lunar facilities, 58
    phases, 51
    von Neumann's contribution, 50
self-similarity, 256-257, 261, 264
    artificial, 258
    recursive, 261
    trees, 263-264
    vs. fractal, 257
self-structuring properties, 311
sensorimotor neurons, 424
Shakey the robot, 174
Shannon, Claude, 245, 171
Shoulders, K. R., 450, 452-453
signal propagation, 27, 30
signals
    acropetal, 234
    basipetal, 234
Simon, G., 67, 171, 388
simulated evolution, 158
simulation, 66, 68, 71
    computer, 66
    going beyond nature, 268
single-step selection, 219
situated action, 421, 439
    approach to artificial intelligence, 439
Sloane, N. J. A., 350
smart memory, 354
Smith, Alvy Ray, 264
social structure, 310
socially regulated clock, 308
Society of Mind model, 415, 417, 429, 431
solitons, 120, 528
sort, 363
species selection, 219
spin glass, 522, 541
spontaneous formation, 125
Stahl, Walter, 19,
Stanford Research Institute Problem Solver, 174

Stanford Artificial Intelligence Laboratory, 175
statistical self-similarity, 265
stickleback, 427
STMs, 446, 503
stochastic device, 365
stochastic re-causalization, 82
strict predecessor, 224
STRIPS, 174
structural coupling, 428
structure-oriented models, 298
structures
    emergent, 312
    geometric, 506
    topological, 506
superconducting devices, 195
superposition principle, 41
supramolecular systems, 454
symbiosis, 478
symbiotic association, 524
symbiotic theory of eukaryote origin, 524
symbol-matter problem, 69
symmetry, 209-215
    bilateral, 206
syntactic inference problem, 244
synthesis, 39, 41
synthetic biology, 125-127
systems, 73
    dataflow, 408
    dynamical vs. non-dynamical, 73
    knowledge-based, 408
    organic vs. mechanical, 507, 511
    paradigm, 299-301
    supramolecular, 454
sythetic biology, 126

T

T4 phage self-assembly & operation, 321
Tamayo, Pablo, 105
tangled hierarchy, 370
Taniguchi, N., 444
Taylor, C. E., 275
tessellation, 50, 344, 361
    localized perturbations, 346
    transformations on, 356
Thatcher, James, 16
theoretical biology, 1
theories
    automata, 13, 105
    clay for the origin of life, 107
    Cairns-Smith, 169

theories (cont'd.)
    cognitive activity, 67
    genetic takeover, 169
    mesurement, 74
Thom, R., 72
Thompson, D'Arcy, 230
Thompson, Richard, 317
Tinbergen, N., 427, 435, 439
top-down environments, 38
top-down specifications, 31
top-down system, 439
topological structures, 506
topology, 506
transport
    channeled, 504
    diffusive, 504
Travers, Michael, 421
tree typology, 264
trees
    branching, 415
    data structures, 261
    dataflow, 417
    decision, 415
Tristram Shandy paradox, 52
tubulin, 525-530, 541, 544
turbulence, 113, 118-120
Turing, Alan, 10, 12, 171, 388
Turing machine, 24
Turing's halting problem, 12
Turing's model for morphogenesis, 108
Turner, S. R., 275
turtles, 399
Typogenetics, 369
    demonstration of reductio strategy, 383
    formal system, 369
    infinitely fertile strands, 380-386
    meta-logical proofs, 383
    paradox, 385
    proof of special derivation format, 379
    propagation, 382-383
        logic for investigating, 370
    regress paradox, 382-383
    self-replication, 382-383

U

Ulam, Stan, 13, 299
ultraminiaturization, 444
    Feynman's, 444
unique path, 349

universal computation, 64
universal constructor, 13, 405
unpredictability, 33
unpredictability of PTYPE from GTYPE, 33
Uribe, R., 131
urstrand, 385
    in nature, 385-386
    in Typogenetics, 386

V

vacuum-tunnel-effect cathode, 451
Varela, F. G., 131
variable structure models, 302-303
Vaucanson, 8
Vaucanson's duck, 8
Vehicles, 422
Vichniac, G. Y., 133
virtual machines, 20
virtual parts, 41
virus self-assembly & operation, 321
viscous fingering, 125
vision, 187
vitalism, 4
Vivarium Project, 407
von Foester, H., 454
von Hippel, A. R., 453

von Neumann, John, 13-15, 50-51, 69, 73, 105-106, 171, 299, 320, 478
von Neumann machines, 477

W

Walter, W. Grey, 18, 171, 184, 399, 436
Ward, J., 457
watershed, 218
Wator, 20
Weismann, 203
White, J., 453
Wiener, Norbert, 15, 170, 388
Wiener's cybernetics, 15
Wilson, Stewart, 157, 436
wire, 351
world model, 184
Wright, Sewall, 514
Wright's genetic landscape, 514

Y

Young, R. D., 457

Z

Zeleny, Milan, 125
Zeltzer, D., 439
Zingsheim, H. P., 454